# APPROXIMATE
# DYNAMIC PROGRAMMING

## THE WILEY BICENTENNIAL–KNOWLEDGE FOR GENERATIONS

Each generation has its unique needs and aspirations. When Charles Wiley first opened his small printing shop in lower Manhattan in 1807, it was a generation of boundless potential searching for an identity. And we were there, helping to define a new American literary tradition. Over half a century later, in the midst of the Second Industrial Revolution, it was a generation focused on building the future. Once again, we were there, supplying the critical scientific, technical, and engineering knowledge that helped frame the world. Throughout the 20th Century, and into the new millennium, nations began to reach out beyond their own borders and a new international community was born. Wiley was there, expanding its operations around the world to enable a global exchange of ideas, opinions, and know-how.

For 200 years, Wiley has been an integral part of each generation's journey, enabling the flow of information and understanding necessary to meet their needs and fulfill their aspirations. Today, bold new technologies are changing the way we live and learn. Wiley will be there, providing you the must-have knowledge you need to imagine new worlds, new possibilities, and new opportunities.

Generations come and go, but you can always count on Wiley to provide you the knowledge you need, when and where you need it!

WILLIAM J. PESCE
PRESIDENT AND CHIEF EXECUTIVE OFFICER

PETER BOOTH WILEY
CHAIRMAN OF THE BOARD

# APPROXIMATE DYNAMIC PROGRAMMING

## Solving the Curses of Dimensionality

**Warren B. Powell**
Princeton University
Princeton, New Jersey

**WILEY-INTERSCIENCE**
**A John Wiley & Sons, Inc., Publication**

Published by John Wiley & Sons, Inc., Hoboken, New Jersey.
Published simultaneously in Canada.



***Library of Congress Cataloging-in-Publication Data:***

Powell, Warren B., 1955–
Approximate dynamic programming : solving the curses of dimensionality / Warren B. Powell.
p. ; cm. (Wiley series in probability and statistics)
Includes bibliographical references.
ISBN 978-0-470-17155-4 (cloth : alk. paper)
1. Dynamic programming. I. Title.
T57 . 83 . P76 2007
519 . 7'03—dc22 2007013724

Printed in the United States of America.

10 9 8 7 6 5

# CONTENTS

# PREFACE

The path to completing this book began in the early 1980's when I first started working on dynamic models arising in the management of fleets of vehicles for the truckload motor carrier industry. It is often said that necessity is the mother of invention, and as with many of my colleagues in this field, the methods that emerged evolved out of a need to solve a problem. The initially ad hoc models and algorithms I developed to solve these complex industrial problems evolved into a sophisticated set of tools supported by an elegant theory within a field which is increasingly being referred to as *approximate dynamic programming*.

The methods in this volume reflect the original motivating applications. We started with elegant models for which academia is so famous, but our work with industry revealed the need to handle a number of complicating factors which were beyond the scope of these models. One of these was a desire from one company to understand the effect of uncertainty on operations, requiring the ability to solve these large-scale optimization problems in the presence of various forms of randomness (but most notably customer demands). This question launched what became a multi-decade search for a modeling and algorithmic strategy that would provide practical, but high-quality, solutions.

This process of discovery took me through multiple fields, including linear and nonlinear programming, Markov decision processes, optimal control, and stochastic programming. It is somewhat ironic that the framework of Markov decision processes, which originally appeared to be limited to toy problems (three trucks moving between five cities), turned out to provide the critical theoretical framework for solving truly industrial-strength problems (thousands of drivers moving between hundreds of locations, each described by complex vectors of attributes).

Our ability to solve these problems required the integration of four major disciplines: dynamic programming (Markov decision processes), math programming (linear, nonlinear and integer programming), simulation and statistics. The desire to bring together the fields of dynamic programming and math programming motivated some fundamental notational choices (in particular, the use of $x$ as a decision variable). There is a heavy dependence on the Monte Carlo methods so widely used in simulation, but a knowledgeable reader will quickly see how much is missing. We cover in some depth a number of important techniques from statistics, but even this presentation only scratches the surface of tools and concepts available from with fields such as nonparametric statistics, signal processing and approximation theory.

## Audience

The book is aimed primarily at an advanced undergraduate/masters audience with no prior background in dynamic programming. The presentation does expect a first course in probability and statistics. Some topics require an introductory course in linear programming. A major goal of the book is the clear and precise presentation of dynamic problems, which means there is an emphasis on modeling and notation.

The body of every chapter focuses on models and algorithms with a minimum of the mathematical formalism that so often makes presentations of dynamic programs inaccessible to a broader audience. Using numerous examples, each chapter emphasizes the presentation of algorithms that can be directly applied to a variety of applications. The book contains dozens of algorithms that are intended to serve as a starting point in the design of practical solutions for real problems. Material for more advanced graduate students (with measure-theoretic training and an interest in theory) is contained in sections marked with **.

The book can also be used quite effectively in a Ph.D. course. Several chapters include "Why does it work" sections at the end which present proofs at an advanced level (these are all marked with **). This material can be easily integrated into the teaching of the material within the chapter.

## Pedagogy

The book is roughly organized into three parts. Part I comprises chapters 1-4, which provide a relatively easy introduction using a simple, discrete representation of states. Part II is the heart of the volume, beginning with a much richer discussion of how to model a dynamic program followed by the most important algorithmic strategies. Part III introduces specific problem classes, including information acquisition and resource allocation, and algorithms that arise in this setting. A number of sections are marked with an *. These can all be skipped when first reading the book without loss of continuity. Sections marked with ** are intended only for advanced graduate students with an interest in the theory behind the techniques.

**Part I** Introduction to dynamic programming using simple state representations - In the first four chapters, we introduce dynamic programming using what is known as a "flat" state representation, which is to say that we assume that we can represent states as $s = 1, 2, \ldots,$. We avoid many of the rich modeling and algorithmic issues that arise in more realistic problems.

**Chapter 1** Here we set the tone for the book, introducing the challenge of the three "curses of dimensionality" that arise in complex systems.

**Chapter 2** Dynamic programs are best taught by example. Here we describe three classes of problems: deterministic problems, stochastic problems and information acquisition problems. Notation is kept simple but precise, and readers see a range of different applications.

**Chapter 3** This is an introduction to classic Markov decision processes. While these models and algorithms are typically dismissed because of "the curse of dimensionality," these ideas represent the foundation of the rest of the book. The proofs in the "why does it work" section are particularly elegant and help provide a deep understanding of this material.

**Chapter 4** This chapter provides an initial introduction to approximate dynamic programming, including the use of the post-decision state variable. Readers see basic algorithms without being exposed to the rich set of modeling and algorithmic issues that have to be addressed in most applications. Value functions are modeled using simple lookup-table representations.

**Part II** Approximate dynamic programming - These chapters represent the most important dimensions of approximate dynamic programming: modeling real applications, the interface with stochastic approximation methods, techniques for approximating general value functions, and a more in-depth presentation of ADP algorithms for finite and infinite horizon applications.

**Chapter 5** A forty page chapter on modeling? This hints at the richness of dynamic problems. To help with assimilating this chapter, we encourage readers to skip sections marked with an * the first time they go through the chapter. It is also useful to reread this chapter from time to time as you are exposed to the rich set of modeling issues that arise in real applications.

**Chapter 6** Stochastic approximation methods represent the theoretical foundation for estimating value functions using sample information. Here we introduce the theory behind smoothing random observations using "stepsizes." We provide a fairly thorough review of stepsize recipes, as well as a section on optimal stepsizes. The more advanced reader should try reading the proofs in the "why does it work" section.

**Chapter 7** Perhaps the most important challenge in ADP is approximating value functions by taking advantage of the structure of the state variable. This chapter introduces a range of techniques, although it is by no means comprehensive. We have access to the entire field of statistics as well as the subfield of approximation theory.

**Chapter 8** This chapter is effectively a more in-depth repeat of chapter 4. We cover a wider range of techniques, including methods from the reinforcement-learning community. In addition, we no longer assume lookup-table representations for value functions, but instead depend on the techniques introduced in chapter 7. The presentation focuses on finite horizon problems, both because of their inherent importance (especially in the ADP arena) and also because they reinforce the modeling of time.

**Chapter 9** Here we make the transition to infinite horizon problems. For the most part, this is simply an adaptation of methods for finite horizon problems, but the infinite horizon setting introduces some subtleties.

**Part III** Additional topics - By now, we have most of the fundamentals to start solving a range of problems. The remainder of the volume focuses on special topics and applications that bring out the richness of approximate dynamic programming.

**Chapter 10** Information acquisition not only is an interesting problem in its own right, but it also arises throughout the design of ADP algorithms. We need to estimate a value function at (or near) a state, but to obtain information about the value of being in a state, we have to visit the state. As a result, we have to make the tradeoff between visiting a state because we think this represents the best decision ("exploitation") and visiting a state just to obtain information about the value of being in a state ("exploration").

**Chapter 11** This chapter introduces a class of approximation techniques that are particularly useful in the resource allocation problems that we present in chapter 12.

**Chapter 12** Dynamic resource allocation problems arise in a vast array of applications. These tend to be high-dimensional problems, with state variables that can easily have thousands or even millions of dimensions. But they also exhibit significant structure which we can exploit.

**Chapter 13** We close with a discussion of a number of more practical issues that arise in the development and testing of ADP algorithms.

This material is best covered in order. Depending on the length of the course and the nature of the class, an instructor will want to skip some sections, or to weave in some of the theoretical material in the "why does it work" sections. Additional material (exercises, solutions, datasets, errata) will be made available over time at the website http://www.castlelab.princeton.edu/adp.htm (this can also be accessed from the CASTLE Lab website at http://www.castlelab.princeton.edu/).

There are two faces of approximate dynamic programming, and we try to present both of them. The first emphasizes models and algorithms, with an emphasis on applications and computation. It is virtually impossible to learn this material without writing software to test and compare algorithms. The other face is a deeply theoretical one that focuses on proofs of convergence and rate of convergence. This material is advanced and accessible primarily to students with training in probability and stochastic processes at a measure-theoretic level.

Approximate dynamic programming is also a field that has emerged from several disciplines. We have tried to expose the reader to the many dialects of ADP, reflecting its origins in artificial intelligence, control theory, and operations research. In addition to the diversity of words and phrases which mean the same thing (but often with different connotations), we also had to make difficult notational choices.

We have found that different communities offer unique insights into different dimensions of the problem. Our finding is that the control theory community has the most thorough understanding of the meaning of a state variable. The artificial intelligence community has the most experience with deeply nested problems (which require numerous steps before earning a reward). The operations research community has evolved a set of tools that are well suited for high-dimensional resource allocation, contributing both math programming and a culture of careful modeling.

WARREN B. POWELL

*Princeton, New Jersey*
*July, 2007*

# ACKNOWLEDGMENTS

The work in this book reflects the contributions of many. Perhaps most important are the problems that motivated the development of this material. This work would not have been possible without the corporate sponsors who posed these problems in the first place. I would like to give special recognition to Schneider National, the largest truckload carrier in the U.S., Yellow Freight System, the largest less-than-truckload carrier, and Norfolk Southern Railroad, one of the four major railroads that serves the U.S. These three companies not only posed difficult problems, but also provided years of research funding that allowed us to work on the development of tools that became the foundation of this book. This work would never have progressed without the thousands of hours of my two senior professional staff members, Hugo Simão and Belgacem Bouzaiëne-Ayari, who have written hundreds of thousands of lines of code to solve industrial-strength problems. It is their efforts working with our corporate sponsors that brought out the richness of real applications, and therefore the capabilities that our tools needed to possess.

While our industrial sponsors provided the problems, without the participation of the graduate students, we would simply have a set of ad hoc procedures. It is the work of my graduate students that provided most of the fundamental insights and algorithms, and virtually all of the convergence proofs. In the order in which they joined by research program, the students are Linos Frantzeskakis, Raymond Cheung, Tassio Carvalho, Zhi-Long Chen, Greg Godfrey, Joel Shapiro, Mike Spivey, Huseyin Topaloglu, Katerina Papadaki, Arun Marar, Tony Wu, Abraham George, Juliana Nascimento, and Peter Frazier, all of whom are my current and former students, have contributed directly to the material presented in this volume. My undergraduate senior thesis advisees provided many colorful applications of dynamic programming, and they contributed their experiences with their computational work.

The presentation has benefited from numerous conversations with professionals in this community. I am particularly grateful to Erhan Çinlar, who taught me the language of stochastic processes that played a fundamental role in guiding my notation in the modeling of information. I am also grateful for many conversations with Ben van Roy, Dimitri Bertsekas, Andy Barto, Mike Fu, Dan Adelman, Lei Zhao and Diego Klabjan. I would also like to thank Paul Werbos at NSF for introducing me to the wonderful neural net community in IEEE which contributed what for me was a fresh perspective on dynamic problems. Jennie Si, Don Wunsch, and George Lendaris all helped educate me in the language and concepts of the control theory community.

This research was first funded by the National Science Foundation, but the bulk of our research in this volume was funded by the Air Force Office of Scientific Research, and I am particularly grateful to Dr. Neal Glassman for supporting us through these early years.

Many people have assisted with the editing of this volume through numerous comments. Mary Fan, Tamas Papp, and Hugo Simão all read various drafts cover to cover. Diego Klabjan and his dynamic programming classes at the University of Illinois provided numerous comments and corrections. Special thanks are due to the students in my own undergraduate and graduate dynamic programming classes who had to survive the very early editions of this volume.

Of course, the preparation of this book required tremendous patience from my wife Shari and my children Elyse and Danny, who had to tolerate my ever-present laptop at home. Without their support, this project could never have been completed.

W. B. P.

# CHAPTER 1

# THE CHALLENGES OF DYNAMIC PROGRAMMING

The optimization of problems over time arises in many settings, ranging from the control of heating systems to managing entire economies. In between are examples including landing aircraft, purchasing new equipment, managing blood inventories, scheduling fleets of vehicles, selling assets, investing money in portfolios, or just playing a game of tic-tac-toe or backgammon. These problems involve making decisions, then observing information, after which we make more decisions, and then more information, and so on. Known as *sequential decision problems*, they can be straightforward (if subtle) to formulate, but solving them is another matter.

Dynamic programming has its roots in several fields. Engineering and economics tend to focus on problems with continuous states and decisions (these communities refer to decisions as controls), which might be quantities such as location, speed, and temperature. By contrast, the fields of operations research and artificial intelligence work primarily with discrete states and decisions (or actions). Problems that are modeled with continuous states and decisions (and typically in continuous time) are often addressed under the umbrella of "control theory," whereas problems with discrete states and decisions, modeled in discrete time, are studied at length under the umbrella of "Markov decision processes." Both of these subfields set up recursive equations that depend on the use of a state variable to capture history in a compact way. There are many high-dimensional problems such as those involving the allocation of resources that are generally studied using the tools of mathematical programming. Most of this work focuses on deterministic problems using tools such as linear, nonlinear, or integer programming, but there is a subfield known as

*Approximate Dynamic Programming*. By Warren B. Powell

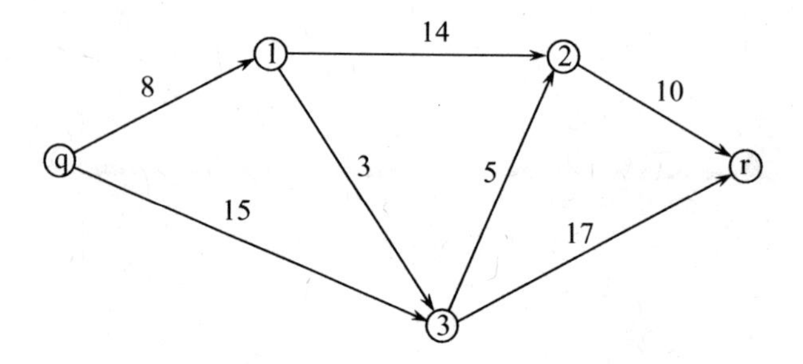


**Figure 1.1** Illustration of a shortest path problem from origin $q$ to destination $r$.

stochastic programming which incorporates uncertainty. Our presentation spans all of these fields.

## 1.1 A DYNAMIC PROGRAMMING EXAMPLE: A SHORTEST PATH PROBLEM

Perhaps one of the best0-known applications of dynamic programming is that faced by a driver choosing a path in a transportation network. For simplicity (and this is a real simplification for this application), we assume that the driver has to decide at each node (or intersection) which link to traverse next (we are not going to get into the challenges of left turns versus right turns). Let $\mathcal{I}$ be the set of intersections. If the driver is at intersection $i$, he can go to a subset of intersections $\mathcal{I}_i^+$ at a cost $c_{ij}$. He starts at the origin node $q \in \mathcal{I}$ and has to find his way to the destination node $r \in \mathcal{I}$ at the least cost. An illustration is shown in figure 1.1.

The problem can be easily solved using dynamic programming. Let

$$v_i \quad = \quad \text{The cost to get from intersection } i \in \mathcal{I} \text{ to the destination node } r.$$

We assume that $v_r = 0$. Initially, we do not know $v_i$, and so we start by setting $v_i = M$, where "$M$" is known as "big M" and represents a large number. We can solve the problem by iteratively computing

$$v_i \leftarrow \min\left\{v_i, \min_{j \in \mathcal{I}^+}\{c_{ij} + v_j\}\right\} \quad \text{for all } i \in \mathcal{I}. \tag{1.1}$$

Equation (1.1) has to be solved iteratively, where at each iteration, we loop over all the nodes $i$ in the network. We stop when none of the values $v_i$ change. It should be noted that this is not a very efficient way of solving a shortest path problem. For example, in the early iterations, it may well be the case that $v_j = M$ for all $j \in \mathcal{I}^+$. However, we use the method to illustrate dynamic programming.

Table 1.1 illustrates the algorithm, assuming that we always traverse the nodes in the order $(q, 1, 2, 3, r)$. Note that we handle node 2 before node 3, which is the reason why, even in the first pass, we learn that the path cost from node 3 to node $r$ is 15 (rather than 17). We are done after iteration 3, but we require iteration 4 to verify that nothing has changed.

Shortest path problems arise in a variety of settings that have nothing to do with transportation or networks. Consider, for example, the challenge faced by a college freshman trying to plan her schedule to graduation. By graduation, she must take 32 courses overall,

| Iteration | Cost from node q | 1 | 2 | 3 | r |
|---|---|---|---|---|---|
| | 100 | 100 | 100 | 100 | 0 |
| 1 | 100 | 100 | 10 | 15 | 0 |
| 2 | 30 | 18 | 10 | 15 | 0 |
| 3 | 26 | 18 | 10 | 15 | 0 |
| 4 | 26 | 18 | 10 | 15 | 0 |

**Table 1.1** Path cost from each node to node $r$ after each node has been visited

including eight departmentals, two math courses, one science course, and two language courses. We can describe the state of her academic program in terms of how many courses she has taken under each of these five categories. Let $S_{tc}$ be the number of courses she has taken by the end of semester $t$ in category $c = \{\text{Total courses, Departmentals, Math, Science, Language}\}$, and let $S_t = (S_{tc})_c$ be the state vector. Based on this state, she has to decide which courses to take in the next semester. To graduate, she has to reach the state $S_8 = (32, 8, 2, 1, 2)$. We assume that she has a measurable desirability for each course she takes, and that she would like to maximize the total desirability of all her courses.

The problem can be viewed as a shortest path problem from the state $S_0 = (0, 0, 0, 0, 0)$ to $S_8 = (32, 8, 2, 1, 2)$. Let $S_t$ be her current state at the beginning of semester $t$ and let $x_t$ represent the decisions she makes while determining what courses to take. We then assume we have access to a *transition function* $S^M(S_t, x_t)$ which tells us that if she is in state $S_t$ and makes decision $x_t$, she will land in state $S_{t+1}$, which we represent by simply using

$$S_{t+1} = S^M(S_t, x_t).$$

In our transportation problem, we would have $S_t = i$ if we are at intersection $i$, and $x_t$ would be the decision to "go to $j$," leaving us in the state $S_{t+1} = j$.

Finally, let $C_t(S_t, x_t)$ be the contribution or reward she generates from being in state $S_t$ and making the decision $x_t$. The value of being in state $S_t$ is defined by the equation

$$V_t(S_t) = \max_{x_t} \{C_t(S_t, x_t) + V_{t+1}(S_{t+1})\} \quad \forall s_t \in \mathcal{S}_t,$$

where $S_{t+1} = S^M(S_t, x_t)$ and where $\mathcal{S}_t$ is the set of all possible (discrete) states that she can be in at the beginning of the year.

## 1.2 THE THREE CURSES OF DIMENSIONALITY

All dynamic programs can be written in terms of a recursion that relates the value of being in a particular state at one point in time to the value of the states that we are carried into at the next point in time. For deterministic problems, this equation can be written

$$V_t(S_t) = \max_{x_t} \big(C_t(S_t, x_t) + V_{t+1}(S_{t+1})\big). \tag{1.2}$$

where $S_{t+1}$ is the state we transition to if we are currently in state $S_t$ and take action $x_t$. Equation (1.2) is known as Bellman's equation, or the Hamilton-Jacobi equation, or increasingly, the Hamilton-Jacobi-Bellman equation (HJB for short). Some textbooks (in control theory) refer to them as the "functional equation" of dynamic programming (or the "recurrence equation"). We primarily use the term "optimality equation" in our presentation, but often use the term "Bellman equation" since this is so widely used in the dynamic programming community.

Most of the problems that we address in this volume involve some form of uncertainty (prices, travel times, equipment failures, weather). For example, in a simple inventory problem, we may have $S_t$ DVD players in stock. We might then order $x_t$ new DVD players, (after which we satisfy a random demand $\hat{D}_{t+1}$ which follows some probability distribution. The state variable would be described by the transition equation

$$S_{t+1} = \max\{0, S_t + x_t - \hat{D}_{t+1}\}.$$

Assume that $C_t(S_t, x_t)$ is the contribution we earn at time $t$, given by

$$C_t(S_t, x_t, \hat{D}_{t+1}) = p_t \min\{S_t + x_t, \hat{D}_{t+1}\} - cx_t.$$

To find the best decision, we need to maximize the contribution we receive from $x_t$ plus the expected value of the state that we end up at (which is random). That means we need to solve

$$V_t(S_t) = \max_{x_t}\{C_t(S_t, x_t, \hat{D}_{t+1}) + V_{t+1}(S_{t+1})|S_t\}. \tag{1.3}$$

This problem is not too hard to solve. Assume we know $V_{t+1}(S_{t+1}$ for each state $S_{t+1}$. We just have to compute (1.3) for each value of $S_t$, which then gives us $V_t(S_t)$. We can keep stepping backward in time to compute all the value functions.

For the vast majority of problems, the state of the system is a vector. For example, if we have to track the inventory of $N$ different products, where we might have $0, 1, \ldots, M-1$ units of inventory of each product, then we would have $M^N$ different states. As we can see, the size of the state space grows very quickly as the number of dimensions grows. This is the widely known "curse of dimensionality" of dynamic programming and is the most often-cited reason why dynamic programming cannot be used.

In fact, there are many applications where there are three curses of dimensionality. Consider the problem of managing blood inventories. There are eight blood types (AB+, AB-, A+, A-, B+, B-, O+, O-), which means we have eight types of blood supplies and eight types of blood demands. Let $B_{ti}$ be the supply of blood type $i$ at time $t$ ($i = 1, 2, \ldots, 8$) and let $D_{ti}$ be the demand for blood type $i$ at time $t$. Our state variable is given by $S_t = (B_t, D_t)$ where $B_t = (B_{ti})_{i=1}^8$ ($D_t$ is defined similarly).

In each time period, there are two types of randomness: random blood donations and random demands. Let $\hat{B}_{ti}$ be the random new donations of blood of type $i$ in week $t$, and let $\hat{D}_{ti}$ be the random new demands for blood of type $i$ in week $t$. We are going to let $W_t = (\hat{B}_t, \hat{D}_t)$ be the vector of random information (new supplies and demands) that becomes known in week $t$.

Finally let $x_{tij}$ be the amount of blood type $i$ used to satisfy a demand for blood of type $j$. There are rules that govern what blood types can substitute for different demand types, shown in figure 1.2.

We can quickly see that $S_t$ and $W_t$ have 16 dimensions each. If we have up to 100 units of blood of any type, then our state space has $100^{16} = 10^{32}$ states. If we have up to 20 units of blood being donated or needed in any week, then $W_t$ has $20^{16} = 6.55 \times 10^{20}$ outcomes. We would need to evaluate 16 nested summations to evaluate the expectation. Finally, $x_t$ has 27 dimensions (there are 27 feasible substitutions of blood types for demand types). Needless to say, evaluating all possible values of $x_t$ is completely intractable.

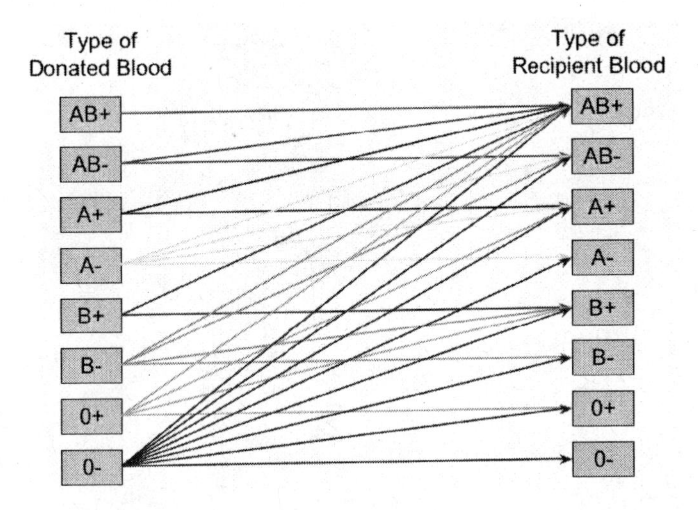


**Figure 1.2** The different substitution possibilities between donated blood and patient types (from Cant (2006)).

This problem illustrates what is, for many applications, the three curses of dimensionality:

1. The state space - If the state variable $S_t = (S_{t1}, S_{t2}, \ldots, S_{ti}, \ldots, S_{tI})$ has $I$ dimensions, and if $S_{ti}$ can take on $L$ possible values, then we might have up to $L^I$ different states.

2. The outcome space - The random variable $W_t = (W_{t1}, W_{t2}, \ldots, W_{tj}, \ldots, W_{tJ})$ might have $J$ dimensions. If $W_{tj}$ can take on $M$ outcomes, then our outcome space might take on up to $M^J$ outcomes.

3. The action space - The decision vector $x_t = (x_{t1}, x_{t2}, \ldots, x_{tk}, \ldots, x_{tK})$ might have $K$ dimensions. If $x_{tk}$ can take on $N$ outcomes, we might have up to $N^K$ outcomes.

By the time we get to chapter 12, we will be able to produce high-quality, implementable solutions not just to the blood problem (see section 12.2), but for problems that are far larger. The techniques that we are going to describe have produced production quality solutions to plan the operations of some of the largest transportation companies in the country. These problems require state variables with millions of dimensions, with very complex dynamics. We will show that these same algorithms converge to optimal solutions for special cases. For these problems, we will produce solutions that are within one percent of optimality in a small fraction of the time required to find the optimal solution using classical techniques. However, we will also describe algorithms for problems with unknown convergence properties, which produce solutions of uncertain quality and with behaviors that can range from the frustrating to the mystifying. This is a very young field.

Not all problems suffer from the three curses of dimensionality. Many problems have small sets of actions (do we buy or sell?), easily computable expectations (did a customer arrive or not?), and small state spaces (the nodes of a network). The field of dynamic programming has identified many problems, some with genuine industrial applications, which avoid the curses of dimensionality. The position of this book will be that these

**Figure 1.3** Windmills are one form of alternative energy resources (from http://www.nrel.gov/data/pix/searchpix.cgi ).

are special cases. We take the position that we are working on a standard problem which exhibits all three curses of dimensionality, and unless otherwise stated, we will develop methods that work for these more general problems.

## 1.3 SOME REAL APPLICATIONS

Most of the applications in this book can be described as resource allocation problems of some form. These arise in a broad range of applications. Resources can be physical objects including people, equipment such as aircraft, trucks or electric power transformers, commodities such as oil and food, and fixed facilities such as buildings and power generators.

As our energy environment changes, we have to plan new energy resources. A challenging dynamic problem requires determining when to acquire or install new energy resources (windmills, ethanol plants, nuclear plants, new types of coal plants) and where to put them. These decisions have to be made considering uncertainty in the demand, prices, and the underlying technologies for creating, storing, and using energy. For example, adding ethanol capacity has to consider the possibility that oil prices will drop (reducing the demand for ethanol) or that government regulations may favor alternative fuels (increasing the demand).

**Figure 1.4** The major railroads in the United States have to manage complex assets such as boxcars, locomotives and the people who operate them. Courtesy Norfolk Southern.

An example of a very complex resource allocation problem arises in railroads (figure 1.4). In North America, there are six major railroads (known as "Class I" railroads) which operate thousands of locomotives, many of which cost over $1 million. Decisions have to be made now to assign locomotives to trains, taking into account how the locomotives will be used at the destination. For example, a train may be going to a location that needs additional power. Or a locomotive may have to be routed to a maintenance facility, and the destination of a train may or may not offer good opportunities for getting the locomotive to the shop. There are many types of locomotives, and different types of locomotives are suited to different types of trains (for example, trains moving coal, grain, or merchandise). Other applications of dynamic programming include the management of freight cars, where decisions about when, where and how many to move have to be made in the presence of numerous sources of uncertainty, including customer demands, transit times and equipment problems.

The military faces a broad range of operational challenges that require positioning resources to anticipate future demands. The problems may be figuring out when and where to position tankers for mid-air refueling (figure 1.5), or whether a cargo aircraft should be modified to carry passengers. The air mobility command needs to think about not only what aircraft is best to move a particular load of freight, but also the value of aircraft in the future (are there repair facilities near the destination?). The military is also interested in the value of more reliable aircraft and the impact of last-minute requests. Dynamic programming provides a means to produce robust decisions, allowing the military to respond to last-minute requests.

Managing the electric power grid requires evaluating the reliability of equipment such as the transformers that convert high-voltage power to the voltages used by homes and businesses. Figure 1.6 shows the high-voltage backbone network managed by PJM Interconnections which provides electricity to the northeastern United States. To ensure

**Figure 1.5** Mid-air refueling is a major challenge for air operations, requiring that tankers be positioned in anticipation of future needs (from http://www.amc.af.mil/photos/).

the reliability of the grid, PJM helps utilities maintain an appropriate inventory of spare transformers. They cost five million dollars each, weigh over 200 tons, and require at least a year to deliver. We must make decisions about how many to buy, how fast they should be delivered (fast delivery costs more) and where to put them when they do arrive. If a transformer fails, the electric power grid may have to purchase power from more expensive

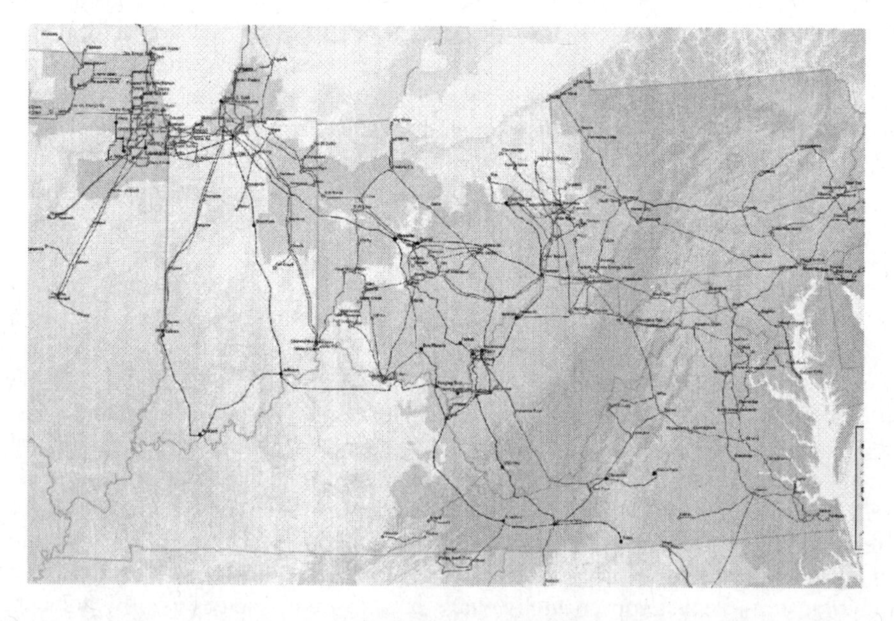

**Figure 1.6** The high-voltage backbone network managed by PJM Interconnections provides electricity to the northeastern United States. Courtesy PJM Interconnections.

**Figure 1.7** Schneider National, the largest truckload motor carrier in the United States, manages a fleet of over 15,000 drivers. Courtesy Schneider National.

utilities to avoid a bottleneck, possibly costing millions of dollars per month. As a result, it is not possible to wait until problems happen. Utilities also face the problem of pricing their energy in a dynamic market, and purchasing commodities such as coal and natural gas in the presence of fluctuating prices.

Similar issues arise in the truckload motor carrier industry, where drivers are assigned to move loads that arise in a highly dynamic environment. Large companies manage fleets of thousands of drivers, and the challenge at any moment in time is to find the best driver (figure 1.7 is from Schneider National, the largest truckload carrier in the United States). There is much more to the problem than simply finding the closest driver; each driver is characterized by attributes such as his or her home location and equipment type as well as his or her skill level and experience. There is a need to balance decisions that maximize profits now versus those that produce good long run behavior. Approximate dynamic programming produced the first accurate model of a large truckload operation. Modeling this large-scale problem produces some of the advances described in this volume.

A separate set of problems can be found in the management of financial assets. These problems involve the management of stocks, bonds, cash, money market certificates, futures, options, and other financial instruments such as derivatives. Since physical objects and money are often interchangeable (money can be used to purchase a physical object; the physical object can be sold and turned into money), the financial community will talk about real assets as opposed to financial assets. Physical resources such as jets, locomotives and people tend to be much more complex than financial assets, but financial models tend to exhibit much more complex information processes.

A third problem class is the acquisition of information. Consider the problem faced by the government which is interested in researching a new technology such as fuel cells or converting coal to hydrogen. There may be dozens of avenues to pursue, and the challenge is to determine the projects in which the government should invest. The state of the system is the set of estimates of how well different components of the technology work. The government funds research to collect information. The result of the research may be the anticipated improvement, or the results may be disappointing. The government wants to plan a research program to maximize the likelihood that a successful technology is developed within a reasonable time frame (say, 20 years). Depending on time and budget constraints, the government may wish to fund competing technologies in the event that one does not work. Alternatively, it may be more effective to fund one promising technology and then switch to an alternative if the first does not work out.

## 1.4 PROBLEM CLASSES

Most of the problems that we use as examples in this book can be described as involving the management of physical, financial, or informational resources. Sometimes we use the term assets which carries the connotation of money or valuable resources (aircraft, real estate, energy commodities). But in some settings, even these terms may seem inappropriate, for example, training machines to play a game such as tic-tac-toe, where it will be more natural to think in terms of managing an "entity." Regardless of the term, there are a number of major problem classes we consider in our presentation:

**The budgeting problem -** Here we face the problem of allocating a fixed resource over a set of activities that returns a reward that is a function of how much we invest in the activity. For example, drug companies have to decide how much to invest in different research projects or how much to spend on advertising for different drugs. Oil exploration companies have to decide how much to spend exploring potential sources of oil. Political candidates have to decide how much time to spend campaigning in different states.

**Asset acquisition with concave costs -** A company can raise capital by issuing stock or floating a bond. There are costs associated with these financial instruments independent of how much money is being raised. Similarly, an oil company purchasing oil will be given quantity discounts (or it may face the fixed cost of purchasing a tanker-load of oil). Retail outlets get a discount if they purchase a truckload of an item. All of these are instances of acquiring assets with a concave (or, more generally, nonconvex) cost function, which means there is an incentive for purchasing larger quantities.

**Asset acquisition with lagged information processes -** We can purchase commodity futures that allow us to purchase a product in the future at a lower cost. Alternatively, we may place an order for memory chips from a factory in southeast Asia with one- to two-week delivery times. A transportation company has to provide containers for a shipper who may make requests several days in advance or at the last minute. All of these are asset acquisition problems with *lagged information processes*.

**Buying/selling an asset -** In this problem class, the process stops when we either buy an asset when it looks sufficiently attractive or sell an asset when market conditions warrant. The game ends when the transaction is made. For these problems, we tend to focus on the price (the purchase price or the sales price), and our success depends on our ability to trade off current value with future price expectations.

**General resource allocation problems -** This class encompasses the problem of managing reusable and substitutable resources over time (equipment, people, products, commodities). Applications abound in transportation and logistics. Railroads have to move locomotives and boxcars to serve different activities (moving trains, moving freight) over time. An airline has to move aircraft and pilots in order to move passengers. Consumer goods have to move through warehouses to retailers to satisfy customer demands.

**Demand management -** There are many applications where we focus on managing the demands being placed on a process. Should a hospital admit a patient? Should a trucking company accept a request by a customer to move a load of freight?

**Shortest paths -** In this problem class, we typically focus on managing a single, discrete entity. The entity may be someone playing a game, a truck driver we are trying to route to return him home, a driver who is trying to find the best path to his destination or a locomotive we are trying to route to its maintenance shop. Shortest path problems, however, also represent a general mathematical structure that applies to a broad range of dynamic programs.

**Dynamic assignment -** Consider the problem of managing multiple entities, such as computer programmers, to perform different tasks over time (writing code or fixing bugs). Each entity and task is characterized by a set of attributes that determines the cost (or contribution) from assigning a particular resource to a particular task.

All of these problems focus on the problem of managing physical or financial resources (or assets, or entities). They provide an idea of the diversity of applications that can be studied. In each case, we have focused on the question of how to manage the resource. In addition, there are three other classes of questions that arise for each application:

**Pricing -** Often the question being asked is, What price should be paid for an asset? The right price for an asset depends on how it is managed, so it should not be surprising that we often find asset prices as a byproduct from determining how to best manage the asset.

**Information collection -** Since we are modeling sequential information and decision processes, we explicitly capture the information that is available when we make a decision, allowing us to undertake studies that change the information process. For example, the military uses unmanned aerial vehicles (UAV's) to collect information about targets in a military setting. Oil companies drill holes to collect information about underground geologic formations. Travelers try different routes to collect information about travel times. Pharmaceutical companies use test markets to experiment with different pricing and advertising strategies.

**Technology switching -** The last class of questions addresses the underlying technology that controls how the physical process evolves over time. For example, when should a power company upgrade a generating plant (e.g., to burn oil and natural gas)? Should an airline switch to aircraft that fly faster or more efficiently? How much should a communications company invest in a technology given the likelihood that better technology will be available in a few years?

Most of these problems arise in both discrete and continuous versions. Continuous models would be used for money, physical products such as oil, grain, and coal, or discrete products that occur in large volume (most consumer products). In other settings, it is important to retain the integrality of the resources being managed (people, aircraft, locomotives, trucks, and expensive items that come in small quantities). For example, how do we position emergency response units around the country to respond to emergencies (bioterrorism, major oil spills, failure of certain components in the electric power grid)?

What makes these problems hard? With enough assumptions, none of these problems are inherently difficult. But in real applications, a variety of issues emerge that can make all of them intractable. These include:

- Evolving information processes - We have to make decisions now before we know the information that will arrive later. This is the essence of stochastic models, and this property quickly turns the easiest problems into computational nightmares.

- High-dimensional problems - Most problems are easy if they are small enough. In real applications, there can be many types of resources, producing decision vectors of tremendous size.

- Measurement problems - Normally, we assume that we look at the state of our system and from this determine what decision to make. In many problems, we cannot measure the state of our system precisely. The problem may be delayed information (stock prices), incorrectly reported information (the truck is in the wrong location), misreporting (a manager does not properly add up his total sales), theft (retail inventory), or deception (an equipment manager underreports his equipment so it will not be taken from him).

- Unknown models (information, system dynamics) - We can anticipate the future by being able to say something about what might happen (even if it is with uncertainty) or the effect of a decision (which requires a model of how the system evolves over time).

- Missing information - There may be costs that simply cannot be computed and that are instead ignored. The result is a consistent model bias (although we do not know when it arises).

- Comparing solutions - Primarily as a result of uncertainty, it can be difficult comparing two solutions to determine which is better. Should we be better on average, or are we interested in the best and worst solution? Do we have enough information to draw a firm conclusion?

## 1.5 THE MANY DIALECTS OF DYNAMIC PROGRAMMING

Dynamic programming arises from the study of sequential decision processes. Not surprisingly, these arise in a wide range of applications. While we do not wish to take anything from Bellman's fundamental contribution, the optimality equations are, to be quite honest, somewhat obvious. As a result, they were discovered independently by the different communities in which these problems arise.

The problems arise in a variety of engineering problems, typically in continuous time with continuous control parameters. These applications gave rise to what is now referred to as control theory. While uncertainty is a major issue in these problems, the formulations tend to focus on deterministic problems (the uncertainty is typically in the estimation of the state or the parameters that govern the system). Economists adopted control theory for a variety of problems involving the control of activities from allocating single budgets or managing entire economies (admittedly at a very simplistic level). Operations research (through Bellman's work) did the most to advance the theory of controlling stochastic problems, thereby producing the very rich theory of Markov decision processes. Computer scientists, especially those working in the realm of artificial intelligence, found that dynamic programming was a useful framework for approaching certain classes of machine learning problems known as reinforcement learning.

As different communities discovered the same concepts and algorithms, they invented their own vocabularies to go with them. As a result, we can solve the Bellman equations, the Hamiltonian, the Jacobian, the Hamilton-Jacobian, or the all-purpose Hamilton-Jacobian-Bellman equations (typically referred to as the HJB equations). In our presentation, we prefer the term "optimality equations."

There is an even richer vocabulary for the types of algorithms that are the focal point of this book. Everyone has discovered that the backward recursions required to solve the optimality equations in section 1.1 do not work if the state variable is multidimensional. For example, instead of visiting node $i$ in a network, we might visit state $S_t = (S_{t1}, S_{t2}, \ldots, S_{tB})$ where $S_{tb}$ is the amount of blood on hand of type $b$. A variety of authors have independently discovered that an alternative strategy is to step forward through time, using iterative algorithms to help estimate the value function. This general strategy has been referred to as forward dynamic programming, iterative dynamic programming, adaptive dynamic programming, heuristic dynamic programming, reinforcement learning, and neuro-dynamic programming. The term that is being increasingly adopted is *approximate dynamic programming*, although perhaps it is convenient that the initials, ADP, apply equally well to "adaptive dynamic programming." However, the artificial intelligence community continues to use "reinforcement learning," while a substantial audience in the control theory community (primarily in engineering where neural networks are popular) uses the term "neuro-dynamic programming." While these may all be viewed as different terms describing the same field, the notation and applications tend to be different, as are the most popular algorithms (which tend to reflect the characteristics of the problems each community works on).

The use of iterative algorithms that are the basis of most approximate dynamic programming procedures also have their roots in a field known as stochastic approximation methods. Again, authors tended to discover the technique and only later learn of its relationship to the field of stochastic approximation methods. Unfortunately, this relationship was sometimes discovered only after certain terms became well established.

Throughout the presentation, students need to appreciate that many of the techniques in the fields of approximate dynamic programming and stochastic approximation methods are fundamentally quite simple. The proofs of convergence and some of the algorithmic strategies can become quite difficult, but the basic strategies often represent what someone would do with no training in the field. As a result, the techniques frequently have a very natural feel to them, and the algorithmic challenges we face often parallel problems we encounter in every day life.

As of this writing, the relationship between control theory (engineering and economics), Markov decision processes (operations research), and reinforcement learning (computer science/artificial intelligence) is well understood by the research community. The relationship between iterative techniques (reviewed in chapter 4) and the field of stochastic approximations is also well established.

There is, however, a separate community that evolved from the field of deterministic math programming, which focuses on very high-dimensional problems. As early as the

1950's, this community was trying to introduce uncertainty into mathematical programs. The resulting subcommunity is called stochastic programming and uses a vocabulary that is quite distinct from that of dynamic programming. The relationship between dynamic programming and stochastic programming has not been widely recognized, despite the fact that Markov decision processes are considered standard topics in graduate programs in operations research.

Our treatment will try to bring out the different dialects of dynamic programming, although we will tend toward a particular default vocabulary for important concepts. Students need to be prepared to read books and papers in this field that will introduce and develop important concepts using a variety of dialects. The challenge is realizing when authors are using different words to say the same thing.

## 1.6 WHAT IS NEW IN THIS BOOK?

As of this writing, dynamic programming has enjoyed a relatively long history, with many superb books. Within the operations research community, the original text by Bellman (Bellman (1957)) was followed by a sequence of books focusing on the theme of Markov decision processes. Of these, the current high-water mark is *Markov Decision Processes* by Puterman, which played an influential role in the writing of chapter 3. This field offers a powerful theoretical foundation, but the algorithms are limited to problems with very low-dimensional state and action spaces.

This volume focuses on a field that is coming to be known as *approximate dynamic programming* that emphasizes modeling and computation for much harder classes of problems. The problems may be hard because they are large (for example, large state spaces), or because we lack a model of the underlying process that the field of Markov decision processes takes for granted. Two major references precede this volume. *Neuro-Dynamic Programming* by Bertsekas and Tsitsiklis was the first book to appear that summarized a vast array of strategies for approximating value functions for dynamic programming. *Reinforcement Learning* by Sutton and Barto presents the strategies of approximate dynamic programming in a very readable format, with an emphasis on the types of applications that are popular in the computer science/artificial intelligence community.

This volume focuses on models of problems that can be broadly described as "resource management," where we cover physical, financial, and informational resources. Many of these applications involve very high-dimensional decision vectors (referred to as controls or actions in other communities) that can only be solved using the techniques from the field of mathematical programming. As a result, we have adopted a notational style that makes the relationship to the field of math programming quite transparent. A major goal of this volume is to lay the foundation for solving these very large and complex problems.

There are several major differences between this volume and the major works that precede it.

- This is the first book to bring together the fields of approximate dynamic programming with mathematical programming, with its tools for handling high-dimensional decision vectors. Prior to this volume, approximate dynamic programming had already synthesized dynamic programming with statistics (estimating value functions) and simulation (the process of stepping forward through time). However, these techniques were not applicable to the types of high-dimensional resource allocation problems that are classically solved using linear, nonlinear and integer programming.

- Our presentation is the first book to highlight the power of using the post-decision state variable as a fundamental algorithmic device in approximate dynamic programming.

- Many authors have spoken about the "curse of dimensionality." We explicitly identify three curses of dimensionality that arise in many resource management problems, and introduce an approximation strategy based on using the post-decision state variable.

- We show how approximate dynamic programming, when coupled with mathematical programming, can solve (approximately) deterministic or stochastic optimization problems that are far larger than anything that could be solved using existing techniques. The result is a new class of algorithms for solving complex resource allocation problems that exhibit state and action (decision) vectors that are effectively infinite-dimensional. The notation is chosen to facilitate the link between dynamic programming and math programming.

- We focus much more heavily on modeling than any of the existing books. Emphasis is placed throughout on the proper representation of exogenous information processes and system dynamics. Partly for this reason, we present finite-horizon models first since it requires more careful modeling of time than is needed for steady-state models.

- Examples are drawn primarily from the classical problems of resource management that arise in operations research. We make a critical distinction between single resource problems (when to sell an asset, how to fly a plane from one location to another, playing a game of backgammon) and problems with multiple resources and resource classes (how to manage a fleet of aircraft, purchasing different types of equipment, managing money in different forms of investments) by introducing specific notation for each.

- The theoretical foundations of this material can be deep and rich, but our presentation is aimed at undergraduate or masters level students with introductory courses in statistics, probability and, for chapter 12, linear programming. For more advanced students, proofs are provided in "Why does it work" sections. The presentation is aimed primarily at students in engineering interested in taking real, complex problems, developing proper mathematical models and producing computationally tractable algorithms.

Our presentation integrates the fields of Markov decision processes, math programming, statistics and simulation. The use of statistics to estimate value functions dates back to Bellman & Dreyfus (1959). Although the foundations for proving convergence of special classes of these algorithms traces its origins to the seminal paper on stochastic approximation theory (Robbins & Monro (1951)), the use of this theory (in a more modern form) to prove convergence of special classes of approximate dynamic programming algorithms did not occur until 1994 (Tsitsiklis (1994), Jaakkola et al. (1994)). The first book to bring these themes together is Bertsekas & Tsitsiklis (1996), although an importance reference within the computer science community is Sutton & Barto (1998). This volume is the first book to combine these techniques with math programming, a step that is not practical without the use of the post-decision state variable.

## 1.7 BIBLIOGRAPHIC NOTES

In the operations research community, Bellman's seminal text (Bellman (1957)) is viewed as the foundation of the field of dynamic programming. Numerous books followed using the framework established by Bellman, each making important contributions to the evolution of the field. Selected highlights include Howard (1971), Derman (1970), Ross (1983), and Heyman & Sobel (1984). As of this writing, the best overall treatment of what has become known as the field of Markov decision processes is given in Puterman (1994).

This book provides a brief introduction to this classical material. The material is included partly because of its historical importance, but more significantly because it provides important theoretical and algorithmic insights into the behavior of algorithms. Our presentation departs from this classical material because it is extremely difficult to apply. During the 1980's, as computers became more accessible, a number of researchers and practitioners evolved a series of computational techniques with a focus on practical applications. Unlike classical Markov decision processes, which traces its roots to Bellman's work in the 1950's, this computational tradition has many roots and, as a result has evolved under many names and computational styles reflecting the nature of the applications.

The study of approximation methods in dynamic programming, designed to overcome "the curse of dimensionality," is almost as old as dynamic programming itself. Bellman & Dreyfus (1959) introduces methods for approximating value functions. Samuel (1959) independently uses approximation techniques to train a computer to play checkers, helping to launch the field that would become known in artificial intelligence as reinforcement learning. Judd (1998) provides a very nice treatment of computational methods for solving continuous problems using a variety of approximation methods, where even relatively low-dimensional problems can be computationally problematic if we attempt to discretize the problem.

This line of research, with its emphasis on computation, made a dramatic step forward in visibility with the emergence of two key books. The first was Bertsekas & Tsitsiklis (1996), which used the name "neuro-dynamic programming." This book synthesized for the first time the relationship between dynamic programming and a field known as stochastic approximation methods, invented in 1950's, which provided the theoretical foundation for proving convergence of algorithms based on randomly sampled information. The second, Sutton & Barto (1998), synthesized the relationship between dynamic programming and reinforcement learning. This book has helped feed an explosion of research in approximate dynamic programming (under the umbrella of "reinforcement learning") in the computer science community.

While these books have received the greatest visibility, special recognition is due to a series of workshops funded by the National Science Foundation under the leadership of Paul Werbos, some of which have been documented in several edited volumes (Miller et al. (1990), White & Sofge (1992) and Si et al. (2004*a*)). These workshops have played a significant role in bringing different communities together, the effect of which can be found throughout this volume.

# CHAPTER 2

# SOME ILLUSTRATIVE MODELS

Dynamic programming is one of those incredibly rich fields that has filled the careers of many. But it is also a deceptively easy idea to illustrate and use. This chapter presents a series of applications that illustrate the modeling of dynamic programs. The goal of the presentation is to teach dynamic programming by example. The applications range from problems that can be solved analytically, to those that can be solved using fairly simple numerical algorithms, to very large scale problems that will require carefully designed applications. The examples in this chapter effectively communicate the range of applications that the techniques in this book are designed to solve.

It is possible, after reading this chapter, to conclude that "dynamic programming is easy" and to wonder "why do I need the rest of this book?" The answer is: sometimes dynamic programming *is* easy and requires little more than the understanding gleaned from these simple problems. But there is a vast array of problems that are quite difficult to model, and where standard solution approaches are computationally intractable.

We present three classes of examples. 1) Deterministic problems, where everything is known; 2) stochastic problems, where some information is unknown but which are described by a known probability distribution; and 3) information acquisition problems, where we have uncertainty described by an unknown distribution. In the last problem class, the focus is on collecting information so that we can better estimate the distribution.

These illustrations are designed to teach by example. The careful reader will pick up subtle modeling decisions, in particular the indexing with respect to time. We defer to chapter 5 a more complete explanation of our choices, where we provide an in-depth treatment of how to model a dynamic program.

## 2.1 DETERMINISTIC PROBLEMS

Dynamic programming is widely used in many deterministic problems as a technique for breaking down what might be a very large problem into a sequence of much smaller problems. The focus of this book is on stochastic problems, but the value of dynamic programming for solving some large, deterministic problems should never be lost.

### 2.1.1 The shortest path problem

Perhaps one of the most popular dynamic programming problems is known as the shortest path problem. Although it has a vast array of applications, it is easiest to describe in terms of the problem faced by every driver when finding a path from one location to the next over a road network. Let

$$\begin{aligned}
\mathcal{I} &= \text{The set of nodes (intersections) in the network,} \\
\mathcal{L} &= \text{The set of links } (i,j) \text{ in the network,} \\
c_{ij} &= \text{The cost (typically the time) to drive from node } i \text{ to node } j, \\
&\quad\; i,j \in \mathcal{I}, (i,j) \in \mathcal{L}, \\
\mathcal{I}_i^+ &= \text{The set of nodes } j \text{ for which there is a link } (i,j) \in \mathcal{L}, \\
\mathcal{I}_j^- &= \text{The set of nodes } i \text{ for which there is a link } (i,j) \in \mathcal{L}.
\end{aligned}$$

We assume that a traveler at node $i$ can choose to traverse any link $(i,j)$, where $j \in \mathcal{I}_i^+$. Assume our traveler is starting at some node $q$ and needs to get to a destination node $r$ at the least cost. Let

$$v_j = \text{The minimum cost required to get from node } j \text{ to node } r.$$

We do not know $v_j$, but we do know that $v_r = 0$. Let $v_j^n$ be our estimate, at iteration $n$, of the cost to get from $j$ to $r$. We can find the optimal costs, $v_j$, by initially setting $v_j^0$ to a large number for $j \neq r$ and then iteratively looping over all the nodes, finding the best link to traverse out of an intersection $i$ by minimizing the sum of the outbound link cost $c_{ij}$ plus our current estimate of the downstream value $v_j^{n-1}$. The complete algorithm is summarized in figure 2.1. This algorithm has been proven to converge to the optimal set of node values.

There is a substantial literature on solving shortest path problems. Because they arise in so many applications, there is tremendous value in solving them very quickly. Our basic algorithm is not very efficient because we are often solving equation (2.1) for an intersection $i$ where $v_i^{n-1} = M$, and where $v_j^{n-1} = M$ for all $j \in \mathcal{I}_i^+$. A more standard strategy is to maintain a candidate list of nodes $\mathcal{C}$ that consists of an ordered list $i_1, i_2, \ldots$. Initially the list will consist only of the destination node $r$ (since we are solving the problem of finding paths into the destination node $r$). As we reach over links into node $i$ in the candidate list, we may find a better path from some node $j$ which is then added to the candidate list (if it is not already there).

This is often referred to as Bellman's algorithm, although the algorithm in figure 2.1 is a purer form of Bellman's equation for dynamic programming. A very effective variation of the algorithm in 2.2 is to keep track of nodes that have already been in the candidate list. If a node is added to the candidate list that was previously in the candidate list, a very effective strategy is to add this node to the top of the list. This variant is known as Pape's algorithm (pronounced "papa's"). Another powerful variation, called Dijkstra's algorithm

---

**Step 0.** Let

$$v_j^0 = \begin{cases} M & j \neq r, \\ 0 & j = r. \end{cases}$$

where "$M$" is known as "big-M" and represents a large number. Let $n = 1$.

**Step 1.** Solve for all $i \in \mathcal{I}$,

$$v_i^n = \min_{j \in \mathcal{I}_i^+} \left( c_{ij} + v_j^{n-1} \right). \tag{2.1}$$

**Step 2.** If $v_i^n < v_i^{n-1}$ for any $i$, let $n = n + 1$ and return to step 1. Else stop.

---

**Figure 2.1** A basic shortest path algorithm.

---

**Step 0.** Let

$$v_j = \begin{cases} M & j \neq r. \\ 0 & j = r. \end{cases}$$

Let $n = 1$. Set the candidate list $\mathcal{C} = \{q\}$.

**Step 1.** Choose node $j \in \mathcal{C}$ from the top of the candidate list.

**Step 2.** For all nodes $i \in \mathcal{I}_j^-$ do:

**Step 2a.**

$$\hat{v}_i = c_{ij} + v_j. \tag{2.2}$$

**Step 2b.** If $\hat{v}_i < v_i$, then set $v_i = \hat{v}_i$. If $i \notin \mathcal{C}$, add $i$ to the candidate list: $\mathcal{C} = \mathcal{C} \cup \{i\}$ ($i$ is assumed to be put at the bottom of the list).

**Step 3.** Drop node $j$ from the candidate list. If the candidate list $\mathcal{C}$ is not empty, return to step 1.

---

**Figure 2.2** A more efficient shortest path algorithm.

(pronounced "Dike-stra"), chooses the node from the candidate list with the smallest value of $v_i^n$.

Almost any (deterministic) discrete dynamic program can be viewed as a shortest path problem. We can view each node $i$ as representing a particular discrete state of the system. The origin node $q$ is our starting state, and the ending state $r$ might be any state at an ending time $T$. We can also have shortest path problems defined over infinite horizons, although we would typically include a discount factor.

### 2.1.2 The discrete budgeting problem

Assume we have to allocate a budget of size $R$ to a series of tasks $\mathcal{T}$. Let $x_t$ be the amount of money allocated to task $t$, and let $C_t(x_t)$ be the contribution (or reward) that we receive from this allocation. We would like to maximize our total contribution

$$\max_x \sum_{t \in \mathcal{T}} C_t(x_t) \tag{2.3}$$

subject to the constraint on our available resources

$$\sum_{t \in \mathcal{T}} x_t = R. \tag{2.4}$$

In addition, we cannot allocate negative resources to any task, so we include

$$x_t \geq 0. \tag{2.5}$$

We refer to (2.3)-(2.5) as the *budgeting problem* (other authors refer to it as the "resource allocation problem," a term we find too general for such a simple problem). In this example, all data are deterministic. There are a number of algorithmic strategies for solving this problem that depend on the structure of the contribution function, but we are going to show how it can be solved without any assumptions.

We will approach this problem by first deciding how much to allocate to task 1, then to task 2, and so on, until the last task, $T$. In the end, however, we want a solution that optimizes over all tasks. Let

$V_t(R_t)$ = The value of having $R_t$ resources remaining to allocate to task $t$ and later tasks.

Implicit in our definition of $V_t(R_t)$ is that we are going to solve the problem of allocating $R_t$ over tasks $t, t+1, \ldots, T$ in an optimal way. Imagine that we somehow know the function $V_{t+1}(R_{t+1})$. The relationship between $R_{t+1}$ and $R_t$ is given by

$$R_{t+1} = R_t - x_t. \tag{2.6}$$

In the language of dynamic programming, $R_t$ is known as the *state variable*, which captures all the information we need to model the system forward in time (we provide a more careful definition in chapter 5). Equation (2.6) is the *transition function* which relates the state at time $t$ to the state at time $t+1$. Sometimes we need to explicitly refer to the transition function (rather than just the state at time $t+1$), in which case we use

$$R^M(R_t, x_t) = R_t - x_t. \tag{2.7}$$

Equation (2.7) is referred to in some communities as the *system model*, since it models the physics of the system over time (hence our use of the superscript $M$).

The relationship between $V_t(R_t)$ and $V_{t+1}(R_{t+1})$ is given by

$$V_t(R_t) = \max_{0 \leq x_t \leq R_t} \left( C_t(x_t) + V_{t+1}(R^M(R_t, x_t)) \right). \tag{2.8}$$

Equation (2.8) is the optimality equation and represents the foundational equation for dynamic programming. It says that the value of having $R_t$ resources for task $t$ is the value of optimizing the contribution from task $t$ plus the value of then having $R^M(R_t, x_t) = R_{t+1} = R_t - x_t$ resources for task $t+1$ (and beyond). It forces us to balance the contribution from task $t$ against the value that we would receive from all future tasks (which is captured in $V_{t+1}(R_t - x_t)$). One way to solve (2.8) is to assume that $x_t$ is discrete. For example, if our budget is $R = \$10$ million, we might require $x_t$ to be in units of \$100,000 dollars. In this case, we would solve (2.8) simply by searching over all possible values of $x_t$ (since it is a scalar, this is not too hard). The problem is that we do not know what $V_{t+1}(R_{t+1})$ is.

The simplest strategy for solving our dynamic program in (2.8) is to start by using $V_{T+1}(R) = 0$ (for any value of $R$). Then we would solve

$$V_T(R_T) = \max_{0 \leq x_T \leq R_T} C_T(x_T) \tag{2.9}$$

for $0 \leq R_T \leq R$. Now we know $V_T(R_T)$ for any value of $R_T$ that might actually happen. Next we can solve

$$V_{T-1}(R_{T-1}) = \max_{0 \leq x_{T-1} \leq R_{T-1}} \left( C_{T-1}(x_{T-1}) + V_T(R_{T-1} - x_{T-1}) \right). \tag{2.10}$$

Clearly, we can play this game recursively, solving (2.8) for $t = T-1, T-2, \ldots, 1$. Once we have computed $V_t$ for $t = (1, 2, \ldots, T)$, we can then start at $t = 1$ and step forward in time to determine our optimal allocations.

This strategy is simple, easy, and optimal. It has the nice property that we do not need to make any assumptions about the shape of $C_t(x_t)$, other than finiteness. We do not need concavity or even continuity; we just need the function to be defined for the discrete values of $x_t$ that we are examining.

### 2.1.3 The continuous budgeting problem

It is usually the case that dynamic programs have to be solved numerically. In this section, we introduce a form of the budgeting problem that can be solved analytically. Assume that the resources we are allocating are continuous (for example, how much money to assign to various activities), which means that both $R_t$ and $x_t$ are continuous. We are then going to assume that the contribution from allocating $x_t$ dollars to task $t$ is given by

$$C_t(x_t) = \sqrt{x_t}.$$

This function assumes that there are diminishing returns from allocating additional resources to a task, as is common in many applications. We can solve this problem exactly using dynamic programming. We first note that if we have $R_T$ dollars left for the last task, the value of being in this state is

$$V_T(R_T) = \max_{x_T \leq R_T} \sqrt{x_T}.$$

Since the contribution increases monotonically with $x_T$, the optimal solution is $x_T = R_T$, which means that $V_T(R_T) = \sqrt{R_T}$. Now consider the problem at time $t = T-1$. The value of being in state $R_{T-1}$ would be

$$V_{T-1}(R_{T-1}) = \max_{x_{T-1} \leq R_{T-1}} \left( \sqrt{x_{T-1}} + V_T(R_T(x_{T-1})) \right) \tag{2.11}$$

where $R_T(x_{T-1}) = R_{T-1} - x_{T-1}$ is the money left over from time period $T-1$. Since we know $V_T(R_T)$ we can rewrite (2.11) as

$$V_{T-1}(R_{T-1}) = \max_{x_{T-1} \leq R_{T-1}} \left( \sqrt{x_{T-1}} + \sqrt{R_{T-1} - x_{T-1}} \right). \tag{2.12}$$

We solve (2.12) by differentiating with respect to $x_{T-1}$ and setting the derivative equal to zero (we are taking advantage of the fact that we are maximizing a continuously differentiable, concave function). Let

$$F_{T-1}(R_{T-1}, x_{T-1}) = \sqrt{x_{T-1}} + \sqrt{R_{T-1} - x_{T-1}}.$$

Differentiating $F_{T-1}(R_{T-1}, x_{T-1})$ and setting this equal to zero gives

$$\begin{aligned} \frac{\partial F_{T-1}(R_{T-1}, x_{T-1})}{\partial x_{T-1}} &= \frac{1}{2}(x_{T-1})^{-\frac{1}{2}} - \frac{1}{2}(R_{T-1} - x_{T-1})^{-\frac{1}{2}} \\ &= 0. \end{aligned}$$

This implies

$$x_{T-1} = R_{T-1} - x_{T-1}$$

which gives

$$x^*_{T-1} = \frac{1}{2} R_{T-1}.$$

We now have to find $V_{T-1}$. Substituting $x^*_{T-1}$ back into (2.12) gives

$$\begin{aligned} V_{T-1}(R_{T-1}) &= \sqrt{R_{T-1}/2} + \sqrt{R_{T-1}/2} \\ &= 2\sqrt{R_{T-1}/2}. \end{aligned}$$

We can continue this exercise, but there seems to be a bit of a pattern forming (this is a common trick when trying to solve dynamic programs analytically). It seems that a general formula might be

$$V_{T-t+1}(R_{T-t+1}) = t\sqrt{R_{T-t+1}/t} \tag{2.13}$$

or, equivalently,

$$V_t(R_t) = (T-t+1)\sqrt{R_t/(T-t+1)}. \tag{2.14}$$

How do we determine if this guess is correct? We use a technique known as proof by induction. We assume that (2.13) is true for $V_{T-t+1}(R_{T-t+1})$ and then show that we get the same structure for $V_{T-t}(R_{T-t})$. Since we have already shown that it is true for $V_T$ and $V_{T-1}$, this result would allow us to show that it is true for all $t$.

Finally, we can determine the optimal solution using the value function in equation (2.14). The optimal value of $x_t$ is found by solving

$$\max_{x_t} \left( \sqrt{x_t} + (T-t)\sqrt{(R_t - x_t)/(T-t)} \right). \tag{2.15}$$

Differentiating and setting the result equal to zero gives

$$\frac{1}{2}(x_t)^{-\frac{1}{2}} - \frac{1}{2}\left(\frac{R_t - x_t}{T-t}\right)^{-\frac{1}{2}} = 0.$$

This implies that

$$x_t = (R_t - x_t)/(T-t).$$

Solving for $x_t$ gives

$$x^*_t = R_t/(T-t+1).$$

This gives us the very intuitive result that we want to evenly divide the available budget among all remaining tasks. This is what we would expect since all the tasks produce the same contribution.

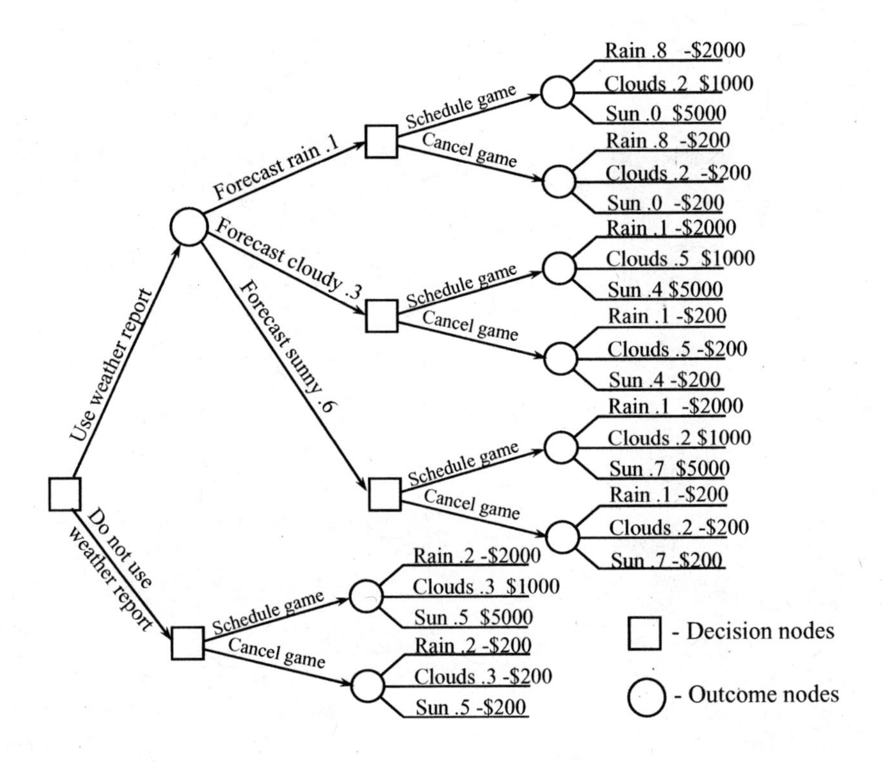


**Figure 2.3** Decision tree showing decision nodes and outcome nodes.

## 2.2 STOCHASTIC PROBLEMS

Dynamic programming can be a useful algorithmic strategy for deterministic problems, but it is often an essential strategy for stochastic problems. In this section, we illustrate a number of stochastic problems, with the goal of illustrating the challenge of modeling the flow of information (essential to any stochastic problem). We also make the transition from problems with fairly simple state variables to problems with extremely large state spaces.

### 2.2.1 Decision trees

One of the most effective ways of communicating the process of making decisions under uncertainty is to use decision trees. Figure 2.3 illustrates a problem facing a Little League baseball coach trying to schedule a playoff game under the threat of bad weather. The coach first has to decide if he should check the weather report. Then he has to decide if he should schedule the game. Bad weather brings a poor turnout that reduces revenues from tickets and the concession stand. There are costs if the game is scheduled (umpires, food, people to handle parking and the concession stand) that need to be covered by the revenue the game might generate.

In figure 2.3, squares denote decision nodes where we have to choose an action (Does he check the weather report? Does he schedule the game?), while circles represent outcome

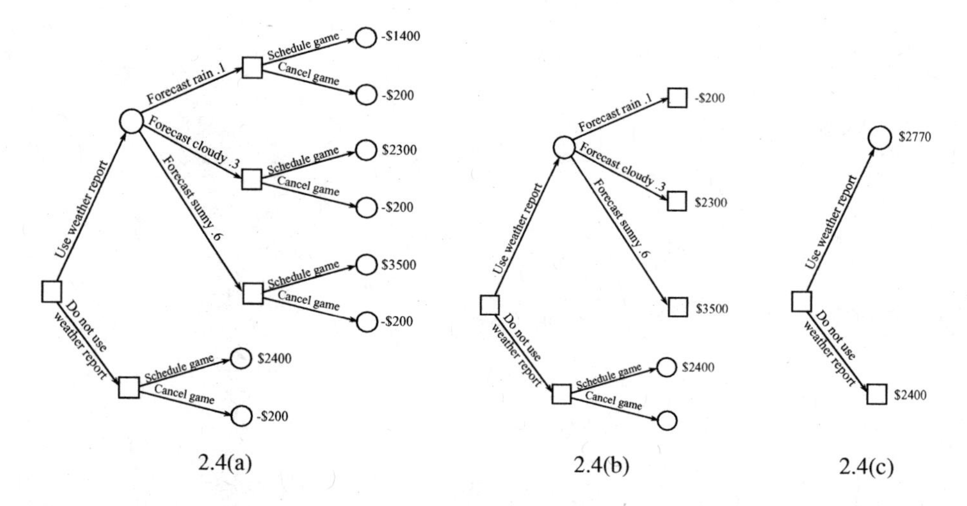


**Figure 2.4** Evaluating a decision tree. (a) Evaluating the final outcome nodes. (b) Evaluating the final decision nodes. (c) Evaluating the first outcome nodes.

nodes where new (and random) information arrives (What will the weather report say? What will the weather be?). We can "solve" the decision tree (that is, find the best decision given the information available), by rolling backward through the tree. In figure 2.4(a), we have found the expected value of being at each of the end outcome nodes. For example, if we check the weather report and see a forecast of rain, the probability it will actually rain is 0.80, producing a loss of \$2000; the probability that it will be cloudy is 0.20, producing a profit of \$1000; the probability it will be sunny is zero (if it were sunny, we would make a profit of \$5000). The expected value of scheduling the game, when the weather forecast is rain, is $(.8)(-\$2000) + (.2)(\$1000) + (0)(\$5000) = -\$1400$. Repeating this calculation for each of the ending outcome nodes produces the results given in figure 2.4(a).

At a decision node, we get to choose an action, and of course we choose the action with the highest expected profit. The results of this calculation are given in figure 2.4(b). Finally, we have to determine the expected value of checking the weather report by again by multiplying the probability of each possible weather forecast (rainy, cloudy, sunny) times the expected value of each outcome. Thus, the expected value of checking the weather report is $(.1)(-\$200) + (.3)(\$2300) + (.6)(\$3500) = \$2770$, shown in figure 2.4(c). The expected value of making decisions without the weather report is \$2400, so the analysis shows that we should check the weather report. Alternatively, we can interpret the result as telling us that we would be willing to pay up to \$300 for the weather report.

Almost any dynamic program with discrete states and actions can be modeled as a decision tree. The problem is that they are not practical when there are a large number of states and actions.

### 2.2.2 A stochastic shortest path problem

We are often interested in shortest path problems where there is uncertainty in the cost of traversing a link. For our transportation example, it is natural to view the travel time on a link as random, reflecting the variability in traffic conditions on each link. There are two ways we can handle the uncertainty. The simplest is to assume that our driver has to make a decision before seeing the travel time over the link. In this case, our updating equation would look like

$$v_i^n = \min_{j \in \mathcal{I}_i^+} \mathbb{E}\{c_{ij}(W) + v_j^{n-1}\}$$

where $W$ is some random variable that contains information about the network (such as travel times). This problem is identical to our original problem; all we have to do is to let $c_{ij} = \mathbb{E}\{c_{ij}(W)\}$ be the expected cost on an arc.

An alternative model is to assume that we know the travel time on a link from $i$ to $j$ as soon as we arrive at node $i$. In this case, we would have to solve

$$v_i^n = \mathbb{E}\left\{ \min_{j \in \mathcal{I}_i^+} \left(c_{ij}(W) + v_j^{n-1}\right) \right\}.$$

Here, the expectation is outside of the min operator that chooses the best decision, capturing the fact that now the decision itself is random.

Note that our notation is ambiguous, in that with the same notation, we have two very different models. In chapter 5, we are going to refine our notation so that it will be immediately apparent when a decision "sees" the random information and when the decision has to be made before the information becomes available.

### 2.2.3 The gambling problem

A gambler has to determine how much of his capital he should bet on each round of a game, where he will play a total of $N$ rounds. He will win a bet with probability $p$ and lose with probability $q = 1 - p$ (assume $q < p$). Let $s^n$ be his total capital after $n$ plays, $n = 1, 2, \ldots, N$, with $s^0$ being his initial capital. For this problem, we refer to $s^n$ as the state of the system. Let $x^n$ be the amount he bets in round $n$, where we require that $x^n \leq s^{n-1}$. He wants to maximize $\ln s^N$ (this provides a strong penalty for ending up with a small amount of money at the end and a declining marginal value for higher amounts).

Let

$$W^n = \begin{cases} 1 & \text{if the gambler wins the } n^{th} \text{ game,} \\ 0 & \text{otherwise.} \end{cases}$$

The system evolves according to

$$S^n = S^{n-1} + x^n W^n - x^n(1 - W^n).$$

Let $V^n(S^n)$ be the value of having $S^n$ dollars at the end of the $n^{th}$ game. The value of being in state $S^n$ at the end of the $n^{th}$ round can be written

$$\begin{aligned} V^n(S^n) &= \max_{0 \leq x^{n+1} \leq S^n} \mathbb{E}\{V^{n+1}(S^{n+1})|S^n\} \\ &= \max_{0 \leq x^{n+1} \leq S^n} \mathbb{E}\{V^{n+1}(S^n + x^{n+1}W^{n+1} - x^{n+1}(1 - W^{n+1}))|S^n\}. \end{aligned}$$

Here, we claim that the value of being in state $S^n$ is found by choosing the decision that maximizes the expected value of being in state $S^{n+1}$ given what we know at the end of the $n^{th}$ round.

We solve this by starting at the end of the $N^{th}$ trial, and assuming that we have finished with $S^N$ dollars. The value of this is

$$V^N(S^N) = \ln S^N.$$

Now step back to $n = N - 1$, where we may write

$$\begin{aligned} V^{N-1}(S^{N-1}) &= \max_{0 \le x^N \le S^{N-1}} \mathbb{E}\{V^N(S^{N-1} + x^N W^N - x^N(1 - W^N))|S^{N-1}\} \\ &= \max_{0 \le x^N \le S^{N-1}} \left[p \ln(S^{N-1} + x^N) + (1-p)\ln(S^{N-1} - x^N)\right]. \end{aligned} \quad (2.16)$$

Let $V^{N-1}(S^{N-1}, x^N)$ be the value within the max operator. We can find $x^N$ by differentiating $V^{N-1}(S^{N-1}, x^N)$ with respect to $x^N$, giving

$$\begin{aligned} \frac{\partial V^{N-1}(S^{N-1}, x^N)}{\partial x^N} &= \frac{p}{S^{N-1} + x^N} - \frac{1-p}{S^{N-1} - x^N} \\ &= \frac{2S^{N-1}p - S^{N-1} - x^N}{(S^{N-1})^2 - (x^N)^2}. \end{aligned}$$

Setting this equal to zero and solving for $x^N$ gives

$$x^N = (2p - 1)S^{N-1}.$$

The next step is to plug this back into (2.16) to find $V^{N-1}(s^{N-1})$ using

$$\begin{aligned} V^{N-1}(S^{N-1}) &= p\ln(S^{N-1} + S^{N-1}(2p-1)) + (1-p)\ln(S^{N-1} - S^{N-1}(2p-1)) \\ &= p\ln(S^{N-1}2p) + (1-p)\ln(S^{N-1}2(1-p)) \\ &= p\ln S^{N-1} + (1-p)\ln S^{N-1} + \underbrace{p\ln(2p) + (1-p)\ln(2(1-p))}_{K} \\ &= \ln S^{N-1} + K, \end{aligned}$$

where $K$ is a constant with respect to $S^{N-1}$. Since the additive constant does not change our decision, we may ignore it and use $V^{N-1}(S^{N-1}) = \ln S^{N-1}$ as our value function for $N - 1$, which is the same as our value function for $N$. Not surprisingly, we can keep applying this same logic backward in time and obtain

$$V^n(S^n) = \ln S^n \; ( +K^N)$$

for all $n$, where again, $K^n$ is some constant that can be ignored. This means that for all $n$, our optimal solution is

$$x^n = (2p - 1)S^{n-1}.$$

The optimal strategy at each iteration is to bet a fraction $\beta = (2p - 1)$ of our current money on hand. Of course, this requires that $p > .5$.

### 2.2.4 Asset pricing

Imagine you are holding an asset that you can sell at a price that fluctuates randomly. In this problem we want to determine the best time to sell the asset, and from this, infer the value of the asset. For this reason, this type of problem arises frequently in the context of asset pricing.

Let $\hat{p}_t$ be the price that is revealed in period $t$, at which point you have to make a decision

$$x_t = \begin{cases} 1 & \text{Sell.} \\ 0 & \text{Hold.} \end{cases}$$

For our simple model, we assume that $\hat{p}_t$ is independent of prior prices (a more typical model would assume that the *change* in price is independent of prior history). With this assumption, our system has two states:

$$S_t = \begin{cases} 1 & \text{We are holding the asset,} \\ 0 & \text{We have sold the asset.} \end{cases}$$

Assume that we measure the state immediately after the price $\hat{p}_t$ has been revealed but before we have made a decision. If we have sold the asset, then there is nothing we can do. We want to maximize the price we receive when we sell our asset. Let the scalar $V_t$ be the value of holding the asset at time $t$. This can be written

$$V_t = \max_{x_t \in \{0,1\}} \big( x_t \hat{p}_t + (1 - x_t)\gamma \mathbb{E} V_{t+1} \big).$$

So, we either get the price $\hat{p}_t$ if we sell, or we get the discounted future value of the asset. Assuming the discount factor $\gamma < 1$, we do not want to hold too long simply because the value in the future is worth less than the value now. In practice, we eventually will see a price $\hat{p}_t$ that is greater than the future expected value, at which point we would stop the process and sell our asset.

The time at which we sell our asset is known as a *stopping time*. By definition, $x_\tau = 1$. It is common to think of $\tau$ as the decision variable, where we wish to solve

$$\max_{\tau} \mathbb{E} \hat{p}_\tau. \tag{2.17}$$

Equation (2.17) is a little tricky to interpret. Clearly, the choice of when to stop is a random variable since it depends on the price $\hat{p}_t$. We cannot optimally choose a random variable, so what is meant by (2.17) is that we wish to choose a *function* (or *policy*) that determines when we are going to sell. For example, we would expect that we might use a rule that says

$$X_t(S_t, \bar{p}) = \begin{cases} 1 & \text{if } \hat{p}_t \geq \bar{p} \text{ and } S_t = 1, \\ 0 & \text{otherwise.} \end{cases} \tag{2.18}$$

In this case, we have a function parameterized by $\bar{p}$. In this case, we would write our problem in the form

$$\max_{\bar{p}} \mathbb{E} \sum_{t=1}^{\infty} \gamma^t X_t(S_t, \bar{p}).$$

This formulation raises two questions. First, while it seems very intuitive that our policy would take the form given in equation (2.18), there is the theoretical question of whether this in fact is the structure of an optimal policy. Section 7.5 demonstrates how these questions can be answered in the context of a class of batch replenishment problems. The second question is how to find the best policy within this class. For this problem, that means finding the parameter $\bar{p}$. For problems where the probability distribution of the random process driving prices is (assumed) known, this is a rich and deep theoretical challenge. Alternatively, there is a class of algorithms from stochastic optimization that allows us to find "good" values of the parameter in a fairly simple way.

### 2.2.5 The asset acquisition problem - I

A basic asset acquisition problem arises in applications where we purchase product at time $t$ to be used during time interval $t+1$. We can model the problem using

$$
\begin{aligned}
R_t &= \text{The assets on hand at time } t \text{ before we make a new ordering decision, and} \\
&\quad\ \text{before we have satisfied any demands arising in time interval } t, \\
x_t &= \text{The amount of product purchased at time } t \text{ to be used during time interval} \\
&\quad\ t+1, \\
\hat{D}_t &= \text{The random demands that arise between } t-1 \text{ and } t.
\end{aligned}
$$

We have chosen to model $R_t$ as the resources on hand in period $t$ before demands have been satisfied. It would have been just as natural to define $R_t$ as the resources on hand *after* demands have been satisfied. Our definition here makes it easier to introduce (in the next section) the decision of how much demand we should satisfy.

We assume we purchase new assets at a fixed price $p^p$ and sell them at a fixed price $p^s$. The amount we earn between $t-1$ and $t$, including the decision we make at time $t$, is given by

$$C_t(x_t) = p^s \min\{R_t, \hat{D}_t\} - p^p x_t.$$

Our inventory $R_t$ is described using the equation

$$R_{t+1} = R_t - \min\{R_t, \hat{D}_t\} + x_t.$$

We assume that any unsatisfied demands are lost to the system.

This problem can be solved using Bellman's equation. For this problem, $R_t$ is our state variable. Let $V_t(R_t)$ be the value of being in state $R_t$. Then Bellman's equation tells us that

$$V_t(R_t) = \max_{x_t} \left( C_t(x_t) + \gamma \mathbb{E} V_{t+1}(R_{t+1}) \right).$$

where the expectation is over all the possible realizations of the demands $\hat{D}_{t+1}$.

We note that our reward for maintaining inventory ($p^s$ times the amount of demand we satisfy) at time $t$ is not a function of $x_t$. An alternative formulation of this problem is to write the contribution based on what we will receive between $t$ and $t+1$. In this case, we would write the contribution as

$$C_{t+1}(x_t, \hat{D}_{t+1}) = p^s \min\{(R_t - \min\{R_t, \hat{D}_t\} + x_t), \hat{D}_{t+1}\} - p^p x_t.$$

Now our contribution function a function of $x_t$ at time $t$, but it is also random, so Bellman's equation would be written

$$V_t(R_t) = \max_{x_t} \mathbb{E} \left( C_{t+1}(x_t, \hat{D}_{t+1}) + \gamma V_{t+1}(R_{t+1}) \right).$$

There are many applications where the contribution function is more naturally measured based on what happened between $t-1$ and $t$ (which means it is deterministic at time $t$), while for other applications it is more natural to measure events happening in the next time period. The only difference is in where we place the expectation.

This equation is not too hard to solve if we assume that $R_t$ is discrete. We start at the end of our planning horizon $T$, where we assume that $V_{T+1}(R_{T+1}) = 0$, and simply solve the equation for each $R_t$ while stepping backward in time.

### 2.2.6 The asset acquisition problem - II

Many asset acquisition problems introduce additional sources of uncertainty. The assets we are acquiring could be stocks, planes, energy commodities such as oil, consumer goods, and blood. In addition to the need to satisfy random demands (the only source of uncertainty we considered in our basic asset acquisition problem), we may also have randomness in the prices at which we buy and sell assets. We may also include exogenous changes to the assets on hand due to additions (cash deposits, blood donations, energy discoveries) and subtractions (cash withdrawals, equipment failures, theft of product).

We can model the problem using

$x_t^p$ = Assets purchased (acquired) at time $t$ to be used during time interval $t+1$,

$x_t^s$ = Amount of assets sold to satisfy demands during time interval $t$,

$x_t$ = $(x_t^p, x_t^s)$,

$R_t$ = Resource level at time $t$ before any decisions are made,

$D_t$ = Demands waiting to be served at time $t$.

Of course, we are going to require that $x_t^s \leq \min\{R_t, D_t\}$ (we cannot sell what we do not have, and we cannot sell more than the market demand). We are also going to assume that we buy and sell our assets at market prices that fluctuate over time. These are described using

$p_t^p$ = Market price for purchasing assets at time $t$,

$p_t^s$ = Market price for selling assets at time $t$,

$p_t$ = $(p_t^s, p_t^p)$.

Our system evolves according to several types of exogenous information processes that include random changes to the supplies (assets on hand), demands and prices. We model these using

$\hat{R}_t$ = Exogenous changes to the assets on hand that occur during time interval $t$,

$\hat{D}_t$ = Demand for the resources during time interval $t$,

$\hat{p}_t^p$ = Change in the purchase price that occurs between $t-1$ and $t$,

$\hat{p}_t^s$ = Change in the selling price that occurs between $t-1$ and $t$,

$\hat{p}_t$ = $(\hat{p}_t^p, \hat{p}_t^s)$.

We assume that the exogenous changes to assets, $\hat{R}_t$, occurs before we satisfy demands.

For more complex problems such as this, it is convenient to have a generic variable for exogenous information. We use the notation $W_t$ to represent all the information that first arrives between $t-1$ and $t$, where for this problem, we would have

$$W_t = (\hat{R}_t, \hat{D}_t, \hat{p}_t).$$

The state of our system is described by

$$S_t = (R_t, D_t, p_t).$$

We represent the evolution of our state variable generically using

$$S_{t+1} = S^M(S_t, x_t, W_{t+1}).$$

Some communities refer to this as the "system model," hence our notation. We refer to this as the transition function. More specifically, the equations that make up our transition function would be

$$\begin{aligned} R_{t+1} &= R_t - x_t^s + x_t^p + \hat{R}_{t+1}, \\ D_{t+1} &= D_t - x_t^s + \hat{D}_{t+1}, \\ p_{t+1}^p &= p_t^p + \hat{p}_{t+1}^p, \\ p_{t+1}^s &= p_t^s + \hat{p}_{t+1}^s. \end{aligned}$$

The one-period contribution function is

$$C_t(S_t, x_t) = p_t^s x_t^s - p_t^p x_t.$$

We can find optimal decisions by solving Bellman's equation

$$V_t(S_t) = \max \left(C_t(S_t, x_t) + \gamma \mathbb{E} V_{t+1}(S_{t+1}^M(S_t, x_t, W_{t+1}))|S_t\right). \tag{2.19}$$

This problem allows us to capture a number of dimensions of the modeling of stochastic problems. This is a fairly classical problem, but we have stated it in a more general way by allowing for unsatisfied demands to be held for the future, and by allowing for random purchasing and selling prices.

### 2.2.7 The lagged asset acquisition problem

A variation of the basic asset acquisition problem we introduced in section 2.2.5 arises when we can purchase assets now to be used in the future. For example, a hotel might book rooms at time $t$ for a date $t'$ in the future. A travel agent might purchase space on a flight or a cruise line at various points in time before the trip actually happens. An airline might purchase contracts to buy fuel in the future. In all of these cases, it will generally be the case that assets purchased farther in advance are cheaper, although prices may fluctuate. For this problem, we are going to assume that selling prices are

$x_{tt'}$ = Assets purchased at time $t$ to be used to satisfy demands that become known during time interval between $t'-1$ and $t'$,

$x_t$ = $(x_{t,t+1}, x_{t,t+2}, \ldots,)$,
 = $(x_{tt'})_{t'>t}$,

$\hat{D}_t$ = Demand for the resources that become known during time interval $t$,

$R_{tt'}$ = Total assets acquired on or before time $t$ that may be used to satisfy demands that become known between $t'-1$ and $t'$,

$R_t$ = $(R_{tt'})_{t'\geq t}$.

Now, $R_{tt}$ is the resources on hand in period $t$ that can be used to satisfy demands $\hat{D}_t$ that become known during time interval $t$. In this formulation, we do not allow $x_{tt}$, which would represent purchases on the spot market. If this were allowed, purchases at time $t$ could be used to satisfy unsatisfied demands arising during time interval between $t-1$ and $t$.

The transition function is given by

$$R_{t+1,t'} = \begin{cases} \left(R_{t,t} - \min(R_{t,t}, \hat{D}_t)\right) + x_{t,t+1} + R_{t,t+1}, & t' = t+1, \\ R_{tt'} + x_{tt'}, & t' > t+1. \end{cases}$$

The one-period contribution function (measuring forward in time) is

$$C_t(R_t, \hat{D}_t) = p^s \min(R_{t,t}, \hat{D}_t) - \sum_{t'>t} p^p x_{tt'}.$$

We can again formulate Bellman's equation as in (2.19) to determine an optimal set of decisions. From a computational perspective, however, there is a critical difference. Now, $x_t$ and $R_t$ are vectors with elements $x_{tt'}$ and $R_{tt'}$, which makes it computationally impossible to enumerate all possible states (or actions).

### 2.2.8 The batch replenishment problem

One of the classical problems in operations research is one that we refer to here as the batch replenishment problem. To illustrate the basic problem, assume that we have a single type of resource that is consumed over time. As the reserves of the resource run low, it is necessary to replenish the resources. In many problems, there are economies of scale in this process. It is cheaper (on an average cost basis) to increase the level of resources in one jump (see examples).

---

■ **EXAMPLE 2.1**

A startup company has to maintain adequate reserves of operating capital to fund product development and marketing. As the cash is depleted, the finance officer has to go to the markets to raise additional capital. There are fixed costs of raising capital, so this tends to be done in batches.

■ **EXAMPLE 2.2**

An oil company maintains an aggregate level of oil reserves. As these are depleted, it will undertake exploration expeditions to identify new oil fields, which will produce jumps in the total reserves under the company's control.

---

We address this problem in some depth in chapter 7 (section 7.5). To introduce the core elements, let

$\hat{D}_t$ = Demand for the resources during time interval $t$,

$R_t$ = Resource level at time $t$,

$x_t$ = Additional resources acquired at time $t$ to be used during time interval $t+1$.

The transition function is given by

$$R^M_{t+1}(R_t, x_t, \hat{D}_{t+1}) = \max\{0, (R_t + x_t - \hat{D}_{t+1})\}.$$

Our one period cost function (which we wish to minimize) is given by

$$\begin{aligned}\hat{C}_{t+1}(R_t, x_t, \hat{D}_{t+1}) &= \text{Total cost of acquiring } x_t \text{ units of the resource} \\ &= c^f I_{\{x_t>0\}} + c^p x_t + c^h R^M_{t+1}(R_t, x_t, \hat{D}_{t+1}),\end{aligned}$$

where

$$\begin{aligned}c^f &= \text{The fixed cost of placing an order,} \\ c^p &= \text{The unit purchase cost,} \\ c^h &= \text{The unit holding cost.}\end{aligned}$$

For our purposes, $\hat{C}_{t+1}(R_t, x_t, \hat{D}_{t+1})$ could be any nonconvex function; this is a simple example of one. Since the cost function is nonconvex, it helps to order larger quantities at the same time.

Assume that we have a family of decision functions $X^\pi(R_t)$, $\pi \in \Pi$, for determining $x_t$. For example, we might use a decision rule such as

$$X^\pi(R_t) = \begin{cases} 0 & \text{if } R_t \geq s, \\ Q - R_t & \text{if } R_t < q. \end{cases}$$

where $Q$ and $q$ are specified parameters. In the language of dynamic programming, a decision rule such as $X^\pi(R_t)$ is known as a *policy* (literally, a rule for making decisions). We index policies by $\pi$, and denote the set of policies by $\Pi$. In this example, a combination $(S, s)$ represents a policy, and $\Pi$ would represent all the possible values of $Q$ and $q$.

Our goal is to solve

$$\min_{\pi \in \Pi} \mathbb{E} \left\{ \sum_{t=0}^{T} \gamma^t \hat{C}_{t+1}(R_t, X^\pi(R_t), \hat{D}_{t+1}) \right\}.$$

This means that we want to search over all possible values of $Q$ and $q$ to find the best performance (on average).

The basic batch replenishment problem, where $R_t$ and $x_t$ are scalars, is quite easy (if we know things like the distribution of demand). But there are many real problems where these are vectors because there are different types of resources. The vectors may be small (different types of fuel, raising different types of funds) or extremely large (hiring different types of people for a consulting firm or the military; maintaining spare parts inventories). Even a small number of dimensions would produce a very large problem using a discrete representation.

### 2.2.9 The transformer replacement problem

The electric power industry uses equipment known as transformers to convert the high-voltage electricity that comes out of power generating plants into currents with successively lower voltage, finally delivering the current we can use in our homes and businesses. The largest of these transformers can weigh 200 tons, might cost millions of dollars to replace and may require a year or more to build and deliver. Failure rates are difficult to estimate (the most powerful transformers were first installed in the 1960's and have yet to reach the end of their natural lifetime). Actual failures can be very difficult to predict, as they often depend on heat, power surges, and the level of use.

We are going to build an aggregate replacement model where we only capture the age of the transformers. Let

$$\begin{aligned} a &= \text{The age of a transformer (in units of time periods) at time } t, \\ R_{ta} &= \text{The number of active transformers of age } a \text{ at time } t. \end{aligned}$$

For our model, we assume that age is the best predictor of the probability that a transformer will fail. Let

$$\begin{aligned} \hat{R}_{ta} &= \text{The number of transformers of age } a \text{ that fail between } t-1 \text{ and } t, \\ p_a &= \text{The probability a transformer of age } a \text{ will fail between } t-1 \text{ and } t. \end{aligned}$$

Of course, $\hat{R}_{ta}$ depends on $R_{ta}$ since transformers can only fail if we own them.

It can take a year or two to acquire a new transformer. Assume that we are measuring time, and therefore age, in fractions of a year (say, three months). Normally it can take about six time periods from the time of purchase before a transformer is installed in the network. However, we may pay extra and get a new transformer in as little as three quarters. If we purchase a transformer that arrives in six time periods, then we might say that we have acquired a transformer that is $a = -6$ time periods old. Paying extra gets us a transformer that is $a = -3$ time periods old. Of course, the transformer is not productive until it is at least $a = 0$ time periods old. Let

$$\begin{aligned} x_{ta} &= \text{Number of transformers of age } a \text{ that we purchase at time } t, \\ c_a &= \text{The cost of purchasing a transformer of age } a. \end{aligned}$$

If we have too few transformers, then we incur what are known as "congestion costs," which represent the cost of purchasing power from more expensive utilities because of bottlenecks in the network. To capture this, let

$$\begin{aligned} \bar{R} &= \text{Target number of transformers that we should have available}, \\ R_t^A &= \text{Actual number of transformers that are available at time } t, \\ &= \sum_{a \geq 0} R_{ta}, \\ C_t(R_t^A, \bar{R}) &= \text{Expected congestion costs if } R_t^A \text{ transformers are available}, \\ &= c_0 \left( \frac{\bar{R}}{R_t^A} \right)^{\beta}. \end{aligned}$$

The function $C_t(R_t^A, \bar{R})$ captures the behavior that as $R_t^A$ falls below $\bar{R}$, the congestion costs rise quickly.

Assume that $x_{ta}$ is determined immediately after $R_{ta}$ is measured. The transition function is given by

$$R_{t+1,a} = R_{t,a-1} + x_{t,a-1} - \hat{R}_{t+1,a}.$$

Let $R_t$, $\hat{R}_t$, and $x_t$ be vectors with components $R_{ta}$, $\hat{R}_{ta}$, and $x_{ta}$, respectively. We can write our system dynamics more generally as

$$R_{t+1} = R^M(R_t, x_t, \hat{R}_{t+1}).$$

If we let $V_t(R_t)$ be the value of having a set of transformers with an age distribution described by $R_t$, then, as previously, we can write this value using Bellman's equation

$$V_t(R_t) = \min_{x_t} \left( cx_t + \mathbb{E}V_{t+1}(R^M(R_t, x_t, \hat{R}_{t+1})) \right).$$

For this application, our state variable $R_t$ might have as many as 100 dimensions. If we have, say, 200 transformers, each of which might be as many as 100 years old, then the number of possible values of $R_t$ could be $100^{200}$. Fortunately, we can develop continuous approximations that allow us to approximate problems such as this relatively easily.

### 2.2.10 The dynamic assignment problem

Consider the challenge of managing a group of technicians that help with the installation of expensive medical devices (for example, medical imaging equipment). As hospitals install this equipment, they need technical assistance with getting the machines running and training hospital personnel. The technicians may have different skills (some may be trained to handle specific types of equipment), and as they travel around the country, we may want to keep track not only of their current location, but also how long they have been on the road. We describe the attributes of a technician using

$$a_t = \begin{pmatrix} a_1 \\ a_2 \\ a_3 \end{pmatrix} = \begin{pmatrix} \text{The location of the technician} \\ \text{The type of equipment the technician is trained to handle} \\ \text{The number of days the technician has been on the road} \end{pmatrix}.$$

Since we have more than one technician, we can model the set of all technicians using

$R_{ta}$ = The number of technicians with attribute $a$,

$\mathcal{A}$ = Set of all possible values of the attribute vector $a$,

$R_t$ = $(R_{ta})_{a \in \mathcal{A}}$.

Over time, demands arise for technical services as the equipment is installed. Let

$b$ = The characteristics of a piece of equipment (location, type of equipment),

$\mathcal{B}$ = The set of all possible values of the vector $b$,

$\hat{D}_{tb}$ = The number of new pieces of equipment of type $b$ that were installed between $t-1$ and $t$ (and now need service),

$\hat{D}_t$ = $(\hat{D}_{tb})_{b \in \mathcal{B}}$,

$D_{tb}$ = The total number of pieces of equipment of type $b$ that still need to be installed at time $t$,

$D_t$ = $(D_{tb})_{b \in \mathcal{B}}$.

We next have to model the decisions that we have to make. Assume that at any point in time, we can either assign a technician to handle a new installation, or we can send the technician home. Let

$\mathcal{D}^H$ = The set of decisions to send a technician home, where $d \in \mathcal{D}^H$ represents a particular location,

$\mathcal{D}^D$ = The set of decisions to have a technician serve a demand, where $d \in \mathcal{D}$ represents a decision to serve a demand of type $b_d$,

$d^\phi$ = The decision to "do nothing" with a technician,

$\mathcal{D}$ = $\mathcal{D}^H \cup \mathcal{D}^D \cup d^\phi$.

A decision has the effect of changing the attributes of a technician, as well as possibly satisfying a demand. The impact on the attribute vector of a technician is captured using the attribute transition function, represented using

$$a_{t+1} = a^M(a_t, d).$$

For algebraic purposes, it is useful to define the indicator function

$$\delta_{a'}(a_t, d) = \begin{cases} 1 & \text{for } a^M(a_t, d) = a', \\ 0 & \text{otherwise.} \end{cases}$$

A decision $d \in \mathcal{D}^D$ means that we are serving a piece of equipment described by an attribute vector $b_d$. This is only possible, of course, if $D_{tb} > 0$. Typically, $D_{tb}$ will be 0 or 1, although our model will allow multiple pieces of equipment with the same attributes. We indicate which decisions we have made using

$x_{tad}$ = The number of times we apply a decision of type $d$ to a technician with attribute $a$,

$x_t$ = $(x_{tad})_{a \in \mathcal{A}, d \in \mathcal{D}}$.

Similarly, we define the cost of a decision to be

$c_{tad}$ = The cost of applying a decision of type $d$ to a technician with attribute $a$,

$c_t$ = $(c_{tad})_{a \in \mathcal{A}, d \in \mathcal{D}}$.

We could solve this problem myopically by making what appears to be the best decisions now, ignoring their impact on the future. We would do this by solving

$$\min_{x_t} \sum_{a \in \mathcal{A}} \sum_{d \in \mathcal{D}} c_{tad} x_{tad} \tag{2.20}$$

subject to

$$\sum_{d \in \mathcal{D}} x_{tad} = R_{ta} \tag{2.21}$$

$$\sum_{a \in \mathcal{A}} x_{tad} \leq D_{tb_d} \quad d \in \mathcal{D}^D \tag{2.22}$$

$$x_{tad} \geq 0. \tag{2.23}$$

Equation (2.21) says that we either have to send a technician home, or assign him to a job. Equation (2.22) says that we can only assign a technician to a job of type $b_d$ if there is in fact a job of type $b_d$. Said differently, we cannot assign more than one technician per job. But we do not have to assign a technician to every job (we may not have enough technicians).

The problem posed by equations (2.20)-(2.23) is a linear program. Real problems may involve managing hundreds or even thousands of individual entities. The decision vector $x_t = (x_{tad})_{a \in \mathcal{A}, d \in \mathcal{D}}$ may have over ten thousand dimensions. But commercial linear programming packages handle problems of this size quite easily.

If we make decisions by solving (2.20)-(2.23), we say that we are using a *myopic policy* since we are using only what we know now, and we are ignoring the impact of decisions now on the future. For example, we may decide to send a technician home rather than have him sit in a hotel room waiting for a job. But this ignores the likelihood that another job may suddenly arise close to the technician's current location. Alternatively, we may have two different technicians with two different skill sets. If we only have one job, we might assign what appears to be the closest technician, ignoring the fact that this technician has specialized skills that are best reserved for difficult jobs.

Given a decision vector, the dynamics of our system can be described using

$$R_{t+1,a} = \sum_{a' \in \mathcal{A}} \sum_{d \in \mathcal{D}} x_{ta'd} \delta_a(a', d), \tag{2.24}$$

$$D_{t+1,b_d} = D_{t,b_d} - \sum_{a \in \mathcal{A}} x_{tad} + \hat{D}_{t+1,b_d}, \quad d \in \mathcal{D}^D. \tag{2.25}$$

Equation (2.24) captures the effect of all decisions (including serving demands) on the attributes of a technician. This is easiest to visualize if we assume that all tasks are completed within one time period. If this is not the case, then we simply have to augment the state vector to capture the attribute that we have partially completed a task. Equation (2.25) subtracts from the list of available demands any of type $b_d$ that are served by a decision $d \in \mathcal{D}^D$ (recall that each element of $\mathcal{D}^D$ corresponds to a type of task, which we denote $b_d$).

The state of our system is given by

$$S_t = (R_t, D_t).$$

The evolution of our state variable over time is determined by equations (2.24) and (2.24). We can now set up an optimality recursion to determine the decisions that minimize costs over time using

$$V_t = \min_{x_t \in \mathcal{X}_t} \left( C_t(S_t, x_t) + \gamma \mathbb{E} V_{t+1}(S_{t+1}) \right),$$

where $S_{t+1}$ is the state at time $t+1$ given that we are in state $S_t$ and action $x_t$. $S_{t+1}$ is random because at time $t$, we do not know $\hat{D}_{t+1}$. The feasible region $\mathcal{X}_t$ is defined by equations (2.21)-(2.23).

Needless to say, the state variable for this problem is quite large. The dimensionality of $R_t$ is determined by the number of attributes of our technician, while the dimensionality of $D_t$ is determined by the relevant attributes of a demand. In real applications, these attributes can become fairly detailed. Fortunately, the methods of approximate dynamic programming can handle these complex problems.

## 2.3 INFORMATION ACQUISITION PROBLEMS

Information acquisition is an important problem in many applications where we face uncertainty about the value of an action, but the only way to obtain better estimates of the value is to take the action. For example, a baseball manager may not know how well a particular player will perform at the plate. The only way to find out is to put him in the lineup and let him hit. The only way a mutual fund can learn how well a manager will perform may be to let her manage a portion of the portfolio. A pharmaceutical company does not know how the market will respond to a particular pricing strategy. The only way to learn is to offer the drug at different prices in test markets.

Information acquisition plays a particularly important role in approximate dynamic programming. Assume that a system is in state $i$ and that a particular action might bring the system to state $j$. We may know the contribution of this decision, but we do not know the value of being in state $j$ (although we may have an estimate). The only way to learn is to try making the decision and then obtain a better estimate of being in state $j$ by actually visiting the state. This process of approximating a value function, first introduced in chapter 4, is fundamental to approximate dynamic programming. For this reason, the information acquisition problem is of special importance, even in the context of solving classical dynamic programs.

The information acquisition problem introduces a new dimension to our thinking about dynamic programs. In all the examples we have considered in this chapter, the state of our system is the state of the resources we are managing. In the information acquisition problem, our state variable has to also include our estimates of unknown parameters. We illustrate this idea in the examples that follow.

### 2.3.1 The bandit problem

The classic information acquisition problem is known as the *bandit problem.* Consider the situation faced by a gambler trying to choose which of $K$ slot machines to play. Now assume that the probability of winning may be different for each machine, but the gambler does not know what these probabilities are. The only way to obtain information is to actually play a slot machine. To formulate this problem, let

$$x_k^n = \begin{cases} 1 & \text{if we choose to play the } k^{th} \text{ slot machine in the } n^{th} \text{ trial,} \\ 0 & \text{otherwise.} \end{cases}$$

$W_k^n$ = Winnings from playing the $k^{th}$ slot machine during the $n^{th}$ trial.

$\bar{w}_k^n$ = Our estimate of the expected winnings from playing the $k^{th}$ slot machine after the $n^{th}$ trial.

$(\bar{s}_k^2)^n$ = Our estimate of the variance of the winnings from playing the $k^{th}$ slot machine after the $n^{th}$ trial.

$N_k^n$ = The number of times we have played the $k^{th}$ slot machine after $n$ trials.

If $x_k^n = 1$, then we observe $W_k^n$ and can update $\bar{w}_k^n$ and $(\bar{s}_k^2)^n$ using

$$\bar{w}_k^n = (1 - \frac{1}{N_k^n})\bar{w}_k^{n-1} + \frac{1}{N_k^n} W_k^n, \tag{2.26}$$

$$(\bar{s}_k^2)^n = \frac{N_k^n - 2}{N_k^n - 1}(\bar{s}_k^2)^{n-1} + \frac{1}{N_k^n}(W_k^n - \bar{w}_k^{n-1})^2. \tag{2.27}$$

Equations (2.26) and (2.27) are equivalent to computing averages and variances from sample observations. Let $\mathcal{N}_k^n$ be the iterations where $x_k^n = 1$. Then we can write the mean and variance using

$$\bar{w}_k^n = \frac{1}{N_k^n} \sum_{m \in \mathcal{N}_k^n} W_k^m,$$

$$(\bar{s}_k^2)^n = \frac{1}{N_k^n - 1} \sum_{m \in \mathcal{N}_k^n} (W_k^m - \bar{w}_k^m)^2.$$

The recursive equations ((2.26) - (2.27)) make the transition from state $(\bar{w}_k^{n-1}, (\bar{s}_k^2)^{n-1})$ to $(\bar{w}_k^n, (\bar{s}_k^2)^n)$ more transparent.

All other estimates remain unchanged. For this system, the "state" of our system (measured after the $n^{th}$ trial) is

$$S^n = (\bar{w}_k^n, (\bar{s}_k^2)^n, N_k^n)_{k=1}^K.$$

We have to keep track of $N_k^n$ for each bandit $k$ since these are needed in the updating equations.

It is useful think of $S^n$ as our "state of knowledge" since it literally captures what we know about the system (given by $\bar{w}_k^n$) and how well we know it (given by $(\bar{s}_k^2)^n$). Some authors call this the information state, or the "hyperstate." Equations (2.26)-(2.27) represent the transition equations that govern how the state evolves over time. Note that the state variable has $2K$ elements, which are also continuous (if the winnings are integer, $S^n$ takes on discrete outcomes, but this is of little practical value).

This is a pure information acquisition problem. Normally, we would choose to play the machine with the highest expected winnings. To express this we would write

$$x^n = \arg\max_k \bar{w}_k^{n-1}.$$

Here, "arg max" means the value of the decision variable (in this case $k$) that maximizes the problem. We set $x^n$ equal to the value of $k$ that corresponds to the largest value of $\bar{w}_k^{n-1}$ (ties are broken arbitrarily). This rule ignores the possibility that our *estimate* of the expected winnings for a particular machine might be wrong. Since we use these estimates to help us make better decisions, we might want to try a machine where the current estimated winnings are lower, because we might obtain information that indicates that the true mean might actually be higher.

The problem can be solved, in theory, using Bellman's equation

$$V^n(S^n) = \max_x \mathbb{E}\left(\sum_k W_k^{n+1} x_k + V^{n+1}(S^{n+1})|S^n\right) \tag{2.28}$$

subject to $\sum_k x_k = 1$. Equation (2.28) is hard to compute since the state variable is continuous. Also, we do not know the distribution of the random variables $W^{n+1}$ (if we knew the distributions then we would know the expected reward for each bandit). However, the estimates of the mean and variance, specified by the state $S^n$, allow us to infer the distribution of possible means (and variances) of the true distribution.

There are numerous examples of information acquisition problems. The examples provide some additional illustrations of bandit problems. These problems can be solved optimally using something known as an *index policy*. In the case of the bandit problem, it is possible to compute a single index for each bandit. The best decision is to then choose the bandit with the highest index. The value of this result is that we can solve a $K$-dimensional problem as $K$ one-dimensional problems. See chapter 10 for more on this topic.

---

**■ EXAMPLE 2.1**

Consider someone who has just moved to a new city and who now has to find the best path to work. Let $T_p$ be a random variable giving the time he will experience if he chooses path $p$ from a predefined set of paths $\mathcal{P}$. The only way he can obtain observations of the travel time is to actually travel the path. Of course, he would like to choose the path with the shortest average time, but it may be necessary to try a longer path because it may be that he simply has a poor estimate. The problem is identical to our bandit problem if we assume that driving one path does not teach us anything about a different path (this is a richer form of bandit problem).

■ **EXAMPLE 2.2**

A baseball manager is trying to decide which of four players makes the best designated hitter. The only way to estimate how well they hit is to put them in the batting order as the designated hitter.

■ **EXAMPLE 2.3**

A couple that has acquired some assets over time is looking to find a good money manager who will give them a good return without too much risk. The only way to determine how well a money manager performs is to them actually manage the money for a period of time. The challenge is determining whether to stay with the money manager or to switch to a new money manager.

■ **EXAMPLE 2.4**

A doctor is trying to determine the best blood pressure medication for a patient. Each patient responds differently to each medication, so it is necessary to try a particular medication for a while, and then switch if the doctor feels that better results can be achieved with a different medication.

---

### 2.3.2 An information-collecting shortest path problem

Now assume that we have to choose a path through a network, just as we did in sections 2.1.1 and 2.2.2, but this time we face the problem that we not only do not know the actual travel time on any of the links of the network, we do not even know the mean or variance (we might be willing to assume that the probability distribution is normal). As with the two previous examples, we solve the problem repeatedly, and sometimes we want to try new paths just to collect more information.

There are two significant differences between this simple problem and the two previous problems. First, imagine that you are at a node $i$ and you are trying to decide whether to follow the link from $i$ to $j_1$ or from $i$ to $j_2$. We have an estimate of the time to get from $j_1$ and $j_2$ to the final destination. These estimates may be correlated because they may share common links to the destination. Following the path from $j_1$ to the destination may teach us something about the time to get from $j_2$ to the destination (if the two paths share common links). The second difference is that making the decision to go from node $i$ to node $j$ changes the set of options that we face. In the bandit problem, we always faced the same set of slot machines.

Information-collecting shortest path problems arise in any information collection problem where the decision now affects not only the information you collect, but also the decisions you can make in the future. While we can solve basic bandit problems optimally, this broader problem class remains unsolved.

## 2.4 A SIMPLE MODELING FRAMEWORK FOR DYNAMIC PROGRAMS

Now that we have covered a number of simple examples, it is useful to briefly review the elements of a dynamic program. We are going to revisit this topic in considerably greater depth in chapter 5, but this discussion provides a brief introduction. Our presentation focuses on *stochastic* dynamic programs which exhibit a flow of uncertain information. These problems, at a minimum, consist of the following elements:

**The state variable -** This captures all the information we need to make a decision, as well as the information that we need to describe how the system evolves over time.

**The decision variable -** Decisions represent how we control the process.

**Exogenous information -** This is data that first become known each time period (for example, the demand for product, or the price at which it can be purchased or sold). In addition, we also have to be told the initial state of our system.

**The transition function -** This function determines how the system evolves from the state $S_t$ to the state $S_{t+1}$ given the decision that was made at time $t$ and the new information that arrived between $t$ and $t+1$.

**The contribution function -** This determines the costs incurred or rewards received during each time interval.

**The objective function -** Here we formally state the problem of maximizing the contribution (or minimizing the cost) over a specified time horizon (which might be infinite).

We can illustrate these elements using the simple asset acquisition problem from section 2.2.6.

The state variable is the information we need to make a decision and compute functions that determine how the system evolves into the future. In our asset acquisition problem, we need three pieces of information. The first is $R_t$, the resources on hand before we make any decisions (including how much of the demand to satisfy). The second is the demand itself, denoted $D_t$, and the third is the price $p_t$. We would write our state variable as $S_t = (R_t, D_t, p_t)$.

We have two decisions to make. The first, denoted $x_t^D$, is how much of the demand $D_t$ during time interval $t$ that should be satisfied using available assets, which means that we require $x_t^D \leq R_t$. The second, denoted $x_t^O$, is how many new assets should be acquired at time $t$ which can be used to satisfy demands during time interval $t+1$.

The exogenous information process consists of three types of information. The first is the new demands that arise during time interval $t$, denoted $\hat{D}_t$. The second is the change in the price at which we can sell our assets, denoted $\hat{p}_t$. Finally, we are going to assume that there may be exogenous changes to our available resources. These might be blood donations or cash deposits (producing positive changes), or equipment failures and cash withdrawals (producing negative changes). We denote these changes by $\hat{R}_t$. We often use a generic variable $W_t$ to represent all the new information that is first learned during time interval $t$, which for our problem would be written $W_t = (\hat{R}_t, \hat{D}_t, \hat{p}_t)$. In addition to specifying the types of exogenous information, for stochastic models we also have to specify the likelihood of a particular outcome. This might come in the form of an assumed probability distribution for $\hat{R}_t$, $\hat{D}_t$, and $\hat{p}_t$, or we may depend on an exogenous source for sample realizations (the actual price of the stock or the actual travel time on a path).

Once we have determined what action we are going to take from our decision rule, we compute our contribution $C_t(S_t, x_t)$ which might depend on our current state and the decision $x_t$ that we take at time $t$. For our asset acquisition problem (where the state variable is $R_t$), the contribution function is

$$C_t(S_t, x_t) = p_t x_t^D - c_t x_t^O.$$

In this particular model, $C_t(S_t, x_t)$ is a deterministic function of the state and action. In other applications, the contribution from decision $x_t$ depends on what happens during time $t+1$.

Next, we have to specify how the state variable changes over time. This is done using a *transition function* which we might represent in a generic way using

$$S_{t+1} = S^M(S_t, x_t, W_{t+1}),$$

where $S_t$ is the state at time $t$, $x_t$ is the decision we made at time $t$ and $W_{t+1}$ is our generic notation for the information that arrives between $t$ and $t+1$. We use the notation $S^M(\cdot)$ to denote the transition function, where the superscript $M$ stands for "model" (or "system model" or "plant model," in recognition of vocabulary that has been in place for many years in the engineering community). The transition function for our asset acquisition problem is given by

$$\begin{aligned} R_{t+1} &= R_t - x_t^D + x_t^O + \hat{R}_{t+1}, \\ D_{t+1} &= D_t - x_t^D + \hat{D}_{t+1}, \\ p_{t+1} &= p_t + \hat{p}_{t+1}. \end{aligned}$$

This model assumes that unsatisfied demands are held until the next time period.

Our final step in formulating a dynamic program is to specify the objective function. Assume we are trying to maximize the total contribution received over a finite horizon $t = (0, 1, \ldots, T)$. If we were solving a deterministic problem, we might formulate the objective function as

$$\max_{(x_t)_{t=0}^T} \sum_{t=0}^{T} C_t(S_t, x_t). \tag{2.29}$$

We would have to optimize (2.29) subject to a variety of constraints on the decisions $(x_0, x_1, \ldots, x_T)$.

If we have a stochastic problem, which is to say that there are a number of possible realizations of the exogenous information process $(W_t)_{t=0}^T$, then we have to formulate the objective function in a different way. If the exogenous information process is uncertain, we do not know which state we will be in at time $t$. Since the state $S_t$ is a random variable, then the choice of decision (which depends on the state) is also a random variable.

We get around this problem by formulating the objective in terms of finding the best *policy* (or decision rule) for choosing decisions. A policy tells us what to do for all possible states, so regardless of which state we find ourselves in at some time $t$, the policy will tell us what decision to make. This policy must be chosen to produce the best *expected* contribution over all outcomes. If we let $X^\pi(S_t)$ be a particular decision rule indexed by $\pi$, and let $\Pi$ be a set of decision rules, then the problem of finding the best policy would be written

$$\max_{\pi \in \Pi} \mathbb{E} \sum_{t=0}^{T} C_t(S_t, X^\pi(S_t)). \tag{2.30}$$

Exactly what is meant by finding the best policy out of a set of policies is very problem specific. Our decision rule might be to order $X^\pi(R_t) = S - R_t$ if $R_t < s$ and order $X^\pi(R_t) = 0$ if $R_t \geq s$. The family of policies is the set of all values of the parameters $(s, S)$ for $s < S$ (here, $s$ and $S$ are parameters to be determined, not state variables). If we are selling an asset, we might adopt a policy of selling if the price of the asset $p_t$ falls below some value $\bar{p}$. The set of all policies is the set of all values of $\bar{p}$. However, policies of this sort tend to work only for very special problems.

Equation (2.30) states our problem as one of finding the best policy (or decision rule, or function $X^\pi$) to maximize the expected value of the total contribution over our horizon. There are a number of variations of this objective function. For applications where the horizon is long enough to affect the time value of money, we might introduce a discount factor $\gamma$ and solve

$$\max_{\pi \in \Pi} \mathbb{E} \sum_{t=0}^{T} \gamma^t C_t(S_t, X^\pi(S_t)). \tag{2.31}$$

There is also considerable interest in infinite horizon problems of the form

$$\max_{\pi \in \Pi} \mathbb{E} \sum_{t=0}^{\infty} \gamma^t C_t(S_t, X^\pi(S_t)). \tag{2.32}$$

Equation (2.32) is often used when we want to study the behavior of a system in steady state.

Equations such as (2.30), (2.31), and (2.32) are all easy to write on a sheet of paper. Solving them computationally is a different matter. That challenge is the focus of this book.

A complete specification of a dynamic program requires that we specify both data and functions, as follows:

Data:

- The initial state $S_0$.
- The exogenous information process $W_t$. We need to know what information is arriving from outside the system, and how it is being generated. For example, we might be given the probability of an outcome, or given a process that creates the data for us.

Functions:

- The contribution function $C(S_t, x_t)$. This may be specified in the form $C(S_t, x_t, W_{t+1})$. In this case, we may need to compute the expectation, or we may work directly with this form of the contribution function.
- The transition function $S^M(S_t, x_t, W_{t+1})$.
- The family of decision functions $(X^\pi(S))_{\pi \in \Pi}$.

This description provides only a taste of the richness of sequential decision processes. Chapter 5 describes the different elements of a dynamic program in far greater detail.

## 2.5 BIBLIOGRAPHIC NOTES

Most of the problems in this chapter are fairly classic, in particular the deterministic and stochastic shortest path problems (see Bertsekas (2000)), asset acquisition problem (see Porteus (1990), for example) and the batch replenishment problem (see Puterman (1994), among others).

Section 2.1.1 - The shortest path problem is one of the most widely studied problems in optimization. One of the early treatments of shortest paths is given in the seminal book on network flows by Ford & Fulkerson (1962). It has long been recognized that shortest paths could be solved directly (if inefficiently) using Bellman's equation.

Section 2.2.2 - Many problems in discrete stochastic dynamic programming can at least conceptually be formulated as some form of stochastic shortest path problem. There is an extensive literature on stochastic shortest paths (see, for example, Frank (1969), Sigal et al. (1980), Frieze & Grimmet (1985), Andreatta & Romeo (1988), Psaraftis & Tsitsiklis (1993), Bertsekas & Tsitsiklis (1991)).

Section 2.2.10 - The dynamic assignment problem is based on Spivey & Powell (2004).

Section 2.3.1 - Bandit problems have long been studied as classic exercises in information collection. For good introductions to this material, see Ross (1983) and Whittle (1982*a*). A more detailed discussion of bandit problems is given in chapter 10.

## PROBLEMS

**2.1** Give an example of a sequential decision process from your own experience. Describe the elements of your problem following the framework provided in section 2.4. Then describe the types of rules you might use to make a decision.

**2.2** What is the state variable at each node in the decision tree in figure 2.3?

**2.3** Describe the gambling problem in section 2.2.3 as a decision tree, assuming that we can gamble only 0, 1 or 2 dollars in each round (this is just to keep the decision tree from growing too large).

**2.4** Repeat the gambling problem assuming that the value of ending up with $S^N$ dollars is $\sqrt{S^N}$.

**2.5** Write out the steps of a shortest path algorithm, similar to that shown in figure 2.2, which starts at the destination and works backward to the origin.

**2.6** Carry out the proof by induction described at the end of section 2.1.3.

**2.7** Repeat the derivation in section 2.1.3 assuming that the reward for task $t$ is $c_t\sqrt{x_t}$.

**2.8** Repeat the derivation in section 2.1.3 assuming that the reward for task $t$ is given by $\ln(x)$.

**2.9** Repeat the derivation in section 2.1.3 one more time, but now assume that all you know is that the reward is continuously differentiable, monotonically increasing and concave.

**2.10** What happens to the answer to the budget allocation problem in section 2.1.3 if the contribution is convex instead of concave (for example, $C_t(x_t) = x_t^2$)?

**2.11** Consider three variations of a shortest path problem:

Case I - All costs are known in advance. Here, we assume that we have a real-time network tracking system that allows us to see the cost on each link of the network before we start our trip. We also assume that the costs do not change during the time when we start the trip to when we arrive at the link.

Case II - Costs are learned as the trip progresses. In this case, we assume that we see the actual link costs for links out of node $i$ when we arrive at node $i$.

Case III - Costs are learned after the fact. In this setting, we only learn the cost on each link after the trip is finished.

Let $v_i^I$ be the expected cost to get from node $i$ to the destination for Case I. Similarly, let $v_i^{II}$ and $v_i^{III}$ be the expected costs for cases II and III. Show that $v_i^I \leq v_i^{II} \leq v_i^{III}$.

**2.12** We are now going to do a budgeting problem where the reward function does not have any particular properties. It may have jumps, as well as being a mixture of convex and concave functions. But this time we will assume that $R = \$30$ dollars and that the allocations $x_t$ must be in integers between 0 and 30. Assume that we have $T = 5$ products, with a contribution function $C_t(x_t) = cf(x_t)$ where $c = (c_1, \ldots, c_5) = (3, 1, 4, 2, 5)$ and where $f(x)$ is given by

$$f(x) = \begin{cases} 0, & x \leq 5, \\ 5, & x = 6, \\ 7, & x = 7, \\ 10, & x = 8, \\ 12, & x \geq 9. \end{cases}$$

Find the optimal allocation of resources over the five products.

**2.13** You suddenly realize towards the end of the semester that you have three courses that have assigned a term project instead of a final exam. You quickly estimate how much each one will take to get 100 points (equivalent to an A+) on the project. You then guess that if you invest $t$ hours in a project, which you estimated would need $T$ hours to get 100 points, then for $t < T$ your score will be

$$R = 100\sqrt{t/T}.$$

That is, there are declining marginal returns to putting more work into a project. So, if a project is projected to take 40 hours and you only invest 10, you estimate that your score will be 50 points (100 times the square root of 10 over 40). You decide that you cannot spend more than a total of 30 hours on the projects, and you want to choose a value of $t$ for each project that is a multiple of 5 hours. You also feel that you need to spend at least 5 hours on each project (that is, you cannot completely ignore a project). The time you estimate to get full score on each of the four projects is given by

| Project | Completion time $T$ |
|---|---|
| 1 | 20 |
| 2 | 15 |
| 3 | 10 |

You decide to solve the problem as a dynamic program.

(a) What is the state variable and decision epoch for this problem?

(b) What is your reward function?

(c) Write out the problem as an optimization problem.

(d) Set up the optimality equations.

(e) Solve the optimality equations to find the right time investment strategy.

**2.14** Rewrite the transition function for the asset acquisition problem II (section 2.2.6) assuming that $R_t$ is the resources on hand after we satisfy the demands.

**2.15** Write out the transition equations for the lagged asset acquisition problem in section 2.2.7 when we allow spot purchases, which means that we may have $x_{tt} > 0$. $x_{tt}$ refers to purchases that are made at time $t$ which can be used to serve unsatisfied demands $D_t$ that occur during time interval $t$.

**2.16** You have to send a set of questionnaires to each of $N$ population segments. The size of each population segment is given by $w_i$. You have a budget of $B$ questionnaires to allocate among the population segments. If you send $x_i$ questionnaires to segment $i$, you will have a sampling error proportional to

$$f(x_i) = 1/\sqrt{x_i}.$$

You want to minimize the weighted sum of sampling errors, given by

$$F(x) = \sum_{i=1}^{N} w_i f(x_i)$$

You wish to find the allocation $x$ that minimizes $F(x)$ subject to the budget constraint $\sum_{i=1}^{N} x_i \leq B$. Set up the optimality equations to solve this problem as a dynamic program (needless to say, we are only interested in integer solutions).

**2.17** Identify three examples of problems where you have to try an action to learn about the reward for an action.

# CHAPTER 3

# INTRODUCTION TO MARKOV DECISION PROCESSES

There is a very elegant theory for solving stochastic, dynamic programs if we are willing to live within some fairly limiting assumptions. Assume that we have a discrete state space $\mathcal{S} = (1, 2, \ldots, |\mathcal{S}|)$, where $\mathcal{S}$ is small enough to enumerate. Next assume that there is a relatively small set of decisions or actions. In many books, an action is denoted by $a$ and $\mathcal{A}$ represents the set of possible actions. A major goal of our presentation is to bridge dynamic programming with the field of math programming so that we can solve large-scale problems which need tools such as linear, nonlinear, and integer programming. For this reason, in this chapter and throughout the book we let $x_t$ be our generic decision variable, and using the vocabulary of math programming, we let $\mathcal{X}$ (or $\mathcal{X}_t$ if it depends on time) be the *feasible region*, which means that we want to choose a decision $x_t \in \mathcal{X}_t$. Finally, assume that we are given a transition matrix $p_t(S_{t+1}|S_t, x_t)$ which gives the probability that if we are in state $S_t$ (at time $t$) and make decision $x_t$, then we will next be in state $S_{t+1}$.

There is a vast array of problems where states are continuous, or the state variable is a vector producing a state space that is far too large to enumerate. In addition, computing the one-step transition matrix $p_t(S_{t+1}|S_t, x_t)$ can also be computationally difficult or impossible to compute. So why cover material that is widely acknowledged to work only on small or highly specialized problems? First, some problems have small state and action spaces and can be solved with these techniques. Second, the theory of Markov decision processes can be used to identify structural properties that can dramatically simplify computational algorithms. But far more importantly, this material provides the intellectual foundation for the types of algorithms that we present in later chapters. Using the framework in this chapter, we can prove very powerful results that will provide a guiding hand as we step into richer and more complex problems in many real-world settings. Furthermore, the behavior of these algorithms provide important insights that guide the behavior of algorithms for more general problems.

There is a rich and elegant theory behind Markov decision processes. Even if the algorithms have limited application, the ideas behind these algorithms, which enjoy a rich history, represent the fundamental underpinnings of most of the algorithms in the remainder of this book. As with most of the chapters in the book, the body of this chapter focuses on the algorithms, and the convergence proofs have been deferred to the "Why does it work" section (section 3.9). The intent is to allow the presentation of results to flow more naturally, but serious students of dynamic programming are encouraged to delve into these proofs, which are quite elegant. This is partly to develop a deeper appreciation of the properties of the problem as well as to develop an understanding of the proof techniques that are used in this field.

## 3.1 THE OPTIMALITY EQUATIONS

In the last chapter, we illustrated a number of stochastic applications that involve solving the following objective function

$$\max_{\pi} \mathbb{E} \left\{ \sum_{t=0}^{T} \gamma^t C_t^{\pi}(S_t, X_t^{\pi}(S_t)) \right\}. \tag{3.1}$$

For most problems, solving equation (3.1) is computationally intractable, but it provides the basis for identifying the properties of optimal solutions and finding and comparing "good" solutions.

### 3.1.1 Bellman's equations

With a little thought, we realize that we do not have to solve this entire problem at once. Assume that we are solving a deterministic shortest path problem where $S_t$ is the index of the node in the network where we have to make a decision. If we are in state $S_t = i$ (that is, we are at node $i$ in our network) and make decision $x_t = j$ (that is, we wish to traverse the link from $i$ to $j$), our transition function will tell us that we are going to land in some state $S_{t+1} = S^M(S_t, x_t)$ (in this case, node $j$). What if we had a function $V_{t+1}(S_{t+1})$ that told us the value of being in state $S_{t+1}$ (giving us the value of the path from node $j$ to the destination)? We could evaluate each possible decision $x_t$ and simply choose the decision $x_t$ that has the largest one-period contribution, $C_t(S_t, x_t)$, plus the value of landing in state $S_{t+1} = S^M(S_t, x_t)$ which we represent using $V_{t+1}(S_{t+1})$. Since this value represents the money we receive one time period in the future, we might discount this by a factor $\gamma$. In other words, we have to solve

$$x_t^*(S_t) = \arg\max_{x_t \in \mathcal{X}_t} \big(C_t(S_t, x_t) + \gamma V_{t+1}(S_{t+1})\big),$$

where "$\arg\max$" means that we want to choose the value of $x_t$ that maximizes the expression in parentheses. We also note that $S_{t+1}$ is a function of $S_t$ and $x_t$, meaning that we could write it as $S_{t+1}(S_t, x_t)$. Both forms are fine. It is common to write $S_{t+1}$ by itself, but the dependence on $S_t$ and $x_t$ needs to be understood.

The value of being in state $S_t$ is the value of using the optimal decision $x_t^*(S_t)$. That is

$$\begin{aligned} V_t(S_t) &= \max_{x_t \in \mathcal{X}_t} \left( C_t(S_t, x_t) + \gamma V_{t+1}(S_{t+1}(S_t, x_t)) \right) \\ &= C_t(S_t, x_t^*(S_t)) + \gamma V_{t+1}(S_{t+1}(S_t, x_t^*(S_t))). \end{aligned} \tag{3.2}$$

Equation (3.2) is the optimality equation for deterministic problems.

When we are solving stochastic problems, we have to model the fact that new information becomes available after we make the decision $x_t$. The result can be uncertainty in both the contribution earned, and in the determination of the next state we visit, $S_{t+1}$. For example, consider the problem of managing oil inventories for a refinery. Let the state $S_t$ be the inventory in thousands of barrels of oil at time $t$ (we require $S_t$ to be integer). Let $x_t$ be the amount of oil ordered at time $t$ that will be available for use between $t$ and $t+1$, and let $\hat{D}_{t+1}$ be the demand for oil between $t$ and $t+1$. The state variable is governed by the simple inventory equation

$$S_{t+1}(S_t, x_t) = \max\{0, S_t + x_t - \hat{D}_{t+1}\}.$$

We have written the state $S_{t+1}$ using $S_{t+1}(S_t, x_t)$ to express the dependence on $S_t$ and $x_t$, but it is common to simply write $S_{t+1}$ and let the dependence on $S_t$ and $x_t$ be implicit. Since $\hat{D}_{t+1}$ is random at time $t$ when we have to choose $x_t$, we do not know $S_{t+1}$. But if we know the probability distribution of the demand $\hat{D}$, we can work out the probability that $S_{t+1}$ will take on a particular value. If $\mathbb{P}^D(d) = \mathbb{P}[\hat{D} = d]$ is our probability distribution, then we can find the probability distribution for $S_{t+1}$ using

$$Prob(S_{t+1} = s') = \begin{cases} 0 & \text{if } s' > S_t + x_t, \\ \mathbb{P}^D(S_t + x_t - s') & \text{if } 0 < s' \leq S_t + x_t, \\ \sum_{d=S_t+x_t}^{\infty} \mathbb{P}^D(d) & \text{if } s' = 0. \end{cases}$$

These probabilities depend on $S_t$ and $x_t$, so we write the probability distribution as

$$\mathbb{P}(S_{t+1}|S_t, x_t) = \text{The probability of } S_{t+1} \text{ given } S_t \text{ and } x_t.$$

We can then modify the deterministic optimality equation in (3.2) by simply adding an expectation, giving us

$$V_t(S_t) = \max_{x_t \in \mathcal{X}_t} \left( C_t(S_t, x_t) + \gamma \sum_{s' \in \mathcal{S}} \mathbb{P}(S_{t+1} = s'|S_t, x_t) V_{t+1}(s') \right). \tag{3.3}$$

We refer to this as the *standard form* of Bellman's equations, since this is the version that is used by virtually every textbook on stochastic, dynamic programming. An equivalent form that is more natural for approximate dynamic programming is to write

$$V_t(S_t) = \max_{x_t \in \mathcal{X}_t} \left( C_t(S_t, x_t) + \gamma \mathbb{E}\{V_{t+1}(S_{t+1}(S_t, x_t))|S_t\} \right), \tag{3.4}$$

where we simply use an expectation instead of summing over probabilities. We refer to this equation as the *expectation form* of Bellman's equation. This version forms the basis for our algorithmic work in later chapters.

**Remark:** Equation (3.4) is often written in the slightly more compact form

$$V_t(S_t) = \max_{x_t \in \mathcal{X}_t} \left( C_t(S_t, x_t) + \gamma \mathbb{E}\{V_{t+1}(S_{t+1})|S_t\} \right), \tag{3.5}$$

where the functional relationship $S_{t+1} = S^M(S_t, x_t, W_{t+1})$ is implicit. At this point, however, we have to deal with some subtleties of mathematical notation. In equation (3.4) we have written the expectation explicitly as a function of the decision $x_t$, conditioned on the state $S_t$. This is similar to our indexing of the one-step transition matrix in (3.3) where we write the probability of transitioning to state $s'$ given that we are in state $S_t$ and take action $x_t$. The problem is in the interpretation of $S_t$. Is $S_t$ a specific state, or is it a random variable which might take on a range of values? It is common to assume that $S_t$ is a random variable (remember that we are solving this problem at time $t = 0$). If we wanted to assume that we are in a specific state $S_t = s$, we would write (3.5) as

$$V_t(s) = \max_{x_t \in \mathcal{X}_t} \left( C_t(s, x_t) + \gamma \mathbb{E}\{V_{t+1}(S_{t+1}(s, x_t)) | S_t = s\} \right). \tag{3.6}$$

When we condition on $S_t$ (as we have in equation (3.4) or (3.5)), then it is common to view $S_t$ as a random variable which might take on many outcomes. In this case, we have to realize that when we solve for an action $x_t$ that it depends on $S_t$, and perhaps might be written $x_t(S_t)$. When we specify an action for each state, we refer to this as a *policy*. For this reason, some authors feel that they have to write the expectation $\mathbb{E}\{V_{t+1}(S_{t+1})|S_t\}$ as conditioned on the action, which they might write as $\mathbb{E}\{V_{t+1}(S_{t+1})|S_t, x_t\}$. This might be interpreted as conditioning on a specific (deterministic) action $x_t$ and a random state (yes, this gets pretty subtle). For this reason, some authors like to think of conditioning on a policy $\pi$, which might be written $\mathbb{E}^\pi\{V_{t+1}(S_{t+1})|S_t\}$. Throughout this book, when we write Bellman's equation we always assume the functional relationship $S_{t+1} = S^M(S_t, x_t, W_{t+1})$ which we might write explicitly or implicitly. In this setting, we view $S_t$ deterministically (with apologies to the probability community that thinks that every capital letter is random), with $x_t$ a specific (deterministic) action.

The standard form of Bellman's equation (3.3) has been popular in the research community since it lends itself to elegant algebraic manipulation when we assume we know the transition matrix. It is common to write it in a more compact form. Recall that a policy $\pi$ is a rule that specifies the action $x_t$ given the state $S_t$. In this chapter, it is easiest if we always think of a policy in terms of a rule "when we are in state $s$ we take action $x$." This a form of "lookup-table" representation of a policy that is very clumsy for most real problems, but it will serve our purposes here. The probability that we transition from state $S_t = s$ to $S_{t+1} = s'$ can be written as

$$p_{ss'}(x) = \mathbb{P}(S_{t+1} = s' | S_t = s, x_t = x).$$

We would say that "$p_{ss'}(x)$ is the probability that we end up in state $s'$ if we start in state $s$ at time $t$ when we are taking action $x$." Now assume that we have a function $X_t^\pi(s)$ that determines the action $x$ we should take when in state $s$. It is common to write the transition probability $p_{ss'}(x)$ in the form

$$p_{ss'}^\pi = \mathbb{P}(S_{t+1} = s' | S_t = s, X_t^\pi(s) = x).$$

We can now write this in matrix form

$$P_t^\pi = \text{The one-step transition matrix under policy } \pi \quad ,$$

where $p_{ss'}^\pi$ is the element in row $s$ and column $s'$. There is a different matrix $P^\pi$ for each policy (decision rule) $\pi$.

Now let $c_t^\pi$ be a column vector with element $c_t^\pi(s) = C_t(s, X_t^\pi(s))$, and let $v_{t+1}$ be a column vector with element $V_{t+1}(s)$. Then (3.3) is equivalent to

$$\begin{bmatrix} \vdots \\ v_t(s) \\ \vdots \end{bmatrix} = \max_\pi \left( \begin{bmatrix} \vdots \\ c_t^\pi(s) \\ \vdots \end{bmatrix} + \gamma \begin{bmatrix} \ddots & & \\ & p_{ss'}^\pi & \\ & & \ddots \end{bmatrix} \begin{bmatrix} \vdots \\ v_{t+1}(s') \\ \vdots \end{bmatrix} \right). \tag{3.7}$$

where the maximization is performed for each element (state) in the vector. In matrix/vector form, equation (3.7) can be written

$$v_t = \max_\pi \left( c_t^\pi + \gamma P_t^\pi v_{t+1} \right). \tag{3.8}$$

Equation (3.8) can be solved by finding $x_t$ for each state $s$. The result is a decision vector $x_t^* = (x_t^*(s))_{s \in \mathcal{S}}$, which is equivalent to determining the best policy. This is easiest to envision when $x_t$ is a scalar (how much to buy, whether to sell), but in many applications $x_t(s)$ is itself a vector. For example, assume our problem is to assign individual programmers to different programming tasks, where our state $S_t$ captures the availability of programmers and the different tasks that need to be completed. Of course, computing a vector $x_t$ for each state $S_t$ which is itself a vector is much easier to write than to implement.

It is very easy to lose sight of the relationship between Bellman's equation and the original objective function that we stated in equation (3.1). To bring this out, we begin by writing the expected profits using policy $\pi$ from time $t$ onward

$$F_t^\pi(S_t) = \mathbb{E}\left\{ \sum_{t'=t}^{T-1} C_{t'}(S_{t'}, X_{t'}^\pi(S_{t'})) + C_T(S_T) | S_t \right\}.$$

$F_t^\pi(S_t)$ is the expected total contribution if we are in state $S_t$ in time $t$, and follow policy $\pi$ from time $t$ onward. If $F_t^\pi(S_t)$ were easy to calculate, we would probably not need dynamic programming. Instead, it seems much more natural to calculate $V_t^\pi$ recursively using

$$V_t^\pi(S_t) = C_t(S_t, X_t^\pi(S_t)) + \mathbb{E}\left\{ V_{t+1}^\pi(S_{t+1}) | S_t \right\}.$$

It is not hard to show (by stepping backward in time) that

$$F_t^\pi(S_t) = V_t^\pi(S_t).$$

The proof, given in section 3.9.1, uses a proof by induction: assume it is true for $V_{t+1}^\pi$, and then show that it is true for $V_t^\pi$ (not surprisingly, inductive proofs are very popular in dynamic programming).

With this result in hand, we can then establish the following key result. Let $V_t(S_t)$ be a solution to equation (3.4) (or (3.3)). Then

$$\begin{aligned} F_t^* &= \max_{\pi \in \Pi} F_t^\pi(S_t) \\ &= V_t(S_t). \end{aligned} \tag{3.9}$$

Equation (3.9) establishes the equivalence between (a) the value of being in state $S_t$ and following the optimal policy and (b) the optimal value function at state $S_t$. While these are indeed equivalent, the equivalence is the result of a theorem (established in section 3.9.1). However, it is not unusual to find people who lose sight of the original objective function. Later, we have to solve these equations approximately, and we will need to use the original objective function to evaluate the quality of a solution.

### 3.1.2 Computing the transition matrix

It is very common in stochastic, dynamic programming (more precisely, Markov decision processes) to assume that the one-step transition matrix $P^\pi$ is given as data (remember that there is a different matrix for each policy $\pi$). In practice, we generally can assume we know the transition function $S^M(S_t, x_t, W_{t+1})$ from which we have to derive the one-step transition matrix.

Assume for simplicity that the random information $W_{t+1}$ that arrives between $t$ and $t+1$ is independent of all prior information. Let $\Omega_{t+1}$ be the set of possible outcomes of $W_{t+1}$ (for simplicity, we assume that $\Omega_{t+1}$ is discrete), where $\mathbb{P}(W_{t+1} = \omega_{t+1})$ is the probability of outcome $\omega_{t+1} \in \Omega_{t+1}$. Also define the indicator function

$$1_{\{X\}} = \begin{cases} 1 & \text{if the statement "}X\text{" is true.} \\ 0 & \text{otherwise.} \end{cases}$$

Here, "$X$" represents a logical condition (such as, "is $S_t = 6$?"). We now observe that the one-step transition probability $\mathbb{P}_t(S_{t+1}|S_t, x_t)$ can be written

$$\begin{aligned} \mathbb{P}_t(S_{t+1}|S_t, x_t) &= \mathbb{E}1_{\{s'=S^M(S_t,x_t,W_{t+1})\}} \\ &= \sum_{\omega_{t+1}\in\Omega_{t+1}} \mathbb{P}(\omega_{t+1})1_{\{s'=S^M(S_t,x_t,\omega_{t+1})\}} \end{aligned}$$

So, finding the one-step transition matrix means that all we have to do is to sum over all possible outcomes of the information $W_{t+1}$ and add up the probabilities that take us from a particular state-action pair $(S_t, x_t)$ to a particular state $S_{t+1} = s'$. Sounds easy.

In some cases, this calculation is straightforward (consider our oil inventory example earlier in the section). But in other cases, this calculation is impossible. For example, $W_{t+1}$ might be a vector of prices or demands. In this case, the set of outcomes $\Omega_{t+1}$ can be much too large to enumerate. We can estimate the transition matrix statistically, but in later chapters (starting in chapter 4) we are going to avoid the need to compute the one-step transition matrix entirely. For the remainder of this chapter, we assume the one-step transition matrix is available.

### 3.1.3 Random contributions

In many applications, the one-period contribution function is a deterministic function of $S_t$ and $x_t$, and hence we routinely write the contribution as the deterministic function $C_t(S_t, x_t)$. However, this is not always the case. For example, a car traveling over a stochastic network may choose to traverse the link from node $i$ to node $j$, and only learn the cost of the movement after making the decision. For such cases, the contribution function is random, and we might write it as

$\hat{C}_{t+1}(S_t, x_t, W_{t+1})$ = The contribution received in period $t+1$ given the state $S_t$ and decision $x_t$, as well as the new information $W_{t+1}$ that arrives in period $t+1$.

In this case, we simply bring the expectation in front, giving us

$$V_t(S_t) = \max_{x_t} \mathbb{E}\left\{\hat{C}_{t+1}(S_t, x_t, W_{t+1}) + \gamma V_{t+1}(S_{t+1})|S_t\right\}. \tag{3.10}$$

Now let

$$C_t(S_t, x_t) = \mathbb{E}\{\hat{C}_{t+1}(S_t, x_t, W_{t+1})|S_t\}.$$

Thus, we may view $C_t(S_t, x_t)$ as the expected contribution given that we are in state $S_t$ and take action $x_t$.

### 3.1.4 Bellman's equation using operator notation*

The vector form of Bellman's equation in (3.8) can be written even more compactly using operator notation. Let $\mathcal{M}$ be the "max" (or "min") operator in (3.8) that can be viewed as acting on the vector $v_{t+1}$ to produce the vector $v_t$. Operating on a particular state $s$, $\mathcal{M}$ is defined by

$$\mathcal{M}v(s) = \max_x \big(C_t(s, x) + \gamma \sum_{s' \in \mathcal{S}} \mathbb{P}_t(s'|s, x)v_{t+1}(s')\big).$$

Here, $\mathcal{M}v$ produces a vector, and $\mathcal{M}v(s)$ refers to element $s$ of this vector. In vector form, we would write

$$\mathcal{M}v = \max_\pi \big(c_t^\pi + \gamma P_t^\pi v_{t+1}\big).$$

Now let $\mathcal{V}$ be the space of value functions. Then, $\mathcal{M}$ is a mapping

$$\mathcal{M} : \mathcal{V} \rightarrow \mathcal{V}.$$

We may also define the operator $\mathcal{M}^\pi$ for a particular policy $\pi$ using

$$\mathcal{M}^\pi(v) = c_t^\pi + \gamma P^\pi v \tag{3.11}$$

for some vector $v \in \mathcal{V}$. $\mathcal{M}^\pi$ is known as a *linear operator* since the operations that it performs on $v$ are additive and multiplicative. In mathematics, the function $c_t^\pi + \gamma P^\pi v$ is known as an *affine function*. This notation is particularly useful in mathematical proofs (see in particular some of the proofs in section 3.9), but we will not use this notation when we describe models and algorithms.

We see later in the chapter that we can exploit the properties of this operator to derive some very elegant results for Markov decision processes. These proofs provide insights into the behavior of these systems, which can guide the design of algorithms. For this reason, it is relatively immaterial that the actual computation of these equations may be intractable for many problems; the insights still apply.

## 3.2 FINITE HORIZON PROBLEMS

Finite horizon problems tend to arise in two settings. First, some problems have a very specific horizon. For example, we might be interested in the value of an American option where we are allowed to sell an asset at any time $t \leq T$ where $T$ is the exercise date. Another problem is to determine how many seats to sell at different prices for a particular flight departing at some point in the future. In the same class are problems that require reaching some goal (but not at a particular point in time). Examples include driving to a destination, selling a house, or winning a game.

A second class of problems is actually infinite horizon, but where the goal is to determine what to do right now given a particular state of the system. For example, a transportation company might want to know what drivers should be assigned to a particular set of loads right now. Of course, these decisions need to consider the downstream impact, so models have to extend into the future. For this reason, we might model the problem over a horizon $T$ which, when solved, yields a decision of what to do right now.

When we encounter a finite horizon problem, we assume that we are given the function $V_T(S_T)$ as data. Often, we simply use $V_T(S_T) = 0$ because we are primarily interested in what to do now, given by $x_0$, or in projected activities over some horizon $t = 0, 1, \ldots, T^{ph}$, where $T^{ph}$ is the length of a planning horizon. If we set $T$ sufficiently larger than $T^{ph}$, then we may be able to assume that the decisions $x_0, x_1, \ldots, x_{T^{ph}}$ are of sufficiently high quality to be useful.

Solving a finite horizon problem, in principle, is straightforward. As outlined in figure 3.1, we simply have to start at the last time period, compute the value function for each possible state $s \in \mathcal{S}$, and then step back another time period. This way, at time period $t$ we have already computed $V_{t+1}(S)$. Not surprisingly, this method is often referred to as "backward dynamic programming." The critical element that attracts so much attention is the requirement that we compute the value function $V_t(S_t)$ for all states $S_t \in \mathcal{S}$.

---

**Step 0.** Initialization:

Initialize the terminal contribution $V_T(S_T)$.

Set $t = T - 1$.

**Step 1.** Calculate:

$$V_t(S_t) = \max_{x_t} \left\{ C_t(S_t, x_t) + \gamma \sum_{s' \in \mathcal{S}} \mathbb{P}(s'|S_t, x_t) V_{t+1}(s') \right\}$$

for all $S_t \in \mathcal{S}$.

**Step 2.** If $t > 0$, decrement $t$ and return to step 1. Else, stop.

---

**Figure 3.1** A backward dynamic programming algorithm.

We first saw backward dynamic programming in section 2.2.1 when we described a simple decision tree problem. The only difference between the backward dynamic programming algorithm in figure 3.1 and our solution of the decision tree problem is primarily notational. Decision trees are visual and tend to be easier to understand, whereas in this section the methods are described using notation. However, decision tree problems tend to be always presented in the context of problems with relatively small numbers of states and actions (What job should I take? Should the United States put a blockade around Cuba? Should the shuttle launch have been canceled due to cold weather?).

Another popular illustration of dynamic programming is the discrete asset acquisition problem. Assume that you order a quantity $x_t$ at each time period to be used in the next time period to satisfy a demand $\hat{D}_{t+1}$. Any unused product is held over to the following time period. For this, our state variable $S_t$ is the quantity of inventory left over at the end of the period after demands are satisfied. The transition equation is given by $S_{t+1} = [S_t + x_t - \hat{D}_{t+1}]^+$ where $[x]^+ = \max(x, 0)$. The cost function (which we seek to minimize) is given by $\hat{C}_{t+1}(S_t, x_t) = c^h S_t + c^o I_{\{x_t > 0\}}$, where $I_{\{X\}} = 1$ if $X$ is true and 0 otherwise. Note that the cost function is nonconvex. This does not create problems if we solve our minimization problem by searching over different (discrete) values of $x_t$. Since all of our quantities are scalar, there is no difficulty finding $C_t(S_t, x_t)$.

To compute the one-step transition matrix, let $\Omega$ be the set of possible outcomes of $\hat{D}_t$, and let $\mathbb{P}(\hat{D}_t = \omega)$ be the probability that $\hat{D}_t = \omega$ (if this use of $\omega$ seems weird, get used to it - we are going to use it a lot).

The one-step transition matrix is computed using

$$\mathbb{P}(s'|s,x) = \sum_{\omega \in \Omega} \mathbb{P}(\hat{D}_{t+1} = \omega) 1_{\{s' = [s+x-\omega]^+\}}$$

where $\Omega$ is the set of (discrete) outcomes of the demand $\hat{D}_{t+1}$.

Another example is the shortest path problem with random arc costs. Assume that you are trying to get from origin node $q$ to destination node $r$ in the shortest time possible. As you reach each intermediate node $i$, you are able to observe the time required to traverse each arc out of node $i$. Let $V_j$ be the expected shortest path time from $j$ to the destination node $r$. At node $i$, you see the link time $\hat{\tau}_{ij}$ which represents a random observation of the travel time. Now we choose to traverse arc $i, j^*$ where $j^*$ solves $\min_j(\hat{\tau}_{ij} + V_j)$ ($j^*$ is random since the travel time is random). We would then compute the value of being at node $i$ using $V_i = \mathbb{E}\{\min_j(\hat{\tau}_{ij} + V_j)\}$.

## 3.3 INFINITE HORIZON PROBLEMS

Infinite horizon problems arise whenever we wish to study a problem where the parameters of the contribution function, transition function and the process governing the exogenous information process do not vary over time. Often, we wish to study such problems in steady state. More importantly, infinite horizon problems provide a number of insights into the properties of problems and algorithms, drawing off an elegant theory that has evolved around this problem class. Even students who wish to solve complex, nonstationary problems will benefit from an understanding of this problem class.

We begin with the optimality equations

$$V_t(S_t) = \max_{x_t \in \mathcal{X}} \mathbb{E}\left\{C_t(S_t, x_t) + \gamma V_{t+1}(S_{t+1}) | S_t\right\}.$$

We can think of a steady-state problem as one without the time dimension. Letting $V(s) = \lim_{t \to \infty} V_t(S_t)$ (and assuming the limit exists), we obtain the steady-state optimality equations

$$V(s) = \max_{x \in \mathcal{X}} \left\{ C(s,x) + \gamma \sum_{s' \in \mathcal{S}} \mathbb{P}(s'|s,x) V(s') \right\}. \tag{3.12}$$

The functions $V(s)$ can be shown (as we do later) to be equivalent to solving the infinite horizon problem

$$\max_{\pi \in \Pi} \mathbb{E} \left\{ \sum_{t=0}^{\infty} \gamma^t C_t(S_t, X_t^\pi(S_t)) \right\}. \tag{3.13}$$

Now define

$$\begin{aligned} P^{\pi,t} &= t\text{-step transition matrix, over periods } 0, 1, \ldots, t-1, \text{ given policy } \pi \\ &= \Pi_{t'=0}^{t-1} P_{t'}^\pi. \end{aligned} \tag{3.14}$$

We further define $P^{\pi,0}$ to be the identity matrix. As before, let $c_t^\pi$ be the column vector of the expected cost of being in each state given that we choose the action $x_t$ described by policy $\pi$, where the element for state $s$ is $c_t^\pi(s) = C_t(s, X^\pi(s))$. The infinite horizon, discounted value of a policy $\pi$ starting at time $t$ is given by

$$v_t^\pi = \sum_{t'=t}^{\infty} \gamma^{t'-t} P^{\pi,t'-t} c_{t'}^\pi. \tag{3.15}$$

Assume that after following policy $\pi_0$ we follow policy $\pi_1 = \pi_2 = \ldots = \pi$. In this case, equation (3.15) can now be written as (starting at $t = 0$)

$$\begin{aligned} v^{\pi_0} &= c^{\pi_0} + \sum_{t'=1}^{\infty} \gamma^{t'} P^{\pi,t'} c_{t'}^\pi & (3.16) \\ &= c^{\pi_0} + \sum_{t'=1}^{\infty} \gamma^{t'} \left( \Pi_{t''=0}^{t'-1} P_{t''}^\pi \right) c_{t'}^\pi & (3.17) \\ &= c^{\pi_0} + \gamma P^{\pi_0} \sum_{t'=1}^{\infty} \gamma^{t'-1} \left( \Pi_{t''=1}^{t'-1} P_{t''}^\pi \right) c_{t'}^\pi & (3.18) \\ &= c^{\pi_0} + \gamma P^{\pi_0} v^\pi. & (3.19) \end{aligned}$$

Equation (3.19) shows us that the value of a policy is the single period reward plus a discounted terminal reward that is the same as the value of a policy starting at time 1. If our decision rule is stationary, then $\pi_0 = \pi_1 = \ldots = \pi_t = \pi$, which allows us to rewrite (3.19) as

$$v^\pi = c^\pi + \gamma P^\pi v^\pi. \tag{3.20}$$

This allows us to solve for the stationary reward explicitly (as long as $0 \le \gamma < 1$), giving us

$$v^\pi = (I - \gamma P^\pi)^{-1} c^\pi.$$

We can also write an infinite horizon version of the optimality equations using our operator notation. Letting $\mathcal{M}$ be the "max" (or "min") operator, the infinite horizon version of equation (3.11) would be written

$$\mathcal{M}^\pi(v) = c^\pi + \gamma P^\pi v. \tag{3.21}$$

There are several algorithmic strategies for solving infinite horizon problems. The first, value iteration, is the most widely used method. It involves iteratively estimating the value function. At each iteration, the estimate of the value function determines which decisions we will make and as a result defines a policy. The second strategy is *policy iteration*. At every iteration, we define a policy (literally, the rule for determining decisions) and then determine the value function for that policy. Careful examination of value and policy iteration reveals that these are closely related strategies that can be viewed as special cases of a general strategy that uses value and policy iteration. Finally, the third major algorithmic strategy exploits the observation that the value function can be viewed as the solution to a specially structured linear programming problem.

**Step 0.** Initialization:

Set $v^0(s) = 0 \ \forall s \in \mathcal{S}$.

Fix a tolerance parameter $\epsilon > 0$.

Set $n = 1$.

**Step 1.** For each $s \in \mathcal{S}$ compute:

$$v^n(s) = \max_{x \in \mathcal{X}} \left( C(s,x) + \gamma \sum_{s' \in \mathcal{S}} \mathbb{P}(s'|s,x) v^{n-1}(s') \right). \tag{3.22}$$

Let $x^n$ be the decision vector that solves equation (3.22).

**Step 2.** If $\|v^n - v^{n-1}\| < \epsilon(1-\gamma)/2\gamma$, let $\pi^\epsilon$ be the resulting policy that solves (3.22), and let $v^\epsilon = v^n$ and stop; else set $n = n + 1$ and go to step 1.

**Figure 3.2** The value iteration algorithm for infinite horizon optimization

## 3.4 VALUE ITERATION

Value iteration is perhaps the most widely used algorithm in dynamic programming because it is the simplest to implement and, as a result, often tends to be the most natural way of solving many problems. It is virtually identical to backward dynamic programming for finite horizon problems. In addition, most of our work in approximate dynamic programming is based on value iteration.

Value iteration comes in several flavors. The basic version of the value iteration algorithm is given in figure 3.2. The proof of convergence (see section 3.9.2) is quite elegant for students who enjoy mathematics. The algorithm also has several nice properties that we explore below.

It is easy to see that the value iteration algorithm is similar to the backward dynamic programming algorithm. Rather than using a subscript $t$, which we decrement from $T$ back to 0, we use an iteration counter $n$ that starts at 0 and increases until we satisfy a convergence criterion. Here, we stop the algorithm when

$$\|v^n - v^{n-1}\| < \epsilon(1-\gamma)/2\gamma,$$

where $\|v\|$ is the max-norm defined by

$$\|v\| = \max_s |v(s)|.$$

Thus, $\|v\|$ is the largest absolute value of a vector of elements. Thus, we stop if the largest change in the value of being in any state is less than $\epsilon(1-\gamma)/2\gamma$ where $\epsilon$ is a specified error tolerance.

Below, we describe a Gauss-Seidel variant which is a useful method for accelerating value iteration, and a version known as relative value iteration.

### 3.4.1 A Gauss-Seidel variation

A slight variant of the value iteration algorithm provides a faster rate of convergence. In this version (typically called the Gauss-Seidel variant), we take advantage of the fact that when we are computing the expectation of the value of the future, we have to loop over all the states $s'$ to compute $\sum_{s'} \mathbb{P}(s'|s,x)v^n(s')$. For a particular state $s$, we would have already computed $v^{n+1}(\hat{s})$ for $\hat{s} = 1, 2, \ldots, s-1$. By simply replacing $v^n(\hat{s})$ with $v^{n+1}(\hat{s})$ for the states we have already visited, we obtain an algorithm that typically exhibits a noticeably faster rate of convergence. The algorithm requires a change to step 1 of the value iteration, as shown in figure 3.3.

---

Replace Step 1 with

**Step 1'.** For each $s \in \mathcal{S}$ compute

$$v^n(s) = \max_{x \in \mathcal{X}} \left\{ C(s,x) + \gamma \left( \sum_{s' < s} \mathbb{P}(s'|s,x) v^n(s') + \sum_{s' \geq s} \mathbb{P}(s'|s,x) v^{n-1}(s') \right) \right\}$$

---

**Figure 3.3** The Gauss-Seidel variation of value iteration.

### 3.4.2 Relative value iteration

Another version of value iteration is called *relative value iteration*, which is useful in problems that do not have a discount factor or where the optimal policy converges much more quickly than the value function, which may grow steadily for many iterations. The relative value iteration algorithm is shown in 3.4.

In relative value iteration, we focus on the fact that we may be more interested in the convergence of the difference $|v(s) - v(s')|$ than we are in the values of $v(s)$ and $v(s')$. This would be the case if we are interested in the best policy rather than the value function itself (this is not always the case). What often happens is that, especially toward the limit, all the values $v(s)$ start increasing by the same rate. For this reason, we can pick any state (denoted $s^*$ in the algorithm) and subtract its value from all the other states.

To provide a bit of formalism for our algorithm, we define the *span* of a vector $v$ as follows:

$$sp(v) = \max_{s \in \mathcal{S}} v(s) - \min_{s \in \mathcal{S}} v(s).$$

---

**Step 0.** Initialization:

- Choose some $v^0 \in \mathcal{V}$.
- Choose a base state $s^*$ and a tolerance $\epsilon$.
- Let $w^0 = v^0 - v^0(s^*)e$ where $e$ is a vector of ones.
- Set $n = 1$.

**Step 1.** Set

$$v^n = \mathcal{M} w^{n-1},$$
$$w^n = v^n - v^n(s^*)e.$$

**Step 2.** If $sp(v^n - v^{n-1}) < (1-\gamma)\epsilon/\gamma$, go to step 3; otherwise, go to step 1.

**Step 3.** Set $x^\epsilon = \arg\max_{x \in \mathcal{X}} \left( C(x) + \gamma P^\pi v^n \right)$.

---

**Figure 3.4** Relative value iteration.

Note that our use of "span" is different than the way it is normally used in linear algebra. Here and throughout this section, we define the norm of a vector as

$$\|v\| = \max_{s \in \mathcal{S}} v(s).$$

Note that the span has the following six properties:

1) $sp(v) \geq 0$.

2) $sp(u + v) \leq sp(u) + sp(v)$.

3) $sp(kv) = |k| sp(v)$.

4) $sp(v + ke) = sp(v)$.

5) $sp(v) = sp(-v)$.

6) $sp(v) \leq 2\|v\|$.

Property $(4)$ implies that $sp(v) = 0$ does not mean that $v = 0$ and therefore it does not satisfy the properties of a norm. For this reason, it is called a *semi-norm*.

The relative value iteration algorithm is simply subtracting a constant from the value vector at each iteration. Obviously, this does not change the optimal decision, but it does change the value itself. If we are only interested in the optimal policy, relative value iteration often offers much faster convergence, but it may not yield accurate estimates of the value of being in each state.

### 3.4.3 Bounds and rates of convergence

One important property of value iteration algorithms is that if our initial estimate is too low, the algorithm will rise to the correct value from below. Similarly, if our initial estimate is too high, the algorithm will approach the correct value from above. This property is formalized in the following theorem:

**Theorem 3.4.1** *For a vector* $v \in \mathcal{V}$*:*

*(a) If* $v$ *satisfies* $v \geq \mathcal{M}v$*, then* $v \geq v^*$.

*(b) If* $v$ *satisfies* $v \leq \mathcal{M}v$*, then* $v \leq v^*$.

*(c) If* $v$ *satisfies* $v = \mathcal{M}v$*, then* $v$ *is the unique solution to this system of equations and* $v = v^*$.

The proof is given in section 3.9.3. It is a nice property because it provides some valuable information on the nature of the convergence path. In practice, we generally do not know the true value function, which makes it hard to know if we are starting from above or below (although some problems have natural bounds, such as nonnegativity).

The proof of the monotonicity property above also provides us with a nice corollary. If $V(s) = \mathcal{M}V(s)$ for all $s$, then $V(s)$ is the unique solution to this system of equations, which must also be the optimal solution.

This result raises the question: What if some of our estimates of the value of being in some states are too high, while others are too low? This means the values may cycle above and below the optimal solution, although at some point we may find that all the values have increased (decreased) from one iteration to the next. If this happens, then it means that the values are all equal to or below (above) the limiting value.

Value iteration also provides a nice bound on the quality of the solution. Recall that when we use the value iteration algorithm, we stop when

$$\|v^{n+1} - v^n\| < \epsilon(1-\gamma)/2\gamma \tag{3.23}$$

where $\gamma$ is our discount factor and $\epsilon$ is a specified error tolerance. It is possible that we have found the optimal policy when we stop, but it is very unlikely that we have found the optimal value functions. We can, however, provide a bound on the gap between the solution $v^n$ and the optimal values $v^*$ by using the following theorem:

**Theorem 3.4.2** *If we apply the value iteration algorithm with stopping parameter $\epsilon$ and the algorithm terminates at iteration $n$ with value function $v^{n+1}$, then*

$$\|v^{n+1} - v^*\| \leq \epsilon/2. \tag{3.24}$$

*Let $\pi^\epsilon$ be the policy that we terminate with, and let $v^{\pi^\epsilon}$ be the value of this policy. Then*

$$\|v^{\pi^\epsilon} - v^*\| \leq \epsilon.$$

The proof is given in section 3.9.4. While it is nice that we can bound the error, the bad news is that the bound can be quite poor. More important is what the bound teaches us about the role of the discount factor.

We can provide some additional insights into the bound, as well as the rate of convergence, by considering a trivial dynamic program. In this problem, we receive a constant reward $c$ at every iteration. There are no decisions, and there is no randomness. The value of this "game" is quickly seen to be

$$\begin{aligned} v^* &= \sum_{n=0}^{\infty} \gamma^n c \\ &= \frac{1}{1-\gamma} c. \end{aligned} \tag{3.25}$$

Consider what happens when we solve this problem using value iteration. Starting with $v^0 = 0$, we would use the iteration

$$v^n = c + \gamma v^{n-1}.$$

After we have repeated this $n$ times, we have

$$\begin{aligned} v^n &= \sum_{m=0}^{n-1} \gamma^n c \\ &= \frac{1-\gamma^n}{1-\gamma} c. \end{aligned} \tag{3.26}$$

Comparing equations (3.25) and (3.26), we see that

$$v^n - v^* = -\frac{\gamma^n}{1-\gamma} c. \tag{3.27}$$

Similarly, the change in the value from one iteration to the next is given by

$$\begin{aligned}
\|v^{n+1} - v^n\| &= \left|\frac{\gamma^{n+1}}{1-\gamma} - \frac{\gamma^n}{1-\gamma}\right| c \\
&= \gamma^n \left|\frac{\gamma}{1-\gamma} - \frac{1}{1-\gamma}\right| c \\
&= \gamma^n \left|\frac{\gamma - 1}{1-\gamma}\right| c \\
&= \gamma^n c.
\end{aligned}$$

If we stop at iteration $n+1$, then it means that

$$\gamma^n c \leq \epsilon/2 \left(\frac{1-\gamma}{\gamma}\right). \tag{3.28}$$

If we choose $\epsilon$ so that (3.28) holds with equality, then our error bound (from 3.24) is

$$\begin{aligned}
\|v^{n+1} - v^*\| &\leq \epsilon/2 \\
&= \frac{\gamma^{n+1}}{1-\gamma} c.
\end{aligned}$$

From (3.27), we know that the distance to the optimal solution is

$$|v^{n+1} - v^*| = \frac{\gamma^{n+1}}{1-\gamma} c,$$

which matches our bound.

This little exercise confirms that our bound on the error may be tight. It also shows that the error decreases geometrically at a rate determined by the discount factor. For this problem, the error arises because we are approximating an infinite sum with a finite one. For more realistic dynamic programs, we also have the effect of trying to find the optimal policy. When the values are close enough that we have, in fact, found the optimal policy, then we have only a Markov reward process (a Markov chain where we earn rewards for each transition). Once our Markov reward process has reached steady state, it will behave just like the simple problem we have just solved, where $c$ is the expected reward from each transition.

## 3.5 POLICY ITERATION

In policy iteration, we choose a policy and then find the infinite horizon, discounted value of the policy. This value is then used to choose a new policy. The general algorithm is described in figure 3.5. Policy iteration is popular for infinite horizon problems because of the ease with which we can find the value of a policy. As we showed in section 3.3, the value of following policy $\pi$ is given by

$$v^\pi = (I - \gamma P^\pi)^{-1} c^\pi. \tag{3.29}$$

While computing the inverse can be problematic as the state space grows, it is, at a minimum, a very convenient formula.

---

**Step 0.** Initialization:

**Step 0a.** Select a policy $\pi^0$.

**Step 0b.** Set $n = 1$.

**Step 1.** Given a policy $\pi^{n-1}$:

**Step 1a.** Compute the one-step transition matrix $P^{\pi^{n-1}}$.

**Step 1b** Compute the contribution vector $c^{\pi^{n-1}}$ where the element for state $s$ is given by $c^{\pi^{n-1}}(s) = C(s, X^{\pi^{n-1}})$.

**Step 2.** Let $v^{\pi,n}$ be the solution to

$$(I - \gamma P^{\pi^{n-1}})v = c^{\pi^{n-1}}.$$

**Step 3.** Find a policy $\pi^n$ defined by

$$x^n(s) = \arg\max_{x \in \mathcal{X}} \left(C(x) + \gamma P^\pi v^n\right).$$

This requires that we compute an action for each state $s$.

**Step 4.** If $x^n(s) = x^{n-1}(s)$ for all states $s$, then set $x^* = x^n$; otherwise, set $n = n + 1$ and go to step 1.

---

**Figure 3.5** Policy iteration

It is useful to illustrate the policy iteration algorithm in different settings. In the first, consider a batch replenishment problem where we have to replenish resources (raising capital, exploring for oil to expand known reserves, hiring people) where there are economies from ordering larger quantities. We might use a simple policy where if our level of resources $R_t < q$ for some lower limit $q$, we order a quantity $x_t = Q - R_t$. This policy is parameterized by $(q, Q)$ and is written

$$X^\pi(R_t) = \begin{cases} 0, & R_t \geq q, \\ Q - R_t, & R_t < q. \end{cases} \tag{3.30}$$

For a given set of parameters $\pi = (q, Q)$, we can compute a one-step transition matrix $P^\pi$ and a contribution vector $c^\pi$.

Policies come in many forms. For the moment, we simply view a policy as a rule that tells us what decision to make when we are in a particular state. In later chapters, we introduce policies in different forms since they create different challenges for finding the best policy.

Given a transition matrix $P^\pi$ and contribution vector $c^\pi$, we can use equation (3.29) to find $v^\pi$, where $v^\pi(s)$ is the discounted value of started in state $s$ and following policy $\pi$. From this vector, we can infer a new policy by solving

$$x^n(s) = \arg\max_{x \in \mathcal{X}} \left(C(x) + \gamma P^\pi v^n\right) \tag{3.31}$$

for each state $s$. For our batch replenishment example, it turns out that we can show that $x^n(s)$ will have the same structure as that shown in (3.30). So, we can either store $x^n(s)$ for each $s$, or simply determine the parameters $(q, Q)$ that correspond to the decisions produced by (3.31). The complete policy iteration algorithm is described in figure 3.5.

The policy iteration algorithm is simple to implement and has fast convergence when measured in terms of the number of iterations. However, solving equation (3.29) is quite hard if the number of states is large. If the state space is small, we can use $v^\pi = (I - \gamma P^\pi)^{-1} c^\pi$, but the matrix inversion can be computationally expensive. For this reason, we may use a hybrid algorithm that combines the features of policy iteration and value iteration.

## 3.6 HYBRID VALUE-POLICY ITERATION

Value iteration is basically an algorithm that updates the value at each iteration and then determines a new policy given the new estimate of the value function. At any iteration, the value function is not the true, steady-state value of the policy. By contrast, policy iteration picks a policy and then determines the true, steady-state value of being in each state given the policy. Given this value, a new policy is chosen.

It is perhaps not surprising that policy iteration converges faster in terms of the number of iterations because it is doing a lot more work in each iteration (determining the true, steady-state value of being in each state under a policy). Value iteration is much faster per iteration, but it is determining a policy given an approximation of a value function and then performing a very simple updating of the value function, which may be far from the true value function.

A hybrid strategy that combines features of both methods is to perform a somewhat more complete update of the value function before performing an update of the policy. Figure 3.6 outlines the procedure where the steady-state evaluation of the value function in equation (3.29) is replaced with a much easier iterative procedure (step 2 in figure 3.6). This step is run for $M$ iterations, where $M$ is a user-controlled parameter that allows the exploration of the value of a better estimate of the value function. Not surprisingly, it will generally be the case that $M$ should decline with the number of iterations as the overall process converges.

## 3.7 THE LINEAR PROGRAMMING METHOD FOR DYNAMIC PROGRAMS

Theorem 3.4.1 showed us that if

$$v \geq \max_x \left( C(s,x) + \gamma \sum_{s' \in \mathcal{S}} \mathbb{P}(s'|s,x) v(s') \right),$$

then $v$ is an upper bound (actually, a vector of upper bounds) on the value of being in each state. This means that the optimal solution, which satisfies $v^* = c + \gamma P v^*$, is the smallest value of $v$ that satisfies this inequality. We can use this insight to formulate the problem of finding the optimal values as a linear program. Let $\beta$ be a vector with elements $\beta_s > 0$, $\forall s \in \mathcal{S}$. The optimal value function can be found by solving the following linear program

$$\min_v \sum_{s \in \mathcal{S}} \beta_s v(s) \tag{3.32}$$

subject to

$$v(s) \geq C(s,x) + \gamma \sum_{s' \in \mathcal{S}} \mathbb{P}(s'|s,x) v(s') \quad \text{for all } s \text{ and } x, \tag{3.33}$$

---

**Step 0.** Initialization:

- Set $n = 1$.
- Select a tolerance parameter $\epsilon$ and inner iteration limit $M$.
- Select some $v^0 \in \mathcal{V}$.

**Step 1.** Find a decision $x^n(s)$ for each $s$ that satisfies

$$x^n(s) = \arg\max_{x \in \mathcal{X}} \left\{ C(s,x) + \gamma \sum_{s' \in \mathcal{S}} \mathbb{P}(s'|s,x) v^{n-1}(s') \right\},$$

which we represent as policy $\pi^n$.

**Step 2.** Partial policy evaluation.

(a) Set $m = 0$ and let: $u^n(0) = c^\pi + \gamma P^{\pi^n} v^{n-1}$.

(b) If $\|u^n(0) - v^{n-1}\| < \epsilon(1-\gamma)/2\gamma$, go to step 3. Else:

(c) While $m < M$ do the following:

i) $u^n(m+1) = c^{\pi^n} + \gamma P^{\pi^n} u^n(m) = \mathcal{M}^\pi u^n(m)$.

ii) Set $m = m + 1$ and repeat $(i)$.

(d) Set $v^n = u^n(M)$, $n = n + 1$ and return to step 1.

**Step 3.** Set $x^\epsilon = x^{n+1}$ and stop.

---

**Figure 3.6** Hybrid value/policy iteration

The linear program has a $|\mathcal{S}|$-dimensional decision vector (the value of being in each state), with $|\mathcal{S}| \times |\mathcal{X}|$ inequality constraints (equation (3.33)).

This formulation was viewed as primarily a theoretical result for many years, since it requires formulating a linear program where the number of constraints is equal to the number of states and actions. While even today this limits the size of problems it can solve, modern linear programming solvers can handle problems with tens of thousands of constraints without difficulty. This size is greatly expanded with the use of specialized algorithmic strategies which are an active area of research as of this writing. The advantage of the LP method over value iteration is that it avoids the need for iterative learning with the geometric convergence exhibited by value iteration. Given the dramatic strides in the speed of linear programming solvers over the last decade, the relative performance of value iteration over the linear programming method is an unresolved question. However, this question only arises for problems with relatively small state and action spaces. While a linear program with 50,000 constraints is considered large, dynamic programs with 50,000 states and actions often arises with relatively small problems.

## 3.8 MONOTONE POLICIES*

One of the most dramatic success stories from the study of Markov decision processes has been the identification of the structure of optimal policies. A common example of structured policies is what are known as *monotone policies*. Simply stated, a monotone policy is one where the decision gets bigger as the state gets bigger, or the decision gets smaller as the state gets bigger (see examples).

---

■ **EXAMPLE 3.1**

A software company must decide when to ship the next release of its operating system. Let $S_t$ be the total investment in the current version of the software. Let $x_t = 1$ denote the decision to ship the release in time period $t$ while $x_t = 0$ means to keep investing in the system. The company adopts the rule that $x_t = 1$ if $S_t \geq \bar{S}$. Thus, as $S_t$ gets bigger, $x_t$ gets bigger (this is true even though $x_t$ is equal to zero or one).

■ **EXAMPLE 3.2**

An oil company maintains stocks of oil reserves to supply its refineries for making gasoline. A supertanker comes from the Middle East each month, and the company can purchase different quantities from this shipment. Let $R_t$ be the current inventory. The policy of the company is to order $x_t = Q - S_t$ if $S_t < R$. $R$ is the reorder point, and $Q$ is the "order up to" limit. The bigger $S_t$ is, the less the company orders.

■ **EXAMPLE 3.3**

A mutual fund has to decide when to sell its holding in a company. Its policy is to sell the stock when the price $\hat{p}_t$ is greater than a particular limit $\bar{p}$.

---

In each example, the decision of what to do in each state is replaced by a function that determines the decision (otherwise known as a policy). The function typically depends on the choice of a few parameters. So, instead of determining the right action for each possible state, we only have to determine the parameters that characterize the function. Interestingly, we do not need dynamic programming for this. Instead, we use dynamic programming to determine the structure of the optimal policy. This is a purely theoretical question, so the computational limitations of (discrete) dynamic programming are not relevant.

The study of monotone policies is included partly because it is an important part of the field of dynamic programming. It is also useful in the study of approximate dynamic programming because it yields properties of the value function. For example, in the process of showing that a policy is monotone, we also need to show that the value function itself is monotone (that is, it increases or decreases with the state variable). Such properties can be exploited in the estimation of a value function approximation.

To demonstrate the analysis of a monotone policy, we consider a classic batch replenishment policy that arises when there is a random accumulation that is then released in batches. Examples include dispatching elevators or trucks, moving oil inventories away from producing fields in tankers, and moving trainloads of grain from grain elevators.

### 3.8.1 The model

For our batch model, we assume resources accumulate and are then reduced using a batch process. For example, oil might accumulate in tanks before a tanker removes it. Money might accumulate in a cash account before it is swept into an investment.

Our model uses the following parameters:

$$
\begin{aligned}
c^r &= \text{The fixed cost incurred each time we dispatch a new batch.} \\
c^h &= \text{Penalty per time period for holding a unit of the resource.} \\
K &= \text{Maximum size of a batch.}
\end{aligned}
$$

Our exogenous information process consists of

$$
\begin{aligned}
A_t &= \text{Quantity of new arrivals during time interval } t. \\
\mathbb{P}^A(i) &= Prob(A_t = i).
\end{aligned}
$$

Our state variable is

$$
R_t = \text{Resources remaining at time } t \text{ before we have made a decision to send a batch.}
$$

There are two decisions we have to make. The first is whether to dispatch a batch, and the second is how many resources to put in the batch. For this problem, once we make the decision to send a batch, we are going to make the batch as large as possible, so the "decision" of how large the batch should be seems unnecessary. It becomes more important when we later consider multiple resource types. For consistency with the more general problem with multiple resource types, we define

$$
\begin{aligned}
x_t &= \begin{cases} 1 & \text{if a batch is sent at time } t, \\ 0 & \text{otherwise,} \end{cases} \\
y_t &= \text{The number of resources to put in the batch.}
\end{aligned}
$$

In theory, we might be able to put a large number of resources in the batch, but we may face a nonlinear cost that makes this suboptimal. For the moment, we are going to assume that we always want to put as many as we can, so we set

$$
\begin{aligned}
y_t &= x_t \min\{K, R_t\}, \\
X^\pi(R_t) &= \text{The decision function that returns } x_t \text{ and } y_t \text{ given } R_t.
\end{aligned}
$$

The transition function is described using

$$
R_{t+1} = R_t - y_t + A_{t+1}. \tag{3.34}
$$

The objective function is modeled using

$$
\begin{aligned}
C_t(R_t, x_t, y_t) &= \text{The cost incurred in period } t\text{, given state } R_t \text{ and dispatch decision } x_t \\
&= c^r x_t + c^h(R_t - y_t). \qquad (3.35)
\end{aligned}
$$

Our problem is to find the policy $X_t^\pi(R_t)$ that solves

$$
\min_{\pi \in \Pi} \mathbb{E} \left\{ \sum_{t=0}^{T} C_t(R_t, X_t^\pi(R_t)) \right\}. \tag{3.36}
$$

where $\Pi$ is the set of policies. If we are managing a single asset class, then $R_t$ and $y_t$ are scalars and the problem can be solved using standard backward dynamic programming techniques of the sort that were presented in chapter 3 (assuming that we have a probability model for the demand). In practice, many problems involve multiple asset classes, which makes standard techniques impractical. But we can use this simple problem to study the structure of the problem.

If $R_t$ is a scalar, and if we know the probability distribution for $A_t$, then we can solve this using backward dynamic programming. Indeed, this is one of the classic dynamic programming problems in operations research. However, the solution to this problem seems obvious. We should dispatch a batch whenever the level of resources $R_t$ is greater than some number $\bar{r}_t$, which means we only have to find $\bar{r}_t$ (if we have a steady state, infinite horizon problem, then we would have to find a single parameter $\bar{r}$). The remainder of this section helps establish the theoretical foundation for making this argument. While not difficult, the mathematical level of this presentation is somewhat higher than our usual presentation.

### 3.8.2 Submodularity and other stories

In the realm of optimization problems over a continuous set, it is important to know a variety of properties about the objective function (such as convexity/concavity, continuity and boundedness). Similarly, discrete problems require an understanding of the nature of the functions we are maximizing, but there is a different set of conditions that we need to establish.

One of the most important properties that we will need is supermodularity (submodularity if we are minimizing). We assume we are studying a function $g(u), u \in \mathcal{U}$, where $\mathcal{U} \subseteq \Re^n$ is an $n$-dimensional space. Consider two vectors $u_1, u_2 \in \mathcal{U}$ where there is no particular relationship between $u_1$ and $u_2$. Now define

$$\begin{aligned} u_1 \wedge u_2 &= \min\{u_1, u_2\}, \\ u_1 \vee u_2 &= \max\{u_1, u_2\}, \end{aligned}$$

where the min and max are defined elementwise. Let $u^+ = u_1 \wedge u_2$ and $u^- = u_1 \vee u_2$. We first have to ask the question of whether $u^+, u^- \in \mathcal{U}$, since this is not guaranteed. For this purpose, we define the following:

**Definition 3.8.1** *The space $\mathcal{U}$ is a* **lattice** *if for each $u_1, u_2 \in \mathcal{U}$, then $u^+ = u_1 \wedge u_2 \in \mathcal{U}$ and $u^- = u_1 \vee u_2 \in \mathcal{U}$.*

The term "lattice" for these sets arises if we think of $u_1$ and $u_2$ as the northwest and southeast corners of a rectangle. In that case, these corners are $u^+$ and $u^-$. If all four corners fall in the set (for any pair $(u_1, u_2)$), then the set can be viewed as containing many "squares," similar to a lattice.

For our purposes, we assume that $\mathcal{U}$ is a lattice (if it is not, then we have to use a more general definition of the operators "$\vee$" and "$\wedge$"). If $\mathcal{U}$ is a lattice, then a general definition of supermodularity is given by the following:

**Definition 3.8.2** *A function $g(u), u \in \mathcal{U}$ is* **supermodular** *if it satisfies*

$$g(u_1 \wedge u_2) + g(u_1 \vee u_2) \geq g(u_1) + g(u_2) \tag{3.37}$$

Supermodularity is the discrete analog of a convex function. A function is **submodular** if the inequality in equation (3.37) is reversed. There is an alternative definition of supermodular when the function is defined on sets. Let $\mathcal{U}_1$ and $\mathcal{U}_2$ be two sets of elements, and let $g$ be a function defined on these sets. Then we have

**Definition 3.8.3** *A function $g : \mathcal{U} \mapsto \Re^1$ is* **supermodular** *if it satisfies*

$$g(\mathcal{U}_1 \cup \mathcal{U}_2) + g(\mathcal{U}_1 \cap \mathcal{U}_2) \geq g(\mathcal{U}_1) + g(\mathcal{U}_2) \tag{3.38}$$

We may refer to definition 3.8.2 as the vector definition of supermodularity, while definition 3.8.3 as the set definition. We give both definitions for completeness, but our work uses only the vector definition.

In dynamic programming, we are interested in functions of two variables, as in $f(s, x)$ where $s$ is a state variable and $x$ is a decision variable. We want to characterize the behavior of $f(s, x)$ as we change $s$ and $x$. If we let $u = (s, x)$, then we can put this in the context of our definition above. Assume we have two states $s^+ \geq s^-$ (again, the inequality is applied elementwise) and two decisions $x^+ \geq x^-$. Now, form two vectors $u_1 = (s^+, x^-)$ and $u_2 = (s^-, x^+)$. With this definition, we find that $u_1 \vee u_2 = (s^+, x^+)$ and $u_1 \wedge u_2 = (s^-, x^-)$. This gives us the following:

**Proposition 3.8.1** *A function $g(s, x)$ is supermodular if for $s^+ \geq s^-$ and $x^+ \geq x^-$, then*

$$g(s^+, x^+) + g(s^-, x^-) \geq g(s^+, x^-) + g(s^-, x^+). \tag{3.39}$$

For our purposes, equation (3.39) will be the version we will use.

A common variation on the statement of a supermodular function is the equivalent condition

$$g(s^+, x^+) - g(s^-, x^+) \geq g(s^+, x^-) - g(s^-, x^-) \tag{3.40}$$

In this expression, we are saying that the incremental change in $s$ for larger values of $x$ is greater than for smaller values of $x$. Similarly, we may write the condition as

$$g(s^+, x^+) - g(s^+, x^-) \geq g(s^-, x^+) - g(s^-, x^-) \tag{3.41}$$

which states that an incremental change in $x$ increases with $s$.

Some examples of supermodular functions include

(a) If $g(s, x) = g_1(s) + g_2(x)$, meaning that it is separable, then (3.39) holds with equality.

(b) $g(s, x) = h(s + x)$ where $h(\cdot)$ is convex and increasing.

(c) $g(s, x) = sx, \; s, x \in \Re^1$.

A concept that is related to supermodularity is *superadditivity*, defined by the following:

**Definition 3.8.4** *A superadditive function $f : \Re^n \to \Re^1$ satisfies*

$$f(x) + f(y) \;\; \leq \;\; f(x + y). \tag{3.42}$$

Some authors use superadditivity and supermodularity interchangeably, but the concepts are not really equivalent, and we need to use both of them.

### 3.8.3 From submodularity to monotonicity

It seems intuitively obvious that we should dispatch a batch if the state $R_t$ (the resources waiting to be served in a batch) is greater than some number (say, $\bar{r}_t$). The dispatch rule that says we should dispatch if $R_t \geq \bar{r}_t$ is known as a *control limit structure*. Similarly, we might be holding an asset and we feel that we should sell it if the price $p_t$ (which is the state of our asset) is over (or perhaps under) some number $\bar{p}_t$. A question arises: when is an optimal policy monotone? The following theorem establishes sufficient conditions for an optimal policy to be monotone.

**Theorem 3.8.1** *Assume that we are maximizing total discounted contribution and that*

*(a)* $C_t(R, x)$ *is supermodular on* $\mathcal{R} \times \mathcal{X}$.

*(b)* $\sum_{R' \in \mathcal{R}} \mathbb{P}(R'|R, x) v_{t+1}(R')$ *is supermodular on* $\mathcal{R} \times \mathcal{X}$.

*Then there exists a decision rule* $X^\pi(R)$ *that is nondecreasing on* $\mathcal{R}$.

The proof of this theorem is provided in section 3.9.6.

In the presentation that follows, we need to show submodularity (instead of supermodularity) because we are minimizing costs rather than maximizing rewards.

It is obvious that $C_t(R, x)$ is nondecreasing in $R$. So it remains to show that $C_t(R, x)$ satisfies

$$C_t(R^+, 1) - C_t(R^-, 1) \leq C_t(R^+, 0) - C_t(R^-, 0). \tag{3.43}$$

Substituting equation (3.35) into (3.43), we must show that

$$c^r + c^h(R^+ - K)^+ - c^r - c^h(R^- - K)^+ \leq c^h R^+ - c^h R^-.$$

This simplifies to

$$(R^+ - K)^+ - (R^- - K)^+ \leq R^+ - R^-. \tag{3.44}$$

Since $R^+ \geq R^-$, $(R^+ - K)^+ = 0 \Rightarrow (R^- - K)^+ = 0$. This implies there are three possible cases for equation (3.44):

**Case 1:** $(R^+ - K)^+ > 0$ and $(R^- - K)^+ > 0$. In this case, (3.44) reduces to $R^+ - R^- = R^+ - R^-$.

**Case 2:** $(R^+ - K)^+ > 0$ and $(R^- - K)^+ = 0$. Here, (3.44) reduces to $R^- \leq K$, which follows since $(R^- - K)^+ = 0$ implies that $R^- \leq K$.

**Case 3:** $(R^+ - K)^+ = 0$ and $(R^- - K)^+ = 0$. Now, (3.44) reduces to $R^- \leq R^+$, which is true by construction.

Now we have to show submodularity of $\sum_{R'=0}^{\infty} \mathbb{P}(R'|R, x) V(R')$. We will do this for the special case that the batch capacity is so large that we never exceed it. A proof is available for the finite capacity case, but it is much more difficult.

Submodularity requires that for $R^- \leq R^+$ we have

$$\begin{aligned}\sum_{R'=0}^{\infty} \mathbb{P}(R'|R^+, 1)V(R') - \sum_{R'=0}^{\infty} \mathbb{P}(R'|R^+, 0)V(R') \quad &\leq \quad \sum_{R'=0}^{\infty} \mathbb{P}(R'|R^-, 1)V(R') \\ &\quad - \sum_{R'=0}^{\infty} \mathbb{P}(R'|R^-, 0)V(R')\end{aligned}$$

For the case that $R^-, R^+ \leq K$ we have

$$\sum_{R'=0}^{\infty} \mathbb{P}^A(R')V(R') - \sum_{R'=R^+}^{\infty} \mathbb{P}^A(R'-R^+)V(R') \leq \sum_{R'=0}^{\infty} \mathbb{P}^A(R')V(R') - \sum_{R'=R^-}^{\infty} \mathbb{P}^A(R'-R^-)V(R'),$$

which simplifies to

$$\sum_{R'=0}^{\infty} \mathbb{P}^A(R')V(R') - \sum_{R'=0}^{\infty} \mathbb{P}^A(R')V(R'+R^+) \leq \sum_{R'=0}^{\infty} \mathbb{P}^A(R')V(R') - \sum_{R'=0}^{\infty} \mathbb{P}^A(R')V(R'+R^-).$$

Since $V$ is nondecreasing we have $V(R'+R^+) \geq V(R'+R^-)$, proving the result.

## 3.9 WHY DOES IT WORK?**

The theory of Markov decision processes is especially elegant. While not needed for computational work, an understanding of why they work will provide a deeper appreciation of the properties of these problems.

Section 3.9.1 provides a proof that the optimal value function satisfies the optimality equations. Section 3.9.2 proves convergence of the value iteration algorithm. Section 3.9.3 then proves conditions under which value iteration increases or decreases monotonically to the optimal solution. Then, section 3.9.4 proves the bound on the error when value iteration satisfies the termination criterion given in section 3.4.3. Section 3.9.5 closes with a discussion of deterministic and randomized policies, along with a proof that deterministic policies are always at least as good as a randomized policy.

### 3.9.1 The optimality equations

Until now, we have been presenting the optimality equations as though they were a fundamental law of some sort. To be sure, they can easily look as though they were intuitively obvious, but it is still important to establish the relationship between the original optimization problem and the optimality equations. Since these equations are the foundation of dynamic programming, it seems beholden on us to work through the steps of proving that they are actually true.

We start by remembering the original optimization problem:

$$F_t^\pi(S_t) = \mathbb{E}\left\{\sum_{t'=t}^{T-1} C_{t'}(S_{t'}, X_{t'}^\pi(S_{t'})) + C_T(S_T)|S_t\right\}. \tag{3.45}$$

Since (3.45) is, in general, exceptionally difficult to solve, we resort to the optimality equations

$$V_t^\pi(S_t) = C_t(S_t, X_t^\pi(S_t)) + \mathbb{E}\left\{V_{t+1}^\pi(S_{t+1})|S_t\right\}. \tag{3.46}$$

Our challenge is to show that these are the same. In order to establish this result, it is going to help if we first prove the following:

**Lemma 3.9.1** *Let $S_t$ be a state variable that captures the relevant history up to time $t$, and let $F_{t'}(S_{t+1})$ be some function measured at time $t' \geq t+1$ conditioned on the random variable $S_{t+1}$. Then*

$$\mathbb{E}\left[\mathbb{E}\{F_{t'}|S_{t+1}\}|S_t\right] = \mathbb{E}\left[F_{t'}|S_t\right]. \tag{3.47}$$

**Proof:** This lemma is variously known as the law of iterated expectations or the tower property. Assume, for simplicity, that $F_{t'}$ is a discrete, finite random variable that takes outcomes in $\mathcal{F}$. We start by writing

$$\mathbb{E}\{F_{t'}|S_{t+1}\} = \sum_{f \in \mathcal{F}} f\mathbb{P}(F_{t'} = f|S_{t+1}). \tag{3.48}$$

Recognizing that $S_{t+1}$ is a random variable, we may take the expectation of both sides of (3.48), conditioned on $S_t$ as follows:

$$\mathbb{E}\left[\mathbb{E}\{F_{t'}|S_{t+1}\}|S_t\right] = \sum_{S_{t+1} \in \mathcal{S}} \sum_{f \in \mathcal{F}} f\mathbb{P}(F_{t'} = f|S_{t+1}, S_t)\mathbb{P}(S_{t+1} = S_{t+1}|S_t). \tag{3.49}$$

First, we observe that we may write $\mathbb{P}(F_{t'} = f|S_{t+1}, S_t) = \mathbb{P}(F_{t'} = f|S_{t+1})$, because conditioning on $S_{t+1}$ makes all prior history irrelevant. Next, we can reverse the summations on the right-hand side of (3.49) (some technical conditions have to be satisfied to do this, but these are satisfied if the random variables are discrete and finite). This means

$$\begin{aligned}
\mathbb{E}\left[\mathbb{E}\{F_{t'}|S_{t+1} = S_{t+1}\}|S_t\right] &= \sum_{f \in \mathcal{F}} \sum_{S_{t+1} \in \mathcal{S}} f\mathbb{P}(F_{t'} = f|S_{t+1}, S_t)\mathbb{P}(S_{t+1} = S_{t+1}|S_t) \\
&= \sum_{f \in \mathcal{F}} f \sum_{S_{t+1} \in \mathcal{S}} \mathbb{P}(F_{t'} = f, S_{t+1}|S_t) \\
&= \sum_{f \in \mathcal{F}} f\mathbb{P}(F_{t'} = f|S_t) \\
&= \mathbb{E}\left[F_{t'}|S_t\right],
\end{aligned}$$

which proves our result. Note that the essential step in the proof occurs in the first step when we add $S_t$ to the conditioning. □

We are now ready to show the following:

**Proposition 3.9.1** $F_t^\pi(S_t) = V_t^\pi(S_t)$.

**Proof:** To prove that (3.45) and (3.46) are equal, we use a standard trick in dynamic programming: proof by induction. Clearly, $F_T^\pi(S_T) = V_T^\pi(S_T) = C_T(S_T)$. Next, assume that it holds for $t+1, t+2, \ldots, T$. We want to show that it is true for $t$. This means that we can write

$$V_t^\pi(S_t) = C_t(S_t, X_t^\pi(S_t)) + \mathbb{E}\left[\underbrace{\mathbb{E}\left\{\sum_{t'=t+1}^{T-1} C_{t'}(S_{t'}, X_{t'}^\pi(S_{t'})) + C_t(S_T(\omega))\middle| S_{t+1}\right\}}_{F_{t+1}^\pi(S_{t+1})}\middle| S_t\right].$$

We then use lemma 3.9.1 to write $\mathbb{E}\left[\mathbb{E}\left\{\ldots|S_{t+1}\right\}|S_t\right] = \mathbb{E}\left[\ldots|S_t\right]$. Hence,

$$V_t^\pi(S_t) = C_t(S_t, X_t^\pi(S_t)) + \mathbb{E}\left[\sum_{t'=t+1}^{T-1} C_{t'}(S_{t'}, X_{t'}^\pi(S_{t'})) + C_t(S_T)|S_t\right].$$

When we condition on $S_t$, $X_t^\pi(S_t)$ (and therefore $C_t(S_t, X_t^\pi(S_t))$) is deterministic, so we can pull the expectation out to the front giving

$$\begin{aligned} V_t^\pi(S_t) &= \mathbb{E}\left[\sum_{t'=t}^{T-1} C_{t'}(S_{t'}, y_{t'}(S_{t'})) + C_t(S_T)|S_t\right] \\ &= F_t^\pi(S_t), \end{aligned}$$

which proves our result. □

Using equation (3.46), we have a backward recursion for calculating $V_t^\pi(S_t)$ for a given policy $\pi$. Now that we can find the expected reward for a given $\pi$, we would like to find the best $\pi$. That is, we want to find

$$F_t^*(S_t) = \max_{\pi\in\Pi} F_t^\pi(S_t).$$

If the set $\Pi$ is infinite, we replace the "max" with "sup". We solve this problem by solving the optimality equations. These are

$$V_t(S_t) = \max_{x\in\mathcal{X}} \left(C_t(S_t, x) + \sum_{s'\in\mathcal{S}} p_t(s'|S_t, x)V_{t+1}(s')\right). \tag{3.50}$$

We are claiming that if we find the set of $V's$ that solves (3.50), then we have found the policy that optimizes $F_t^\pi$. We state this claim formally as:

**Theorem 3.9.1** *Let $V_t(S_t)$ be a solution to equation (3.50). Then*

$$\begin{aligned} F_t^* &= V_t(S_t) \\ &= \max_{\pi\in\Pi} F_t^\pi(S_t). \end{aligned}$$

**Proof:** The proof is in two parts. First, we show by induction that $V_t(S_t) \geq F_t^*(S_t)$ for all $S_t \in \mathcal{S}$ and $t = 0, 1, \ldots, T-1$. Then, we show that the reverse inequality is true, which gives us the result.

Part 1:

We resort again to our proof by induction. Since $V_T(S_T) = C_t(S_T) = F_T^\pi(S_T)$ for all $S_T$ and all $\pi \in \Pi$, we get that $V_T(S_T) = F_T^*(S_T)$.

Assume that $V_{t'}(S_{t'}) \geq F_{t'}^*(S_{t'})$ for $t' = t+1, t+2, \ldots, T$, and let $\pi$ be an arbitrary policy. For $t' = t$, the optimality equation tells us

$$V_t(S_t) = \max_{x\in\mathcal{X}} \left(C_t(S_t, x) + \sum_{s'\in\mathcal{S}} p_t(s'|S_t, x)V_{t+1}(s')\right).$$

By the induction hypothesis, $F_{t+1}^*(s) \leq V_{t+1}(s)$, so we get

$$V_t(S_t) \geq \max_{x\in\mathcal{X}} \left(C_t(S_t, x) + \sum_{s'\in\mathcal{S}} p_t(s'|S_t, x)F_{t+1}^*(s')\right).$$

Of course, we have that $F^*_{t+1}(s) \geq F^\pi_{t+1}(s)$ for an arbitrary $\pi$. Also let $X^\pi(S_t)$ be the decision that would be chosen by policy $\pi$ when in state $S_t$. Then

$$\begin{aligned} V_t(S_t) &\geq \max_{x\in\mathcal{X}}\left(C_t(S_t,x) + \sum_{s'\in\mathcal{S}} p_t(s'|S_t,x)F^\pi_{t+1}(s')\right) \\ &\geq C_t(S_t,X^\pi(S_t)) + \sum_{s'\in\mathcal{S}} p_t(s'|S_t,X^\pi(S_t))F^\pi_{t+1}(s') \\ &= F^\pi_t(S_t). \end{aligned}$$

This means

$$V_t(S_t) \geq F^\pi_t(S_t) \quad \text{for all } \pi \in \Pi,$$

which proves part 1.
Part 2:

Now we are going to prove the inequality from the other side. Specifically, we want to show that for any $\epsilon > 0$ there exists a policy $\pi$ that satisfies

$$F^\pi_t(S_t) + (T-t)\epsilon \geq V_t(S_t). \tag{3.51}$$

To do this, we start with the definition

$$V_t(S_t) = \max_{x\in\mathcal{X}}\left(C_t(S_t,x) + \sum_{s'\in\mathcal{S}} p_t(s'|S_t,x)V_{t+1}(s')\right). \tag{3.52}$$

We may let $x_t(S_t)$ be the decision rule that solves (3.52). This rule corresponds to the policy $\pi$. In general, the set $\mathcal{X}$ may be infinite, whereupon we have to replace the "max" with a "sup" and handle the case where an optimal decision may not exist. For this case, we know that we can design a decision rule $x_t(S_t)$ that returns a decision $x$ that satisfies

$$V_t(S_t) \leq C_t(S_t,x) + \sum_{s'\in\mathcal{S}} p_t(s'|S_t,x)V_{t+1}(s') + \epsilon. \tag{3.53}$$

We can prove (3.51) by induction. We first note that (3.51) is true for $t = T$ since $F^\pi_T(S_t) = V_T(S_T)$. Now assume that it is true for $t' = t+1, t+2, \ldots, T$. We already know that

$$F^\pi_t(S_t) = C_t(S_t,X^\pi(S_t)) + \sum_{s'\in\mathcal{S}} p_t(s'|S_t,X^\pi(S_t))F^\pi_{t+1}(s').$$

We can use our induction hypothesis which says $F^\pi_{t+1}(s') \geq V_{t+1}(s') - (T-(t+1))\epsilon$ to get

$$\begin{aligned} F^\pi_t(S_t) &\geq C_t(S_t,X^\pi(S_t)) + \sum_{s'\in\mathcal{S}} p_t(s'|S_t,X^\pi(S_t))[V_{t+1}(s') - (T-(t+1))\epsilon] \\ &= C_t(S_t,X^\pi(S_t)) + \sum_{s'\in\mathcal{S}} p_t(s'|S_t,X^\pi(S_t))V_{t+1}(s') \\ &\quad - \sum_{s'\in\mathcal{S}} p_t(s'|S_t,X^\pi(S_t))\left[(T-t-1)\epsilon\right] \\ &= \left\{C_t(S_t,X^\pi(S_t)) + \sum_{s'\in\mathcal{S}} p_t(s'|S_t,X^\pi(S_t))V_{t+1}(s') + \epsilon\right\} - (T-t)\epsilon. \end{aligned}$$

Now, using equation (3.53), we replace the term in brackets with the smaller $V_t(S_t)$ (equation (3.53)):

$$F_t^\pi(S_t) \geq V_t(S_t) - (T-t)\epsilon,$$

which proves the induction hypothesis. We have shown that

$$F_t^*(S_t) + (T-t)\epsilon \geq F_t^\pi(S_t) + (T-t)\epsilon \geq V_t(S_t) \geq F_t^*(S_t).$$

This proves the result. □

Now we know that solving the optimality equations also gives us the optimal value function. This is our most powerful result because we can solve the optimality equations for many problems that cannot be solved any other way.

### 3.9.2 Convergence of value iteration

We now undertake the proof that the basic value function iteration converges to the optimal solution. This is not only an important result, it is also an elegant one that brings some powerful theorems into play. The proof is also quite short. However, we will need some mathematical preliminaries:

**Definition 3.9.1** *Let $\mathcal{V}$ be a set of (bounded, real-valued) functions and define the norm of $v$ by:*

$$\|v\| = \sup_{s\in\mathcal{S}} v(s)$$

*where we replace the "$sup$" with a "$max$" when the state space is finite. Since $\mathcal{V}$ is closed under addition and scalar multiplication and has a norm, it is a* **normed linear space**.

**Definition 3.9.2** $T : \mathcal{V} \rightarrow \mathcal{V}$ *is a* **contraction mapping** *if there exists a $\gamma$, $0 \leq \gamma < 1$ such that:*

$$\|Tv - Tu\| \leq \gamma\|v - u\|.$$

**Definition 3.9.3** *A sequence $v^n \in \mathcal{V}$, $n = 1, 2, \ldots$ is said to be a* **Cauchy sequence** *if for all $\epsilon > 0$, there exists $N$ such that for all $n, m \geq N$:*

$$\|v^n - v^m\| < \epsilon.$$

**Definition 3.9.4** *A normed linear space is* **complete** *if every Cauchy sequence contains a limit point in that space.*

**Definition 3.9.5** *A* **Banach space** *is a complete normed linear space.*

**Definition 3.9.6** *We define the norm of a matrix $Q$ as*

$$\|Q\| = \max_{s\in\mathcal{S}} \sum_{j\in\mathcal{S}} |q(j|s)|,$$

*that is, the largest row sum of the matrix. If $Q$ is a one-step transition matrix, then $\|Q\| = 1$.*

**Definition 3.9.7** *The* **triangle inequality** *means that given two vectors $a, b \in \Re^n$:*

$$\|a + b\| \leq \|a\| + \|b\|.$$

The triangle inequality is commonly used in proofs because it helps us establish bounds between two solutions (and in particular, between a solution and the optimum).

We now state and prove one of the famous theorems in applied mathematics and then use it immediately to prove convergence of the value iteration algorithm.

**Theorem 3.9.2** *(Banach Fixed-Point Theorem) Let $\mathcal{V}$ be a Banach space, and let $T : \mathcal{V} \rightarrow \mathcal{V}$ be a contraction mapping. Then:*

*(a) There exists a unique $v^* \in \mathcal{V}$ such that $Tv^* = v^*$.*

*(b) For an arbitrary $v^0 \in \mathcal{V}$, the sequence $v^n$ defined by: $v^{n+1} = Tv^n = T^{n+1}v^0$ converges to $v^*$.*

**Proof:** We start by showing that the distance between two vectors $v^n$ and $v^{n+m}$ goes to zero for sufficiently large $n$ and by writing the difference $v^{n+m} - v^n$ using

$$\begin{aligned} v^{n+m} - v^n &= v^{n+m} - v^{n+m-1} + v^{n+m-1} - \cdots - v^{n+1} + v^{n+1} - v^n \\ &= \sum_{k=0}^{m-1} (v^{n+k+1} - v^{n+k}). \end{aligned}$$

Taking norms of both sides and invoking the triangle inequality gives

$$\begin{aligned} \|v^{n+m} - v^n\| &= \|\sum_{k=0}^{m-1} (v^{n+k+1} - v^{n+k})\| \\ &\leq \sum_{k=0}^{m-1} \|(v^{n+k+1} - v^{n+k})\| \\ &= \sum_{k=0}^{m-1} \|(T^{n+k}v^1 - T^{n+k}v^0)\| \\ &\leq \sum_{k=0}^{m-1} \gamma^{n+k}\|v^1 - v^0\| \\ &= \frac{\gamma^n(1-\gamma^m)}{(1-\gamma)}\|v^1 - v^0\|. \end{aligned} \tag{3.54}$$

Since $\gamma < 1$, for sufficiently large $n$ the right-hand side of (3.54) can be made arbitrarily small, which means that $v^n$ is a Cauchy sequence. Since $\mathcal{V}$ is *complete*, it must be that $v^n$ has a limit point $v^*$. From this we conclude

$$\lim_{n \to \infty} v^n \rightarrow v^*. \tag{3.55}$$

We now want to show that $v^*$ is a fixed point of the mapping $T$. To show this, we observe

$$\begin{aligned} 0 &\leq \|Tv^* - v^*\| && (3.56) \\ &= \|Tv^* - v^n + v^n - v^*\| && (3.57) \\ &\leq \|Tv^* - v^n\| + \|v^n - v^*\| && (3.58) \\ &= \|Tv^* - Tv^{n-1}\| + \|v^n - v^*\| && (3.59) \\ &\leq \gamma\|v^* - v^{n-1}\| + \|v^n - v^*\|. && (3.60) \end{aligned}$$

Equation (3.56) comes from the properties of a norm. We play our standard trick in (3.57) of adding and subtracting a quantity (in this case, $v^n$), which sets up the triangle inequality in (3.58). Using $v^n = Tv^{n-1}$ gives us (3.59). The inequality in (3.60) is based on the assumption of the theorem that $T$ is a contraction mapping. From (3.55), we know that

$$\lim_{n\to\infty} \|v^* - v^{n-1}\| = \lim_{n\to\infty} \|v^n - v^*\| = 0. \tag{3.61}$$

Combining (3.56), (3.60), and (3.61) gives

$$0 \leq \|Tv^* - v^*\| \leq 0$$

from which we conclude

$$\|Tv^* - v^*\| = 0,$$

which means that $Tv^* = v^*$.

We can prove uniqueness by contradiction. Assume that there are two limit points that we represent as $v^*$ and $u^*$. The assumption that $T$ is a contraction mapping requires that

$$\|Tv^* - Tu^*\| \leq \gamma\|v^* - u^*\|.$$

But, if $v^*$ and $u^*$ are limit points, then $Tv^* = v^*$ and $Tu^* = u^*$, which means

$$\|v^* - u^*\| \leq \gamma\|v^* - u^*\|.$$

Since $\gamma < 1$, this is a contradiction, which means that it must be true that $v^* = u^*$. □

We can now show that the value iteration algorithm converges to the optimal solution if we can establish that $\mathcal{M}$ is a contraction mapping. So we need to show the following:

**Proposition 3.9.2** *If $0 \leq \gamma < 1$, then $\mathcal{M}$ is a contraction mapping on $\mathcal{V}$.*

**Proof:** Let $u, v \in \mathcal{V}$ and assume that $\mathcal{M}v \geq \mathcal{M}u$ where the inequality is applied element-wise. For a particular state $s$ let

$$x_s^*(v) \in \arg\max_{x\in\mathcal{X}} \left( C(s,x) + \gamma \sum_{s'\in\mathcal{S}} \mathbb{P}(s'|s,x)v(s') \right)$$

where we assume that a solution exists. Then

$$
\begin{aligned}
0 &\leq \mathcal{M}v(s) - \mathcal{M}u(s) & (3.62)\\
&= C(s, x_s^*(v)) + \gamma \sum_{s' \in \mathcal{S}} \mathbb{P}(s'|s, x_s^*(v))v(s') \\
&\quad - \left( C(s, x_s^*(u)) + \gamma \sum_{s' \in \mathcal{S}} \mathbb{P}(s'|s, x_s^*(u))u(s') \right) & (3.63)\\
&\leq C(s, x_s^*(v)) + \gamma \sum_{s' \in \mathcal{S}} \mathbb{P}(s'|s, x_s^*(v))v(s') \\
&\quad - \left( C(s, x_s^*(v)) + \gamma \sum_{s' \in \mathcal{S}} \mathbb{P}(s'|s, x_s^*(v))u(s') \right) & (3.64)\\
&= \gamma \sum_{s' \in \mathcal{S}} \mathbb{P}(s'|s, x_s^*(v))[v(s') - u(s')] & (3.65)\\
&\leq \gamma \sum_{s' \in \mathcal{S}} \mathbb{P}(s'|s, x_s^*(v))\|v - u\| & (3.66)\\
&= \gamma\|v - u\| \sum_{s' \in \mathcal{S}} \mathbb{P}(s'|s, x_s^*(v)) & (3.67)\\
&= \gamma\|v - u\|. & (3.68)
\end{aligned}
$$

Equation (3.62) is true by assumption, while (3.63) holds by definition. The inequality in (3.64) holds because $x_s^*(v)$ is not optimal when the value function is $u$, giving a reduced value in the second set of parentheses. Equation (3.65) is a simple reduction of (3.64). Equation (3.66) forms an upper bound because the definition of $\|v - u\|$ is to replace all the elements $[v(s) - u(s)]$ with the largest element of this vector. Since this is now a vector of constants, we can pull it outside of the summation, giving us (3.67), which then easily reduces to (3.68) because the probabilities add up to one.

This result states that if $\mathcal{M}v(s) \geq \mathcal{M}u(s)$, then $\mathcal{M}v(s) - \mathcal{M}u(s) \leq \gamma|v(s) - u(s)|$. If we start by assuming that $\mathcal{M}v(s) \leq \mathcal{M}u(s)$, then the same reasoning produces $\mathcal{M}v(s) - \mathcal{M}u(s) \geq -\gamma|v(s) - u(s)|$. This means that we have

$$
|\mathcal{M}v(s) - \mathcal{M}u(s)| \leq \gamma|v(s) - u(s)| \tag{3.69}
$$

for *all* states $s \in \mathcal{S}$. From the definition of our norm, we can write

$$
\begin{aligned}
\sup_{s \in \mathcal{S}} |\mathcal{M}v(s) - \mathcal{M}u(s)| &= \|\mathcal{M}v - \mathcal{M}u\| \\
&\leq \gamma\|v - u\|.
\end{aligned}
$$

This means that $\mathcal{M}$ is a contraction mapping, which means that the sequence $v^n$ generated by $v^{n+1} = \mathcal{M}v^n$ converges to a unique limit point $v^*$ that satisfies the optimality equations.
□

### 3.9.3 Monotonicity of value iteration

Infinite horizon dynamic programming provides a compact way to study the theoretical properties of these algorithms. The insights gained here are applicable to problems even when we cannot apply this model, or these algorithms, directly.

We assume throughout our discussion of infinite horizon problems that the reward function is bounded over the domain of the state space. This assumption is virtually always satisfied in practice, but notable exceptions exist. For example, the assumption is violated if we are maximizing a utility function that depends on the log of the resources we have at hand (the resources may be bounded, but the function is unbounded if the resources are allowed to hit zero).

Our first result establishes a monotonicity property that can be exploited in the design of an algorithm.

**Theorem 3.9.3** *For a vector $v \in \mathcal{V}$:*

*(a) If $v$ satisfies $v \geq \mathcal{M}v$, then $v \geq v^*$.*

*(b) If $v$ satisfies $v \leq \mathcal{M}v$, then $v \leq v^*$.*

*(c) If $v$ satisfies $v = \mathcal{M}v$, then $v$ is the unique solution to this system of equations and $v = v^*$.*

**Proof:** Part $(a)$ requires that

$$\begin{aligned} v &\geq \max_{\pi \in \Pi} \{c^\pi + \gamma P^\pi v\} & (3.70) \\ &\geq c^{\pi_0} + \gamma P^{\pi_0} v & (3.71) \\ &\geq c^{\pi_0} + \gamma P^{\pi_0} (c^{\pi_1} + \gamma P^{\pi_1} v) & (3.72) \\ &= c^{\pi_0} + \gamma P^{\pi_0} c^{\pi_1} + \gamma^2 P^{\pi_0} P^{\pi_1} v \end{aligned}$$

Equation (3.70) is true by assumption (part $(a)$ of the theorem) and equation (3.71) is true because $\pi_0$ is some policy that is not necessarily optimal for the vector $v$. Using similar reasoning, equation (3.72) is true because $\pi_1$ is another policy which, again, is not necessarily optimal. Using $P^{\pi,(t)} = P^{\pi_0} P^{\pi_1} \cdots P^{\pi_t}$, we obtain by induction

$$v \geq c^{\pi_0} + \gamma P^{\pi_0} c^{\pi_1} + \cdots + \gamma^{t-1} P^{\pi_0} P^{\pi_1} \cdots P^{\pi_{t-1}} c^{\pi_t} + \gamma^t P^{\pi,(t)} v \quad (3.73)$$

Recall that

$$v^\pi = \sum_{t=0}^{\infty} \gamma^t P^{\pi,(t)} c^{\pi_t} \quad (3.74)$$

Breaking the sum in (3.74) into two parts allows us to rewrite the expansion in (3.73) as

$$v \geq v^\pi - \sum_{t'=t+1}^{\infty} \gamma^{t'} P^{\pi,(t')} c^{\pi_{t'+1}} + \gamma^t P^{\pi,(t)} v \quad (3.75)$$

Taking the limit of both sides of (3.75) as $t \to \infty$ gives us

$$\begin{aligned} v &\geq \lim_{t \to \infty} v^\pi - \sum_{t'=t+1}^{\infty} \gamma^{t'} P^{\pi,(t')} c^{\pi_{t'+1}} + \gamma^t P^{\pi,(t)} v & (3.76) \\ &\geq v^\pi \quad \forall \pi \in \Pi & (3.77) \end{aligned}$$

The limit in (3.76) exists as long as the reward function $c^\pi$ is bounded and $\gamma < 1$. Because (3.77) is true for all $\pi \in \Pi$, it is also true for the optimal policy, which means that

$$\begin{aligned} v &\geq v^{\pi *} \\ &= v^* \end{aligned}$$

which proves part (a) of the theorem. Part $(b)$ can be proved in an analogous way. Parts (a) and (b) mean that $v \geq v^*$ and $v \leq v^*$. If $v = \mathcal{M}v$, then we satisfy the preconditions of both parts (a) and (b), which means they are both true and therefore we must have $v = v^*$. □

This result means that if we start with a vector that is higher than the optimal vector, then we will decline monotonically to the optimal solution (almost – we have not quite proven that we actually get to the optimal). Alternatively, if we start below the optimal vector, we will rise to it. Note that it is not always easy to find a vector $v$ that satisfies either condition $(a)$ or $(b)$ of the theorem. In problems where the rewards can be positive and negative, this can be tricky.

### 3.9.4 Bounding the error from value iteration

We now wish to establish a bound on our error from value iteration, which will establish our stopping rule. We propose two bounds: one on the value function estimate that we terminate with and one for the long-run value of the decision rule that we terminate with. To define the latter, let $\pi^\epsilon$ be the policy that satisfies our stopping rule, and let $v^{\pi^\epsilon}$ be the infinite horizon value of following policy $\pi^\epsilon$.

**Theorem 3.9.4** *If we apply the value iteration algorithm with stopping parameter $\epsilon$ and the algorithm terminates at iteration $n$ with value function $v^{n+1}$, then*

$$\|v^{n+1} - v^*\| \leq \epsilon/2, \tag{3.78}$$

*and*

$$\|v^{\pi^\epsilon} - v^*\| \leq \epsilon. \tag{3.79}$$

**Proof:** We start by writing

$$\begin{aligned} \|v^{\pi^\epsilon} - v^*\| &= \|v^{\pi^\epsilon} - v^{n+1} + v^{n+1} - v^*\| \\ &\leq \|v^{\pi^\epsilon} - v^{n+1}\| + \|v^{n+1} - v^*\|. \end{aligned} \tag{3.80}$$

Recall that $\pi^\epsilon$ is the policy that solves $\mathcal{M}v^{n+1}$, which means that $\mathcal{M}^{\pi^\epsilon} v^{n+1} = \mathcal{M}v^{n+1}$. This allows us to rewrite the first term on the right-hand side of (3.80) as

$$\begin{aligned} \|v^{\pi^\epsilon} - v^{n+1}\| &= \|\mathcal{M}^{\pi^\epsilon} v^{\pi^\epsilon} - \mathcal{M}v^{n+1} + \mathcal{M}v^{n+1} - v^{n+1}\| \\ &\leq \|\mathcal{M}^{\pi^\epsilon} v^{\pi^\epsilon} - \mathcal{M}v^{n+1}\| + \|\mathcal{M}v^{n+1} - v^{n+1}\| \\ &= \|\mathcal{M}^{\pi^\epsilon} v^{\pi^\epsilon} - \mathcal{M}^{\pi^\epsilon} v^{n+1}\| + \|\mathcal{M}v^{n+1} - \mathcal{M}v^n\| \\ &\leq \gamma\|v^{\pi^\epsilon} - v^{n+1}\| + \gamma\|v^{n+1} - v^n\|. \end{aligned}$$

Solving for $\|v^{\pi^\epsilon} - v^{n+1}\|$ gives

$$\|v^{\pi^\epsilon} - v^{n+1}\| \leq \frac{\gamma}{1-\gamma}\|v^{n+1} - v^n\|.$$

We can use similar reasoning applied to the second term in equation (3.80) to show that

$$\|v^{n+1} - v^*\| \leq \frac{\gamma}{1-\gamma}\|v^{n+1} - v^n\|. \tag{3.81}$$

The value iteration algorithm stops when $\|v^{n+1} - v^n\| \leq \epsilon(1-\gamma)/2\gamma$. Substituting this in (3.81) gives

$$\|v^{n+1} - v^*\| \leq \frac{\epsilon}{2}. \tag{3.82}$$

Recognizing that the same bound applies to $\|v^{\pi^\epsilon} - v^{n+1}\|$ and combining these with (3.80) gives us

$$\|v^{\pi^\epsilon} - v^*\| \leq \epsilon,$$

which completes our proof. □

### 3.9.5 Randomized policies

We have implicitly assumed that for each state, we want a single action. An alternative would be to choose a policy probabilistically from a family of policies. If a state produces a single action, we say that we are using a *deterministic policy*. If we are randomly choosing an action from a set of actions probabilistically, we say we are using a *randomized policy*.

Randomized policies may arise because of the nature of the problem. For example, you wish to purchase something at an auction, but you are unable to attend yourself. You may have a simple rule ("purchase it as long as the price is under a specific amount") but you cannot assume that your representative will apply the same rule. You can choose a representative, and in doing so you are effectively choosing the probability distribution from which the action will be chosen.

Behaving randomly also plays a role in two-player games. If you make the same decision each time in a particular state, your opponent may be able to predict your behavior and gain an advantage. For example, as an institutional investor you may tell a bank that you not willing to pay any more than \$14 for a new offering of stock, while in fact you are willing to pay up to \$18. If you always bias your initial prices by \$4, the bank will be able to guess what you are willing to pay.

When we can only influence the likelihood of an action, then we have an instance of a randomized MDP. Let

$q_t^\pi(x|S_t) =$ The probability that decision $x$ will be taken at time $t$ given state $S_t$ and policy $\pi$ (more precisely, decision rule $X^\pi$).

In this case, our optimality equations look like

$$V_t^*(S_t) = \max_{\pi \in \Pi^{MR}} \sum_{x \in \mathcal{X}} \left[ q_t^\pi(x|S_t) \left( C_t(S_t, x) + \sum_{s' \in \mathcal{S}} p_t(s'|S_t, x) V_{t+1}^*(s') \right) \right]. \tag{3.83}$$

Now let us consider the single best action that we could take. Calling this $x^*$, we can find it using

$$x^* = \arg\max_{x \in \mathcal{X}} \left[ C_t(S_t, x) + \sum_{s' \in \mathcal{S}} p_t(s'|S_t, x) V_{t+1}^*(s') \right].$$

This means that

$$C_t(S_t, x^*) + \sum_{s' \in \mathcal{S}} p_t(s'|S_t, x^*) V_{t+1}^*(s') \geq C_t(S_t, x) + \sum_{s' \in \mathcal{S}} p_t(s'|S_t, x) V_{t+1}^*(s') \tag{3.84}$$

for all $x \in \mathcal{X}$. Substituting (3.84) back into (3.83) gives us

$$\begin{aligned} V_t^*(S_t) &= \max_{\pi \in \Pi^{MR}} \sum_{x \in \mathcal{X}} \left\{ q_t^\pi(x|S_t) \left( C_t(S_t, x) + \sum_{s' \in \mathcal{S}} p_t(s'|S_t, x) V_{t+1}^*(s') \right) \right\} \\ &\leq \max_{\pi \in \Pi^{MR}} \sum_{x \in \mathcal{X}} \left\{ q_t^\pi(x|S_t) \left( C_t(S_t, x^*) + \sum_{s' \in \mathcal{S}} p_t(s'|S_t, x^*) V_{t+1}^*(s') \right) \right\} \\ &= C_t(S_t, x^*) + \sum_{s' \in \mathcal{S}} p_t(s'|S_t, x^*) V_{t+1}^*(s'). \end{aligned}$$

What this means is that if you have a choice between picking exactly the action you want versus picking a probability distribution over potentially optimal and nonoptimal actions, you would always prefer to pick exactly the best action. Clearly, this is not a surprising result.

The value of randomized policies arise primarily in two-person games, where one player tries to anticipate the actions of the other player. In such situations, part of the state variable is the estimate of what the other play will do when the game is in a particular state. By randomizing his behavior, a player reduces the ability of the other player to anticipate his moves.

### 3.9.6 Optimality of monotone policies

The foundational result that we use is the following technical lemma:

**Lemma 3.9.2** *If a function $g(s, x)$ is supermodular, then*

$$x^*(s) = \max \left\{ x' \in \arg\max_x g(s, x) \right\} \tag{3.85}$$

*is monotone and nondecreasing in $s$.*

If the function $g(s, x)$ has a unique, optimal $x^*(s)$ for each value of $s$, then we can replace (3.85) with

$$x^*(s) = \max_x g(s, x). \tag{3.86}$$

**Discussion:** The lemma is saying that if $g(s, x)$ is supermodular, then as $s$ grows larger, the optimal value of $x$ given $s$ will grow larger. When we use the version of supermodularity given in equation (3.41), we see that the condition implies that as the state becomes larger, the value of increasing the decision also grows. As a result, it is not surprising that the condition produces a decision rule that is monotone in the state vector.

**Proof of the lemma:** Assume that $s^+ \geq s^-$, and choose $x \leq x^*(s^-)$. Since $x^*(s)$ is, by definition, the best value of $x$ given $s$, we have

$$g(s^-, x^*(s^-)) - g(s^-, x) \geq 0. \tag{3.87}$$

The inequality arises because $x^*(s^-)$ is the best value of $x$ given $s^-$. Supermodularity requires that

$$g(s^-, x) + g(s^+, x^*(s^-)) \geq g(s^-, x^*(s^-)) + g(s^+, x) \tag{3.88}$$

Rearranging (3.88) gives us

$$g(s^+, x^*(s^-)) \geq \underbrace{\{g(s^-, x^*(s^-)) - g(s^-, x)\}}_{\geq 0} + g(s^+, x) \quad \forall x \leq x^*(s^-) \tag{3.89}$$

$$\geq \; g(s^+, x) \qquad \forall x \leq x^*(s^-) \tag{3.90}$$

We obtain equation (3.90) because the term in brackets in (3.89) is nonnegative (from (3.87)).

Clearly

$$g(s^+, x^*(s^+)) \;\; \geq \;\; g(s^+, x^*(s^-))$$

because $x^*(s^+)$ optimizes $g(s^+, x)$. This means that $x^*(s^+) \geq x^*(s^-)$ since otherwise, we would simply have chosen $x = x^*(s^-)$.

Just as the sum of concave functions is concave, we have the following:

**Proposition 3.9.3** *The sum of supermodular functions is supermodular.*

The proof follows immediately from the definition of supermodularity, so we leave it as one of those proverbial exercises for the reader.

The main theorem regarding monotonicity is relatively easy to state and prove, so we will do it right away. The conditions required are what make it a little more difficult.

**Theorem 3.9.5** *Assume that:*

*(a) $C_t(s, x)$ is supermodular on $\mathcal{S} \times \mathcal{X}$.*

*(b) $\sum_{s' \in \mathcal{S}} \mathbb{P}(s'|s, x) v_{t+1}(s')$ is supermodular on $\mathcal{S} \times \mathcal{X}$.*

*Then there exists a decision rule $x(s)$ that is nondecreasing on $\mathcal{S}$.*

**Proof:** Let

$$w(s, x) \;\; = \;\; C_t(s, x) + \sum_{s' \in \mathcal{S}} \mathbb{P}(s'|s, x) v_{t+1}(s') \tag{3.91}$$

The two terms on the right-hand side of (3.91) are assumed to be supermodular, and we know that the sum of two supermodular functions is supermodular, which tells us that $w(s, x)$ is supermodular. Let

$$x^*(s) \;\; = \;\; \arg\max_{x \in \mathcal{X}} w(s, x)$$

From Lemma 3.9.2, we obtain the result that the decision $x^*(s)$ increases monotonically over $\mathcal{S}$, which proves our result.

The proof that the one-period reward function $C_t(s, x)$ is supermodular must be based on the properties of the function for a specific problem. Of greater concern is establishing the conditions required to prove condition $(b)$ of the theorem because it involves the property of the value function, which is not part of the basic data of the problem.

In practice, it is sometimes possible to establish condition $(b)$ directly based on the nature of the problem. These conditions usually require conditions on the monotonicity of the reward function (and hence the value function) along with properties of the one-step transition matrix. For this reason, we will start by showing that if the one-period reward function is nondecreasing (or nonincreasing), then the value functions are nondecreasing (or nonincreasing). We will first need the following technical lemma:

**Lemma 3.9.3** *Let $p_j, p'_j, j \in \mathcal{J}$ be probability mass functions defined over $\mathcal{J}$ that satisfy*

$$\sum_{j=j'}^{\infty} p_j \geq \sum_{j=j'}^{\infty} p'_j \quad \forall j' \in \mathcal{J} \tag{3.92}$$

*and let $v_j, j \in \mathcal{J}$ be a nondecreasing sequence of numbers. Then*

$$\sum_{j=0}^{\infty} p_j v_j \geq \sum_{j=0}^{\infty} p'_j v_j \tag{3.93}$$

We would say that the distribution represented by $\{p_j\}_{j\in\mathcal{J}}$ *stochastically dominates* the distribution $\{p'_j\}_{j\in\mathcal{J}}$. If we think of $p_j$ as representing the probability a random variable $V = v_j$, then equation (3.93) is saying that $E^p V \geq E^{p'} V$. Although this is well known, a more algebraic proof is as follows:

**Proof:** Let $v_{-1} = 0$ and write

$$\begin{aligned}
\sum_{j=0}^{\infty} p_j v_j &= \sum_{j=0}^{\infty} p_j \sum_{i=0}^{j} (v_i - v_{i-1}) && (3.94)\\
&= \sum_{j=0}^{\infty} (v_j - v_{j-1}) \sum_{i=j}^{\infty} p_i && (3.95)\\
&= \sum_{j=1}^{\infty} (v_j - v_{j-1}) \sum_{j=i}^{\infty} p_i + v_0 \sum_{i=0}^{\infty} p_i && (3.96)\\
&\geq \sum_{j=1}^{\infty} (v_j - v_{j-1}) \sum_{i=j}^{\infty} p'_j + v_0 \sum_{i=0}^{\infty} p'_j && (3.97)\\
&= \sum_{j=0}^{\infty} p'_j v_j && (3.98)
\end{aligned}$$

In equation (3.94), we replace $v_j$ with an alternating sequence that sums to $v_j$. Equation (3.95) involves one of those painful change of variable tricks with summations. Equation (3.96) is simply getting rid of the term that involves $v_{-1}$. In equation (3.97), we replace the cumulative distributions for $p_j$ with the distributions for $p'_j$, which gives us the inequality. Finally, we simply reverse the logic to get back to the expectation in (3.98). □

We stated that lemma 3.9.3 is true when the sequences $\{p_j\}$ and $\{p'_j\}$ are probability mass functions because it provides an elegant interpretation as expectations. For example, we may use $v_j = j$, in which case equation (3.93) gives us the familiar result that when one probability distribution stochastically dominates another, it has a larger mean. If we use an increasing sequence $v_j$ instead of $j$, then this can be viewed as nothing more than the same result on a transformed axis.

In our presentation, however, we need a more general statement of the lemma, which follows:

**Lemma 3.9.4** *Lemma 3.9.3 holds for any real valued, nonnegative (bounded) sequences $\{p_j\}$ and $\{p'_j\}$.*

The proof involves little more than realizing that the proof of lemma 3.9.3 never required that the sequences $\{p_j\}$ and $\{p'_j\}$ be probability mass functions.

**Proposition 3.9.4** *Suppose that:*

*(a)* $C_t(s,x)$ *is nondecreasing (nonincreasing) in s for all* $x \in \mathcal{X}$ *and* $t \in \mathcal{T}$.

*(b)* $C_T(s)$ *is nondecreasing (nonincreasing) in s.*

*(c)* $q_t(\bar{s}|s,x) = \sum_{s' \geq \bar{s}} \mathbb{P}(s'|s,x)$, *the reverse cumulative distribution function for the transition matrix, is nondecreasing in s for all* $s \in \mathcal{S}, x \in \mathcal{X}$ *and* $t \in \mathcal{T}$.

*Then,* $v_t(s)$ *is nondecreasing (nonincreasing) in s for* $t \in \mathcal{T}$.

**Proof:** As always, we use a proof by induction. We will prove the result for the nondecreasing case. Since $v_T(s) = C_t(s)$, we obtain the result by assumption for $t = T$. Now, assume the result is true for $v_{t'}(s)$ for $t' = t+1, t+2, \ldots, T$. Let $x_t^*(s)$ be the decision that solves:

$$\begin{aligned} v_t(s) &= \max_{x \in \mathcal{X}} C_t(s,x) + \sum_{s' \in \mathcal{S}} \mathbb{P}(s'|s,x) v_{t+1}(s') \\ &= C_t(s, x_t^*(s)) + \sum_{s' \in \mathcal{S}} \mathbb{P}(s'|s, x_t^*(s)) v_{t+1}(s') \end{aligned} \tag{3.99}$$

Let $\hat{s} \geq s$. Condition $(c)$ of the proposition implies that:

$$\sum_{s' \geq s} \mathbb{P}(s'|s,x) \leq \sum_{s' \geq s} \mathbb{P}(s'|\hat{s},x) \tag{3.100}$$

Lemma 3.9.4 tells us that when (3.100) holds, and if $v_{t+1}(s')$ is nondecreasing (the induction hypothesis), then:

$$\sum_{s' \in \mathcal{S}} \mathbb{P}(s'|s,x) v_{t+1}(s') \leq \sum_{s' \in \mathcal{S}} \mathbb{P}(s'|\hat{s},x) v_{t+1}(s') \tag{3.101}$$

Combining equation (3.101) with condition (a) of proposition 3.9.4 into equation (3.99) gives us

$$\begin{aligned} v_t(s) &\leq C_t(\hat{s}, x^*(s)) + \sum_{s' \in \mathcal{S}} \mathbb{P}(s'|\hat{s}, x^*(s)) v_{t+1}(s') \\ &\leq \max_{x \in \mathcal{X}} C_t(\hat{s}, x) + \sum_{s' \in \mathcal{S}} \mathbb{P}(s'|\hat{s}, x) v_{t+1}(s') \\ &= v_t(\hat{s}), \end{aligned}$$

which proves the proposition. □

With this result, we can establish condition (b) of theorem 3.9.5:

**Proposition 3.9.5** *If*

*(a)* $q_t(\bar{s}|s,x) = \sum_{s' \geq \bar{s}} \mathbb{P}(s'|s,x)$ *is supermodular on* $\mathcal{S} \times \mathcal{X}$ *and*

*(b)* $v(s)$ *is nondecreasing in s,*

*then* $\sum_{s' \in \mathcal{S}} \mathbb{P}(s'|s,x)v(s')$ *is supermodular on* $\mathcal{S} \times \mathcal{X}$.

**Proof:** Supermodularity of the reverse cumulative distribution means:

$$\sum_{s' \geq \bar{s}} \mathbb{P}(s'|s^+, x^+) + \sum_{s' \geq \bar{s}} \mathbb{P}(s'|s^-, x^-) \geq \sum_{s' \geq \bar{s}} \mathbb{P}(s'|s^+, x^-) + \sum_{s' \geq \bar{s}} \mathbb{P}(s'|s^-, x^+)$$

We can apply Lemma 3.9.4 using $p_{\bar{s}} = \sum_{s' \geq \bar{s}} \mathbb{P}(s'|s^+, x^+) + \sum_{s' \geq \bar{s}} \mathbb{P}(s'|s^-, x^-)$ and $p'_{\bar{s}} = \sum_{s' \geq \bar{s}} \mathbb{P}(s'|s^+, x^-) + \sum_{s' \geq \bar{s}} \mathbb{P}(s'|s^-, x^+)$, which gives

$$\sum_{s' \in \mathcal{S}} \left(\mathbb{P}(s'|s^+, x^+) + \mathbb{P}(s'|s^-, x^-)\right) v(s') \geq \sum_{s' \in \mathcal{S}} \left(\mathbb{P}(s'|s^+, x^-) + \mathbb{P}(s'|s^-, x^+)\right) v(s')$$

which implies that $\sum_{s' \in \mathcal{S}} \mathbb{P}(s'|s,x)v(s')$ is supermodular. □

**Remark:** Supermodularity of the reverse cumulative distribution $\sum_{s' \in \mathcal{S}} \mathbb{P}(s'|s,x)$ may seem like a bizarre condition at first, but a little thought suggests that it is often satisfied in practice. As stated, the condition means that

$$\sum_{s' \in \mathcal{S}} \mathbb{P}(s'|s^+, x^+) - \sum_{s' \in \mathcal{S}} \mathbb{P}(s'|s^+, x^-) \geq \sum_{s' \in \mathcal{S}} \mathbb{P}(s'|s^-, x^+) - \sum_{s' \in \mathcal{S}} \mathbb{P}(s'|s^-, x^-)$$

Assume that the state $s$ is the water level in a dam, and the decision $x$ controls the release of water from the dam. Because of random rainfalls, the amount of water behind the dam in the next time period, given by $s'$, is random. The reverse cumulative distribution gives us the probability that the amount of water is greater than $s^+$ (or $s^-$). Our supermodularity condition can now be stated as: "If the amount of water behind the dam is higher one month ($s^+$), then the effect of the decision of how much water to release ($x$) has a greater impact than when the amount of water is initially at a lower level ($s^-$)." This condition is often satisfied because a control frequently has more of an impact when a state is at a higher level than a lower level.

For another example of supermodularity of the reverse cumulative distribution, assume that the state represents a person's total wealth, and the control is the level of taxation. The effect of higher or lower taxes is going to have a bigger impact on wealthier people than on those who are not as fortunate (but not always: think about other forms of taxation that affect less affluent people more than the wealthy, and use this example to create an instance of a problem where a monotone policy may not apply).

We now have the result that if the reward function $C_t(s,x)$ is nondecreasing in $s$ for all $x \in \mathcal{X}$ and the reverse cumulative distribution $\sum_{s' \in \mathcal{S}} \mathbb{P}(s'|s,x)$ is supermodular, then $\sum_{s' \in \mathcal{S}} \mathbb{P}(s'|s,x)v(s')$ is supermodular on $\mathcal{S} \times \mathcal{X}$. Combine this with the supermodularity of the one-period reward function, and we obtain the optimality of a nondecreasing decision function.

## 3.10 BIBLIOGRAPHIC NOTES

This chapter presents the classic view of Markov decision processes, for which the literature is extensive. Beginning with the seminal text of Bellman (Bellman (1957)), there has been numerous, significant textbooks on the subject, including Howard (1971), Nemhauser (1966), White (1969), Derman (1970), Bellman (1971), Dreyfus & Law (1977), Dynkin (1979), Denardo (1982), Ross (1983) and Heyman & Sobel (1984). As of this writing, the current high-water mark for textbooks in this area is the landmark volume by Puterman (1994). Most of this chapter is based on Puterman (1994), modified to our notational style.

Section 3.7 - The linear programming method was first proposed in Manne (1960) (see subsequent discussions in Derman (1962) and Puterman (1994)). The so-called "linear programming method" was ignored for many years because of the large size of the linear programs that were produced, but the method has seen a resurgence of interest using approximation techniques. Recent research into algorithms for solving problems using this method are discussed in section 9.2.5.

Section 3.9.6 - In addition to Puterman (1994), see also Topkins (1978).

## PROBLEMS

**3.1** A classical inventory problem works as follows: Assume that our state variable $R_t$ is the amount of product on hand at the end of time period $t$ and that $D_t$ is a random variable giving the demand during time interval $(t-1, t)$ with distribution $p_d = \mathbb{P}(D_t = d)$. The demand in time interval $t$ must be satisfied with the product on hand at the beginning of the period. We can then order a quantity $x_t$ at the end of period $t$ that can be used to replenish the inventory in period $t+1$.

(a) Give the transition function that relates $R_{t+1}$ to $R_t$ if the order quantity is $x_t$ (where $x_t$ is fixed for all $R_t$).

(b) Give an algebraic version of the one-step transition matrix $P^\pi = \{p^\pi_{ij}\}$ where $p^\pi_{ij} = \mathbb{P}(R_{t+1} = j | R_t = i, X^\pi = x_t)$.

**3.2** Repeat the previous exercise, but now assume that we have adopted a policy $\pi$ that says we should order a quantity $x_t = 0$ if $R_t \geq s$ and $x_t = Q - R_t$ if $R_t < q$ (we assume that $R_t \leq Q$). Your expression for the transition matrix will now depend on our policy $\pi$ (which describes both the structure of the policy and the control parameter $s$).

**3.3** We are going to use a very simple Markov decision process to illustrate how the initial estimate of the value function can affect convergence behavior. In fact, we are going to use a Markov reward process to illustrate the behavior because our process does not have any decisions. Assume we have a two-stage Markov chain with one-step transition matrix

$$P = \begin{bmatrix} 0.7 & 0.3 \\ 0.05 & 0.95 \end{bmatrix}.$$

The contribution from each transition from state $i \in \{1, 2\}$ to state $j \in \{1, 2\}$ is given by the matrix

$$\begin{bmatrix} 10 & 30 \\ 30 & 5 \end{bmatrix}.$$

That is, a transition from state 1 to state 2 returns a contribution of 30. Apply the value iteration algorithm for an infinite horizon problem (note that you are not choosing a decision so there is no maximization step). The calculation of the value of being in each state will depend on your previous estimate of the value of being in each state. The calculations can be easily implemented in a spreadsheet. Assume that your discount factor is .8.

(a) Plot the value of being in state 1 as a function of the number of iterations if your initial estimate of the value of being in each state is 0. Show the graph for 50 iterations of the algorithm.

(b) Repeat this calculation using initial estimates of 100.

(c) Repeat the calculation using an initial estimate of the value of being in state 1 of 100, and use 0 for the value of being in state 2. Contrast the behavior with the first two starting points.

**3.4** Show that $\mathbb{P}(S_{t+\tau}|S_t)$, given that we are following a policy $\pi$ (for stationary problems), is given by (3.14). [Hint: first show it for $\tau = 1, 2$ and then use inductive reasoning to show that it is true for general $\tau$.]

**3.5** Apply policy iteration to the problem given in exercise 3.3. Plot the average value function (that is, average the value of being in each state) after each iteration alongside the average value function found using value iteration after each iteration (for value iteration, initialize the value function to zero). Compare the computation time for one iteration of value iteration and one iteration of policy iteration.

**3.6** Now apply the hybrid value-policy iteration algorithm to the problem given in exercise 3.3. Show the average value function after each major iteration (update of $n$) with $M = 1, 2, 3, 5, 10$. Compare the convergence rate to policy iteration and value iteration.

**3.7** An oil company will order tankers to fill a group of large storage tanks. One full tanker is required to fill an entire storage tank. Orders are placed at the beginning of each four week accounting period but do not arrive until the end of the accounting period. During this period, the company may be able to sell 0, 1 or 2 tanks of oil to one of the regional chemical companies (orders are conveniently made in units of storage tanks). The probability of a demand of 0, 1 or 2 is 0.40, 0.40 and 0.20, respectively.

A tank of oil costs \$1.6 million (M) to purchase and sells for \$2M. It costs \$0.020M to store a tank of oil during each period (oil ordered in period $t$, which cannot be sold until period $t + 1$, is not charged any holding cost in period $t$). Storage is only charged on oil that is in the tank at the beginning of the period and remains unsold during the period. It is possible to order more oil than can be stored. For example, the company may have two full storage tanks, order three more, and then only sell one. This means that at the end of the period, they will have four tanks of oil. Whenever they have more than two tanks of oil, the company must sell the oil directly from the ship for a price of \$0.70M. There is no penalty for unsatisfied demand.

An order placed in time period $t$ must be paid for in time period $t$ even though the order does not arrive until $t + 1$. The company uses an interest rate of 20 percent per accounting period (that is, a discount factor of 0.80).

(a) Give an expression for the one-period reward function $r(s, d)$ for being in state $s$ and making decision $d$. Compute the reward function for all possible states (0, 1, 2) and all possible decisions (0, 1, 2).

(b) Find the one-step probability transition matrix when your action is to order one or two tanks of oil. The transition matrix when you order zero is given by

| From-To | 0 | 1 | 2 |
|---|---|---|---|
| 0 | 1 | 0 | 0 |
| 1 | 0.6 | 0.4 | 0 |
| 2 | 0.2 | 0.4 | 0.4 |

(c) Write out the general form of the optimality equations and solve this problem in steady state.

(d) Solve the optimality equations using the value iteration algorithm, starting with $V(s) = 0$ for $s = 0, 1$ and 2. You may use a programming environment, but the problem can be solved in a spreadsheet. Run the algorithm for 20 iterations. Plot $V^n(s)$ for $s = 0, 1, 2$, and give the optimal action for each state at each iteration.

(e) Give a bound on the value function after each iteration.

**3.8** Every day, a salesman visits $N$ customers in order to sell the $R$ identical items he has in his van. Each customer is visited exactly once and each customer buys zero or one item. Upon arrival at a customer location, the salesman quotes one of the prices $0 < p_1 \leq p_2 \leq \ldots \leq p_m$. Given that the quoted price is $p_i$, a customer buys an item with probability $r_i$. Naturally, $r_i$ is decreasing in $i$. The salesman is interested in maximizing the total expected revenue for the day. Show that if $r_i p_i$ is increasing in $i$, then it is always optimal to quote the highest price $p_m$.

**3.9** You need to decide when to replace your car. If you own a car of age $y$ years, then the cost of maintaining the car that year will be $c(y)$. Purchasing a new car (in constant dollars) costs $P$ dollars. If the car breaks down, which it will do with probability $b(y)$ (the breakdown probability), it will cost you an additional $K$ dollars to repair it, after which you immediately sell the car and purchase a new one. At the same time, you express your enjoyment with owning a new car as a negative cost $-r(y)$ where $r(y)$ is a declining function with age. At the beginning of each year, you may choose to purchase a new car ($z = 1$) or to hold onto your old one ($z = 0$). You anticipate that you will actively drive a car for another $T$ years.

(a) Identify all the elements of a Markov decision process for this problem.

(b) Write out the objective function which will allow you to find an optimal decision rule.

(c) Write out the one-step transition matrix.

(d) Write out the optimality equations that will allow you to solve the problem.

**3.10** You are trying to find the best parking space to use that minimizes the time needed to get to your restaurant. There are 50 parking spaces, and you see spaces $1, 2, \ldots, 50$ in order. As you approach each parking space, you see whether it is full or empty. We assume, somewhat heroically, that the probability that each space is occupied follows an independent Bernoulli process, which is to say that each space will be occupied with probability $p$, but will be free with probability $1 - p$, and that each outcome is independent of the other.

It takes 2 seconds to drive past each parking space and it takes 8 seconds to walk past. That is, if we park in space n, it will require $8(50 - n)$ seconds to walk to the restaurant. Furthermore, it would have taken you $2n$ seconds to get to this space. If you get to the last space without finding an opening, then you will have to drive into a special lot down the block, adding 30 seconds to your trip.

We want to find an optimal strategy for accepting or rejecting a parking space.

(a) Give the sets of state and action spaces and the set of decision epochs.

(b) Give the expected reward function for each time period and the expected terminal reward function.

(c) Give a formal statement of the objective function.

(d) Give the optimality equations for solving this problem.

(e) You have just looked at space 45, which was empty. There are five more spaces remaining (46 through 50). What should you do? Using $p = 0.6$, find the optimal policy by solving your optimality equations for parking spaces 46 through 50.

f) Give the optimal value of the objective function in part (e) corresponding to your optimal solution.

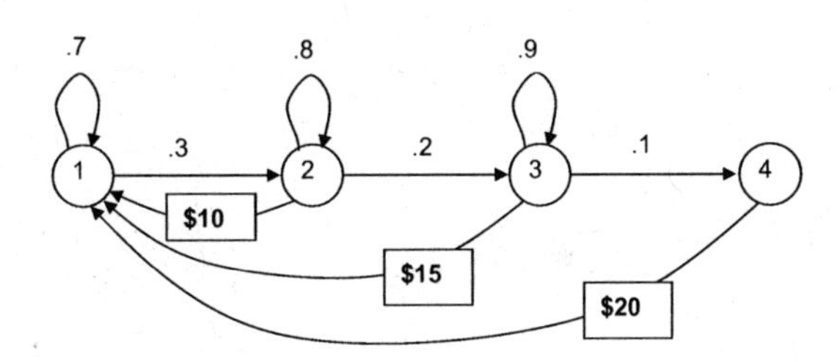


**3.11** We have a four-state process (shown in the figure). In state 1, we will remain in the state with probability 0.7 and will make a transition to state 2 with probability 0.3. In states 2 and 3, we may choose between two policies: Remain in the state waiting for an upward transition or make the decision to return to state 1 and receive the indicated reward. In state 4, we return to state 1 immediately and receive $20. We wish to find an optimal long run policy using a discount factor $\gamma = .8$. Set up and solve the optimality equations for this problem.

**3.12** Assume that you have been applying value iteration to a four-state Markov decision process, and that you have obtained the values over iterations 8 through 12 shown in the table below (assume a discount factor of 0.90). Assume you stop after iteration 12. Give the tightest possible (valid) bounds on the optimal value of being in each state.

**3.13** In the proof of theorem 3.9.3 we showed that if $v \geq \mathcal{M}v$, then $v \geq v^*$. Go through the steps of proving the converse, that if $v \leq \mathcal{M}v$, then $v \leq v^*$.

**3.14** Theorem 3.9.3 states that if $v \leq \mathcal{M}v$, then $v \leq v^*$. Show that if $v^n \leq v^{n+1} = \mathcal{M}v^n$, then $v^{m+1} \geq v^m$ for all $m \geq n$.

**3.15** Consider a finite-horizon MDP with the following properties:

| State | Iteration | | | | |
| --- | --- | --- | --- | --- | --- |
| | 8 | 9 | 10 | 11 | 12 |
| 1 | 7.42 | 8.85 | 9.84 | 10.54 | 11.03 |
| 2 | 4.56 | 6.32 | 7.55 | 8.41 | 9.01 |
| 3 | 11.83 | 13.46 | 14.59 | 15.39 | 15.95 |
| 4 | 8.13 | 9.73 | 10.85 | 11.63 | 12.18 |

- $\mathcal{S} \in \Re^n$, the action space $\mathcal{X}$ is a compact subset of $\Re^n$, $\mathcal{X}(s) = \mathcal{X}$ for all $s \in \mathcal{S}$.
- $C_t(S_t, x_t) = c_t S_t + g_t(x_t)$, where $g_t(\cdot)$ is a known scalar function, and $C_T(S_T) = c_T S_T$.
- If action $x_t$ is chosen when the state is $S_t$ at time $t$, the next state is

$$S_{t+1} = A_t S_t + f_t(x_t) + \omega_{t+1},$$

where $f_t(\cdot)$ is scalar function, and $A_t$ and $\omega_t$ are respectively $n \times n$ and $n \times 1$-dimensional random variables whose distributions are independent of the history of the process prior to $t$.

(a) Show that the optimal value function is linear in the state variable.

(b) Show that there exists an optimal policy $\pi^* = (x_1^*, \ldots, x_{T-1}^*)$ composed of constant decision functions. That is, $X_t^{\pi^*}(s) = x_t^*$ for all $s \in \mathcal{S}$ for some constant $x_t^*$.

**3.16** Assume that you have invested $R_0$ dollars in the stock market which evolves according to the equation

$$R_t = \gamma R_{t-1} + \varepsilon_t$$

where $\varepsilon_t$ is a discrete, positive random variable that is independent and identically distributed and where $0 < \gamma < 1$. If you sell the stock at the end of period $t$, it will earn a riskless return $r$ until time $T$, which means it will evolve according to

$$R_t = (1 + r)R_{t-1}.$$

You have to sell the stock, all on the same day, some time before $T$.

(a) Write a dynamic programming recursion to solve the problem.

(b) Show that there exists a point in time $\tau$ such that it is optimal to sell for $t \geq \tau$, and optimal to hold for $t < \tau$.

(c) How does your answer to (b) change if you are allowed to sell only a portion of the assets in a given period? That is, if you have $R_t$ dollars in your account, you are allowed to sell $x_t \leq R_t$ at time $t$.

# CHAPTER 4

# INTRODUCTION TO APPROXIMATE DYNAMIC PROGRAMMING

In chapter 3, we saw that we could solve intractably complex optimization problems of the form

$$\max_{\pi} \mathbb{E} \left\{ \sum_{t=0}^{T} \gamma^t C_t^{\pi}(S_t, X_t^{\pi}(S_t)) \right\} \tag{4.1}$$

by recursively computing the optimality equations

$$V_t(S_t) = \max_{x_t} \left( C_t(S_t, x_t) + \gamma \mathbb{E}\{V_{t+1}(S_{t+1})|S_t\} \right) \tag{4.2}$$

backward through time. As before, we use the shorthand notation $S_{t+1} = S^M(S_t, x_t, W_{t+1})$ to express the dependency of state $S_{t+1}$ on the previous state $S_t$, the action $x_t$ and the new information $W_{t+1}$. Equation (4.1) can be computationally intractable even for very small problems. The optimality equations (4.2) give us a mechanism for solving these stochastic optimization problems in a simple and elegant way. Unfortunately, in a vast array of applications, the optimality equations are themselves computationally intractable.

Approximate dynamic programming offers a powerful set of strategies for problems that are hard because they are large, but this is not the only application. Even small problems may be hard because we lack a formal model of the information process, or we may not know the transition function. For example, we may have observations of changes in prices in an asset, but we do not have a mathematical model that describes these changes. In this case, we are not able to compute the expectation. Alternatively, consider the problem of modeling how the economy of a small country responds to loans from the International

*Approximate Dynamic Programming*. By Warren B. Powell

Monetary Fund. If we do not now how the economy responds to the size of a loan, then this means that we do not know the transition function.

There is also a vast range of problems which are intractable simply because of size. Imagine a multi-skill call center which handles questions about personal computers. People who call in identify the nature of their question through the options they select in the phone menu. Calls are then routed to people with expertise in those areas. Assigning phone calls to technicians is a dynamic assignment problem involving multiattribute resources, creating a dynamic program with a state space that is effectively infinite.

We begin our presentation by revisiting the "curses of dimensionality." The remainder of the chapter provides an overview of the basic principles and vocabulary of approximate dynamic programming. As with the previous chapters, we retain the basic notation that we have a system in a state $S_t \in \mathcal{S} = \{1, 2, \ldots, s, \ldots\}$. In chapter 5, we introduce a much richer notational framework that makes it easier to develop approximation strategies.

## 4.1 THE THREE CURSES OF DIMENSIONALITY (REVISITED)

The concept of the backward recursion of dynamic programming is so powerful that we have to remind ourselves again why its usefulness can be so limited for many problems. Consider again our simple asset acquisition problem, but this time assume that we have multiple asset classes. For example, we may wish to purchase stocks each month for our retirement account using the money that is invested in it each month. Our on-line brokerage charges \$50 for each purchase order we place, so we have an incentive to purchase stocks in reasonably sized lots. Let $p_{tk}$ be the purchase price of asset type $k \in \mathcal{K}$ in period $t$, and let $x_{tk}$ be the number of shares we purchase in period $t$. The total purchase cost of our order is

$$C^p(x_t) = \sum_{k \in \mathcal{K}} -\left(50 I_{\{x_{tk}>0\}} + p_{tk} x_{tk}\right).$$

where $I_{\{x_{tk}>0\}} = 1$ if $x_{tk} > 0$ is true, and 0 otherwise. Each month, we have \$2000 to invest in our retirement plan. We have to make sure that $\sum_{k \in \mathcal{K}} p_{tk} x_{tk} \leq \$2000$ and, of course, $x_{tk} \geq 0$. While it is important to diversify the portfolio, transaction costs rise as we purchase a wider range of stocks, which reduces how much we can buy.

Let $R_{tk}$ be the number of shares of stock $k$ that we have on hand at the end of period $t$. The value of the portfolio reflects the prices $p_{tk}$ which fluctuate over time according to

$$p_{t+1,k} = p_{t,k} + \hat{p}_{t+1,k},$$

where $\hat{p}_{t,k}$ is the exogenous change in the price of stock $k$ between $t-1$ and $t$. In addition, each stock returns a random dividend $\hat{d}_{tk}$ (between $t-1$ and $t$) given in dollars per share which are reinvested. Thus, the information we receive in each time period is

$$W_t = (\hat{p}_t, \hat{d}_t).$$

If we were to reinvest our dividends, then our transition function would be

$$R_{t+1,k} = R_{tk} + \frac{\hat{d}_{t+1,k}}{p_{t+1,k}} R_{tk} + x_{tk}.$$

The state of our system is given by

$$S_t = (R_t, p_t).$$

Let the total value of the portfolio be

$$Y_t = \sum_{k \in \mathcal{K}} p_{tk} R_{tk},$$

which we evaluate using a concave utility function $U(Y_t)$. Our goal is to maximize total expected discounted utility over a specified horizon.

We now have a problem where our state $S_t$ variable has $2|\mathcal{K}|$ (the resource allocation $R_t$ and the prices $p_t$), the decision variable $x_t$ has $|\mathcal{K}|$ dimensions, and our information process $W_t$ has $2|\mathcal{K}|$ dimensions (the changes in prices and the dividends). To illustrate how bad this gets, assume that we are restricting our attention to 50 different stocks and that we always purchase shares in blocks of 100, but that we can own as many as 20,000 shares (or 200 blocks) of any one stock at a given time. This means that the number of possible portfolios $R_t$ is $200^{50}$. Now assume that we can discretize each price into 100 different prices. This means that we have up to $100^{50}$ different price vectors $p_t$. Thus, we would have to loop over equation (4.2) $200^{50} \times 100^{50}$ times to compute $V_t(S_t)$ for all possible values of $S_t$. This is the classical "curse of dimensionality" that is widely cited as the Achilles heel of dynamic programming.

As we look at our problem, we see the situation is much worse. For each of the $200^{50} \times 100^{50}$ states, we have to compute the expectation in (4.2). Since our random variables might not be independent (the prices generally will not be) we could conceive of finding a joint probability distribution and performing $2 \times 50$ nested summations to complete the expectation (we have 50 prices and 50 dividends). It seems that this can be avoided if we work with the one-step transition matrix and use the form of the optimality equations expressed in (3.3). This form of the recursion, however, only hides the expectation (the one-step transition matrix is, itself, an expectation).

We are not finished. This expectation has to be computed for each action $x$. Assume we are willing to purchase up to 10 blocks of 100 shares of any one of the 50 stocks. Since we can also purchase 0 shares of a stock, we have upwards of $11^{50}$ different combinations of $x_t$ that we might have to consider (the actual number is smaller because we have a budget constraint of \$2000 to spend).

So, we see that we have three curses of dimensionality if we wish to work in discrete quantities. In other words, vectors really cause problems. In this chapter, we are going to provide a brief introduction to approximate dynamic programming, although the techniques in this chapter will not, by themselves, solve this particular problem.

## 4.2 THE BASIC IDEA

The foundation of approximate dynamic programming is based on an algorithmic strategy that steps *forward* through time (earning it the name "forward dynamic programming" in some communities). If we wanted to solve this problem using classical dynamic programming, we would have to find the value function $V_t(S_t)$ using

$$\begin{aligned} V_t(S_t) &= \max_{x_t \in \mathcal{X}_t} \left( C(S_t, x_t) + \gamma \mathbb{E}\{V_{t+1}(S_{t+1})|S_t\} \right) \\ &= \max_{x_t \in \mathcal{X}_t} \left( C(S_t, x_t) + \gamma \sum_{s' \in \mathcal{S}} \mathbb{P}(s'|S_t, x_t) V_{t+1}(s') \right) \end{aligned}$$

for each value of $S_t$. We have written our maximization problem as one of choosing the best $x_t \in \mathcal{X}_t$. $\mathcal{X}_t$ is known as the *feasible region* and captures the constraints on our decisions.

In our portfolio example above, the feasible region would be $\mathcal{X}_t = \{x_t | \sum_{k\in\mathcal{K}} p_{tk}x_{tk} \leq 2000, x_{tk} \geq 0\}$. Note that $\mathcal{X}_t$ depends on our state variable. For notational simplicity, we indicate the dependence on $S_t$ through the subscript $t$.

This problem cannot be solved using the techniques we presented in chapter 3 which require that we loop over all possible states (in fact, we usually have three nested loops, two of which require that we enumerate all states while the third enumerates all actions). With approximate dynamic programming, we step forward in time. In order to simulate the process forward in time, we need to solve two problems. The first is that we need a way to randomly generate a sample of what *might* happen (in terms of the various sources of random information). The second is that we need a way to make decisions. We start with the problem of making decisions first, and then turn to the problem of simulating random information.

### 4.2.1 Making decisions (approximately)

When we used exact dynamic programming, we stepped backward in time, exactly computing the value function which we then used to produce optimal decisions. When we step forward in time, we have not computed the value function, so we have to turn to an approximation in order to make decisions.

Let $\bar{V}_t(S_t)$ be an approximation of the value function. This is easiest to understand if we assume that we have an estimate $\bar{V}_t(S_t)$ for each state $S_t$, but since it is an approximation, we may use any functional form we wish. For our portfolio problem above, we might create an approximation $\bar{V}_t(R_t)$ that depends only on the number of shares we own, rather than the prices. In fact, we might even use a separable approximation of the form

$$V_t(S_t) \approx \sum_{k\in\mathcal{K}} \bar{V}_{tk}(R_{tk})$$

where $\bar{V}_{tk}(R_{tk})$ is a nonlinear function giving the value of holding $R_{tk}$ shares of stock $k$. Obviously assuming such a structure (where we are both ignoring the price vector $p_t$ as well as assuming separability) will introduce errors. Welcome to the field of approximate dynamic programming. The challenge, of course, is finding approximations that are good enough for the purpose at hand.

Approximate dynamic programming proceeds by estimating the approximation $\bar{V}_t(S_t)$ iteratively. We assume we are given an initial approximation $\bar{V}_t^0$ for all $t$ (we often use $\bar{V}_t^0 = 0$ when we have no other choice). Let $\bar{V}_t^{n-1}$ be the value function approximation after $n-1$ iterations, and consider what happens during the $n^{th}$ iteration. We are going to start at time $t = 0$ in state $S_0$. We determine $x_0$ by solving

$$\begin{aligned} x_0 &= \arg\max_{x\in\mathcal{X}_0} \left(C(S_0, x) + \gamma\mathbb{E}\{\bar{V}_1(S_1)|S_0\}\right) \\ &= \arg\max_{x\in\mathcal{X}_0} \left(C(S_0, x) + \gamma \sum_{s'\in\mathcal{S}} \mathbb{P}_0(s'|S_0, x)\bar{V}_1(s')\right). \end{aligned} \tag{4.3}$$

where $x_0$ is the value of $x$ that maximizes the right-hand side of (4.3), and $\mathbb{P}(s'|S_0, x)$ is the one-step transition matrix (which we temporarily assume is given). Assuming this can be accomplished (or at least approximated) we can use (4.3) to determine $x_0$.

For the moment, we are going to assume that we now have a way of making decisions.

### 4.2.2 Stepping forward through time

We are now going to step forward in time from $S_0$ to $S_1$. This starts by making the decision $x_0$, which we are going to do using our approximate value function. Given our decision, we next need to know the information that arrived between $t = 0$ and $t = 1$. At $t = 0$, this information is unknown, and therefore random. Our strategy will be to simply pick a sample realization of the information (for our example, this would be $\hat{d}_1$ and $\hat{p}_1$) at random, a process that is often referred to as *Monte Carlo simulation.* Monte Carlo simulation refers to the popular practice of generating random information, using some sort of artificial process to pick a random observation from a population. For example, imagine that we think a price will be uniformly distributed between 60 and 70. Most computer languages have a random number generator that produces a number that is uniformly distributed between 0 and 1. If we let rand() denote this function, the statement

$$U = \text{rand()}$$

produces a value for $U$ that is some number between 0 and 1 (say, 0.804663917). We can then create our random price using the equation

$$\begin{aligned} \hat{p}_t &= 60 + 10 * U \\ &= 68.04663917. \end{aligned}$$

While this is not the only way to obtain random samples of information, it is one of the most popular. We revisit this technique in chapter 6.

Using our random sample of new information, we can now compute the next state we visit, given by $S_1$. We do this by assuming that we have been given a *transition function* which we represent using

$$S_{t+1} = S^M(S_t, x_t, W_{t+1}).$$

where $W_{t+1}$ is a set of random variables representing the information that arrived between $t$ and $t+1$. We first introduced this notation in chapter 2, which provides a number of examples.

Keeping in mind that we assume that we have been given $\bar{V}_t$ for all $t$, we simply repeat the process of making a decision by solving

$$x_1 = \arg\max_{x \in \mathcal{X}_1} \left( C(S_1, x) + \gamma \sum_{s' \in \mathcal{S}} \mathbb{P}_1(s'|S_1, x) \bar{V}_2(s') \right). \tag{4.4}$$

Once we determine $x_1$, we again sample the new information $(\hat{p}_2, \hat{d}_2)$, compute $S_2$ and repeat the process. Whenever we make a decision based on a value function approximation (as we do in equation (4.4)), we refer to this as a *greedy strategy.* This term, which is widely used in dynamic programming, can be somewhat misleading since the value function approximation is trying to produce decisions that balance rewards now with rewards in the future.

Fundamental to approximate dynamic programming is the idea of following a *sample path.* A sample path refers to a particular sequence of exogenous information. To illustrate, assume we have a problem with two assets, where the price of each asset at time $t$ is given by $(p_{t1}, p_{t2})$. Assume that these changes occur randomly over time according to

$$p_{ti} = p_{t-1,i} + \hat{p}_{ti}, \quad i = 1, 2.$$

| | $t=1$ | $t=2$ | $t=3$ | $t=4$ | $t=5$ | $t=6$ | $t=7$ | $t=8$ |
|---|---|---|---|---|---|---|---|---|
| $\omega$ | $\hat{p}_1$ | $\hat{p}_2$ | $\hat{p}_3$ | $\hat{p}_4$ | $\hat{p}_5$ | $\hat{p}_6$ | $\hat{p}_7$ | $\hat{p}_8$ |
| 1 | (-14,-26) | (7,16) | (-10,20) | (41,7) | (15,-11) | (8,-29) | (-6,-50) | (13,-36) |
| 2 | (11,-33) | (-36,-50) | (-23,0) | (3,-50) | (7,-23) | (12,-10) | (35,46) | (-18,-10) |
| 3 | (-50,-12) | (-2,-18) | (-15,12) | (-31,3) | (5,-2) | (-24,33) | (10,2) | (38,-19) |
| 4 | (1,-13) | (41,20) | (32,-2) | (3,-4) | (-46,34) | (-15,14) | (38,16) | (48,-49) |
| 5 | (2,-34) | (-24,34) | (-9,25) | (-19,-40) | (1,-28) | (34,-7) | (36,-3) | (-46,-35) |
| 6 | (41,-49) | (-24,-33) | (-5,-25) | (16,36) | (8,47) | (-17,-4) | (-29,45) | (-48,-30) |
| 7 | (44,37) | (7,-19) | (49,-40) | (-13,5) | (38,37) | (-30,45) | (-48,-47) | (19,41) |
| 8 | (-19,37) | (-50,-35) | (-28,32) | (-13,-17) | (2,-2) | (-10,-22) | (-2,47) | (2,-24) |
| 9 | (-13,48) | (-48,25) | (-37,39) | (-2,30) | (-28,33) | (-35,-49) | (-44,-13) | (-6,-18) |
| 10 | (48,5) | (37,-39) | (43,34) | (-13,-6) | (28,-37) | (-47,-12) | (13,28) | (26,-35) |

**Table 4.1** Illustration of a set of sample paths.

Further assume that from history, we have constructed 10 potential realizations of the price changes $\hat{p}_t$, $t = 1, 2, \ldots, 8$, which we have shown in table 4.1. Each sample path is a particular set of outcomes of the vector $\hat{p}_t$ for all time periods. For historical reasons, we index each potential set of outcomes by $\omega$, and let $\Omega$ be the set of all sample paths where, for our example, $\Omega = \{1, 2, \ldots, 10\}$. Let $W_t$ be a generic random variable representing the information arriving at time $t$, where for this example we would have $W_t = (\hat{p}_{t1}, \hat{p}_{t2})$. Then we would let $W_t(\omega)$ be a particular sample realization of the prices in time period $t$. For example, using the data in table 4.1, we see that $W_3(5) = (-9, 25)$. In an iterative algorithm, we might choose $\omega^n$ in iteration $n$, and then follow the sample path $W_1(\omega^n), W_2(\omega^n), \ldots, W_t(\omega^n)$,. In each iteration, we have to choose a new outcome $\omega^n$.

The notation $W_t(\omega)$ matches how we have laid out the data in table 4.1. We go to row $\omega$, column $t$ to get the information we need. This layout seems to imply that we have all the random data generated in advance, and we just pick out what we want. In practice, this may not be how the code is actually written. It is often the case that we are randomly generating information as the algorithm progresses. Readers need to realize that when we write $W_t(\omega)$, we mean a random observation of the information that arrived at time $t$, regardless of whether we just generated the data, looked it up in a table, or were given the data from an exogenous process (perhaps an actual price realization from the internet).

Our sample path notation is useful because it is quite general. For example, it does not require that the random observations be independent across time (by contrast, Bellman's equation explicitly assumes that the distribution of $W_{t+1}$ depends only on $S_t$). In approximate dynamic programming, we do not impose any restrictions of this sort. ADP allows us to use a state variable that does not include all the relevant history (but remember - this is *approximate*!).

### 4.2.3 An ADP algorithm

Of course, stepping forward through time following a single set of sample realizations would not produce anything of value. The price of using approximate value functions is that we have to do this over and over, using a fresh set of sample realizations each time.

---

**Step 0.** Initialization:

Step 0a. Initialize $\bar{V}_t^0(S_t)$ for all states $S_t$.

Step 0b. Choose an initial state $S_0^1$.

Step 0c. Set $n = 1$.

**Step 1.** Choose a sample path $\omega^n$.

**Step 2.** For $t = 0, 1, 2, \ldots, T$ do:

**Step 2a.** Solve

$$\hat{v}_t^n = \max_{x_t \in \mathcal{X}_t^n} \left( C_t(S_t^n, x_t) + \gamma \sum_{s' \in \mathcal{S}} \mathbb{P}(s'|S_t^n, x_t)\bar{V}_{t+1}^{n-1}(s') \right)$$

and let $x_t^n$ be the value of $x_t$ that solves the maximization problem.

**Step 2b.** Update $\bar{V}_t^{n-1}(S_t)$ using

$$\bar{V}_t^n(S_t) = \begin{cases} \hat{v}_t^n & S_t = S_t^n \\ \bar{V}_t^{n-1}(S_t) & \text{otherwise.} \end{cases}$$

**Step 2c.** Compute $S_{t+1}^n = S^M(S_t^n, x_t^n, W_{t+1}(\omega^n))$.

**Step 3.** Let $n = n + 1$. If $n < N$, go to step 1.

---

**Figure 4.1** An approximate dynamic programming algorithm using the one-step transition matrix.

Using our vocabulary of the previous section, we would say that we follow a new sample path each time.

When we run our algorithm iteratively, we index everything by the iteration counter $n$. We would let $\omega^n$ represent the specific value of $\omega$ that we sampled for iteration $n$. At time $t$, we would be in state $S_t^n$, and make decision $x_t^n$ using the value function approximation $\bar{V}^{n-1}$. The value function is indexed $n - 1$ because it was computed using information from iteration $n - 1$. After finding $x_t^n$, we would observe the information $W_{t+1}(\omega^n)$ to obtain $S_{t+1}^n$. After reaching the end of our horizon, we would increment $n$ and start over again (we defer until chapter 9 dealing with infinite horizon problems).

A basic approximate dynamic programming algorithm is summarized in figure 4.1. This is very similar to backward dynamic programming (see figure 3.1), except that we are stepping forward in time. We no longer have the dangerous "loop over all states" requirement, but we have introduced a fresh set of challenges. As this book should make clear, this basic strategic is exceptionally powerful but introduces a number of technical challenges that have to be overcome (indeed, this is what fills up the remainder of this book).

This algorithmic strategy introduces several problems we have to solve.

- Forward dynamic programming avoids the problem of looping over all possible states, but it still requires the use of a one-step transition matrix, with the equally dangerous $\sum_{s' \in \mathcal{S}} \mathbb{P}(s'|S_t^n, x_t)[\ldots]$.

- We only update the value of states we visit, but we need the value of states that we *might* visit. We need a way to estimate the value of being in states that we have not visited.

- We might get caught in a circle where states we have never visited look bad (relative to states we have visited), so we never visit them. We may never discover that a state that we have never visited is actually quite good!

### 4.2.4 Approximating the expectation

There are a number of problems where computing the expectation is easy, but these are special cases (but we have to be aware that these special cases do exist). For a vast range of applications, computing the expectation is often computationally intractable. One way to get around this is to randomly generate a sample of outcomes $\hat{\Omega}_{t+1}$ of what $W_{t+1}$ might be. Let $p_{t+1}(\omega_{t+1})$ be the probability of outcome $\hat{\omega}_{t+1} \in \hat{\Omega}_{t+1}$, where if we choose $N$ observations in the set $\hat{\Omega}_{t+1}$, it might be the case that $p_{t+1}(\omega_{t+1}) = 1/N$. We could then approximate the expectation using

$$\sum_{\omega_{t+1} \in \hat{\Omega}_{t+1}} p_{t+1}(\omega_{t+1}) \bar{V}_{t+1}^{n-1}(S^M(S_t, x_t, \omega_{t+1})).$$

In section 4.4, we show how we can avoid this clumsy operation.

An outline of an approximate dynamic programming algorithm is provided in figure 4.2. In contrast with the algorithm in figure 4.1, we now have completely eliminated any step which requires looping over all the states in the state space. In theory at least, we can now tackle problems with state spaces that are infinitely large. We face only the practical problem of filling in the details so that the algorithm actually works on a particular application.

### 4.2.5 Updating the value function approximation

One area that will receive considerable attention in this volume is the challenge of approximating the value function. There is a wide range of methods for doing this which depend largely on the specific form of the value function approximation. In the simplest case, we update the value of being in a particular state by first computing

$$\hat{v}_t^n = \max_{x_t \in \mathcal{X}_t} \left( C_t(S_t^n, x_t) + \gamma \sum_{\hat{\omega} \in \hat{\Omega}_{t+1}} p_{t+1}(\hat{\omega}) \bar{V}_{t+1}^{n-1}(S_{t+1}) \right), \tag{4.6}$$

which is an estimate of the value of being in state $S_t^n$. $\hat{v}_t^n$ should be viewed as a random observation from what we would like to think of as the true distribution of the value of being in state $S_t^n$ at time $t$. By contrast, $\bar{V}_t^{n-1}(S_t^n)$ is our current estimate of the mean of this distribution. The difference

$$\varepsilon_t^n = \hat{v}_t^n - \bar{V}_t^{n-1}(S_t^n)$$

is referred to as the *Bellman error*. Of course, $\hat{v}_t^n$ is a biased observation of the value of being in state $S_t^n$ since it depends on the value function approximation $\bar{V}_{t+1}^{n-1}(S_{t+1})$. Since there is typically some noise in our estimate of $\hat{v}_t^n$ (because of the random sampling needed to construct $\hat{\Omega}_{t+1}$), we perform the following smoothing operation to update $\bar{V}_t^{n-1}(S_t^n)$, which is our current estimate of the value of being in state $S_t^n$, to produce an updated estimate

$$\bar{V}_t^n(S_t^n) = (1 - \alpha_{n-1})\bar{V}_t^{n-1}(S_t^n) + \alpha_{n-1}\hat{v}_t^n. \tag{4.7}$$

---

**Step 0.** Initialization:

Step 0a. Initialize $\bar{V}_t^0(S_t)$ for all states $S_t$.

Step 0b. Choose an initial state $S_0^1$.

Step 0c. Set $n = 1$.

**Step 1.** Choose a sample path $\omega^n$.

**Step 2.** For $t = 0, 1, 2, \ldots, T$ do:

**Step 2a.** Choose a random sample of outcomes $\hat{\Omega}_{t+1}^n \subset \Omega$ representing possible realizations of the information arriving between $t$ and $t+1$.

**Step 2b.** Solve

$$\hat{v}_t^n = \max_{x_t \in \mathcal{X}_t^n} \left( C_t(S_t^n, x_t) + \gamma \sum_{\hat{\omega} \in \hat{\Omega}_{t+1}^n} p_{t+1}(\hat{\omega}) \bar{V}_{t+1}^{n-1}(S_{t+1}) \right) \quad (4.5)$$

where $S_{t+1} = S^M(S_t^n, x_t, W_{t+1}(\hat{\omega}))$. Let $x_t^n$ be the value of $x_t$ that solves the maximization problem.

**Step 2c.** Update $\bar{V}_t^{n-1}(S_t^n)$:

$$\bar{V}_t^n(S_t) = \begin{cases} (1-\alpha_{n-1})\bar{V}_t^{n-1}(S_t^n) + \alpha_{n-1}\hat{v}_t^n & S_t = S_t^n \\ \bar{V}_t^{n-1}(S_t) & \text{Otherwise.} \end{cases}$$

**Step 2d.** Compute $S_{t+1}^n = S^M(S_t^n, x_t^n, W_{t+1}(\omega^n))$.

**Step 3.** Let $n = n + 1$. If $n < N$, go to step 1.

---

**Figure 4.2** A forward dynamic programming algorithm approximating the expectation

Here, $\alpha_{n-1}$ is known as a "stepsize" (among other things), and generally takes on values between 0 and 1 (see chapter 6 for an in-depth discussion of equation (4.7)). Equation (4.7) is an operation that we see a lot of in approximate dynamic programming. It is variously described as "smoothing," (or "exponential smoothing"), a "linear filter" or "stochastic approximation." Sometimes we use a constant stepsize (say $\alpha_n = .1$), a deterministic formula (such as $\alpha_n = 1/n$) or one that adapts to the data. All are trying to do the same thing: use observations of noisy data ($\hat{v}_t^n$) to approximate the mean of the distribution from which the observations are being drawn. $\bar{V}_t^n(S_t^n)$ is viewed as the best estimate of the value of being in state $S_t^n$ after iteration $n$.

The smoothing is needed only because of the randomness in $\hat{v}_t^n$ due to the way we approximated the expectation. If we were able to compute the expectation exactly, we would compute

$$\bar{V}_t^n(S_t^n) = \max_{x_t \in \mathcal{X}_t^n} \left( C(S_t^n, x_t) + \gamma \sum_{s' \in \mathcal{S}} \mathbb{P}(s'|S_t^n, x_t) V_{t+1}^{n-1}(s') \right).$$

This is similar to equation (4.7) using $\alpha_{n-1} = 1$. It is also the same as our value iteration algorithm (section 3.4), except that now we are updating a single state $S_t^n$ rather than all states.

Equation (4.7) assumes we are using what is known as a "lookup-table" version of the value function. That is, for each state $S_t^n$ we have an estimate $\bar{V}_t^n(S_t^n)$ of the value of being in this state, which means we have to estimate a value for each state (!!). It is important to emphasize that the idea of stepping forward through states and updating estimates of the value of being in only the states that you visit does not solve the problem of high-dimensional state spaces. Instead, it replaces the computational burden of looping over every state with the statistical problem of estimating the value of being in each state that we *might* visit. However, there are a number of strategies we can use if we exploit the structure of the problem, something that backward dynamic programming does not really allow us to do. As you gain experience with these methods, you will find that some understanding of the structure of the problem tends to be critical. Chapter 7 describes methods for approximation value functions in much greater detail.

## 4.3 SAMPLING RANDOM VARIABLES

Our ADP algorithm depends on having access to a sequence of sample realizations of our random variable. How this is done depends on the setting. There are three ways that we can obtain random samples:

1) **The real world** - Random realizations may come from real physical processes. For example, we may be trying to estimate average demand using sequences of actual demands. We may also be trying to estimate prices, costs, travel times, or other system parameters from real observations.

2) **Computer simulations** - The realization may be a calculation from a computer simulation of a complex process. The simulation may be of a physical system such as a supply chain or an asset allocation model. Some simulation models can require extensive calculations (a single sample realization could take hours or days on a computer).

3) **Sampling from a known distribution** - This is the easiest way to sample a random variable. We can use existing tools available in most software languages or spreadsheet packages to generate samples from standard probability distributions. These tools can be used to generate many thousands of random observations extremely quickly.

The ability of ADP algorithms to work with real data (or data coming from a complex computer simulation) means that we can solve problems without actually knowing the underlying probability distribution. We may have multiple random variables that exhibit complex correlations. For example, observations of interest rates or currency exchange rates can exhibit complex interdependencies that are difficult to estimate. As a result, it is possible that we cannot compute an expectation because we do not know the underlying probability distribution. That we can still solve such a problem (approximately) greatly expands the scope of problems that we can address.

When we do have a probability model describing our information process, we can use the power of computers to generate random observations from a distribution using a process that is generally referred to as Monte Carlo sampling. Although most software tools come with functions to generate observations from major distributions, it is often necessary to customize tools to handle more general distributions. When this is the case, we often find ourselves relying on sampling techniques that depend on functions for generating random variables that are uniformly distributed between 0 and 1, which we denote by $U$ (often called RAND() in many computer languages), or for generating random variables that are normally distributed with mean 0 and variance 1 (the standard normal distribution), which we denote by $Z$. Using the computer to generate random variables in this way is known as *Monte Carlo simulation*. A "Monte Carlo sample" (or "Monte Carlo realization") refers to a particular observation of a random variable. Not surprisingly, it is much easier to use a single observation of a random variable than the entire distribution.

If we need a random variable $X$ that is uniformly distributed between $a$ and $b$, we use the internal random number generator to produce $U$ and then calculate

$$X = a + (b - a)U.$$

Similarly, if we wish to randomly generate a random variable that is normally distributed with mean $\mu$ and variance $\sigma^2$, we can usually depend on an internal random number generator to compute a random variable $Z$ which has mean 0 and variance 1. We can then perform the transformation

$$X = \mu + \sigma Z$$

to obtain a random variable with mean $\mu$ and variance $\sigma^2$.

If we need a random variable $X$ with a cumulative distribution function $F_X(x) = \mathbb{P}(X \leq x)$, then we can obtain observations of $X$ using a simple property. Let $Y = F_x(X)$ be a random variable that we compute by taking the original random variable $X$, and then computing $F_X(X)$. It is possible to show that $Y$ is a random variable with a uniform distribution between 0 and 1. This implies that $F_X^{-1}(U)$ is a random variable that has the same distribution as $X$, which means that we can generate observations of $X$ using

$$X = F_X^{-1}(U).$$

For example, consider the case of an exponential density function $\lambda e^{-\lambda x}$ with cumulative distribution function $1 - e^{-\lambda x}$. Setting $U = 1 - e^{-\lambda x}$ and solving for $x$ gives

$$X = -\frac{1}{\lambda} \ln(1 - U).$$

Since $1 - U$ is also uniformly distributed between 0 and 1, we can use

$$X = -\frac{1}{\lambda} \ln(U).$$

Figure 4.3 illustrates using the inverse cumulative-distribution method to generate both uniformly distributed and exponentially distributed random numbers. After generating a uniformly distributed random number in the interval [0,1] (denoted $U(0, 1)$ in the figure), we then map this number from the vertical axis to the horizontal axis. If we want to find a random number that is uniformly distributed between $a$ and $b$, the cumulative distribution simply stretches (or compresses) the uniform (0,1) distribution over the range $(a, b)$.

There is an extensive literature on generating Monte Carlo random variables that goes well beyond the scope of this book. This section provides only a brief introduction.

## 4.4 ADP USING THE POST-DECISION STATE VARIABLE

In the practical application of approximate dynamic programming, one of the more difficult steps can be computing (or approximating) the expectation within the max (or min) operator (see equations (4.2) or (4.5)). Most textbooks on dynamic programming seem to avoid this by assuming that the one-step transition matrix is given as data. In fact, the one-step transition matrix requires that we compute the expectation. In a vast range of practical applications, finding the one-step transition matrix is computationally intractable.

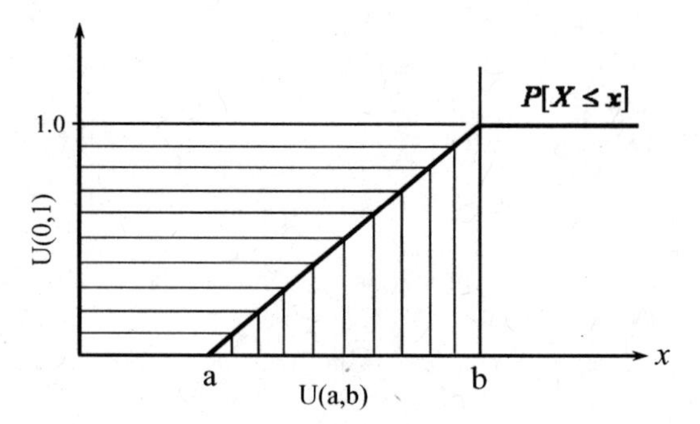


4.3a: Generating uniform random variables.

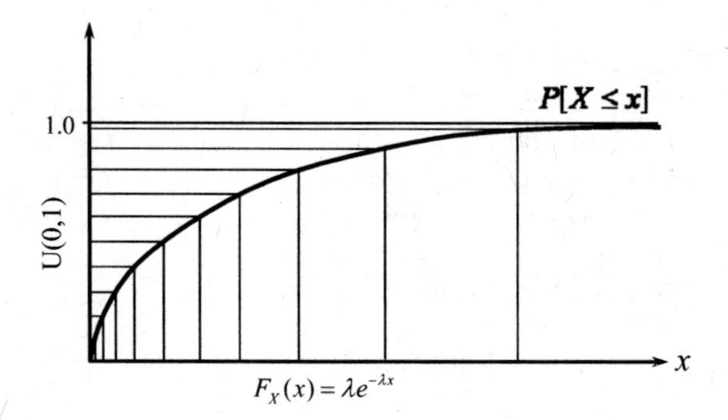


4.3a: Generating exponentially-distributed random variables.

**Figure 4.3** Generating uniformly and exponentially distributed random variables using the inverse cumulative distribution method.

We can circumvent this step by using the idea of the post-decision state variable. We begin by deriving the optimality equations around the post-decision state variable. We then show how this changes our basic ADP strategy. This strategy is one of the defining characteristics of our presentation, as we will use this approach through the remainder of the volume.

### 4.4.1 Finding the post-decision state variable

For many problems, it is possible to break down the effect of decisions and information on the state variable. For example, we might write the transition function for an inventory problem as

$$R_{t+1} = \max\{R_t + x_t - \hat{D}_{t+1}, 0\}.$$

We can break this into two steps, given by

$$\begin{aligned} R_t^x &= R_t + x_t, \\ R_{t+1} &= \max\{R_t^x - \hat{D}_{t+1}, 0\}. \end{aligned}$$

$R_t^x$ captures the pure effect of the decision $x_t$ to order additional assets, while $R_{t+1}$ captures the effect of serving demands $\hat{D}_{t+1}$.

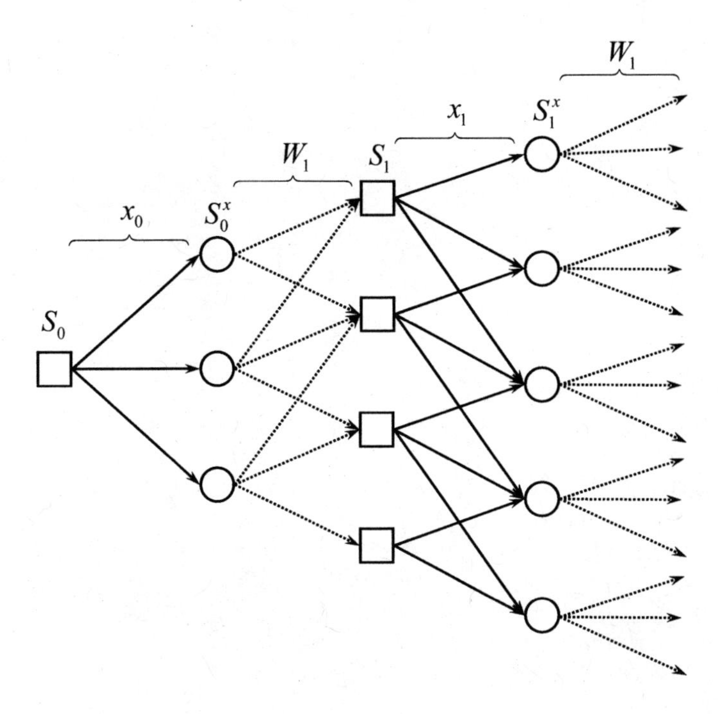


**Figure 4.4** A generic decision tree, showing decision nodes (squares) and outcome nodes (circles). Solid lines are decisions, and dotted lines are random outcomes.

If this is possible, we can break our original transition function

$$S_{t+1} = S^M(S_t, x_t, W_{t+1})$$

into the two steps

$$\begin{aligned} S_t^x &= S^{M,x}(S_t, x_t), \\ S_{t+1} &= S^{M,W}(S_t^x, W_{t+1}). \end{aligned}$$

$S_t$ is the state of the system immediately before we make a decision, while $S_t^x$ is the state immediately after we make a decision (which is why it is indexed by $t$). For this reason, we sometimes refer to $S_t$ as the *pre-decision state variable* while $S_t^x$ is the *post-decision state variable*. Whenever we refer to a "state variable" we always mean the pre-decision state $S_t$. We refer to the pre-decision state only when there is a specific need to avoid confusion.

We first saw pre- and post-decision state variables in section 2.2.1 on decision trees. Figure 4.4 illustrates a generic decision tree with decision nodes (squares) and outcome nodes (circles). The information available at a decision node is the pre-decision state, and the information available at an outcome node is the post-decision state. The function $S^{M,x}(S_t, x_t)$ takes us from a decision node (pre-decision state) to an outcome node (post-decision state). The function $S^{M,W}(S_t^x, W_{t+1})$ takes us from an outcome node to a decision node.

We revisit the use of post-decision state variables repeatedly through this book. Its value is purely computational. The benefits are problem specific, but in some settings they are significant.

### 4.4.2 The optimality equations using the post-decision state variable

Just as we earlier defined $V_t(S_t)$ to be the value of being in state $S_t$ (just before we made a decision), let $V_t^x(S_t^x)$ be the value of being in state $S_t^x$ immediately after we made a decision. There is a simple relationship between $V_t(S_t)$ and $V_t^x(S_t^x)$ that is summarized as follows

$$V_{t-1}^x(S_{t-1}^x) = \mathbb{E}\left\{V_t(S_t)|S_{t-1}^x\right\}, \quad (4.8)$$

$$V_t(S_t) = \max_{x_t \in \mathcal{X}_t} \left(C_t(S_t, x_t) + \gamma V_t^x(S_t^x)\right), \quad (4.9)$$

$$V_t^x(S_t^x) = \mathbb{E}\left\{V_{t+1}(S_{t+1})|S_t^x\right\}. \quad (4.10)$$

where $S_t = S^{M,W}(S_{t-1}^x, W_t)$ in (4.8), $S_t^x = S^{M,x}(S_t, x_t)$ in equation (4.9), and $S_{t+1} = S^{M,W}(S_t^x, W_{t+1})$ in equation (4.10). Equation (4.8) writes $V_{t-1}^x(S_{t-1}^x)$ as a function of $V_t(S_t)$. Equation (4.10) does the same for the next time period, while equation (4.9) writes $V_t(S_t)$ as a function of $V_t^x(S_t^x)$. If we substitute (4.10) into (4.9), we obtain the standard form of Bellman's equation

$$V_t(S_t) = \max_{x_t \in \mathcal{X}_t} \left(C_t(S_t, x_t) + \gamma \mathbb{E}\left\{V_{t+1}(S_{t+1})|S_t\right\}\right).$$

By contrast, if we substitute (4.9) into (4.8), we obtain the optimality equations around the post-decision state variable

$$V_{t-1}^x(S_{t-1}^x) = \mathbb{E}\left\{ \max_{x_t \in \mathcal{X}_t} \left(C_t(S_t, x_t) + \gamma V_t^x(S_t^x)\right) \Big| S_{t-1}^x \right\}. \quad (4.11)$$

What should immediately stand out is that the expectation is now outside of the max operator. While this should initially appear to be a much more difficult equation to solve (we have to solve an optimization problem within the expectation), this is going to give us a tremendous computational advantage. Equation (4.9) is a deterministic optimization problem which can still be quite challenging for some problem classes (especially those involving the management of discrete resources), but there is a vast research base that we can draw on to solve these problems. By contrast, the expectation in equation (4.8) or (4.10) may be easy, but is often computationally intractable. This is the step that typically requires the use of approximation methods.

Writing the value function in terms of the post-decision state variable provides tremendous computational advantages (as well as considerable simplicity). Since Bellman's equation, written around the pre-decision state variable, is absolutely standard even in the field of approximate dynamic programming, this represents a significant point of departure of our treatment of approximate dynamic programming.

### 4.4.3 An ADP algorithm

Assume, as we did in section 4.2, that we have found a suitable approximation $\bar{V}_t(S_t^x)$ for the value function around the post-decision state $S_t^x$.

As before, we are going to run our algorithm iteratively. Assume that we are in iteration $n$ and that at time $t-1$ that we are in post-decision state $S_{t-1}^{x,n}$. We are then going to use a sample realization of the information that might arrive between $t-1$ and $t$, which we have represented using $W_t(\omega^n)$, to compute the next pre-decision state using

$$S_t^n = S^{M,W}(S_{t-1}^x, W_t(\omega^n)).$$

We use the notation $S_t^n$ to indicate that we depend on sample realization $\omega^n$ (this is more compact than writing $S_t(\omega^n)$). We also note that the feasible region $\mathcal{X}_t$ depends on the state $S_t^n$. Again, rather than write $\mathcal{X}_t(S_t^n)$ or $\mathcal{X}_t(\omega^n)$, we let $\mathcal{X}_t^n$ be the feasible region given that we are in state $S_t^n$ (the subscript $t$ indicates that we are at time $t$ with access to the information in $S_t$, while the superscript $n$ indicates that we are depending on the $n^{th}$ sample realization).

Our optimization problem at time $t$ can now be written

$$\hat{v}_t^n = \max_{x_t \in \mathcal{X}_t^n} \left( C_t(S_t^n, x_t) + \gamma \bar{V}_t^{n-1}(S^{M,x}(S_x^n, x_t)) \right), \tag{4.12}$$

where we let $x_t^n$ be the value of $x_t$ that solves this problem. What is very important about this problem is that it is deterministic. We do not require directly computing or even approximating an expectation as we did in equation (4.6) (the expectation is already captured in the deterministic function $\bar{V}_t^{n-1}(S_t^x)$). For large-scale problems which we consider later, this will prove critical.

We next need to update our value function approximation. $\hat{v}_t^n$ is a sample realization of the value of being in state $S_t^n$, so we could update our value function approximation using

$$\bar{V}_t^n(S_t^n) = (1 - \alpha_{n-1})\bar{V}_t^{n-1}(S_t^n) + \alpha_{n-1}\hat{v}_t^n.$$

There is nothing wrong with this expression, except that it will give us an estimate of the value of being in the pre-decision state $S_t$. When we are solving the decision problem at time $t$ (in iteration $n$), we would use $\bar{V}_{t+1}^{n-1}(S_{t+1})$, but since $S_{t+1}$ is a random variable at time $t$, we have to compute (or at least approximate) the expectation.

A much more effective alternative is to use $\hat{v}_t^n$ to update the value of being in the post-decision state $S_{t-1}^{x,n}$. Keep in mind that the previous decision $x_{t-1}^n$ put us in state $S_{t-1}^{x,n}$, after which a random outcome, $W_t(\omega^n)$, put us in state $S_t^n$. So, while $\hat{v}_t^n$ is a sample of the value of being in state $S_t^n$, it is also a sample of the value of the decision that put us in state $S_{t-1}^{x,n}$. Thus, we can update our post-decision value function approximation using

$$\bar{V}_{t-1}^n(S_{t-1}^{x,n}) = (1 - \alpha_{n-1})\bar{V}_{t-1}^{n-1}(S_{t-1}^{x,n}) + \alpha_{n-1}\hat{v}_t^n. \tag{4.13}$$

Note that we have chosen not to use a superscript "$x$" for the value function approximation $\bar{V}_{t-1}^n(S_{t-1}^{x,n})$. In the remainder of this volume, we are going to use the value function around the post-decision state almost exclusively, so we suppress the superscript "$x$" notation to reduce notational clutter. By contrast, we have to retain the superscript for our state variable since we will use both pre-decision and post-decision state variables throughout.

The smoothing in equation (4.13) is where we are taking our expectation. If we write

$$V_t^x(S_t^x) = \mathbb{E}\left\{V_{t+1}(S_{t+1})|S_t^x\right\},$$

then the classical form of Bellman's equation becomes

$$V_t(S_t) = \max_{x_t} \left( C_t(S_t, x_t) + \gamma V_t^x(S_t^x) \right).$$

The key step is that we are writing $\mathbb{E}\left\{V_{t+1}(S_{t+1})|S_t^x\right\}$ as a function of $S_t^x$ rather than $S_t$, where we take advantage of the fact that $S_t^x$ is a deterministic function of $x_t$.

A complete sketch of an ADP algorithm using the post-decision state variable is given in figure 4.5.

An important benefit of the post-decision state variable is that it is often simpler, and sometimes dramatically so, simplifying the challenge of creating a good approximation.

---

**Step 0.** Initialization:

**Step 0a.** Initialize $\bar{V}_t^0, \quad t \in \mathcal{T}$.

**Step 0b.** Set $n = 1$.

**Step 0c.** Initialize $S_0^1$.

**Step 1.** Choose a sample path $\omega^n$.

**Step 2.** Do for $t = 0, 1, 2, \ldots, T$:

**Step 2a.** Solve:

$$\hat{v}_t^n = \max_{x_t \in \mathcal{X}_t^n} \left( C_t(S_t^n, x_t) + \gamma \bar{V}_t^{n-1}(S^{M,x}(S_t^n, x_t)) \right)$$

and let $x_t^n$ be the value of $x_t$ that solves the maximization problem.

**Step 2b.** If $t > 0$, update $\bar{V}_{t-1}^{n-1}$ using

$$\bar{V}_{t-1}^n(S_{t-1}^{x,n}) = (1 - \alpha_{n-1})\bar{V}_{t-1}^{n-1}(S_{t-1}^{x,n}) + \alpha_{n-1}\hat{v}_t^n.$$

**Step 2c.** Find the post-decision state

$$S_t^{x,n} = S^{M,x}(S_t^n, x_t^n)$$

and the next pre-decision state

$$S_{t+1}^n = S^{M,W}(S_t^{x,n}, W_{t+1}(\omega^n)).$$

**Step 3.** Increment $n$. If $n \leq N$ go to Step 1.

**Step 4.** Return the value functions $(\bar{V}_t^N)_{t=0}^T$.

---

**Figure 4.5** Forward dynamic programming using the post-decision state variable.

Consider the problem of assigning resources to demands (customer requests), where demands become known during time period $t$ at which time we have to decide which resources should serve each demand. Further assume that any demands that are not served are discarded. We can let $R_{ti}$ be the number of resources of type $i \in \mathcal{I}$ that are available at time $t$, while $D_{tj}$ is the number of demands of type $j \in \mathcal{J}$ that have to be served. The pre-decision state variable is $R_t = (R_t, D_t)$ (with $|\mathcal{I}| + |\mathcal{J}|$ dimensions), while the post decision state variable describes only the modified resources (with $|\mathcal{I}|$ dimensions).

Forward dynamic programming using the post-decision state variable is more elegant because it avoids the need to approximate the expectation explicitly within the optimization problem. It also gives us another powerful device. Since the decision function "sees" $\bar{V}_t(S_t^x)$ directly (rather than indirectly through the approximation of the expectation), we are able to control the structure of $\bar{V}_t(S_t^x)$. This feature is especially useful when the myopic problem $\max_{x_t \in \mathcal{X}_t} C_t(S_t, x_t)$ is an integer program or a difficult linear or nonlinear program that requires special structure. This feature comes into play in the context of more complex problems, such as those discussed in chapter 12.

Although it may not be entirely apparent now, this basic strategy makes it possible to produce production quality models for large-scale, industrial problems which are described by state variables with millions of *dimensions*. The process of stepping forward in time uses all the tools of classical simulation, where it is possible to capture an arbitrarily high level of detail as we simulate forward in time. Our interest in this volume is obtaining the highest quality decisions as measured by our objective function. The key here is creating approximate value functions that accurately capture the impact of decisions now on the future. For problems with possibly high-dimensional decision vectors, the optimization problem

$$\max_{x_t \in \mathcal{X}_t^n} \left( C_t(S_t^n, x_t) + \gamma \bar{V}_t^{n-1}(S^{M,x}(S_t^n, x_t)) \right)$$

will have to be solved using the various tools of mathematical programming (linear programming, nonlinear programming and integer programming). In order to use these tools, the value function approximation may need to have certain properties which may play a role in the design of the approximation strategy.

### 4.4.4 A perspective of the post-decision state variable

The decision of when to estimate the value function around a post-decision state variable is fairly simple. If you can easily compute the expectation $\mathbb{E}V_{t+1}(S_{t+1})$, you should use the pre-decision state variable. If the expectation is hard (the "second curse of dimensionality"), you should approximate the value function around the post-decision state since it does not require any additional work and is generally easier (the post-decision state is never more complicated than the pre-decision state and may be simpler). If you can compute an expectation exactly, taking advantage of the structure of the problem, you are going to produce a more reliable estimate than using Monte Carlo-based methods.

The vast majority of papers in the ADP community use the pre-decision state variable reflecting, we believe, the classical development of Markov decision processes where the ability to take expectations is taken for granted. A simple example illustrates the power and ease of the post-decision state variable. Imagine that we are trying to predict unemployment for year $t$ as a function of interest rates in year $t$. If $U_t$ is the unemployment rate in year $t$ and $I_t$ is the interest rate, we might create a model of the form

$$U_t = \theta_0 + \theta_1 (I_t)^2 \tag{4.14}$$

where we would estimate $\theta_0$ and $\theta_1$ using historical data on unemployment and interest rates. Now imagine that we wish to use our model to predict unemployment next year. We would first have to predict interest rates and then use our model to predict unemployment.

Now imagine that we use the same data to estimate the model

$$U_t = \theta_0 + \theta_1 (I_{t-1})^2.$$

Estimating this model requires no additional work, but now we can use interest rates this year to predict next year's unemployment. Not surprisingly, this is a much easier model to use.

Approximating a value function around the pre-decision state $S_{t+1}$ is comparable to using equation (4.14). At time $t$, we first have to forecast $S_{t+1}$ and then compute $\bar{V}_{t+1}(S_{t+1})$. Approximating the value function around the post-decision state variable is comparable to using (4.15), which allows us to compute a variable in the future ($U_{t+1}$) based on what we know now ($I_t$).

## 4.5 LOW-DIMENSIONAL REPRESENTATIONS OF VALUE FUNCTIONS

Classical dynamic programming typically assumes a discrete representation of the value function. This means that for every state $s \in \mathcal{S}$, we have to estimate a parameter $v_s$ that gives the value of being in state $s$. Forward dynamic programming may eliminate the loop over all states that is required in backward dynamic programming, but it does not solve the classic "curse of dimensionality" in the state space. Forward dynamic programming focuses attention on the states that we actually visit, but it also requires that we have some idea of the value of being in a state that we *might* visit (we need this estimate to conclude that we should not visit the state).

Virtually every large-scale problem in approximate dynamic programming will focus on determining how to approximate the value function with a smaller number of parameters. In backward discrete dynamic programming, we have one parameter per state, and we want to avoid searching over a large number of states. In forward dynamic programming, we depend on Monte Carlo sampling, and the major issue is statistical error. It is simply easier to estimate a function that is characterized by fewer parameters.

In practice, approximating value functions always requires understanding the structure of the problem. However, there are general strategies that emerge. Below we discuss two of the most popular.

### 4.5.1 Aggregation

In the early days of dynamic programming, aggregation was quickly viewed as a way to provide a good approximation with a smaller state space, allowing the tools described in chapter 3 to be used. A major problem with aggregation when we use the framework in chapter 3 is that we have to solve the entire problem in the aggregated state space. With approximate dynamic programming, we only have to aggregate for the purpose of computing the value function approximation. It is extremely valuable that we can retain the full state variable for the purpose of computing the transition function, costs, and constraints. In the area of approximate dynamic programming, aggregation provides a mechanism for reducing statistical error. Thus, while it may introduce structural error, it actually makes a model more accurate by improving statistical robustness.

---

■ **EXAMPLE 4.1**

Consider the problem of evaluating a basket of securities where $\hat{p}_{ti}$ is the price of the $i^{th}$ security at time $t$, where the state variable $S_t = (\hat{p}_{ti})_i$ is the vector of all the prices. We can discretize the prices, but we can dramatically simplify the problem of estimating $\bar{V}_t(S_t)$ if we use a fairly coarse discretization of the prices. We can discretize prices for the purpose of estimating the value function without changing how we represent prices of each security as we step forward in time.

■ **EXAMPLE 4.2**

A trucking company that moves loads over long distances has to consider the profitability of assigning a driver to a particular load. Estimating the value of the load requires estimating the value of a driver at the destination of the load. We can estimate the value of the driver at the level of the 5-digit zip code of the location, or the 3-digit level, or at the level of a region (companies typically represent the United States using about 100 regions, which is far more aggregate than a 3-digit zip). The value of a driver in a location may also depend on his home domicile, which can also be represented at several levels of aggregation.

---

Aggregation is particularly powerful when there are no other structural properties to exploit, which often arises when the state contains dimensions that are categorical rather than numerical. When this is the case, we typically find that the state space $\mathcal{S}$ does not have any metric to provide a "distance" between two attributes. For example, we have an intuitive sense that a disk drive company and a company that makes screens for laptops both serve the personal computer industry. We would expect that valuations in these two segments would be more closely correlated than they would be with a textile manufacturer. But we do not have a formal metric that measures this relationship.

In chapter 10 we investigate the statistics of aggregation in far greater detail.

### 4.5.2 Continuous value function approximations

In many problems, the state variable does include continuous elements. Consider a resource allocation problem where $R_{ti}$ is the number of resources of type $i \in \mathcal{I}$. $R_{ti}$ may be either continuous (how much money is invested in a particular asset class, and how long has it been invested there) or discrete (the number of people with a particular skill set), but it is always numerical. The number of possible values of a vector $R_t = (R_{ti})_i$ can be huge (the curse of dimensionality). Now consider what happens when we replace the value function $V_t(R_t)$ with a linear approximation of the form

$$\bar{V}_t = \sum_{i \in \mathcal{I}} \bar{v}_{ti} R_{ti}.$$

Instead of having to estimate the value $V_t(R_t)$ for each possible vector $R_t$, we have only to estimate $\bar{v}_{ti}$, one for each value of $i \in \mathcal{I}$. For some problems, this reduces the size of the problem from $10^{100}$ or greater to one with several hundred or several thousand parameters.

Not all problems lend themselves to linear-in-the-resource approximations, but other approximations may emerge. Early in the development of dynamic programming, Bellman realized that a value function could be represented using statistical models and estimated with regression techniques. To illustrate let $\bar{V}(R|\theta)$ be a statistical model where the elements of $R_t$ are used to create the independent variables, and $\theta$ is the set of parameters. For example, we might specify:

$$\bar{V}(R|\theta) = \sum_{i \in \mathcal{I}} \left( \theta_{1i} R_i + \theta_{2i} (R_i)^2 \right).$$

The formulation and estimation of continuous value function approximations is one of the most powerful tools in approximate dynamic programming. The fundamentals of this approach is presented in considerably more depth in chapters 7 and 11.

### 4.5.3 Algorithmic issues

The design of an approximation strategy involves two algorithmic challenges. First, we have to be sure that our value function approximation does not unnecessarily complicate the solution of the myopic problem. We have to assume that the myopic problem is solvable in a reasonable period of time. If we are choosing our decision $x_t$ by enumerating all possible decisions (a "lookup-table" decision function), then this is not an issue, but if $x_t$ is a vector, then this is not going to be possible. If our myopic problem is a linear or nonlinear program, it is usually impossible to consider a value function that is of the discrete, lookup-table variety. If our myopic problem is continuously differentiable and concave, we do not want to introduce a potentially nonconcave value function. By contrast, if our myopic problem is a discrete scheduling problem that is being solved with a search heuristic, then lookup-table value functions can work just fine.

Once we have decided on the structure of our functional approximation, we have to devise an updating strategy. Value functions are basically statistical models that are updated using classical statistical techniques. However, it is very convenient when our updating algorithm is in a recursive form. A strategy that fits a set of parameters by using a sequence of observations using standard regression techniques may be too expensive for many applications.

## 4.6 SO JUST WHAT IS APPROXIMATE DYNAMIC PROGRAMMING?

Approximate dynamic programming can be viewed from three very different perspectives. Depending on the problem you are trying to solve, ADP can be viewed as an algorithmic strategy for solving complex dynamic programs, a way of making classical simulations more intelligent, and a decomposition technique for large-scale mathematical programs.

### 4.6.1 ADP for solving complex dynamic programs

In the research community, approximate dynamic programming is primarily viewed as a way of solving dynamic programs which suffer from "the curse of dimensionality" (in this book, the three curses). Faced with the need to solve Bellman's equation, we have to overcome the classical problem of computing the value function $V(S)$ when the number of states $S$ is too large to enumerate. We also consider problems where the expectation cannot be computed, and where the action space is too large to enumerate. The remainder of this volume presents modeling and algorithmic strategies for overcoming these problems.

### 4.6.2 ADP as an "optimizing simulator"

Part of the power of approximate dynamic programming is that it is basically a form of classical simulation with more intelligent decision rules. Almost any ADP algorithm looks something like the algorithm depicted in figure 4.6, which includes two major steps: an optimization step, where we choose a decision, and a simulation step, where we capture the effects of random information. Viewed in this way, ADP is basically an "optimizing-simulator." The complexity of the state variable $S_t$ and the underlying physical processes, captured by the transition functions $S^{M,x}(S_t, x_t)$ and $S^{M,W}(S_t^x, W_{t+1})$, are virtually unlimited (as is the case with any simulation model). We are, as a rule, limited in the complexity of the value function approximation $\bar{V}_t(S_t^x)$. If $S_t^x$ is too complex, we have to design a functional approximation that uses only the most important features. But it is critical in many applications that we do not have to similarly simplify the state variable used to compute the transition function.

There are many complex, industrial applications where it is very important to capture the physical process at a high level of detail, but where the quality of the decisions (that is, how close the decisions are to the optimal decisions) is much harder to measure. A modest relaxation in the optimality of the decisions may be difficult or impossible to measure, whereas simplifications in the state variable are quickly detected because the system simply does not evolve correctly. It has been our repeated experience in many

---

**Step 0.** Given an initial state $S_0^1$ and value function approximations $\bar{V}_t^0(S_t)$ for all $S_t$ and $t$, set $n = 1$.

**Step 1.** Choose a sample path $\omega^n$.

**Step 2.** For $t = 0, 1, 2, \ldots, T$ do:

> **Step 2a.** Optimization: Compute a decision $x_t^n = X_t^\pi(S_t)$ and find the post-decision state $S_t^{x,n} = S^{M,x}(S_t^n, x_t^n)$.
>
> **Step 2b.** Simulation: Find the next pre-decision state using $S_t^n = S^{M,W}(S_t^{x,n}, W_{t+1}(\omega^n))$.

**Step 3.** Update the value function approximation to obtain $\bar{V}_t^n(S_t)$ for all $t$.

**Step 4.** If we have not met our stopping rule, let $n = n + 1$ and go to step 1.

---

**Figure 4.6** A generic approximate dynamic programming algorithm.

industrial applications that it is far more important to capture a high degree of realism in the transition function than it is to produce truly optimal decisions.

### 4.6.3 ADP as a decomposition technique for large-scale math programs

Our foray into approximate dynamic programming began with our efforts to solve very large-scale linear (and integer) programs that arise in a range of transportation applications. Expressed as a linear program, these would look like

$$\max_{(x_t)_t} \sum_{t=0}^{T} c_t x_t$$

subject to

$$\begin{aligned} A_0 x_0 &= R_0, \\ A_t x_t - B_{t-1} x_{t-1} &= \hat{R}_t \quad t = 1, \ldots, T, \\ D_t x_t &\leq u_t, \\ x &\geq 0 \text{ and integer.} \end{aligned}$$

Here, $R_0$ is the initial inventories of vehicles and drivers (or crews or pilots), while $\hat{R}_t$ is the (deterministic) net inflow or outflow of resources to or from the system. $A_t$ captures flow conservation constraints, while $B_{t-1}$ tells us where resources that were moved at time $t-1$ end up in the future. $D_t$ tells us which elements of $x_t$ are limited by an upper bound $u_t$.

In industrial applications, $x_t$ might easily have 10,000 elements, defined over 50 time periods. $A_t$ might have thousands of rows, or hundreds of thousands (depending on whether we were modeling equipment or drivers).

Formulated as a single, large linear (or integer) program (over 50 time periods), we obtain mathematical programs with hundreds of thousands of rows and upwards of millions of columns. As this volume is being written, there are numerous researchers attacking these problems using a wide range of heuristic algorithms and decomposition strategies. Yet, our finding has been that even these large-scale models ignored a number of operating issues (even if we ignore uncertainty).

ADP would solve this problem by solving sequences of problems that look like

$$\max_{x_t} \left( c_t x_t + \bar{V}_{t+1}(R_{t+1}(x_t)) \right)$$

subject to for $t = 1, \ldots, T,,$

$$\begin{aligned} A_t x_t &= R_t + \hat{R}_t, \\ R_{t+1} - B_t x_t &= 0, \\ D_t x_t &\leq u_t, \\ x &\geq 0 \text{ and integer}, \end{aligned}$$

where $R_t = B_{t-1}x_{t-1}$ and $\bar{V}_{t+1}(R_{t+1})$ is some sort of approximation that captures the impact of decision at time $t$ on the future. After solving the problem at time $t$, we would compute $R_{t+1}$, increment $t$ and solve the problem again. We would run repeated simulations over the time horizon while improving our approximation $\bar{V}_t(R_t)$.

In this setting, it is quite easy to introduce uncertainty, but the real value is that instead of solving one extremely large linear program over many time periods (something which even modern optimization packages struggle with), we solve the problem one time period at a time. Our finding is that commercial solvers handle these problems quite easily, even for the largest and most complex industrial applications. The price is that we have to design an effective approximation for $\bar{V}_t(R_t)$.

## 4.7 EXPERIMENTAL ISSUES

Once we have developed an approximation strategy, we have the problem of testing the results. Two issues tend to dominate the experimental side: rate of convergence and solution quality. The time required to execute a forward pass can range between microseconds and hours, depending on the underlying application. You may be able to assume that you can run millions of iterations, or you may have to work with a few dozen. The answer depends on both your problem and the setting. Are you able to run for a week on a large parallel processor to get the very best answer? Or do you need real-time response on a purchase order or to run a "what if" simulation with a user waiting at the terminal?

Within our time constraints, we usually want the highest-quality solution possible. Here, the primary interest tends to be in the rate of convergence. Monte Carlo methods are extremely flexible, but the price of this flexibility is a slow rate of convergence. If we are estimating a parameter from statistical observations drawn from a stationary process, the quality of the solution tends to improve at a rate proportional to the square root of the number of observations. Unfortunately, we typically do not enjoy the luxury of stationarity. We usually face the problem of estimating a value function which is steadily increasing or decreasing (it might well be increasing for some states and decreasing for others) as the learning process evolves.

### 4.7.1 The initialization problem

All of the strategies above depend on setting an initial estimate $\bar{V}^0$ for the value function approximation. In chapter 3, the initialization of the value function did not affect our ability to find the optimal solution; it only affected the rate of convergence. In approximate dynamic programming, it is often the case that the value function approximation $\bar{V}^{n-1}$ used in iteration $n$ affects the choice of the action taken and therefore the next state visited.

Consider the shortest path problem illustrated in figure 4.7a, where we want to find the best path from node 1 to node 7.

Assume that at any node $i$, we choose the link that goes to node $j$ which minimizes the cost $c_{ij}$ from $i$ to $j$, plus the cost $\bar{v}_j$ of getting from node $j$ to the destination. If

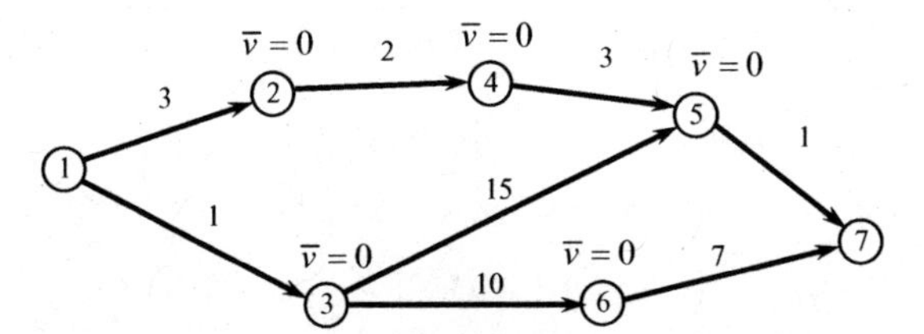


4.7a: Shortest path problem with low initial estimates for the value at each node.

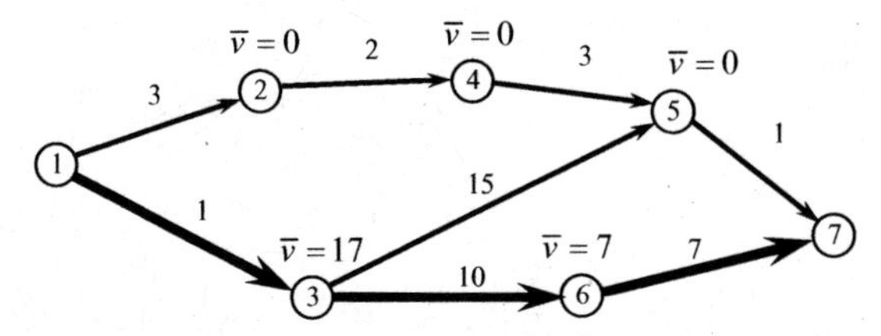


4.7b: After one pass; update cost to destination along this path.

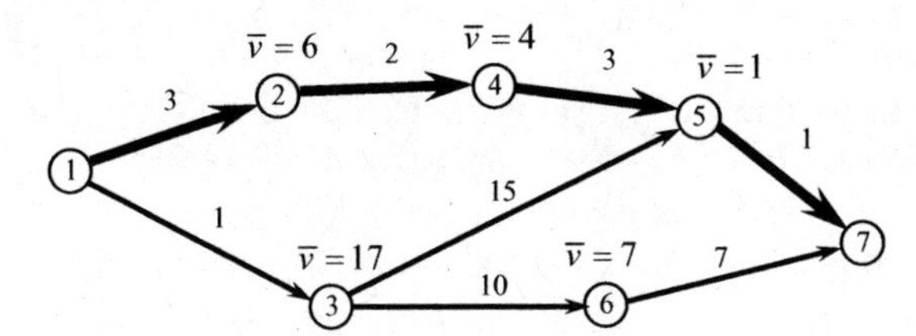


4.7c: After second pass; finds optimal path.

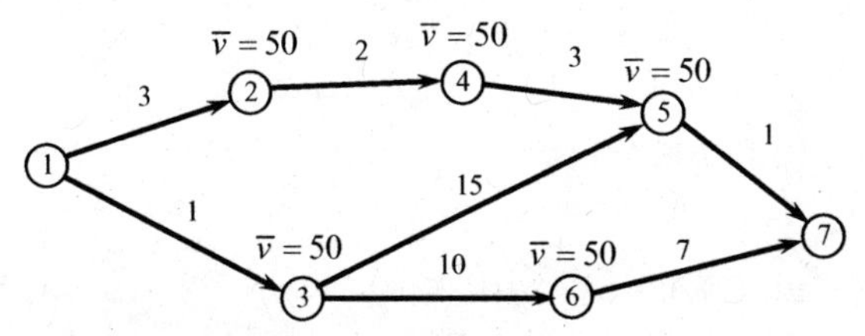


4.7d: Shortest path problem with high initial estimates for the value at each node.

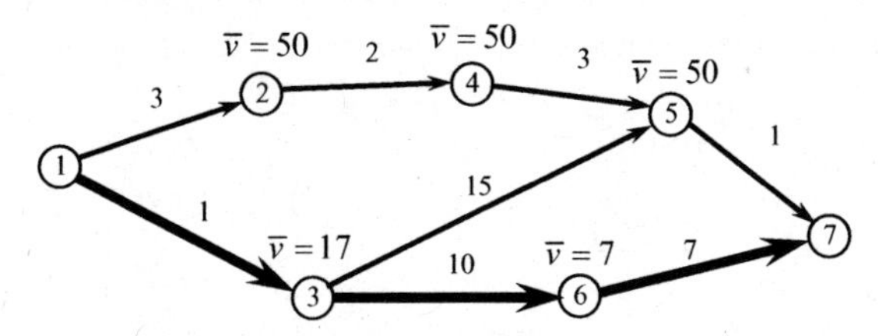


4.7e: After one pass, finds cost along more expensive path, but never finds best path.

**Figure 4.7** Shortest path problems with low (a, b, c) and high (d, e) initial estimates of the cost from each node to the destination.

$\hat{v}_i = \min_j(c_{ij} + \bar{v}_j)$ is the value of the best decision out of node $i$, then we will set $\bar{v}_i = \hat{v}_i$. If we start with a low initial estimate for each $\bar{v}$ (figure 4.7a), we first explore a more expensive path (figure 4.7b) before finding the best path (figure 4.7c) where the algorithm stops.

---

**Step 0.** Initialize an approximation for the value function $\bar{V}^0(S)$ for all states $S$. Let $n = 1$.

**Step 1.** For each state $s \in \mathcal{S}$, do:

$$\hat{v}^n(s) = \max_{x \in \mathcal{X}} \left( C(s,x) + \gamma \sum_{s' \in \mathcal{S}} \mathbb{P}(s'|s,x)\bar{V}^{n-1}(s') \right)$$

**Step 2.** Use the estimate $\hat{v}^n(s)$ to update $\bar{V}^{n-1}(s)$, giving us $\bar{V}^n(s)$.

**Step 3.** Let $n = n + 1$. If $n < N$, go to step 1.

---

**Figure 4.8** Synchronous dynamic programming

Now assume we start with values for $\bar{v}_j$ that are too high, as illustrated in figure 4.7d. We first explore the lower path (figure 4.7e), but we never find the better path.

If we have a deterministic problem and start with an optimistic estimate of the value of being in a state (too low if we are minimizing, too high if we are maximizing), then we are guaranteed to eventually find the best solution. This is a popular algorithm in the artificial intelligence community known as the A* algorithm A* algorithm (pronounced "A star"). However, we generally can never guarantee this in the presence of uncertainty. Just the same, the principle of starting with optimistic estimates remains the same. If we have an optimistic estimate of the value of being in a state, we are more likely to explore that state. The problem is that if the estimates are too optimistic, we may end up exploring too many states.

### 4.7.2 State sampling strategies

There are several strategies for sampling states in dynamic programming. In chapter 3, we used *synchronous* updating, because we updated the value of all states at the same time. The term is derived from parallel implementations of the algorithm, where different processors might independently (but at the same time, hence synchronously) update the value of being in each state. In approximate dynamic programming, synchronous updating would imply that we are going to loop over all the states, updating the value of being in each state (potentially on different processors). An illustration of synchronous approximate dynamic programming is given in figure 4.8 for an infinite horizon problem. Note that at iteration $n$, all decisions are made using $\bar{V}^{n-1}(s)$.

*Asynchronous* dynamic programming, on the other hand, assumes that we are updating one state at a time, after which we update the entire value function. In asynchronous ADP, we might choose states at random, allowing us to ensure that we sample all states infinitely often. The term "asynchronous" is based on parallel implementations where different processors might be updating different states without any guarantee of the order or timing with which states are being updated. The procedure is illustrated in figure 4.9.

A third variant is known in the literature as *real-time dynamic programming*, abbreviated RTDP. RTDP is motivated by a physical system where our action determines the next state we visit. By contrast, in synchronous and asynchronous dynamic programming, we may use an exogenous process to determine which states are sampled (with synchronous DP, we sample all the states all the time). The idea behind RTDP is that we have a physical system, where the decision (and the physical process) determines the state that we next visit. In

---

**Step 0.** Initialize an approximation for the value function $\bar{V}^0(S)$ for all states $S$. Let $n = 1$.

**Step 1.** Randomly choose a state $s^n$.

**Step 2.** Solve:

$$\hat{v}^n = \max_{x \in \mathcal{X}} \left( C(s^n, x) + \gamma \sum_{s' \in \mathcal{S}} \mathbb{P}(s'|s^n, x) \bar{V}^{n-1}(s') \right)$$

**Step 3.** Use $\hat{v}^n$ to update the approximation $\bar{V}^{n-1}(s)$ for all $s$.

**Step 4.** Let $n = n + 1$. If $n < N$, go to step 1.

---

**Figure 4.9** Asynchronous approximate dynamic programming

---

**Step 0.** Initialization:

Step 0a. Initialize $\bar{V}_t^0(S_t)$ for all states $S_t$.

Step 0b. Choose an initial state $S_0^1$.

Step 0c. Set $n = 1$.

**Step 1.** Choose a sample path $\omega^n$.

**Step 2.** For $t = 0, 1, 2, \ldots, T$ do:

**Step 2a.** Solve

$$\hat{v}_t^n = \max_{x_t \in \mathcal{X}_t^n} \left( C_t(S_t^n, x_t) + \gamma \sum_{s' \in \mathcal{S}} \mathbb{P}(s'|S_t^n, x_t) \bar{V}_{t+1}^{n-1}(s') \right),$$

and let $x_t^n$ be the value of $x_t$ that solves the maximization problem.

**Step 2b.** Update $\bar{V}_t^{n-1}(S_t)$ using

$$\bar{V}_t^n(S_t) = \begin{cases} \hat{v}_t^n & S_t = S_t^n \\ \bar{V}_t^{n-1}(S_t) & \text{otherwise.} \end{cases}$$

**Step 2c.** Compute $S_{t+1}^n = S^M(S_t^n, x_t^n, W_{t+1}(\omega^n))$.

**Step 3.** Let $n = n + 1$. If $n < N$, go to step 1.

---

**Figure 4.10** An illustrative real-time dynamic programming algorithm.

RTDP, if we are in state $S_t^n$, we choose an action $x_t^n$ by solving

$$x_t^n = \arg\max_{x_t \in \mathcal{X}_t^n} \left( C(S_t^n, x_t) + \gamma \sum_{s'} \mathbb{P}(s'|S_t^n, x_t) V_{t+1}(s') \right).$$

after which we choose the next state using

$$S_{t+1}^n = S^M(S_t^n, x_t^n, W_{t+1}(\omega^n). \tag{4.15}$$

Papers describing RTDP say that we choose state $S_{t+1}^n = s'$ with probability $\mathbb{P}(s'|S_t^n, x_t)$. This is mathematically equivalent to using the transition equation in (4.15). A version of an RTDP algorithm (there are different variations) is given in figure 4.10.

For the large-scale problems that motivate most applications of approximate dynamic programming, looping over all states (synchronous DP) is out of the question. For problems with high-dimensional state variables, randomly sampling states may also offer little value. What is the point of randomly sampling one hundred thousand or a million states out of a total population of $10^{100}$ states? For this reason, we have adopted the RTDP framework throughout this book as the default strategy. However, we do not use "real-time dynamic programming" to describe the strategy since there are many uses of this approach that have nothing to do with real-time applications.

### 4.7.3 Exploration vs. exploitation

Closely related to the initialization problem is the classical issue of "exploration" vs. "exploitation." Exploration implies visiting states specifically to obtain better information about the value of being in a state, regardless of whether the decision appears to be the best given the current value function approximation. Exploitation means using our best estimate of the contributions and values, and making decisions that seem to be the best given the information we have (we are "exploiting" our estimates of the value function). Of course, exploiting our value function to visit a state is also a form of exploration, but the literature typically uses the term "exploration" to mean that we are visiting a state in order to improve our estimate of the value of being in that state.

There are two ways to explore. After we update the value of one state, we might then choose another state at random, without regard to the state that we just visited (or the action we chose). The second is to randomly choose an action (even it is not the best) which leads us to a somewhat randomized state. The value of the latter is that it constrains our search to states that can be reasonably reached, avoiding the problem of visiting states that can never be reached.

There is a larger issue when considering different forms of exploration. We have introduced several problems (the blood management problem in section 1.2, the dynamic assignment problem in section 2.2.10, and the asset acquisition problem in section 4.1) where both the state variable and the decision variable are multidimensional vectors. For these problems, the state space and the action space are effectively infinite. Randomly sampling a state or even an action for these problems makes little sense.

Determining the right balance between exploring states just to estimate their values, along with using current value functions to visit the states that appear to be the most profitable, represents one of the great unsolved problems in approximate dynamic programming. If we only visit states that appear to be the most profitable given the current estimates (a pure exploitation strategy) we run the risk of landing in local optima unless the problem has special properties. There are strategies that help minimize the likelihood of being trapped in a local optima but at a cost of very slow convergence. Finding good strategies with fairly fast convergence appears to depend on taking advantage of natural problem structure. This topic is covered in considerably more depth in chapter 10.

### 4.7.4 Evaluating policies

A common question is whether a policy $X^{\pi_1}$ is better than another policy $X^{\pi_2}$. Assume we are facing a finite horizon problem that can be represented by the objective function

$$F^{\pi} = \mathbb{E}\left\{\sum_{t=0}^{T} \gamma^t C_t(S_t, X_t^{\pi}(S_t))\right\}.$$

Since we cannot compute the expectation, we might choose a sample $\hat{\Omega} \subseteq \Omega$ and then calculate

$$\begin{aligned} \hat{F}^\pi(\omega) &= \sum_{t=0}^{T} \gamma^t C_t(X_t^\pi(S_t(\omega))), \\ \bar{F}^\pi &= \sum_{\omega \in \hat{\Omega}} \hat{p}(\omega)\hat{F}^\pi(\omega), \end{aligned}$$

where $\hat{p}(\omega)$ is the probability of the outcome $\omega \in \hat{\Omega}$. If we have chosen the outcomes in $\hat{\Omega}$ at random from within $\Omega$, then, letting $N = |\hat{\Omega}|$ be our sample size, we would use

$$\hat{p}(\omega) = \frac{1}{|\hat{\Omega}|} = \frac{1}{N}.$$

Alternatively, we may choose $\hat{\Omega}$ so that we control the types of outcomes in $\Omega$ that are represented in $\hat{\Omega}$. Such sampling strategies fall under names such as stratified sampling or importance sampling. They require that we compute the sample probability distribution $\hat{p}$ to reflect the proper frequency of an outcome $\omega \in \hat{\Omega}$ within the larger sample space $\Omega$.

The choice of the size of $\hat{\Omega}$ should be based on a statistical analysis of $\bar{F}^\pi$. For a given policy $\pi$, it is possible to compute the variance of $\bar{F}^\pi(\hat{\Omega})$ using

$$(\bar{\sigma}^\pi)^2 = \frac{1}{N}\left(\frac{1}{N-1}\right)\sum_{\omega \in \hat{\Omega}} \left(\hat{F}^\pi(\omega) - \bar{F}^\pi\right)^2.$$

This formula assumes that we are sampling outcomes "at random" from $\Omega$, which means that they should be equally weighted. More effective strategies will use sampling strategies that will overrepresent certain types of outcomes.

In most applications, it is reasonable to assume that $(\bar{\sigma}^\pi)^2$ is independent of the policy, allowing us to use a single policy to estimate the variance of our estimate. If we treat the estimates of $\bar{F}^{\pi_1}$ and $\bar{F}^{\pi_2}$ as independent random variables the variance of the difference is $2(\bar{\sigma}^\pi)^2$. If we are willing to assume normality, we can then compute a confidence interval on the difference using

$$\left[(\bar{F}^{\pi_1} - \bar{F}^{\pi_2}) - z_{\alpha/2}\sqrt{2(\bar{\sigma}^\pi)^2}, (\bar{F}^{\pi_1} - \bar{F}^{\pi_2}) + z_{\alpha/2}\sqrt{2(\bar{\sigma}^\pi)^2}\right] \quad (4.16)$$

where $z_{\alpha/2}$ is the standard normal deviate for a confidence level $\alpha$.

Typically, we can obtain a much tighter confidence interval by using the same sample $\hat{\Omega}$ to test both policies. In this case, $\bar{F}^{\pi_1}$ and $\bar{F}^{\pi_2}$ will not be independent and may, in fact, be highly correlated (in a way we can use to our benefit). Instead of computing an estimate of the variance of the value of each policy, we should compute a sample realization of the difference

$$\hat{\Delta}^{\pi_1,\pi_2}(\omega) = \hat{F}^{\pi_1}(\omega) - \hat{F}^{\pi_2}(\omega)$$

from which we can compute an estimate of the difference

$$\begin{aligned} \bar{\Delta}^{\pi_1,\pi_2} &= \sum_{\omega \in \hat{\Omega}} \hat{p}(\omega)\hat{\Delta}^{\pi_1,\pi_2}(\omega) \\ &= \bar{F}^{\pi_1} - \bar{F}^{\pi_2}. \end{aligned}$$

When comparing two policies, it is very important to compute the variance of the estimate of the difference to see if it is statistically significant. If we evaluate each policy using a different set of random outcomes (say, $\omega_1$ and $\omega_2$), the variance of the difference would be given by

$$Var\left[\hat{\Delta}^{\pi_1,\pi_2}\right] = Var\left[\hat{F}^{\pi_1}\right] + Var\left[\hat{F}^{\pi_2}\right]. \tag{4.17}$$

This is generally not the best way to estimate the variance of the difference between two policies. It is better to evaluate two policies using the same random sample for each policy. In this case, $\hat{F}^{\pi_1}(\omega)$ and $\hat{F}^{\pi_2}(\omega)$ are usually correlated, which means the variance would be

$$Var\left[\hat{\Delta}^{\pi_1,\pi_2}\right] = Var\left[\hat{F}^{\pi_1}\right] + Var\left[\hat{F}^{\pi_2}\right] - 2Cov\left[\hat{F}^{\pi_1}, \hat{F}^{\pi_2}\right].$$

The covariance is typically positive, so this estimate of the variance will be smaller (and possibly much smaller). One way to estimate the variance is to compute $\hat{\Delta}^{\pi_1,\pi_2}(\omega)$ for each $\omega \in \hat{\Omega}$ and then compute

$$\bar{\sigma}^{\pi_1,\pi_2} = \frac{1}{N}\left(\frac{1}{N-1}\right)\sum_{\omega\in\hat{\Omega}}\left(\hat{\Delta}^{\pi_1,\pi_2}(\omega) - \bar{\Delta}^{\pi}\right)^2.$$

In general, $\bar{\sigma}^{\pi_1,\pi_2}$ will be much smaller than $2(\bar{\sigma}^{\pi})^2$ which we would obtain if we chose independent estimates.

For some large-scale experiments, it will be necessary to perform comparisons using a single sample realization $\omega$. In fact, this is the strategy that would typically be used if we were solving a steady-state problem. However, the strategy can be used for any problem where the horizon is sufficiently long that the variance of an estimate of the objective function is not too large. We again emphasize that the variance of the difference in the estimates of the contribution of two different policies may be much smaller than would be expected if we used multiple sample realizations to compute the variance of an estimate of the value of a policy.

## 4.8 DYNAMIC PROGRAMMING WITH MISSING OR INCOMPLETE MODELS

Some physical systems are so complex that we cannot describe them with mathematical models, but we are able to observe behaviors directly. Such applications arise in operation settings where a model is running in production, allowing us to observe exogenous outcomes and state transitions from physical processes rather than depending on mathematical equations.

There are three key mathematical models used in dynamic programming. In engineering communities, "model" refers to the transition function (also known as the system model or plant model). Model-free dynamic programming refers to applications where we do not have an explicit transition function (the physical problem may be quite complex). In such problems, we may make a decision but then have to observe the results of the decision from a physical process.

The second "model" refers to the exogenous information process, where we may not have a probability law describing the outcomes. The result is that we will not be able to compute a one-step transition matrix, but it also means that we cannot even run simulations

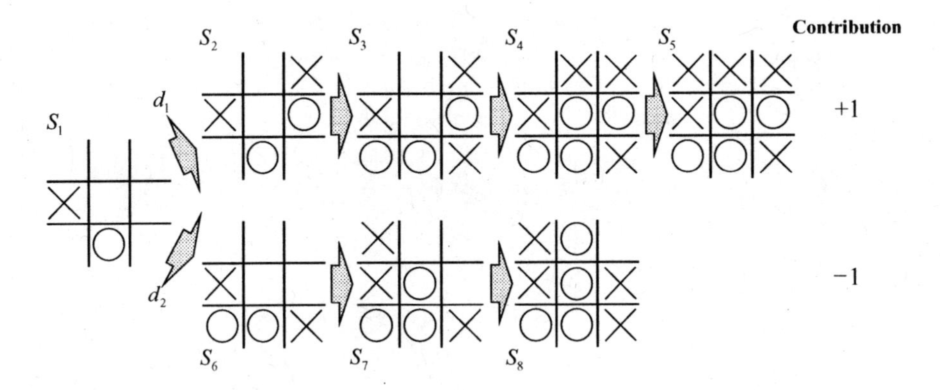


**Figure 4.11** Two outcomes from playing lose tic-tac-toe, where opponent is playing O.

because we do not know the likelihood of an outcome. In such settings, we assume that we have an exogenous process generating outcomes (e.g. stock prices or demands).

The third use of the term model refers to the cost or contribution function. For example, we may have a process where decisions are being made by a human maximizing an unknown utility function. We may have a system where we observe the behavior of a human and infer what action we should be taking given a state.

In an approximate dynamic programming algorithm, we can depend on an exogenous process to give us sample realizations, the cost or contribution function, and the next state we visit. This is possible because the forward nature of approximate dynamic programming algorithms closely matches the physical process. The importance of the issue of missing models depends on the application area. In operations research, it is common to assume that we know the transition function, but we often do not know the probability law for the exogenous information process.

## 4.9 RELATIONSHIP TO REINFORCEMENT LEARNING

A different path of development for dynamic programming emerged from within the artificial intelligence community that found the dynamic programming paradigm a useful framework for thinking about mechanisms for learning human behavior. Consider the problem of training a robot to execute a series of steps to insert a part in a machine or teaching a computer to play a game of backgammon. A rule (or policy) may specify that when the system is in state $s$ (the positions of the pieces on the game board), we should make decision $d$ (given the role of the dice, what move should we make). After a series of moves, we win or lose the game.

We illustrate the process using the game of "tic-tac-toe" where the goal is to get three of your symbol in a row. There are three outcomes: win, lose and draw, which we assign contributions of 1, $-1$ and 0, respectively. Figure 4.11 illustrates two outcomes of the game for player A playing X (who plays first), where his first move brings him to state $S_1$ in the figure. After player A makes his move, we show the response of the opposing player B (as O). We represent the state of the board as the state of each of the nine positions, each of which has three outcomes (X, O, or blank).

From state $S_1$, player A experiments with two different decisions. Initially, he has no basis to choose between decisions $d_1$ and $d_2$ (the value of states $S_2$ and $S_6$ initially are both zero). After this initial decision, the game follows a deterministic path leading to a win (+1) or a loss (−1). Assume that he starts by choosing $d_2$, which eventually leads to a loss for player A (value −1). This means he visits states $S_1, S_6, S_7$ and $S_8$. Let $W^n$ be the winnings in the $n^{th}$ game ($W^1 = -1$). We would update the value of each state using

$$\begin{aligned}
\bar{V}^1(S_8) &= (1-\alpha)\bar{V}^0(S_8) + \alpha W^1 \\
&= (0.5)(0) + (0.5)(-1) \\
&= -0.5, \\
\bar{V}^1(S_7) &= (1-\alpha)\bar{V}^0(S_7) + \alpha \bar{V}^0(S_8) \\
&= 0, \\
\bar{V}^1(S_6) &= (1-\alpha)\bar{V}^0(S_6) + \alpha \bar{V}^0(S_7) \\
&= 0, \\
\bar{V}^1(S_1) &= (1-\alpha)\bar{V}^0(S_1) + \alpha \bar{V}^0(S_6) \\
&= 0.
\end{aligned}$$

In equation (4.18), we smooth the previous estimate of the value of being in state $S_8$, given by $\bar{V}^0(S_8)$, with our observation of our winnings $W^1$. In equation (4.18), we update our estimate of the value of being in state $S_7$ with our current estimate of the value of going to state $S_8$, $\bar{V}^0(S_8)$.

Throughout this exercise, we use a stepsize $\alpha = 0.5$. As shown in table 4.2, we repeat this sequence three times before we finally learn that the value of state $S_5$ is worse than the value of state $S_2$. Starting in iteration 4, we switch to decision $d_1$, after which we no longer update states $S_6$, $S_7$, and $S_8$. After 20 iterations, we have learned that the probability of winning (the value of being in state $S_1$) is approaching 1.0. In a real game, our opponent would behave more randomly, introducing more variability in the updates.

The artificial intelligence community used this strategy for many years to estimate the probability of winning from a particular state. It was only in the 1980's that the probability of winning was viewed as the value of being in a state for an appropriately defined Markov decision process. A unique feature of this class of problems is that the contribution from an action is zero until a particular state is reached (win, lose or draw). This creates challenges in the updating of the value function that we address more thoroughly in future chapters (in chapters 6, 7, and 8).

## 4.10 BUT DOES IT WORK?

The technique of stepping forward through time using Monte Carlo sampling is a powerful strategy, but it effectively replaces the challenge of looping over all possible states with the problem of statistically estimating the value of being in "important states." Furthermore, it is not enough just to get reasonable estimates for the value of being in a state. We have to get reasonable estimates for the value of being in states we *might* want to visit.

As of this writing, formal proofs of convergence are limited to a small number of very specialized algorithms. For lookup-table representations of the value function, it is not possible to obtain convergence proofs without a guarantee that all states will be visited infinitely often. This precludes pure exploitation algorithms where we only use the decision that appears to be the best (given the current value function approximation).

| | | Value of being in state | | | | | | | | |
|---|---|---|---|---|---|---|---|---|---|---|
| Iteration | Decision | $S_1$ | $S_2$ | $S_3$ | $S_4$ | $S_5$ | $S_6$ | $S_7$ | $S_8$ | W |
| 0 | | 0.000 | 0.000 | 0.000 | 0.000 | 0.000 | 0.000 | 0.000 | 0.000 | |
| 1 | $d_2$ | 0.000 | 0.000 | 0.000 | 0.000 | 0.000 | 0.000 | 0.000 | -0.500 | -1 |
| 2 | $d_2$ | 0.000 | 0.000 | 0.000 | 0.000 | 0.000 | 0.000 | -0.250 | -0.500 | -1 |
| 3 | $d_2$ | 0.000 | 0.000 | 0.000 | 0.000 | 0.000 | -0.125 | -0.375 | -0.500 | -1 |
| 4 | $d_1$ | 0.000 | 0.000 | 0.000 | 0.000 | 0.500 | -0.125 | -0.375 | -0.500 | 1 |
| 5 | $d_1$ | 0.000 | 0.000 | 0.000 | 0.250 | 0.750 | -0.125 | -0.375 | -0.500 | 1 |
| 6 | $d_1$ | 0.000 | 0.000 | 0.125 | 0.500 | 0.875 | -0.125 | -0.375 | -0.500 | 1 |
| 7 | $d_1$ | 0.000 | 0.063 | 0.313 | 0.688 | 0.938 | -0.125 | -0.375 | -0.500 | 1 |
| 8 | $d_1$ | 0.031 | 0.188 | 0.500 | 0.813 | 0.969 | -0.125 | -0.375 | -0.500 | 1 |
| 9 | $d_1$ | 0.109 | 0.344 | 0.656 | 0.891 | 0.984 | -0.125 | -0.375 | -0.500 | 1 |
| 10 | $d_1$ | 0.227 | 0.500 | 0.773 | 0.938 | 0.992 | -0.125 | -0.375 | -0.500 | 1 |
| 11 | $d_1$ | 0.363 | 0.637 | 0.855 | 0.965 | 0.996 | -0.125 | -0.375 | -0.500 | 1 |
| 12 | $d_1$ | 0.500 | 0.746 | 0.910 | 0.980 | 0.998 | -0.125 | -0.375 | -0.500 | 1 |
| 13 | $d_1$ | 0.623 | 0.828 | 0.945 | 0.989 | 0.999 | -0.125 | -0.375 | -0.500 | 1 |
| 14 | $d_1$ | 0.726 | 0.887 | 0.967 | 0.994 | 1.000 | -0.125 | -0.375 | -0.500 | 1 |
| 15 | $d_1$ | 0.806 | 0.927 | 0.981 | 0.997 | 1.000 | -0.125 | -0.375 | -0.500 | 1 |
| 16 | $d_1$ | 0.867 | 0.954 | 0.989 | 0.998 | 1.000 | -0.125 | -0.375 | -0.500 | 1 |
| 17 | $d_1$ | 0.910 | 0.971 | 0.994 | 0.999 | 1.000 | -0.125 | -0.375 | -0.500 | 1 |
| 18 | $d_1$ | 0.941 | 0.982 | 0.996 | 1.000 | 1.000 | -0.125 | -0.375 | -0.500 | 1 |
| 19 | $d_1$ | 0.962 | 0.989 | 0.998 | 1.000 | 1.000 | -0.125 | -0.375 | -0.500 | 1 |
| 20 | $d_1$ | 0.975 | 0.994 | 0.999 | 1.000 | 1.000 | -0.125 | -0.375 | -0.500 | 1 |

**Table 4.2** Learning the probability of winning in tic-tac-toe.

Compounding this lack of proofs is experimental work that illustrate cases in which the methods simply do not work. What has emerged from the various laboratories doing experimental work are two themes:

- The functional form of an approximation has to reasonably capture the true value function.
- For large problems, it is essential to exploit the structure of the problem so that a visit to one state provides improved estimates of the value of visiting a large number of other states.

For example, a discrete lookup-table function will always capture the general shape of a (discrete) value function, but it does little to exploit what we have learned from visiting one state in terms of updated estimates of the value of visiting other states. As a result, it is quite easy to design an approximate dynamic programming strategy (using, for example, a lookup-table value function) that either does not work at all, or provides a suboptimal solution that is well below the optimal solution.

At the same time, approximate dynamic programming has proven itself to be an exceptionally powerful tool in the context of specific problem classes. This chapter has illustrated ADP in the context of very simple problems, but it has been successfully applied to very complex resource allocation problems that arise in some of the largest transportation companies. Approximate dynamic programming (and reinforcement learning, as it is still called in artificial intelligence) has proven to be very effective in problems ranging from playing games (such as backgammon) to solving engine control problems (such as managing fuel mixture ratios in engines).

It is our belief that general-purpose results in approximate dynamic programming will be few and far between. Instead, our experience suggests that most results will involve taking advantage of the structure of a particular problem class. Identifying a value function approximation, along with a sampling and updating strategy, that produces a high-quality solution represents a major contribution to the field in which the problem arises. The best we can offer in a general textbook on the field is to provide guiding principles and general tools, allowing domain experts to devise the best possible solution for a particular problem class. We suspect that an ADP strategy applied to a problem context is probably a patentable invention.

## 4.11 BIBLIOGRAPHIC NOTES

Section 4.2 - Research in approximate dynamic programming, which uses statistical methods to approximate value functions, dates to the 1950's when the "curse of dimensionality" was first recognized in the operations research community (Bellman & Dreyfus (1959)). The problems are so important that people independently pursued these methods from within different fields. One of the earliest uses of approximate dynamic programming was training a computer to play a game of checkers (Samuel (1959) and Samuel (1967)). Schweitzer & Seidmann (1985) describes the use of basis functions in the context of the linear programming method (see sections 3.7 and 9.2.5), value iteration and policy iteration. The edited volume by White & Sofge (1992) summarizes an extensive literature in the engineering control community on approximate dynamic programming. Bertsekas & Tsitsiklis (1996) is also written largely from a control perspective, although the influence of problems from operations research and artificial intelligence are apparent. A separate and important line of research grew out of the artificial intelligence community which is nicely summarized in the review by Kaelbling et al. (1996) and the introductory textbook by Sutton & Barto (1998). More recently Si et al. (2004*a*) brought together papers from the engineering controls community, artificial intelligence, and operations research. Reviews are also given in Tsitsiklis & Van Roy (1996) and Van Roy (2001).

Section 4.3 - There is a very rich literature on generating random variables. An easy introduction to simulation is given in Ross (2002). More extensive treatments are given in Banks et al. (1996) and Law & Kelton (2000).

Section 4.4 - Virtually every textbook on dynamic programming (or approximate dynamic programming) sets up Bellman's equations around the pre-decision state variable. However, a number of authors have found that some problems are naturally formulated around the post-decision state variable. The first use of the term "post-decision state variable" that we have found is in Van Roy et al. (1997), although uses of the post-decision state variable date as far back as Bellman (1957). The first presentation describing the post-decision state as a general solution strategy (as opposed to a technique for a special problem class) appears to be Powell & Van Roy (2004). This volume is the first book to systematically use the post-decision state variable as a way of approximating dynamic programs.

Section 4.7 - The A* algorithm is presented in a number of books on artificial intelligence. See, for example, Pearl (1984).

Section 4.7.2 - The concept of synchronous and asynchronous dynamic programming is based on Bertsekas (1982) and in particular Bertsekas & Tsitsiklis (1989). Real-time dynamic programming was introduced in Barto et al. (1995).

## PROBLEMS

**4.1** Let $U$ be a random variable that is uniformly distributed between 0 and 1. Let $R = -\frac{1}{\lambda} \ln U$. Show that $\mathbb{P}[R \leq x] = 1 - e^{-\lambda x}$, which shows that $R$ has an exponential distribution.

**4.2** Let $R = U_1 + U_2$ where $U_1$ and $U_2$ are independent, uniformly distributed random variables between 0 and 1. Derive the probability density function for $R$.

**4.3** Let $Z$ be a normally distributed random variable with mean 0 and variance 1, and let $U$ be a uniform $[0, 1]$ random variable. Let $\Phi(z) = \mathbb{P}(Z \leq z)$ be the cumulative distribution and let $\Phi^{-1}(u)$ be the inverse cumulative. Then $R = \Phi^{-1}(U)$ is also normally distributed with mean 0 and variance 1. Use a spreadsheet to randomly generate 10,000 observations of $U$ and the associated observations of $R$. Let $N(z)$ be the number of observations of $R$ that are less than $z$. Compare $N(z)$ to $10000\Phi(z)$ for $z = -2, -1, -.5, 0, .5, 1, 2$. [Note that in a spreadsheet, $Rand()$ generates a uniform $[0, 1]$ random variable, $\Phi(z) = NORMSDIST(z)$ and $\Phi^{-1}(u) = NORMSINV(u)$.]

**4.4** Let $X$ be a continuous random variable. Let $F(x) = 1 - \mathbb{P}[X \leq x]$, and let $F^{-1}(y)$ be its inverse (that is, if $y = F(x)$, then $F^{-1}(y) = x$). Show that $F^{-1}(U)$ has the same distribution as $X$.

**4.5** You are holding an asset that can be sold at time $t = 1, 2, \ldots, 10$ at a price $\hat{p}_t$ which fluctuates from period to period. Assume that $\hat{p}_t$ is uniformly distributed between 0 and 10. Further assume that $\hat{p}_t$ is independent of prior prices. You have 10 opportunities to sell the asset and you must sell it by the end of the $10^{th}$ time period. Let $x_t$ be the decision variable, where $x_t = 1$ means sell and $x_t = 0$ means hold. Assume that the reward you receive is the price that you receive when selling the asset ($\sum_{t=1}^{10} x_t p_t$).

(a) Define the pre- and post-decision state variables for this problem. Plot the shape of the value function around the pre- and post-decision state variables.

(b) Set up the optimality equations and show that there exists a price $\bar{p}_t$ where we will sell if $p_t > \bar{p}_t$. Also show that $\bar{p}_t \geq \bar{p}_{t+1}$.

(c) Use backward dynamic programming to compute the optimal value function for each time period.

(d) Use an approximate dynamic programming algorithm with a pre-decision state variable to estimate the optimal value function, and give your estimate of the optimal policy (that is, at each price, should we sell or hold). Note that even if you sell the asset at time $t$, you need to stay in the state that you are holding the asset for $t + 1$ (we are not interested in estimating the value of being in the state that we have sold the asset). Run the algorithm for 1000 iterations using a stepsize of $\alpha_{n-1} = 1/n$, and compare the results to that obtained using a constant stepsize $\alpha = 0.20$.

**4.6** Assume that we have an asset selling problem where you can sell at price $p_t$ that evolves according to

$$p_{t+1} = p_t + .5(120 - p_t) + \hat{p}_{t+1}$$

where $p_0 = 100$ and $\hat{p}_{t+1}$ is uniformly distributed between -10 and +10. Assume that we have to sell the asset within the first 10 time periods. Solve the problem using approximate dynamic programming. Unlike 4.5, now $p_t$ is part of the state variable, and we have to estimate $V_t(p_t)$ (the value of holding the asset when the price is $p_t$). Since $p_t$ is continuous, define your value function approximation by discretizing $p_t$ to obtain $\bar{p}_t$ (e.g. rounded to the nearest dollar or nearest five dollars). After training the value function for 10,000 iterations, run 1000 samples (holding the value function fixed) and determine when the model decided to sell the asset. Plot the distribution of times that the asset was held over these 1000 realizations. If $\delta$ is your discretization parameter (e.g., $\delta = 5$ means rounding to the nearest 5 dollars), compare your results for $\delta = 1, 5$ and 10.

**4.7** Here we are going to solve a variant of the asset selling problem using a post-decision state variable. We assume we are holding a real asset and we are responding to a series of offers. Let $\hat{p}_t$ be the $t^{th}$ offer, which is uniformly distributed between 500 and 600 (all prices are in thousands of dollars). We also assume that each offer is independent of all prior offers. You are willing to consider up to 10 offers, and your goal is to get the highest possible price. If you have not accepted the first nine offers, you must accept the $10^{th}$ offer.

(a) Write out the decision function you would use in an approximate dynamic programming algorithm in terms of a Monte Carlo sample of the latest price and a current estimate of the value function approximation.

(b) Use the knowledge that $\hat{p}_t$ is uniform between 500 and 600 and derive the exact value of holding the asset after each offer.

(c) Write out the updating equations (for the value function) you would use after solving the decision problem for the $t^{th}$ offer using Monte Carlo sampling.

(d) Implement an approximate dynamic programming algorithm using *synchronous* state sampling (you sample both states at every iteration). Using 100 iterations, write out your estimates of the value of being in each state immediately after each offer.

(e) From your value functions, infer a decision rule of the form "sell if the price is greater than $\bar{p}_t$."

**4.8** We are going to use approximate dynamic programming to estimate

$$F^T = \mathbb{E}\sum_{t=0}^{T} \gamma^t R_t$$

where $R_t$ is a random variable that is uniformly distributed between 0 and 100 and $\gamma = 0.7$. We assume that $R_t$ is independent of prior history. We can think of this as a single-state Markov decision process with no decisions.

(a) Using the fact that $\mathbb{E}R_t = 50$, give the exact value for $F^{20}$.

(b) Let $\hat{v}_t^n = \sum_{t'=t}^{T} \gamma^{t'-t} R_{t'}(\omega^n)$, where $\omega^n$ represents the $n^{th}$ sample path and $R_{t'}^n = R_{t'}(\omega^n)$ is the realization of the random variable $R_{t'}$ for the $n^{th}$ sample path. Show that $\hat{v}_t^n = R_t^n + \gamma \hat{v}_{t+1}^n$ which means that $\hat{v}_0^n$ is a sample realization of $R^T$.

(c) Propose an approximate dynamic programming algorithm to estimate $F^T$. Give the value function updating equation, using a stepsize $\alpha_{n-1} = 1/n$ for iteration $n$.

(d) Perform 100 iterations of the approximate dynamic programming algorithm to produce an estimate of $F^{20}$. How does this compare to the true value?

(e) Repeat part (d) 10 times, but now use a discount factor of 0.9. Average the results to obtain an averaged estimate. Now use these 10 calculations to produce an estimate of the standard deviation of your average.

(f) From your answer to part (e), estimate how many iterations would be required to obtain an estimate where 95 percent confidence bounds would be within two percent of the true number.

**4.9** Consider a batch replenishment problem where we satisfy demand in a period from the available inventory at the beginning of the period and then order more inventory at the end of the period. Define both the pre- and post-decision state variables, and write the pre-decision and post-decision forms of the transition equations.

**4.10** A mutual fund has to maintain a certain amount of cash on hand for redemptions. The cost of adding funds to the cash reserve is \$250 plus \$0.005 per dollar transferred into the reserve. The demand for cash redemptions is uniformly distributed between \$5,000 and \$20,000 per day. The mutual fund manager has been using a policy of transferring in \$500,000 whenever the cash reserve goes below \$250,000. He thinks he can lower his transaction costs by transferring in \$650,000 whenever the cash reserve goes below \$250,000. However, he pays an opportunity cost of \$0.00005 per dollar per day for cash in the reserve account.

(a) Use a spreadsheet to set up a simulation over 1000 days to estimate total transaction costs plus opportunity costs for the policy of transferring in \$500,000, assuming that you start with \$500,000 in the account. Perform this simulation 10 times, and compute the sample average and variance of these observations. Then compute the variance and standard deviation of your sample average.

(b) Repeat (a), but now test the policy of transferring in \$650,000. Repeat this 1000 times and compute the sample average and the standard deviation of this average. Use a new set of observations of the demands for capital.

(c) Can you conclude which policy is better at a 95 percent confidence level? Estimate the number of observations of each that you think you would need to conclude there is a difference at a 95 percent confidence level.

(d) How does your answer to (d) change if instead of using a fresh set of observations for each policy, you use the same set of random demands for each policy?

**4.11** Let $\mathcal{M}$ be the dynamic programming "max" operator:

$$\mathcal{M}v(s) = \max_x \left( C(s,x) + \gamma \sum_{s'} \mathbb{P}(s'|s,x)v(s') \right).$$

Let

$$\bar{c} = \max_s \left( \mathcal{M}v(s) - v(s) \right)$$

and define a policy $\pi(v)$ using

$$x(s) = \arg\max_x \left( C(s,x) + \gamma \mathbb{E} \sum_{s'} \mathbb{P}(s'|s,x) v(s') \right).$$

Let $v^\pi$ be the value of this policy. Show that

$$v^\pi \leq v + \frac{\bar{c}}{1-\gamma} e,$$

where $e$ is a vector of 1's.

**4.12** As a broker, you sell shares of a company you think your customers will buy. Each day, you start with $s_{t-1}$ shares left over from the previous day, and then buyers for the brokerage add $r_t$ new shares to the pool of shares you can sell that day ($r_t$ fluctuates from day to day). Each day, customer $i \in \mathcal{I}$ calls in asking to purchase $q_{ti}$ units at price $p_{ti}$, but you do not confirm orders until the end of the day. Your challenge is to determine $x_t$ which is the number of shares you wish to sell to the market on that day. You will collect all the orders. Then, at the end of the day, you must choose $x_t$, which you will then allocate among the customers willing to pay the highest price. Any shares you do not sell on one day are available for sale the next day. Let $p_t = (p_{ti})_{i \in \mathcal{I}}$ and $q_t = (q_{ti})_{i \in \mathcal{I}}$ be the vectors of price/quantity offers from all the customers on a given day.

(a) What is the exogenous information process for this system? What is a history for the process?

(b) Give a general definition of a state variable. Is $s_t$ a valid state variable given your definition?

(c) Set up the optimality equations for this problem using the post-decision state variable. Be precise!

(d) Set up the optimality equations for this problem using the pre-decision state variable. Contrast the two formulations from a computational perspective.

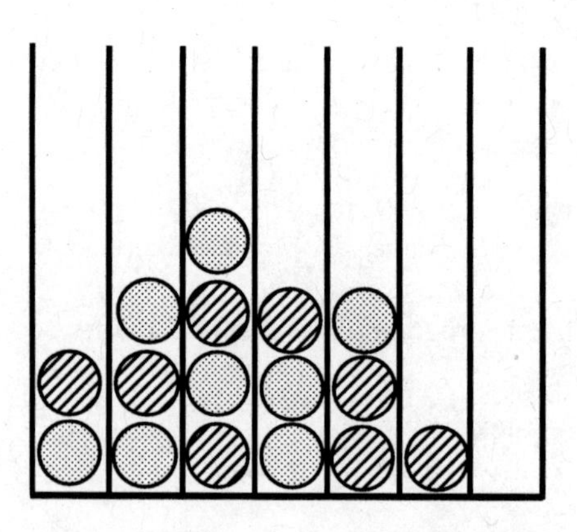

**Figure 4.12** Connect 4 playing grid

**4.13 Connect-4 project** - Connect-4 is a game played where players alternatively drop in chips with two colors (one for each player), as illustrated in figure 4.12. The grid is six high by seven wide. Chips are dropped into the top of a column where they then fall to the last available spot. The goal is to get four in a row (vertically, horizontally or diagonally) of the same color. Since this is a two-player game, you will have to train two value functions - one for the player that starts first, and one for the player who starts second. Once these are trained, you should be able to play against it.

Training a value function for board games will require playing the game a few times, and identifying specific patterns which should be captured. These patterns become your features (also known as *basis functions*). These could be indicator variables (does your opponent have three in a row) or a numerical quantity (how many pieces are in a given row or column). The simplest representation is a separate indicator variable for each cell, producing a value function with 42 parameters.

You must write software that allows a human to play your model after your value functions have been trained. Allow a person to select which column to place his piece (no need for fancy graphics - we can keep track of the game board on a sheet of paper or using a real game piece).

CHAPTER 5

# MODELING DYNAMIC PROGRAMS

Perhaps one of the most important skills to develop in approximate dynamic programming is the ability to write down a model of the problem. Everyone who wants to solve a linear program learns to write out

$$\min_x c^T x$$

subject to

$$\begin{aligned} Ax &= b, \\ x &\geq 0. \end{aligned}$$

This standard modeling framework allows people around the world to express their problem in a standard format.

Stochastic, dynamic problems are much richer than a linear program, and require the ability to model the flow of information and complex system dynamics. Just the same, there is a standard framework for modeling dynamic programs. We provided a taste of this framework in chapter 2, but that chapter only hinted at the richness of the problem class.

In chapters 2, 3 and 4, we used fairly standard notation, and have avoided discussing some important subtleties that arise in the modeling of stochastic, dynamic systems. We intentionally overlooked trying to define a state variable, which we have viewed as simply $S_t$, where the set of states was given by the indexed set $\mathcal{S} = \{1, 2, \ldots, |\mathcal{S}|\}$. We have avoided discussions of how to properly model time or more complex information processes. This style has facilitated introducing some basic ideas in dynamic programming, but would severely limit our ability to apply these methods to real problems.

The goal of this chapter is to describe a standard modeling framework for dynamic programs, providing a vocabulary that will allow us to take on a much wider set of applications. Notation is not as critical for simple problems, as long as it is precise and consistent. But what seems like benign notational decisions for a simple problem can cause unnecessary difficulties, possibly making the model completely intractable as problems become more complex. Complex problems require considerable discipline in notation because they combine the details of the original physical problem with the challenge of modeling sequential information and decision processes. The modeling of time can be particularly subtle. In addition to a desire to model problems accurately, we also need to be able to understand and exploit the structure of the problem, which can become lost in a sea of complex notation.

Good modeling begins with good notation. The choice of notation has to balance traditional style with the needs of a particular problem class. Notation is easier to learn if it is mnemonic (the letters look like what they mean) and compact (avoiding a profusion of symbols). Notation also helps to bridge communities. For example, it is common in dynamic programming to refer to actions using "$a$" (where $a$ is discrete); in control theory a decision (control) is "$u$" (which may be continuous). For high-dimensional problems, it is essential to draw on the field of mathematical programming, where decisions are typically written as "$x$" and resource constraints are written in the standard form $Ax = b$. In this text, many of our problems involve managing resources where we are trying to maximize or minimize an objective subject to constraints. For this reason, we adopt, as much as possible, the notation of math programming to help us bridge the fields of math programming and dynamic programming.

Sections 5.1 to 5.3 provide some foundational material. Section 5.1 begins by describing some basic guidelines for notational style. Section 5.2 addresses the critical question of modeling time, and section 5.3 provides notation for modeling resources that we will use throughout the remainder of the volume.

The general framework for modeling a dynamic program is covered in sections 5.4 to 5.9. There are six elements to a dynamic program, consisting of the following:

**1)** State variables - These describe what we need to know at a point in time (section 5.4).

**2)** Decision variables - These are the variables we control. Choosing these variables ("making decisions") represents the central challenge of dynamic programming (section 5.5).

**3)** Exogenous information processes - These variables describe information that arrives to us exogenously, representing the sources of randomness (section 5.6).

**4)** Transition function - This is the function that describes how the state evolves from one point in time to another (section 5.7).

**5)** Contribution function - We are either trying to maximize a contribution function (profits, revenues, rewards, utility) or minimize a cost function. This function describes how well we are doing at a point in time (section 5.8).

**6)** Objective function - This is what we are maximizing or minimizing over time (section 5.9).

This chapter describes modeling in considerable depth, and as a result it is quite long. A number of sections are marked with a '*', indicating that these can be skipped on a first read. There is a single section marked with a '**' which, as with all sections marked this way, is material designed for readers with more advanced training in probability and stochastic processes.

## 5.1 NOTATIONAL STYLE

Notation is a language: the simpler the language, the easier it is to understand the problem. As a start, it is useful to adopt notational conventions to simplify the style of our presentation. For this reason, we adopt the following notational conventions:

Variables - Variables are *always* a single letter. We would never use, for example, $CH$ for "holding cost."

Modeling time - We always use $t$ to represent a point in time, while we use $\tau$ to represent an interval over time. When we need to represent different points in time, we might use $t, t', \bar{t}, t^{max}$, and so on.

Indexing vectors - Vectors are almost always indexed in the subscript, as in $x_{ij}$. Since we use discrete time models throughout, an activity at time $t$ can be viewed as an element of a vector. When there are multiple indices, they should be ordered from outside in the general order over which they might be summed (think of the outermost index as the most detailed information). So, if $x_{tij}$ is the flow from $i$ to $j$ at time $t$ with cost $c_{tij}$, we might sum up the total cost using $\sum_t \sum_i \sum_j c_{tij} x_{tij}$. Dropping one or more indices creates a vector over the elements of the missing indices to the right. So, $x_t = (x_{tij})_{\forall i, \forall j}$ is the vector of all flows occurring at time $t$. If we write $x_{ti}$, this would be the vector of flows out of $i$ at time $t$ to all destinations $j$. Time, when present, is always the innermost index.

Indexing time - If we are modeling activities in discrete time, then $t$ is an index and should be put in the subscript. So $x_t$ would be an activity at time $t$, with the vector $x = (x_1, x_2, \ldots, x_t, \ldots, x_T)$ giving us all the activities over time. When modeling problems in continuous time, it is more common to write $t$ as an argument, as in $x(t)$. $x_t$ is notationally more compact (try writing a complex equation full of variables written as $x(t)$ instead of $x_t$).

Flavors of variables - It is often the case that we need to indicate different flavors of variables, such as holding costs and order costs. These are always indicated as superscripts, where we might write $c^h$ or $c^{hold}$ as the holding cost. Note that while variables must be a single letter, superscripts may be words (although this should be used sparingly). We think of a variable like "$c^h$" as a single piece of notation. It is better to write $c^h$ as the holding cost and $c^p$ as the purchasing cost than to use $h$ as the holding cost and $p$ as the purchasing cost (the first approach uses a single letter $c$ for cost, while the second approach uses up two letters - the roman alphabet is a scarce resource). Other ways of indicating flavors is hats ($\hat{x}$), bars ($\bar{x}$), tildes ($\tilde{x}$) and primes ($x'$).

Iteration counters - We place iteration counters in the superscript, and we primarily use $n$ as our iteration counter. So, $x^n$ is our activity at iteration $n$. If we are using a descriptive superscript, we might write $x^{h,n}$ to represent $x^h$ at iteration $n$. Sometimes algorithms require inner and outer iterations. In this case, we use $n$ to index the outer iteration and $m$ for the inner iteration. While this will prove to be the most natural way to index iterations, there is potential for confusion where it may not be clear if the superscript $n$ is an index (as we view it) or raising a variable to the $n^{th}$ power. We make one notable exception to our policy of indexing iterations in the superscript. In approximate dynamic programming, we make wide use of a parameter known as a stepsize $\alpha$ where $0 \leq \alpha \leq 1$. We often make the stepsize vary with the iterations. However, writing $\alpha^n$ looks too much like raising the stepsize to the power of $n$. Instead, we write $\alpha_n$ to indicate the stepsize in iteration $n$. This is our only exception to this rule.

Sets are represented using capital letters in a calligraphic font, such as $\mathcal{X}, \mathcal{F}$ or $\mathcal{I}$. We generally use the lowercase roman letter as an element of a set, as in $x \in \mathcal{X}$ or $i \in \mathcal{I}$.

Exogenous information - Information that first becomes available (from outside the system) at time $t$ is denoted using hats, for example, $\hat{D}_t$ or $\hat{p}_t$. Our only exception to this rule is $W_t$ which is our generic notation for exogenous information (since $W_t$ *always* refers to exogenous information, we do not use a hat).

Statistics - Statistics computed using exogenous information are generally indicated using bars, for example $\bar{x}_t$ or $\bar{V}_t$. Since these are functions of random variables, they are also random.

Index variables - Throughout, $i, j, k, l, m$ and $n$ are always scalar indices.

Of course, there are exceptions to every rule. It is extremely common in the transportation literature to model the flow of a type of resource (called a commodity and indexed by $k$) from $i$ to $j$ using $x_{ij}^k$. Following our convention, this should be written $x_{kij}$. Authors need to strike a balance between a standard notational style and existing conventions.

## 5.2 MODELING TIME

A survey of the literature reveals different styles toward modeling time. When using discrete time, some authors start at 1 while others start at zero. When solving finite horizon problems, it is popular to index time by the number of time periods remaining, rather than elapsed time. Some authors index a variable, say $S_t$, as being a function of information up through $t-1$, while others assume it includes information up through time $t$. $t$ may be used to represent when a physical event actually happens, or when we first know about a physical event.

The confusion over modeling time arises in large part because there are two processes that we have to capture: the flow of information, and the flow of physical and financial resources. There are many applications of dynamic programming to deterministic problems where the flow of information does not exist (everything is known in advance). Similarly, there are many models where the arrival of the information about a physical resource, and when the information takes effect in the physical system, are the same. For example, the time at which a customer physically arrives to a queue is often modeled as being the same as when the information about the customer first arrives. Similarly, we often assume that we can sell a resource at a market price as soon as the price becomes known.

There is a rich collection of problems where the information process and physical process are different. A buyer may purchase an option now (an information event) to buy a commodity in the future (the physical event). Customers may call an airline (the

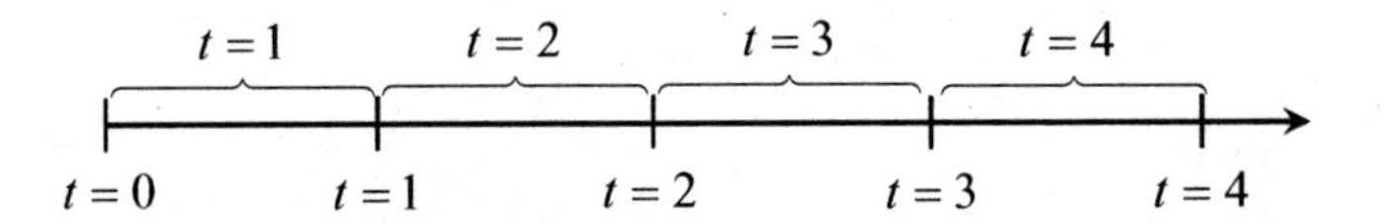


5.1a: Information processes

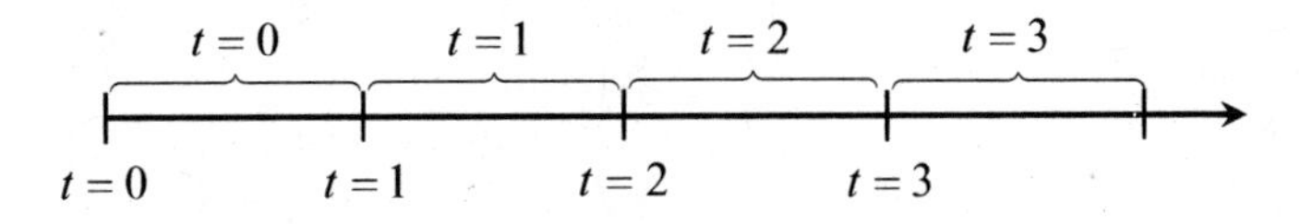


5.1b: Physical processes

**Figure 5.1** Relationship between discrete and continuous time for information processes (5.1a) and physical processes (5.1b).

information event) to fly on a future flight (the physical event). An electric power company has to purchase equipment now to be used one or two years in the future. All of these problems represent examples of *lagged information processes* and force us to explicitly model the informational and physical events (see section 2.2.7 for an illustration).

Notation can easily become confused when an author starts by writing down a deterministic model of a physical process, and then adds uncertainty. The problem arises because the proper convention for modeling time for information processes is different than what should be used for physical processes.

We begin by establishing the relationship between discrete and continuous time. All of the models in this book assume that decisions are made in discrete time (sometimes referred to as *decision epochs*). However, the flow of information, and many of the physical processes being modeled, are best viewed in continuous time. A common error is to assume that when you model a dynamic program in discrete time then all events (information events and physical events) are also occurring in discrete time (in some applications, this is the case). Throughout this volume, decisions are made in discrete time, while all other activities occur in continuous time.

The relationship of our discrete time approximation to the real flow of information and physical resources is depicted in figure 5.1. Above the line, "$t$" refers to a time interval while below the line, "$t$" refers to a point in time. When we are modeling information, time $t = 0$ is special; it represents "here and now" with the information that is available at the moment. The discrete time $t$ refers to the time interval from $t-1$ to $t$ (illustrated in figure 5.1a). This means that the first new information arrives during time interval 1. This notational style means that any variable indexed by $t$, say $S_t$ or $x_t$, is assumed to have access to the information that arrived up to time $t$, which means up through time interval $t$. This property will dramatically simplify our notation in the future. For example, assume that $f_t$ is our forecast of the demand for electricity. If $\hat{D}_t$ is the observed demand during time interval $t$, we would write our updating equation for the forecast using

$$f_t = (1-\alpha)f_{t-1} + \alpha\hat{D}_t. \tag{5.1}$$

We refer to this form as the *informational representation*. Note that the forecast $f_t$ is written as a function of the information that became available during time interval $t$.

When we are modeling a physical process, it is more natural to adopt a different convention (illustrated in figure 5.1b): discrete time $t$ refers to the time interval between $t$ and $t+1$. This convention arises because it is most natural in deterministic models to use time to represent when something is happening or when a resource can be used. For example, let $R_t$ be our cash on hand that we can use during day $t$ (implicitly, this means that we are measuring it at the beginning of the day). Let $\hat{D}_t$ be the demand for cash during the day, and let $x_t$ represent additional cash that we have decided to add to our balance (to be used during day $t$). We can model our cash on hand using

$$R_{t+1} = R_t + x_t - \hat{D}_t. \tag{5.2}$$

We refer to this form as the *actionable representation*. Note that the left-hand side is indexed by $t+1$, while all the quantities on the right-hand side are indexed by $t$. This equation makes perfect sense when we interpret time $t$ to represent when a quantity can be used. For example, many authors would write our forecasting equation (5.1) as

$$f_{t+1} = (1-\alpha)f_t + \alpha\hat{D}_t. \tag{5.3}$$

This equation is correct if we interpret $f_t$ as the forecast of the demand that will happen in time interval $t$.

A review of the literature quickly reveals that both modeling conventions are widely used. It is important to be aware of the two conventions and how to interpret them. We handle the modeling of informational and physical processes by using two time indices, a form that we refer to as the "$(t, t')$" notation. For example,

$\hat{D}_{tt'}$ = The demands that first become known during time interval $t$ to be served during time interval $t'$.

$f_{tt'}$ = The forecast for activities during time interval $t'$ made using the information available up through time $t$.

$R_{tt'}$ = The resources on hand at time $t$ that cannot be used until time $t'$.

$x_{tt'}$ = The decision to purchase futures at time $t$ to be exercised during time interval $t'$.

For each variable, $t$ indexes the information content (literally, when the variable is measured or computed), while $t'$ represents the time at which the activity takes place. Each of these variables can be written as vectors, such as

$$\begin{aligned}
\hat{D}_t &= (\hat{D}_{tt'})_{t' \geq t}, \\
f_t &= (f_{tt'})_{t' \geq t}, \\
x_t &= (x_{tt'})_{t' \geq t}, \\
R_t &= (R_{tt'})_{t' \geq t}.
\end{aligned}$$

Note that these vectors are now written in terms of the information content. For stochastic problems, this style is the easiest and most natural. If we were modeling a deterministic problem, we would drop the first index "$t$" and model the entire problem in terms of the second index "$t'$."

Each one of these quantities is computed at the end of time interval $t$ (that is, with the information up through time interval $t$) and represents a quantity that can be used at time

$t'$ in the future. We could adopt the convention that the first time index uses the indexing system illustrated in figure 5.1a, while the second time index uses the system in figure 5.1b. While this convention would allow us to easily move from a natural deterministic model to a natural stochastic model, we suspect most people will struggle with an indexing system where time interval $t$ in the information process refers to time interval $t-1$ in the physical process. Instead, we adopt the convention to model information in the most natural way, and live with the fact that product arriving at time $t$ can only be used during time interval $t+1$.

Using this convention it is instructive to interpret the special case where $t = t'$. $\hat{D}_{tt}$ is simply demands that arrive during time interval $t$, where we first learn of them when they arrive. $f_{tt}$ makes no sense, because we would never forecast activities during time interval $t$ after we have this information. $R_{tt}$ represents resources that we know about during time interval $t$ and which can be used during time interval $t$. Finally, $x_{tt}$ is a decision to purchase resources to be used during time interval $t$ given the information that arrived during time interval $t$. In financial circles, this is referred to as purchasing on the spot market.

The most difficult notational decision arises when first starting to work on a problem. It is natural at this stage to simplify the problem (often, the problem *appears* simple) and then choose notation that seems simple and natural. If the problem is deterministic and you are quite sure that you will never solve a stochastic version of the problem, then the actionable representation (figure 5.1b) and equation (5.2) is going to be the most natural. Otherwise, it is best to choose the informational format. If you do not have to deal with lagged information processes (e.g., ordering at time $t$ to be used at some time $t'$ in the future) you should be able to get by with a single time index, but you need to remember that $x_t$ may mean purchasing product to be used during time interval $t+1$.

Care has to be used when taking expectations of functions. Consider what happens when we want to know the expected costs to satisfy customer demands $\hat{D}_t$ that arose during time interval $t$ given the decision $x_{t-1}$ we made at time $t-1$. We would have to compute $\mathbb{E}\{C_t(x_{t-1}, \hat{D}_t)\}$, where the expectation is over the random variable $\hat{D}_t$. The function that results from taking the expectation is now a function of information up through time $t-1$. Thus, we might use the notation

$$\bar{C}_{t-1}(x_{t-1}) = \mathbb{E}\{C_t(x_{t-1}, \hat{D}_t)\}.$$

This can take a little getting used to. The costs are incurred during time interval $t$, but now we are indexing the function with time $t-1$. The problem is that if we use a single time index, we are not capturing when the activity is actually happening. An alternative is to switch to a double time index, as in

$$\bar{C}_{t-1,t}(x_{t-1}) = \mathbb{E}\{C_t(x_{t-1}, \hat{D}_t)\},$$

where $\bar{C}_{t-1,t}(x_{t-1})$ is the expected costs that will be incurred during time interval $t$ using the information known at time $t-1$.

## 5.3 MODELING RESOURCES

There is a vast range of problems that can be broadly described in terms of managing "resources." Resources can be equipment, people, money, robots or even games such as backgammon or chess. Depending on the setting, we might use the term asset (financial applications, expensive equipment) or entity (managing a robot, playing a board game). It is very common to start with fairly simple models of these problems, but the challenge is to solve complex problems.

There are four important problem classes that we consider in this volume, each offering unique computational challenges. These can be described along two dimensions: the number of resources or entities being managed, and the complexity of the attributes of each resource or entity. We may be managing a single entity (a robot, an aircraft, an electrical power plant) or multiple entities (a fleet of aircraft, trucks or locomotives, funds in different asset classes, groups of people). The attributes of each entity or resource may be simple (the truck may be described by its location, the money by the asset class into which it has been invested) or complex (all the characteristics of an aircraft or pilot).

The computational implications of each problem class can be quite different. Not surprisingly, different communities tend to focus on specific problem classes, making it possible for people to make the connection between mathematical notation (which can be elegant but vague) and the characteristics of a problem. Problems involving a single, simple entity can usually be solved using the classical techniques of Markov decision processes (chapter 3), although even here some problems can be difficult. The artificial intelligence community often works on problems involving a single, complex entity (games such as Connect-4 and backgammon, moving a robot arm or flying an aircraft). The operations research community has major subcommunities working on problems that involve multiple, simple entities (multicommodity flow problems, inventory problems), while portions of the simulation community and math programming community (for deterministic problems) will work on applications with multiple, complex entities.

In this section, we describe notation that allows us to evolve from simple to complex problems in a natural way. Our goal is to develop mathematical notation that does a better job of capturing which problem class we are working on.

### 5.3.1 Modeling a single, discrete resource

Many problems in dynamic programming involve managing a single resource such as flying an aircraft, planning a path through a network, or planning a game of chess or backgammon. These problems are distinctly different than those involving the management of fleets of vehicles, inventories of blood or groups of people. For this reason, we adopt specific notation for the single entity problem.

If we are managing a single, discrete resource, we find it useful to introduce specific notation to model the attributes of the resource. For this purpose, we use

$$
\begin{aligned}
a_t &= \text{The attribute vector of the resource at time } t \\
&= (a_1, a_2, \ldots, a_N), \\
\mathcal{A} &= \text{Set of all possible attribute vectors.}
\end{aligned}
$$

Attributes might be discrete $(0, 1, 2, \ldots)$, continuous $(0 \leq a_i \leq 1)$ or categorical $(a_i = \text{red})$. We typically assume that the number of dimensions of $a_t$ is not too large. For example, if we are modeling the flow of product through a supply chain, the attribute vector might consist of the product type and location. If we are playing chess, the attribute vector would have 64 dimensions (the piece on each square of the board).

### 5.3.2 Modeling multiple resources

Imagine that we are modeling a fleet of unmanned aerial vehicles (UAV's), which are robotic aircraft used primarily for collecting information. We can let $a_t$ be the attributes of a single UAV at time $t$, but we would like to describe the collective attributes of a fleet of UAV's. There is more than one way to do this, but one way that will prove to be notationally convenient is to define

$$
\begin{aligned}
R_{ta} &= \text{The number of resources with attribute } a \text{ at time } t, \\
R_t &= (R_{ta})_{a \in \mathcal{A}}.
\end{aligned}
$$

$R_t$ is known as the *resource state vector*. If $a$ is a vector, then $|\mathcal{A}|$ may be quite large. It is not hard to create problems where $R_t$ has hundreds of thousands of dimensions. If elements of $a_t$ are continuous, then in theory at least, $R_t$ is infinite-dimensional. It is important to emphasize that in such cases, we would never enumerate the entire vector of elements in $R_t$.

We note that we can use this notation to model a single resource. Instead of letting $a_t$ be our state vector (for the single resource), we let $R_t$ be the state vector, where $\sum_{a \in \mathcal{A}} R_{ta} = 1$. This may seem clumsy, but it offers notational advantages we will exploit from time to time.

### 5.3.3 Illustration: the nomadic trucker

The "nomadic trucker" is a colorful illustration of a multiattribute resource which helps to illustrate some of the modeling conventions being introduced in this chapter. We use this example to illustrate different issues that arise in approximate dynamic programming, leading up to the solution of large-scale resource management problems later in our presentation.

The problem of the nomadic trucker arises in what is known as the truckload trucking industry. In this industry, a truck driver works much like a taxicab. A shipper will call a truckload motor carrier and ask it to send over a truck. The driver arrives, loads up the shipper's freight and takes it to the destination where it is unloaded. The shipper pays for the entire truck, so the carrier is not allowed to consolidate the shipment with freight from other shippers. In this sense, the trucker works much like a taxicab for people. However, as we will soon see, our context of the trucking company adds an additional level of richness that offers some relevant lessons for dynamic programming.

Our trucker runs around the United States, where we assume that his location is one of the 48 contiguous states. When he arrives in a state, he sees the customer demands for loads to move from that state to other states. There may be none, one, or several. He may choose a load to move if one is available; alternatively, he has the option of doing nothing or moving empty to another state (even if a load is available). Once he moves out of a state, all other customer demands (in the form of loads to be moved) are assumed to be picked up by other truckers and are therefore lost. He is not able to see the availability of loads out of states other than where he is located.

Although truckload motor carriers can boast fleets of over 10,000 drivers, our model focuses on the decisions made by a single driver. There are, in fact, thousands of trucking "companies" that consist of a single driver. In chapter 12 we will show that the concepts we develop here form the foundation for managing the largest and most complex versions of this problem. For now, our "nomadic trucker" represents a particularly effective way of illustrating some important concepts in dynamic programming.

### *A basic model*

The simplest model of our nomadic trucker assumes that his only attribute is his location, which we assume has to be one of the 48 contiguous states. We let

$$\mathcal{I} = \text{The set of "states" (locations) at which the driver can be located.}$$

We use $i$ and $j$ to index elements of $\mathcal{I}$. His attribute vector then consists of

$$a = \{i\}.$$

In addition to the attributes of the driver, we also have to capture the attributes of the loads that are available to be moved. For our basic model, loads are characterized only by where they are going. Let

$$\begin{aligned} b &= \text{The vector of characteristics of a load} \\ &= \begin{pmatrix} b_1 \\ b_2 \end{pmatrix} = \begin{pmatrix} \text{The origin of the load.} \\ \text{The destination of the load.} \end{pmatrix}. \end{aligned}$$

We let $\mathcal{A}$ be the set of all possible values of the driver attribute vector $a$, and we let $\mathcal{B}$ be the set of all possible load attribute vectors $b$.

### *A more realistic model*

We need a richer set of attributes to capture some of the realism of an actual truck driver. To begin, we need to capture the fact that at a point in time, a driver may be in the process of moving from one location to another. If this is the case, we represent the attribute vector as the attribute that we expect when the driver arrives at the destination (which is the next point at which we can make a decision). In this case, we have to include as an attribute the time at which we expect the driver to arrive.

Second, we introduce the dimension that the equipment may fail, requiring some level of repair. A failure can introduce additional delays before the driver is available.

A third important dimension covers the rules that limit how much a driver can be on the road. In the United States, drivers are governed by a set of rules set by the Department of Transportation ("DOT"). There are three basic limits: the amount a driver can be behind the wheel in one shift, the amount of time a driver can be "on duty" in one shift (includes time waiting), and the amount of time that a driver can be on duty over any contiguous eight-day period. As of this writing, these rules are as follows: a driver can drive at most 11 hours at a stretch, he may be on duty for at most 14 continuous hours (there are exceptions to this rule), and the driver can work at most 70 hours in any eight-day period. The last clock is reset if the driver is off-duty for 34 successive hours during any stretch (known as the "34-hour reset").

A final dimension involves getting a driver home. In truckload trucking, drivers may be away from home for several weeks at a time. Companies work to assign drivers that get them home in a reasonable amount of time.

If we include these additional dimensions, our attribute vector grows to

$$a_t = \begin{pmatrix} a_1 \\ a_2 \\ a_3 \\ a_4 \\ \\ a_5 \\ \\ a_6 \\ \\ a_7 \\ \\ a_8 \end{pmatrix} = \begin{pmatrix} \text{The current or future location of the driver.} \\ \text{The time at which the driver is expected to arrive at his future location.} \\ \text{The maintenance status of the equipment.} \\ \text{The number of hours a driver has been behind the wheel during his} \\ \text{current shift.} \\ \text{The number of hours a driver has been on-duty during his current shift.} \\ \text{An eight-element vector giving the number of hours the driver was on} \\ \text{duty over each of the previous eight days.} \\ \text{The driver's home domicile.} \\ \\ \text{The number of days a driver has been away from home.} \end{pmatrix}.$$

We note that element $a_6$ is actually a vector that holds the number of hours the driver was on duty during each calendar day over the last eight days.

A single attribute such as location (including the driver's domicile) might have 100 outcomes, or over 1,000. The number of hours a driver has been on the road might be measured to the nearest hour, while the number of days away from home can be as large as 30 (in rare cases). Needless to say, the number of potential attribute vectors is extremely large.

## 5.4 THE STATES OF OUR SYSTEM

The most important quantity in a dynamic program is the state variable. The state variable captures what we need to know, but just as important it is the variable around which we construct value function approximations. Success in developing a good approximation strategy depends on a deep understanding of what is important in a state variable to capture the future behavior of a system.

### 5.4.1 Defining the state variable

Surprisingly, other presentations of dynamic programming spend little time defining a state variable. Bellman's seminal text [Bellman (1957), p. 81] says "... we have a physical system characterized at any stage by a small set of parameters, the *state variables*." In a much more modern treatment, Puterman first introduces a state variable by saying [Puterman (1994), p. 18] "At each decision epoch, the system occupies a *state*." In both cases, the italics are in the original manuscript, indicating that the term "state" is being introduced. In effect, both authors are saying that given a system, the state variable will be apparent from the context.

Interestingly, different communities appear to interpret state variables in slightly different ways. We adopt an interpretation that is fairly common in the control theory community, but offer a definition that appears to be somewhat tighter than what typically appears in the literature. We suggest the following definition:

**Definition 5.4.1** *A* **state variable** *is the minimally dimensioned function of history that is necessary and sufficient to compute the decision function, the transition function, and the contribution function.*

Later in the chapter, we discuss the decision function (section 5.5), the transition function (section 5.7), and the contribution function (section 5.8). In plain English, a state variable is everything you need to know (at time $t$) to model the system from time $t$ onward. Initially, it is easiest to think of the state variable in terms of the physical state of the system (the status of the pilot, the amount of money in our bank account), but ultimately it is necessary to think of it as the "state of knowledge."

This definition provides a very quick test of the validity of a state variable. If there is a piece of data in either the decision function, the transition function, or the contribution function which is not in the state variable, then we do not have a complete state variable. Similarly, if there is information in the state variable that is never needed in any of these three functions, then we can drop it and still have a valid state variable.

We use the term "minimally dimensioned function" so that our state variable is as compact as possible. For example, we could argue that we need the entire history of events up to time $t$ to model future dynamics. But this is not practical. As we start doing computational work, we are going to want $S_t$ to be as compact as possible. Furthermore, there are many problems where we simply do not need to know the entire history. It might be enough to know the status of all our resources at time $t$ (the resource variable $R_t$). But there are examples where this is not enough.

Assume, for example, that we need to use our history to forecast the price of a stock. Our history of prices is given by $(\hat{p}_1, \hat{p}_2, \ldots, \hat{p}_t)$. If we use a simple exponential smoothing model, our estimate of the mean price $\bar{p}_t$ can be computed using

$$\bar{p}_t = (1-\alpha)\bar{p}_{t-1} + \alpha\hat{p}_t,$$

where $\alpha$ is a stepsize satisfying $0 \leq \alpha \leq 1$. With this forecasting mechanism, we do not need to retain the history of prices, but rather only the latest estimate $\bar{p}_t$. As a result, $\bar{p}_t$ is called a *sufficient statistic*, which is a statistic that captures all relevant information needed to compute any additional statistics from new information. A state variable, according to our definition, is always a sufficient statistic.

Consider what happens when we switch from exponential smoothing to an $N$-period moving average. Our forecast of future prices is now given by

$$\bar{p}_t = \frac{1}{N}\sum_{\tau=0}^{N-1} \hat{p}_{t-\tau}.$$

Now, we have to retain the $N$-period rolling set of prices $(\hat{p}_t, \hat{p}_{t-1}, \ldots, \hat{p}_{t-N+1})$ in order to compute the price estimate in the next time period. With exponential smoothing, we could write

$$S_t = \bar{p}_t.$$

If we use the moving average, our state variable would be

$$S_t = (\hat{p}_t, \hat{p}_{t-1}, \ldots, \hat{p}_{t-N+1}). \tag{5.4}$$

Many authors say that if we use the moving average model, we no longer have a proper state variable. Rather, we would have an example of a "history-dependent process" where the state variable needs to be augmented with history. Using our definition of a state variable, the concept of a history-dependent process has no meaning. The state variable is simply the minimal information required to capture what is needed to model future dynamics. Needless to say, having to explicitly retain history, as we did with the moving average model, produces a much larger state variable than the exponential smoothing model.

### 5.4.2 The three states of our system

To set up our discussion, assume that we are interested in solving a relatively complex resource management problem, one that involves multiple (possibly many) different types of resources which can be modified in various ways (changing their attributes). For such a problem, it is necessary to work with three types of states:

**The state of a single resource (or entity) -** As a resource evolves, the state of a resource is captured by its attribute vector $a$.

**The resource state vector -** This is the state of all the different types of resources (or entities) at the same time, given by $R_t = (R_{ta})_{a \in \mathcal{A}}$.

**The information state -** This captures what we know at time $t$, which includes $R_t$ along with estimates of parameters such as prices, times and costs, or the parameters of a function for forecasting demands (or prices or times). An alternative name for this larger state is the *knowledge state*, which carries the connotation of not only capturing what we know, but how well we know it.

In engineering applications, the attribute vector $a_t$, or the resource state vector $R_t$, would be known as the *physical state*. Often, we will use $S_t$ even though we are still referring to just the physical state. In section 2.3, we introduced problems where the state included estimates of parameters that are not properly described by the physical state. There is a distinct lack of agreement how this additional information should be handled. Some authors want to reserve the word "state" for physical state. Bellman himself used the term "hyperstate" to describe a larger state variable to include this additional information. We have found it is best to (a) let "state" refer to all the information we need to model the system, and (b) distinguish between physical state, information and knowledge when we define the state variable.

There are many problems in dynamic programming that involve the management of a single resource or entity (or asset - the best terminology depends on the context), such as using a computer to play backgammon, routing a single aircraft, controlling a power plant, or selling an asset. There is nothing wrong with letting $S_t$ be the state of this entity. When we are managing multiple entities (which often puts us in the domain of "resource management"), it is useful to distinguish between the state of an individual "resource" and the state of all the resources.

We can use $S_t$ to be the state of a single resource (if this is all we are managing), or let $S_t = R_t$ be the state of all the resources we are managing. There are many problems where the state of the system consists only of $a_t$ or $R_t$. We suggest using $S_t$ as a generic state variable when it is not important to be specific, but it must be used when we may wish to include other forms of information. For example, we might be managing resources (consumer products, equipment, people) to serve customer demands $\hat{D}_t$ that become known at time $t$. If $R_t$ describes the state of the resources we are managing, our state variable would consist of $S_t = (R_t, \hat{D}_t)$, where $\hat{D}_t$ represents additional information we need to solve the problem.

Alternatively, other information might include estimates of parameters of the system (costs, speeds, times, prices). To represent this, let

$\bar{\theta}_t$ = A vector of estimates of different problem parameters at time $t$.

$\hat{\theta}_t$ = New information about problem parameters that arrive during time interval $t$.

We can think of $\bar{\theta}_t$ as the state of our information about different problem parameters at time $t$. We can now write a more general form of our state variable as:

$$\begin{aligned} S_t &= \text{The } \textit{information state} \text{ at time } t \\ &= (R_t, \bar{\theta}_t). \end{aligned}$$

In chapter 10, we will show that it is important to include not just the point estimate $\bar{\theta}_t$, but the entire distribution (or the parameters needed to characterize the distribution, such as the variance).

A particularly important version of this more general state variable arises in approximate dynamic programming. Recall that in chapter 4 we used an approximation of the value function to make a decision, as in

$$x_t^n = \arg\max_{x_t \in \mathcal{X}_t^n} \left( C_t(R_t^n, x_t) + \bar{V}_t^{n-1}(R_t^x) \right) \tag{5.5}$$

Here $\bar{V}^{n-1}(\cdot)$ is an estimate of the value function if our decision takes us from resource state $R_t$ to $R_t^x$, and $x_t^n$ is the value of $x_t$ that solves the right-hand side of (5.5). In this case, our state variable would consist of

$$S_t = (R_t, \bar{V}^{n-1}).$$

The idea that the value function is part of our state variable is quite important in approximate dynamic programming.

### 5.4.3 The post-decision state variable

state, post-decision We can view our system as evolving through sequences of new information followed by a decision followed by new information (and so on). Although we have not yet discussed decisions, for the moment let the decisions (which may be a vector) be represented generically using $x_t$ (we discuss our choice of notation for a decision in the next section). In this case, a history of the process might be represented using

$$h_t = (S_0, x_0, W_1, x_1, W_2, x_2, \ldots, x_{t-1}, W_t).$$

$h_t$ contains all the information we need to make a decision $d_t$ at time $t$. As we discussed before, $h_t$ is sufficient but not necessary. We expect our state variable to capture what is needed to make a decision, allowing us to represent the history as

$$h_t = (S_0, x_0, W_1, S_1, x_1, W_2, S_2, x_2, \ldots, x_{t-1}, W_t, S_t). \tag{5.6}$$

The sequence in equation (5.6) defines our state variable as occurring after new information arrives and before a decision is made. For this reason, we call $S_t$ the *pre-decision state variable*. This is the most natural place to write a state variable because the point of capturing information from the past is to make a decision.

For most problem classes, we can design more effective computational strategies using the *post-decision state variable*. This is the state of the system after a decision $x_t$. For this reason, we denote this state variable $S_t^x$, which produces the history

$$h_t = (S_0, x_0, S_0^x, W_1, S_1, x_1, S_1^x, W_2, S_2, x_2, S_2^x, \ldots, x_{t-1}, S_{t-1}^x, W_t, S_t). \tag{5.7}$$

We again emphasize that our notation $S_t^x$ means that this function has access to all the exogenous information up through time $t$, along with the decision $x_t$ (which also has access to the information up through time $t$).

The examples below provide some illustrations of pre- and post-decision states.

---

## EXAMPLE 5.1

A traveler is driving through a network, where the travel time on each link of the network is random. As she arrives at node $i$, she is allowed to see the travel times on each of the links out of node $i$, which we represent by $\hat{\tau}_i = (\hat{\tau}_{ij})_j$. As she arrives at node $i$, her pre-decision state is $S_t = (i, \hat{\tau}_i)$. Assume she decides to move from $i$ to $k$ (we might write $x_{ik} = 1$). Her post-decision state is $S_t^x = (k)$ (note that she is still at node $i$; the post-decision state captures the fact that she will next be at node $k$, and we no longer have to include the travel times on the links out of node $i$).

## EXAMPLE 5.2

The nomadic trucker revisited. Let $R_{ta} = 1$ if the trucker has attribute vector $a$ at time $t$ and 0 otherwise. Now let $D_{tb}$ be the number of customer demands (loads of freight) of type $b$ available to be moved at time $t$. The pre-decision state variable for the trucker is $S_t = (R_t, D_t)$, which tells us the state of the trucker and the demands available to be moved. Assume that once the trucker makes a decision, all the unserved demands in $D_t$ are lost, and new demands become available at time $t+1$. The post-decision state variable is given by $S_t^x = R_t^x$ where $R_{ta}^x = 1$ if the trucker has attribute vector $a$ after a decision has been made.

## EXAMPLE 5.3

Imagine playing backgammon where $R_{ti}$ is the number of your pieces on the $i^{th}$ "point" on the backgammon board (there are 24 points on a board). The transition from $S_t$ to $S_{t+1}$ depends on the player's decision $x_t$, the play of the opposing player, and the next roll of the dice. The post-decision state variable is simply the state of the board after a player moves but before his opponent has moved.

---

The importance of the post-decision state variable, and how to use it, depends on the problem at hand. We saw in chapter 4 that the post-decision state variable allowed us to make decisions without having to compute the expectation within the max or min operator. Later we will see that this allows us to solve some very large scale problems.

There are three ways of finding a post-decision state variable:

***Decomposing decisions and information*** All of the problems with which we have had direct experience have allowed us to separate the pure effect of decisions and information, which means that we can create functions $S^{M,x}(\cdot)$ and $S^{M,W}(\cdot)$ from which we can compute

$$S_t^x = S^{M,x}(S_t, x_t), \tag{5.8}$$

$$S_{t+1} = S^{M,W}(S_t^x, W_{t+1}). \tag{5.9}$$

The structure of these functions is highly problem-dependent. However, there are sometimes significant computational benefits, primarily when we face the problem of approximating the value function. Recall that the state variable captures all the information we need to make a decision, compute the transition function, and compute the contribution

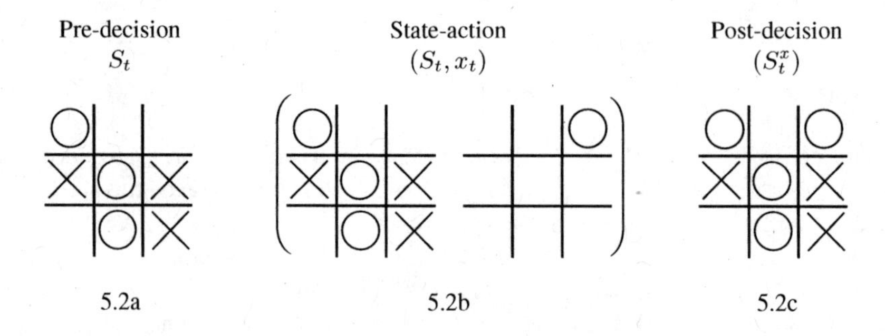


**Figure 5.2** Pre-decision state, augmented state-action, and post-decision state for tic-tac-toe.

function. $S_t^x$ *only* has to carry the information needed to compute the transition function. For some applications, $S_t^x$ has the same dimensionality as $S_t$, but in many settings, $S_t^x$ is dramatically simpler than $S_t$, simplifying the problem of approximating the value function.

***State-action pairs*** A very generic way of representing a post-decision state is to simply write

$$S_t^x = (S_t, x_t).$$

Figure 5.2 provides a nice illustration using our tic-tac-toe example. Figure 5.2a shows a tic-tac-toe board just before player O makes his move. Figure 5.2b shows the augmented state-action pair, where the decision (O decides to place his move in the upper right hand corner) is distinct from the state. Finally, figure 5.2c shows the post-decision state. For this example, the pre- and post-decision state spaces are the same, while the augmented state-action pair is nine times larger.

The augmented state $(S_t, x_t)$ is closely related to the post-decision state $S_t^x$ (not surprising, since we can compute $S_t^x$ deterministically from $S_t$ and $x_t$). But computationally, the difference is significant. If $\mathcal{S}$ is the set of possible values of $S_t$, and $\mathcal{X}$ is the set of possible values of $x_t$, then our augmented state space has size $|\mathcal{S}| \times |\mathcal{X}|$, which is obviously much larger.

The augmented state variable is used in a popular class of algorithms known as $Q$-learning (introduced in chapter 8), where the challenge is to statistically estimate $Q$-factors which give the value of being in state $S_t$ *and* taking action $x_t$. The $Q$-factors are written $Q(S_t, x_t)$, in contrast with value functions $V_t(S_t)$ which provide the value of being in a state. This allows us to directly find the best action by solving $\min_x Q(S_t, x_t)$. This is the essence of $Q$-learning, but the price of this algorithmic step is that we have to estimate $Q(S_t, x_t)$ for each $S_t$ and $x_t$. It is not possible to determine $x_t$ by optimizing a function of $S_t^x$ alone, since we generally cannot determine which action $x_t$ brought us to $S_t^x$.

***The post-decision as a point estimate*** Assume that we have a problem where we can compute a point estimate of future information. Let $\bar{W}_{t,t+1}$ be a point estimate, computed at time $t$, of the outcome of $W_{t+1}$. If $W_{t+1}$ is a numerical quantity, we might use $\bar{W}_{t,t+1} = \mathbb{E}(W_{t+1}|S_t)$ or $\bar{W}_{t,t+1} = 0$. $W_{t+1}$ might be a discrete outcome such as the number of equipment failures. It may not make sense to use an expectation (we may have problems working with 0.10 failures), so in this setting $W_{t+1}$ might be the most likely outcome. Finally, we might simply assume that $W_{t+1}$ is empty (a form of "null" field). For example, a taxi picking up a customer may not know the destination of the customer before the customer gets in the cab. In this case, if $W_{t+1}$ represents the destination, we might use $\bar{W}_{t,t+1} =$ '-'.

If we can create a reasonable estimate $\bar{W}_{t,t+1}$, we can compute post- and pre-decision state variables using

$$\begin{aligned} S_t^x &= S^M(S_t, x_t, \bar{W}_{t,t+1}), \\ S_{t+1} &= S^M(S_t, x_t, W_{t+1}). \end{aligned}$$

Measured this way, we can think of $S_t^x$ as a point estimate of $S_{t+1}$, but this does not mean that $S_t^x$ is necessarily an approximation of the expected value of $S_{t+1}$.

### 5.4.4 Partially observable states*

There are many applications where we are not able to observe (or measure) the state of the system precisely, as illustrated in the examples. These problems are referred to as *partially observable Markov decision processes*, and require introducing a new class of exogenous information representing the difference between the true state and the observed state.

---

**■ EXAMPLE 5.1**

A retailer may have to order inventory without being able to measure the precise current inventory. It is possible to measure sales, but theft and breakage introduce errors.

**■ EXAMPLE 5.2**

A transportation company needs to dispatch a fleet of trucks, but does not know the precise location or maintenance status of each truck.

**■ EXAMPLE 5.3**

The military has to make decisions about sending out aircraft to remove important military targets that may have been damaged in previous raids. These decisions typically have to be made without knowing the precise state of the targets.

---

To model this class of applications, let

$\tilde{S}_t$ = The true state of the system at time $t$.

$\widetilde{W}_t$ = Errors that arise when measuring the state $\tilde{S}_t$.

In this context, we assume that our state variable $S_t$ is the observed state of the system. Now, our history is given by

$$\begin{aligned} h_t = \; & (S_0, x_0, S_0^x, W_1, \tilde{S}_1, \widetilde{W}_1, S_1, x_1, S_1^x, W_2, \tilde{S}_2, \widetilde{W}_2, S_2, x_2, S_2^x, \ldots, \\ & x_{t-1}, S_{t-1}^x, W_t, \tilde{S}_t, \widetilde{W}_t, S_t). \end{aligned}$$

We view our original exogenous information $W_t$ as representing information such as the change in price of a resource, a customer demand, equipment failures, or delays in the completion of a task. By contrast, $\widetilde{W}_t$, which captures the difference between $\tilde{S}_t$ and $S_t$, represents measurement error or the inability to observe information. Examples of measurement error might include differences between actual and calculated inventory of product on a store shelf (due, for example, to theft or breakage), the error in estimating the location of a vehicle (due to errors in the GPS tracking system), or the difference between the actual state of a piece of machine such as an aircraft (which might have a failed part) and the observed state (we do not yet know about the failure). A different form of observational error arises when there are elements we simply cannot observe (for example, we know the location of the vehicle but not its fuel status).

It is important to realize that there are two transition functions at work here. The "real" transition function models the dynamics of the true (unobservable) state, as in

$$\tilde{S}_{t+1} = \tilde{S}^M(\tilde{S}_t, x_t, \widetilde{W}_{t+1}).$$

In practice, not only do we have the problem that we cannot perfectly measure $\tilde{S}_t$, we may not know the transition function $\tilde{S}^M(\cdot)$. Instead, we are working with our "engineered" transition function

$$S_{t+1} = S^M(S_t, x_t, W_{t+1}),$$

where $W_{t+1}$ is capturing some of the effects of the observation error. When building a model where observability is an issue, it is important to try to model $\tilde{S}_t$, the transition function $\tilde{S}^M(\cdot)$ and the observation error $\widetilde{W}_t$ as much as possible. However, anything we cannot measure may have to be captured in our generic noise vector $W_t$.

### 5.4.5 Flat vs. factored state representations*

It is very common in the dynamic programming literature to define a discrete set of states $\mathcal{S} = (1, 2, \ldots, |\mathcal{S}|)$, where $s \in \mathcal{S}$ indexes a particular state. For example, consider an inventory problem where $S_t$ is the number of items we have in inventory (where $S_t$ is a scalar). Here, our state space $\mathcal{S}$ is the set of integers, and $s \in \mathcal{S}$ tells us how many products are in inventory. This is the style we used in chapters 3 and 4.

Now assume that we are managing a set of $K$ product types. The state of our system might be given by $S_t = (S_{t1}, S_{t2}, \ldots, S_{tk}, \ldots)$ where $S_{tk}$ is the number of items of type $k$ in inventory at time $t$. Assume that $S_{tk} \leq M$. Our state space $\mathcal{S}$ would consist of all possible values of $S_t$, which could be as large as $K^M$. A state $s \in \mathcal{S}$ corresponds to a particular vector of quantities $(S_{tk})_{k=1}^K$.

Modeling each state with a single scalar index is known as a flat or unstructured representation. Such a representation is simple and elegant, and produces very compact models that have been popular in the operations research community. The presentation in chapter 3 depends on this representation. However, the use of a single index completely disguises the structure of the state variable, and often produces intractably large state spaces.

In the arena of approximate dynamic programming, it is often essential that we exploit the structure of a state variable. For this reason, we generally find it necessary to use what is known as a *factored* representation, where each factor represents a *feature* of the state variable. For example, in our inventory example we have $K$ factors (or features). It is possible to build approximations that exploit the structure that each dimension of the state variable is a particular quantity.

Our attribute vector notation, which we use to describe a single entity, is an example of a factored representation. Each element $a_i$ of an attribute vector represents a particular feature of the entity. The resource state variable $R_t = (R_{ta})_{a \in \mathcal{A}}$ is also a factored representation, since we explicitly capture the number of resources with a particular attribute. This is useful when we begin developing approximations for problems such as the dynamic assignment problem that we introduced in section 2.2.10.

## 5.5 MODELING DECISIONS

Fundamental to dynamic programs is the characteristic that we are making decisions over time. For stochastic problems, we have to model the sequencing of decisions and information, but there are many uses of dynamic programming that address deterministic problems. In this case, we use dynamic programming because it offers specific structural advantages, such as our budgeting problem in chapter 1. But the concept of sequencing decisions over time is fundamental to a dynamic program.

It is important to model decisions properly so we can scale to high-dimensional problems. This requires that we start with a good fundamental model of decisions for resource management problems. Our choices are nonstandard for the dynamic programming community, but very compatible with the math programming community.

### 5.5.1 Decisions, actions, and controls

A survey of the literature reveals a distressing variety of words used to mean "decisions." The classical literature on Markov decision process talks about choosing an action $a \in \mathcal{A}$ (or $a \in \mathcal{A}_s$, where $\mathcal{A}_s$ is the set of actions available when we are in state $s$) or a policy (a rule for choosing an action). The optimal control community chooses a control $u \in \mathcal{U}_x$ when the system is in state $x$. The math programming community wants to choose a decision represented by the vector $x$, and the simulation community wants to apply a rule.

The proper notation for decisions will depend on the specific application. However, we are often modeling problems that can be broadly described as resource (or asset) allocation problems. For these problems, it is useful to draw on the skills of the math programming community where decisions are typically vectors represented by $x$. This notation is especially well suited for problems that involve buying, selling and/or managing "resources." For this class of applications, we define the following:

$d$ = A type of decision that acts on a resource (or resource type) in some way (buying, selling, or managing),

$\mathcal{D}$ = The set of potential types of decisions that can be used to act on a resource.

$x_{tad}$ = The quantity of resources with attribute vector $a$ acted on with decision $d$ at time $t$,

$x_t$ = $(x_{tad})_{a \in \mathcal{A}, d \in \mathcal{D}}$,

$\mathcal{X}_t$ = Set of allowable decisions given the information available at time $t$.

If $d$ is a decision to purchase a resource, then $x_d$ is the quantity of resources being purchased. If we are moving transportation resources from one location $i$ to another location $j$, then $d$ would represent the decision to move from $i$ to $j$, and $x_{tad}$ (where $a = i$ and $d = (i, j)$) would be the flow of resources. $d$ might be a decision to move a robot arm through a sequence of coordinates, or to make a particular move in a chess game. In these cases, $x_{tad}$ is an indicator variable where $x_{tad} = 1$ if we choose to make decision $d$.

Earlier, we observed that the attribute vector $a$ of a single resource might have as many as 10 or 20 dimensions, but we would never expect the attribute vector to have 100 or more dimensions (problems involving financial assets, for example, might require 0, 1 or 2 dimensions). Similarly, the set of decision types, $\mathcal{D}$, might be one or two (e.g., buying or selling), or on the order of 100 or 1000 for the most complex problems, but we simply would never expect sets with, say, $10^{10}$ types of decisions (that can be used to act on a single resource class). Note that there is a vast array of problems where the size of $\mathcal{D}$ is less than 10, as illustrated in the examples.

---

**EXAMPLE 5.1**

Assume that you are holding a financial asset, where you will receive a price $\hat{p}_t$ if you sell at time $t$. We can think of the decision set as consisting of $\mathcal{D} = (\text{hold}, \text{sell})$. We sell if $x_{sell} = 1$ (if we have more than one asset, $x_{sell}$ can be the quantity that we sell). For the case of a single asset, we would require $\sum_{d \in \mathcal{D}} x_d = 1$ since we must hold or sell.

**EXAMPLE 5.2**

A taxi waiting at location $i$ can serve a customer if one arrives, or it can sit and do nothing, or it can reposition to another location (without a customer) where the chances of finding one seem better. We can let $\mathcal{D}^M$ be the set of locations the cab can move to (without a customer), where the decision $d = i$ represents the decision to hold (at location $i$). Then let $d^s$ be the decision to serve a customer, although this decision can only be made if there is a customer to be served. The complete set of decisions is $\mathcal{D} = d^s \cup \mathcal{D}^M$. $x_d = 1$ if we choose to take decision $d$.

---

It is significant that we are representing a decision $d$ as acting on a single resource or resource type. The field of Markov decision processes represents a decision as an action (typically denoted $a$), but the concept of an action in this setting is equivalent to our decision type $d$. Actions are typically represented as being discrete, whereas our decision vector $x$ can be discrete or continuous. We do, however, restrict our attention to cases where the set $\mathcal{D}$ is discrete and finite.

In some problem classes, we manage a discrete resource which might be someone playing a game, the routing of a single car through traffic, or the control of a single elevator moving up and down in a building. In this case, at any point in time we face the problem of choosing a single decision $d \in \mathcal{D}$. Using our $x$ notation, we would represent this using $x_{ta\hat{d}} = 1$ if we choose decision $\hat{d}$ and $x_{tad} = 0$ for $d \neq \hat{d}$. Alternatively, we could simply drop our "$x$" notation and simply let $d_t$ be the decision we chose at time $t$. While recognizing that the "$d$" notation is perhaps more natural and elegant if we face simply the scalar problem of choosing a single decision, it greatly complicates our transition from simple, scalar problems to the complex, high-dimensional problems.

Our notation represents a difficult choice between the vocabularies of the math programming community (which dominates the field of high-dimensional problems), and the artificial intelligence and control communities (which dominate the field of approximate dynamic programming and reinforcement learning). We use $x$ (which might be a vector) to form a bridge with the math programming community, where $x_d$ is an element of $x$. In most applications, $d$ will represent a type of decision acting on a single entity, whereas $x_d$ tells us how often, or to what degree, we are using this decision. However, there are many applications where we have to control an entity (for example, the acceleration of an aircraft or robot arm); in such cases, we think the classical control variable $u_t$ is entirely appropriate. Rather than use a single, general purpose notation for a decision, control, or action, we encourage the use of notation that reflects the characteristics of the variable (discrete, low-dimensional continuous, or high-dimensional continuous or integer).

### 5.5.2 The nomadic trucker revisited

We return to our nomadic trucker example to review the decisions for this application. There are two classes of decisions the trucker may choose from

$$\mathcal{D}^D = \text{The set of decisions to serve a demand, where each element of } \mathcal{D}^D \text{ corresponds to an element of the set of demand types } \mathcal{B},$$

$$\mathcal{D}^M = \text{The set of decisions to move empty to location } i \in \mathcal{I}, \text{ where each element of } \mathcal{D}^M \text{ corresponds to a location,}$$

$$\mathcal{D} = \mathcal{D}^D \cup \mathcal{D}^M.$$

The trucker may move "empty" to the same state that he is located in, which represents the decision to do nothing. The set $\mathcal{D}^D$ requires a little explanation. Recall that $b$ is the attribute vector of a load to be moved. An element of $\mathcal{D}^D$ represents the decision to move a type of load. Other decision classes that could be modeled include buying and selling trucks, repairing them, or reconfiguring them (for example, adding refrigeration units so a trailer can carry perishable commodities).

### 5.5.3 Policies

When we are solving deterministic problems, our interest is in finding a set of decisions $x_t$ over time. When we are solving stochastic problems (problems with dynamic information processes), the decision $x_t$ for $t \geq 1$ is a random variable. This happens because we do not know (at time $t = 0$) the state of our system $S_t$ at time $t$.

How do we make a decision if we do not know the state of the system? The answer is that instead of finding the best decision, we are going to focus on finding the best rule for making a decision given the information available at the time. This rule is commonly known as a *policy*:

**Definition 5.5.1** *A* **policy** *is a rule (or function) that determines a decision given the available information (in state $S_t$).*

This definition implies that our policy produces a decision deterministically; that is, given a state $S_t$, it produces a single action $x$. There are, however, instances where $S_t$ does not contain all the information needed to make a decision (for example, when the true state $\tilde{S}_t$ is partially observable). In addition, there are special situations (arising in the context of two-player games) where there is value in choosing a decision somewhat randomly. For our computational algorithms, there will be many instances when we want to choose what appears to be a non-optimal decision for the purpose of collecting more information (this idea was introduced in section 2.3).

Policies come in several flavors. Examples include:

1) Lookup-table policies - Let $s \in \mathcal{S}$ be a discrete state, and that we have an action $X(s)$ that gives us the action we should take when we are in state $s$. We might use

this format to store the move we should make when we are playing a board game, where $s$ is the state of the board game. When we use a lookup-table policy, we store an action $x$ for each state $s$.

2) Parameterized policies - Let $S$ be the number of items that we have in inventory. We might use a reorder policy of the form

$$X(s) = \begin{cases} Q - s & \text{if } s < q, \\ 0 & \text{otherwise.} \end{cases}$$

Here, the policy is determined by a functional form and two parameters $(Q, q)$. Finding the best policy means finding the best values of $Q$ and $q$.

3) A regression model - Assume we need to purchase natural gas to put in a storage facility. If the price is lower, we want to purchase more. Assume we observe a human over a number of time periods who chooses to purchase $x_t$ on a day when the price is $p_t$. We decide to express the relationship between $x_t$ and $p_t$ using the regression model

$$x_t = \theta_0 + \theta_1 p_t + \theta_2 p_t^2. \tag{5.10}$$

We can use a set of observed decisions $x_t$ made over time and correlate these decisions against the observed prices at the same point in time to fit our equation. The linear regression determines the structure of the policy which is parameterized by the vector $\theta$.

4) Myopic policies - These can include virtually any policy that ignores the impact of decisions now on the future, but here we assume that we have a contribution function $C(S_t, x_t)$ that returns a contribution given the decision $x_t$ when we are in state $S_t$. The decision function would be

$$X^\pi(S_t) = \arg\max_{x_t \in \mathcal{X}_t} C(S_t, x_t).$$

5) Value function approximations - Imagine that we need to determine how much of our stockpile of vaccines to use at time $t$ from our inventory $R_t$. We might solve a problem of the form

$$X_t^\pi(R_t) = \arg\max_{0 \le x_t \le R_t} \big(C_t(x_t) + \bar{v}_t(R_t - x_t)\big). \tag{5.11}$$

Our "policy" is to find $x_t$ using our linear value function approximation, parameterized by $\bar{v}_t$.

Mathematically, all of these are equivalent in that they each specify a decision given a state. Computationally, however, they are quite different. A lookup-table policy will be completely impractical if the state variable is a vector (it also hard if the decision is a vector). It is much easier to find the best value of two parameters (such as $(Q, q)$) than to find an entire table of actions for each state. A regression model policy such as equation (5.10) requires that we have a source of states and actions to fit our model. A value function policy such as equation (5.11) requires that we have an effective value function approximation and an updating mechanism.

**Remark:** Some authors like to write the decision function in the form

$$X_t^\pi(R_t) \in \arg\max F(x)$$

for some function $F(x)$. The "$\in$" captures the fact that $F(x)$ may have more than one optimal solution, which means $\arg\max F(x)$ is really a set. If this is the case, then we need some rule for choosing which member of the set we are going to use. Most algorithms make this choice at random (typically, we find a best solution, and then replace it only when we find a solution that is even better, rather than just as good). In rare cases, we do care. However, if we use $X_t^\pi(R_t) \in \arg\max$ then we have to introduce additional logic to solve this problem. We assume that the "arg max" operator includes whatever logic we are going to use to solve this problem (since this is typically what happens in practice).

### 5.5.4 Randomized policies*

Assume you need to buy a resource at an auction, and you do not have the time to attend the auction yourself. Your problem is to decide which of your two assistants to send. Assistant A is young and aggressive, and is more likely to bid a higher price (but may also scare off other bidders). Assistant B is more tentative and conservative and might drop out if he thinks the bidding is heading too high.

This is an example of a randomized policy. We are not directly making a decision of what to bid, but we are making a decision that will influence the probability distribution of whether a bid will be made. This is known as a randomized policy.

In section 3.9.5, we showed that given a choice between a deterministic policy and a randomized policy, the deterministic policy will always be at least as good as a randomized policy. But there are situations where we may not have a choice (for example, when we do not have direct control over a system). Randomized policies are particularly important in multi-player game situations. In a two-player game, one player may be able to guess how another player will respond in a particular situation and act accordingly. A player can be more successful by randomizing his behavior, sometimes making moves that will not necessarily be the best just to keep the other player from guessing his moves.

## 5.6 THE EXOGENOUS INFORMATION PROCESS

An important dimension of many of the problems that we address is the arrival of exogenous information, which changes the state of our system. While there are many important deterministic dynamic programs, exogenous information processes represent an important dimension in many problems in resource management.

### 5.6.1 Basic notation for information processes

The central challenge of dynamic programs is dealing with one or more exogenous information processes, forcing us to make decisions before all the information is known. These might be stock prices, travel times, equipment failures, or the behavior of an opponent in a game. There might be a single exogenous process (the price at which we can sell an asset) or a number of processes. For a particular problem, we might model this process using notation that is specific to the application. Here, we introduce generic notation that can handle any problem.

| Sample path | $t=0$ | $t=1$ | | $t=2$ | | $t=3$ | |
|---|---|---|---|---|---|---|---|
| $\omega$ | $p_0$ | $\hat{p}_1$ | $p_1$ | $\hat{p}_2$ | $p_2$ | $\hat{p}_3$ | $p_3$ |
| 1 | 29.80 | 2.44 | 32.24 | 1.71 | 33.95 | -1.65 | 32.30 |
| 2 | 29.80 | -1.96 | 27.84 | 0.47 | 28.30 | 1.88 | 30.18 |
| 3 | 29.80 | -1.05 | 28.75 | -0.77 | 27.98 | 1.64 | 29.61 |
| 4 | 29.80 | 2.35 | 32.15 | 1.43 | 33.58 | -0.71 | 32.87 |
| 5 | 29.80 | 0.50 | 30.30 | -0.56 | 29.74 | -0.73 | 29.01 |
| 6 | 29.80 | -1.82 | 27.98 | -0.78 | 27.20 | 0.29 | 27.48 |
| 7 | 29.80 | -1.63 | 28.17 | 0.00 | 28.17 | -1.99 | 26.18 |
| 8 | 29.80 | -0.47 | 29.33 | -1.02 | 28.31 | -1.44 | 26.87 |
| 9 | 29.80 | -0.24 | 29.56 | 2.25 | 31.81 | 1.48 | 33.29 |
| 10 | 29.80 | -2.45 | 27.35 | 2.06 | 29.41 | -0.62 | 28.80 |

**Table 5.1** A set of sample realizations of prices ($p_t$) and the changes in prices ($\hat{p}_t$)

Consider a problem of tracking the value of an asset. Assume the price evolves according to

$$p_{t+1} = p_t + \hat{p}_{t+1}.$$

Here, $\hat{p}_{t+1}$ is an exogenous random variable representing the chance in the price during time interval $t+1$. At time $t$, $p_t$ is a number, while (at time $t$) $p_{t+1}$ is random. We might assume that $\hat{p}_{t+1}$ comes from some probability distribution. For example, we might assume that it is normally distributed with mean 0 and variance $\sigma^2$. However, rather than work with a random variable described by some probability distribution, we are going to primarily work with sample realizations. Table 5.1 shows 10 sample realizations of a price process that starts with $p_0 = 29.80$ but then evolves according to the sample realization. Following standard mathematical convention, we index each path by the Greek letter $\omega$ (in the example below, $\omega$ runs from 1 to 10). At time $t = 0$, $p_t$ and $\hat{p}_t$ is a random variable (for $t \geq 1$), while $p_t(\omega)$ and $\hat{p}_t(\omega)$ are *sample realizations*. We refer to the sequence

$$p_1(\omega), p_2(\omega), p_3(\omega), \ldots$$

as a *sample path* (for the prices $p_t$).

We are going to use "$\omega$" notation throughout this volume, so it is important to understand what it means. As a rule, we will primarily index exogenous random variables such as $\hat{p}_t$ using $\omega$, as in $\hat{p}_t(\omega)$. $\hat{p}_{t'}$ is a random variable if we are sitting at a point in time $t < t'$. $\hat{p}_t(\omega)$ is not a random variable; it is a sample realization. For example, if $\omega = 5$ and $t = 2$, then $\hat{p}_t(\omega) = -0.73$. We are going to create randomness by choosing $\omega$ at random. To make this more specific, we need to define

$$\begin{aligned} \Omega &= \text{The set of all possible sample realizations (with } \omega \in \Omega\text{)}, \\ p(\omega) &= \text{The probability that outcome } \omega \text{ will occur.} \end{aligned}$$

A word of caution is needed here. We will often work with continuous random variables, in which case we have to think of $\omega$ as being continuous. In this case, we cannot say $p(\omega)$ is the "probability of outcome $\omega$." However, in all of our work, we will use discrete samples. For this purpose, we can define

$$\hat{\Omega} = \text{A set of discrete sample observations of } \omega \in \Omega.$$

In this case, we can talk about $p(\omega)$ being the probability that we sample $\omega$ from within the set $\hat{\Omega}$.

For more complex problems, we may have an entire family of random variables. In such cases, it is useful to have a generic "information variable" that represents all the information that arrives during time interval $t$. For this purpose, we define

$$W_t = \text{The exogenous information becoming available during interval } t.$$

$W_t$ may be a single variable, or a collection of variables (travel times, equipment failures, customer demands). We note that while we use the convention of putting hats on variables representing exogenous information $(\hat{D}_t, \hat{p}_t)$, we do not use a hat for $W_t$ since this is our only use for this variable, whereas $D_t$ and $p_t$ have other meanings. We always think of information as arriving in continuous time, hence $W_t$ is the information arriving during time interval $t$, rather than at time $t$. This eliminates the ambiguity over the information available when we make a decision at time $t$.

The choice of notation $W_t$ as a generic "information function" is not standard, but it is mnemonic (it looks like $\omega_t$). We would then write $\omega_t = W_t(\omega)$ as a sample realization of the information arriving during time interval $t$. This notation adds a certain elegance when we need to write decision functions and information in the same equation.

Some authors use $\omega$ to index a particular sample path, where $W_t(\omega)$ is the information that arrives during time interval $t$. Other authors view $\omega$ as the information itself, as in

$$\omega = (-0.24, 2.25, 1.48).$$

Obviously, both are equivalent. Sometimes it is convenient to define

$$\begin{aligned}
\omega_t &= \text{The information that arrives during time period } t \\
&= W_t(\omega), \\
\omega &= (\omega_1, \omega_2, \ldots).
\end{aligned}$$

We sometimes need to refer to the *history* of our process, for which we define

$$\begin{aligned}
H_t &= \text{The history of the process, consisting of all the information known through time } t, \\
&= (W_1, W_2, \ldots, W_t), \\
\mathcal{H}_t &= \text{The set of all possible histories through time } t, \\
&= \{H_t(\omega) | \omega \in \Omega\}, \\
h_t &= \text{A sample realization of a history}, \\
&= H_t(\omega), \\
\Omega(h_t) &= \{\omega \in \Omega | H_t(\omega) = h_t\}.
\end{aligned}$$

In some applications, we might refer to $h_t$ as the state of our system, but this is usually a very clumsy representation. However, we will use the history of the process for a specific modeling and algorithmic strategy.

### 5.6.2 Outcomes and scenarios

Some communities prefer to use the term *scenario* to refer to a sample realization of random information. For most purposes, "outcome," "sample path," and "scenario" can be used interchangeably (although sample path refers to a sequence of outcomes over time). The term scenario causes problems of both notation and interpretation. First, "scenario" and "state" create an immediate competition for the interpretation of the letter "s." Second, "scenario" is often used in the context of major events. For example, we can talk about the scenario that the Chinese might revalue their currency (a major question in the financial markets at this time). We could talk about two scenarios: (1) the Chinese hold the current relationship between the yuan and the dollar, and (2) they allow their currency to float. For each scenario, we could talk about the fluctuations in the exchange rates between all currencies.

Recognizing that different communities use "outcome" and "scenario" to mean the same thing, we suggest that we may want to reserve the ability to use both terms simultaneously. For example, we might have a set of scenarios that determine if and when the Chinese revalue their currency (but this would be a small set). We recommend denoting the set of scenarios by $\Psi$, with $\psi \in \Psi$ representing an individual scenario. Then, for a particular scenario $\psi$, we might have a set of outcomes $\omega \in \Omega$ (or $\Omega(\psi)$) representing various minor events (currency exchange rates, for example).

---

■ **EXAMPLE 5.1**

Planning spare transformers - In the electric power sector, a certain type of transformer was invented in the 1960's. As of this writing, the industry does not really know the failure rate curve for these units (is their lifetime roughly 50 years? 60 years?). Let $\psi$ be the scenario that the failure curve has a particular shape (for example, where failures begin happening at a higher rate around 50 years). For a given scenario (failure rate curve), $\omega$ represents a sample realization of failures (transformers can fail at any time, although the likelihood they will fail depends on $\psi$).

■ **EXAMPLE 5.2**

Energy resource planning - The federal government has to determine energy sources that will replace fossil fuels. As research takes place, there are random improvements in various technologies. However, the planning of future energy technologies depends on the success of specific options, notably whether we will be able to sequester carbon underground. If this succeeds, we will be able to take advantage of vast stores of coal in the United States. Otherwise, we have to focus on options such as hydrogen and nuclear.

---

In section 11.6, we provide a brief introduction to the field of stochastic programming where "scenario" is the preferred term to describe a set of random outcomes. Often, applications of stochastic programming apply to problems where the number of outcomes (scenarios) is relatively small.

### 5.6.3 Lagged information processes*

There are many settings where the information about a new arrival comes before the new arrival itself as illustrated in the examples.

---

**EXAMPLE 5.1**

An airline may order an aircraft at time $t$ and expect the order to be filled at time $t'$.

**EXAMPLE 5.2**

An orange juice products company may purchase futures for frozen concentrated orange juice at time $t$ that can be exercised at time $t'$.

**EXAMPLE 5.3**

A programmer may start working on a piece of coding at time $t$ with the expectation that it will be finished at time $t'$.

---

This concept is important enough that we offer the following term:

**Definition 5.6.1** *The actionable time of a resource is the time at which a decision may be used to change its attributes (typically generating a cost or reward).*

The actionable time is simply one attribute of a resource. For example, if at time $t$ we own a set of futures purchased in the past with exercise dates of $t+1, t+2, \ldots, t'$, then the exercise date would be an attribute of each futures contract (the exercise dates do not need to coincide with the discrete time instances when decisions are made). When writing out a mathematical model, it is sometimes useful to introduce an index just for the actionable time (rather than having it buried as an element of the attribute vector $a$). Before, we let $R_{ta}$ be the number of resources that we know about at time $t$ with attribute vector $a$. The attribute might capture that the resource is not actionable until time $t'$ in the future. If we need to represent this explicitly, we might write

$$
\begin{aligned}
R_{t,t'a} &= \text{The number of resources that we know about at time } t \text{ that will} \\
&\phantom{=} \text{be actionable with attribute vector } a \text{ at time } t', \\
R_{tt'} &= (R_{t,t'a})_{a \in \mathcal{A}}, \\
R_t &= (R_{t,t'})_{t' \geq t}.
\end{aligned}
$$

It is very important to emphasize that while $t$ is discrete (representing when decisions are made), the actionable time $t'$ may be continuous. When this is the case, it is generally best to simply leave it as an element of the attribute vector.

### 5.6.4 Models of information processes*

Information processes come in varying degrees of complexity. Needless to say, the structure of the information process plays a major role in the models and algorithms used to solve the problem. Below, we describe information processes in increasing levels of complexity.

### *Processes with independent increments*

A large number of problems in resource management can be characterized by what are known as processes with independent increments. What this means is that the *change* in the process is independent of the history of the process, as illustrated in the examples.

---

■ **EXAMPLE 5.1**

A publicly traded index fund has a price process that can be described (in discrete time) as $p_{t+1} = p_t + \sigma\delta$, where $\delta$ is normally distributed with mean $\mu$, variance 1, and $\sigma$ is the standard deviation of the change over the length of the time interval.

■ **EXAMPLE 5.2**

Requests for credit card confirmations arrive according to a Poisson process with rate $\lambda$. This means that the number of arrivals during a period of length $\Delta t$ is given by a Poisson distribution with mean $\lambda\Delta t$, which is independent of the history of the system.

---

The practical challenge we typically face in these applications is that we do not know the parameters of the system. In our price process, the price may be trending upward or downward, as determined by the parameter $\mu$. In our customer arrival process, we need to know the rate $\lambda$ (which can also be a function of time).

### *State-dependent information processes*

The standard dynamic programming models allow the probability distribution of new information (for example, the chance in price of an asset) to be a function of the state of the system (the mean change in the price might be negative if the price is high enough, positive if the price is low enough). This is a more general model than one with independent increments, where the distribution is independent of the state of the system.

---

■ **EXAMPLE 5.1**

Customers arrive at an automated teller machine according to a Poisson process, but as the line grows longer, an increasing proportion decline to join the queue (a property known as *balking* in the queueing literature). The apparent arrival rate at the queue is a process that depends on the length of the queue.

■ **EXAMPLE 5.2**

A market with limited information may respond to price changes. If the price drops over the course of a day, the market may interpret the change as a downward movement, increasing sales and putting further downward pressure on the price. Conversely, upward movement may be interpreted as a signal that people are buying the stock, encouraging more buying behavior.

---

Interestingly, many models of Markov decision processes use information processes that do, in fact, exhibit independent increments. For example, we may have a queueing problem where the state of the system is the number of customers in the queue. The number of arrivals may be Poisson, and the number of customers served in an increment of time is determined primarily by the length of the queue. It is possible, however, that our arrival process is a function of the length of the queue itself (see the examples for illustrations).

State-dependent information processes are more difficult to model and introduce additional parameters that must be estimated. However, from the perspective of dynamic programming, they do not introduce any fundamental complexities. As long as the distribution of outcomes is dependent purely on the state of the system, we can apply our standard models. In fact, approximate dynamic programming algorithms simply need some mechanism to sample information. It does not even matter if the exogenous information depends on information that is not in the state variable (although this will introduce errors).

It is also possible that the information arriving to the system depends on its state, as depicted in the next set of examples.

---

**■ EXAMPLE 5.1**

A driver is planning a path over a transportation network. When the driver arrives at intersection $i$ of the network, he is able to determine the transit times of each of the segments $(i, j)$ emanating from $i$. Thus, the transit times that are observed by the driver depend on the path taken by the driver.

**■ EXAMPLE 5.2**

A private equity manager learns information about a company only by investing in the company and becoming involved in its management. The information arriving to the manager depends on the state of his portfolio.

---

This is a different form of state-dependent information process. Normally, an outcome $\omega$ is assumed to represent *all* information available to the system. A probabilist would insist that this is still the case with our driver; the fact that the *driver* does not know the transit times on all the links is simply a matter of modeling the information the driver uses. However, many will find it more natural to think of the information as depending on the state.

### *More complex information processes*

Now consider the problem of modeling currency exchange rates. The change in the exchange rate between one pair of currencies is usually followed quickly by changes in others. If the Japanese yen rises relative to the U.S. dollar, it is likely that the Euro will also rise relative to it, although not necessarily proportionally. As a result, we have a vector of information processes that are correlated.

In addition to correlations between information processes, we can also have correlations over time. An upward push in the exchange rate between two currencies in one day is likely to be followed by similar changes for several days while the market responds to new information. Sometimes the changes reflect long term problems in the economy of a country. Such processes may be modeled using advanced statistical models which capture correlations between processes as well as over time.

An *information model* can be thought of as a probability density function $\phi_t(\omega_t)$ that gives the density (we would say the probability of $\omega$ if it were discrete) of an outcome $\omega_t$ in time $t$. If the problem has independent increments, we would write the density simply as $\phi_t(\omega_t)$. If the information process is Markovian (dependent on a state variable), then we would write it as $\phi_t(\omega_t|S_{t-1})$.

In some cases with complex information models, it is possible to proceed without any model at all. Instead, we can use realizations drawn from history. For example, we may take samples of changes in exchange rates from different periods in history and assume that these are representative of changes that may happen in the future. The value of using samples from history is that they capture all of the properties of the real system. This is an example of planning a system without a model of an information process.

### 5.6.5 Supervisory processes*

We are often trying to control systems where we have access to a set of decisions from an exogenous source. These may be decisions from history, or they may come from a knowledgeable expert. Either way, this produces a dataset of states $(S^m)_{m=1}^n$ and decisions $(x^m)_{m=1}^n$. In some cases, we can use this information to fit a statistical model which we use to try to predict the decision that would have been made given a state.

The nature of such a statistical model depends very much on the context, as illustrated in the examples.

---

**EXAMPLE 5.1**

Consider our nomadic trucker where we measure his attribute vector $a^n$ (his state) and his decision $d^n$ which we represent in terms of the destination of his next move. We could use a historical file $(a^m, d^m)_{m=1}^n$ to build a probability distribution $\rho(a, d)$ which gives the probability that we make decision $d$ given attribute vector $a$. We can use $\rho(a, d)$ to predict decisions in the future.

**EXAMPLE 5.2**

A mutual fund manager adds $x_t$ dollars in cash at the end of day $t$ (to be used to cover withdrawals on day $t + 1$) when there are $R_t$ dollars in cash left over at the end of the day. We can use a series of observations of $x_{t_m}$ and $R_{t_m}$ on days $t_1, t_2, \ldots, t_m$ to fit a model of the form $X(R) = \theta_0 + \theta_1 R + \theta_2 R^2 + \theta_3 R^3$.

---

We can use supervisory processes to statistically estimate a decision function that forms an initial policy. We can then use this policy in the context of an approximate dynamic programming algorithm to help fit value functions that can be used to improve the decision function. The supervisory process helps provide an initial policy that may not be perfect, but at least is reasonable.

### 5.6.6 Policies in the information process*

The sequence of information $(\omega_1, \omega_2, \ldots, \omega_t)$ is assumed to be driven by some sort of exogenous process. However, we are generally interested in quantities that are functions of both exogenous information as well as the decisions. It is useful to think of decisions as *endogenous information*. But where do the decisions come from? We now see that decisions come from policies. In fact, it is useful to represent our sequence of information and decisions as

$$H_t^\pi = (S_0, X_0^\pi, W_1, S_1, X_1^\pi, W_2, S_2, X_2^\pi, \ldots, X_{t-1}^\pi, W_t, S_t). \tag{5.12}$$

Now our history is characterized by a family of functions: the information variables $W_t$, the decision functions (policies) $X_t^\pi$, and the state variables $S_t$. We see that to characterize a particular history $h_t$, we have to specify both the sample outcome $\omega$ as well as the policy $\pi$. Thus, we might write a sample realization as

$$h_t^\pi = H_t^\pi(\omega).$$

We can think of a complete history $H_\infty^\pi(\omega)$ as an outcome in an expanded probability space (if we have a finite horizon, we would denote this by $H_T^\pi(\omega)$). Let

$$\omega^\pi = H_\infty^\pi(\omega)$$

be an outcome in our expanded space, where $\omega^\pi$ is determined by $\omega$ and the policy $\pi$. Let $\Omega^\pi$ be the set of all outcomes of this expanded space. The probability of an outcome in $\Omega^\pi$ obviously depends on the policy we are following. Thus, computing expectations (for example, expected costs or rewards) requires knowing the policy as well as the set of exogenous outcomes. For this reason, if we are interested, say, in the expected costs during time period $t$, some authors will write $E_t^\pi\{C_t(S_t, x_t)\}$ to express the dependence of the expectation on the policy. However, even if we do not explicitly index the policy, it is important to understand that we need to know how we are making decisions if we are going to compute expectations or other quantities.

## 5.7 THE TRANSITION FUNCTION

The next step in modeling a dynamic system is the specification of the *transition function*. This function describes how the system evolves from one state to another as a result of decisions and information. We begin our discussion of system dynamics by introducing some general mathematical notation. While useful, this generic notation does not provide much guidance into how specific problems should be modeled. We then describe how to model the dynamics of some simple problems, followed by a more general model for complex resources.

### 5.7.1 A general model

The dynamics of our system are represented by a function that describes how the state evolves as new information arrives and decisions are made. The dynamics of a system can be represented in different ways. The easiest is through a simple function that works as follows

$$S_{t+1} = S^M(S_t, X_t^\pi, W_{t+1}). \tag{5.13}$$

The function $S^M(\cdot)$ goes by different names such as "plant model" (literally, the model of a physical production plant), "plant equation," "law of motion," "transfer function," "system dynamics," "system model," "transition law," and "transition function." We prefer "transition function" because it is the most descriptive. We use the notation $S^M(\cdot)$ to reflect that this is the *state* transition function, which represents a *model* of the dynamics of the system. Below, we reinforce the "$M$" superscript with other modeling devices.

The arguments of the function follow standard notational conventions in the control literature (state, action, information), but different authors will follow one of two conventions for modeling time. While equation (5.13) is fairly common, many authors will write the recursion as

$$S_{t+1} = S^M(S_t, X_t^\pi, W_t). \tag{5.14}$$

If we use the form in equation (5.14), we would say "the state of the system at the beginning of time interval $t+1$ is determined by the state at time $t$, plus the decision that is made at time $t$ and the information that arrives during time interval $t$." In this representation, $t$ indexes when we are using the information. We refer to (5.14) as the *actionable representation* since it captures when we can act on the information. This representation is always used for deterministic models, and many authors adopt it for stochastic models as well. We prefer the form in equation (5.13) since we are measuring the information available at time $t$ when we are about to make a decision. If we are making a decision $x_t$ at time $t$, it is natural to index by $t$ all the variables that can be measured at time $t$. We refer to this style as the *informational representation.*

In equation (5.13), we have written the function assuming that the *function* does not depend on time (it does depend on data that depends on time). A common notational error is to write a function, say, $S_t^M(S_t, x_t)$ as if it depends on time, when in fact the function is stationary, but depends on data that depends on time. If the parameters (or structure) of the function depend on time, then we would use $S_t^M(S_t, x_t, W_{t+1})$ (or possibly $S_{t+1}^M(S_t, x_t, W_{t+1})$). If not, the transition function should be written $S^M(S_t, x_t, W_{t+1})$.

This is a very general way of representing the dynamics of a system. In many problems, the information $W_{t+1}$ arriving during time interval $t+1$ depends on the state $S_t$ at the end of time interval $t$, but is conditionally independent of all prior history given $S_t$. For example, a driver moving over a road network may only learn about the travel times on a link from $i$ to $j$ when he arrives at node $i$. When this is the case, we say that we have a Markov information process. When the decisions depend only on the state $S_t$, then we have a Markov decision process. In this case, we can store the system dynamics in the form of a one-step transition matrix using

$$\begin{aligned}
\mathbb{P}(s'|s,x) &= \text{The probability that } S_{t+1} = s' \text{ given } S_t = s \text{ and } X_t^\pi = x, \\
P^\pi &= \text{Matrix of elements where } \mathbb{P}(s'|s,x) \text{ is the element in row } s \text{ and} \\
&\quad\ \text{column } s' \text{ and where the decision } x \text{ to be made in each state is} \\
&\quad\ \text{determined by a policy } \pi.
\end{aligned}$$

There is a simple relationship between the transition function and the one-step transition matrix. Let

$$1_X = \begin{cases} 1 & X \text{ is true} \\ 0 & \text{Otherwise.} \end{cases}$$

Assuming that the set of outcomes $\Omega$ is discrete, the one-step transition matrix can be computed using

$$\begin{aligned} \mathbb{P}(s'|s,x) &= \mathbb{E}\{1_{\{s'=S^M(S_t,x,W_{t+1})\}}|S_t = s\} \\ &= \sum_{\omega_{t+1}\in\Omega_{t+1}} \mathbb{P}(W_{t+1} = \omega_{t+1})1_{\{s'=S^M(S_t,x,W_{t+1})\}}. \end{aligned} \tag{5.15}$$

It is common in the field of Markov decision processes to assume that the one-step transition is given as data. Often, it can be quickly derived (for simple problems) using assumptions about the underlying process. For example, consider a financial asset selling problem with state variable $S_t = (R_t, p_t)$ where

$$R_t = \begin{cases} 1 & \text{We are still holding the asset,} \\ 0 & \text{The asset has been sold.} \end{cases}$$

and where $p_t$ is the price at time $t$. We assume the price process is described by

$$p_t = p_{t-1} + \epsilon_t,$$

where $\epsilon_t$ is a random variable with distribution

$$\epsilon_t = \begin{cases} +1 & \text{with probability } 0.3, \\ 0 & \text{with probability } 0.6, \\ -1 & \text{with probability } 0.1. \end{cases}$$

Assume the prices are integer and range from 1 to 100. We can number our states from 0 to 100 using

$$\mathcal{S} = \{(0,-),(1,1),(1,2),\ldots,(1,100)\}.$$

We propose that our rule for determining when to sell the asset is of the form

$$X^\pi(R_t,p_t) = \begin{cases} \text{Sell asset} & \text{if } p_t < \bar{p}, \\ \text{Hold asset} & \text{if } p_t \geq \bar{p}. \end{cases}$$

Assume that $\bar{p} = 60$. A portion of the one-step transition matrix for the rows and columns corresponding to the state $(0,-)$ and $(1,58),(1,59),(1,60),(1,61),(1,62)$ looks like

$$P^{60} = \begin{matrix} (0,-) \\ (1,58) \\ (1,59) \\ (1,60) \\ (1,61) \\ (1,62) \end{matrix} \begin{bmatrix} 1 & 0 & 0 & 0 & 0 & 0 \\ 1 & 0 & 0 & 0 & 0 & 0 \\ 1 & 0 & 0 & 0 & 0 & 0 \\ 0 & 0 & .1 & .6 & .3 & 0 \\ 0 & 0 & 0 & .1 & .6 & .3 \\ 0 & 0 & 0 & 0 & .1 & .6 \end{bmatrix}.$$

As we saw in chapter 3, this matrix plays a major role in the theory of Markov decision processes, although its value is more limited in practical applications. By representing the system dynamics as a one-step transition matrix, it is possible to exploit the rich theory surrounding matrices in general and Markov chains in particular.

In engineering problems, it is far more natural to develop the transition function first. Given this, it may be possible to compute the one-step transition matrix exactly or estimate it using simulation. The techniques in this book do not, in general, use the one-step transition matrix, but instead use the transition function directly. However, formulations based on the transition matrix provide a powerful foundation for proving convergence of both exact and approximate algorithms.

### 5.7.2 The resource transition function

The vast majority of dynamic programming problems can be modeled in terms of managing "resources" where the state vector is denoted $R_t$. We use this notation when we want to specifically exclude other dimensions that might be in a state variable (for example, the challenge of making decisions to better estimate a quantity, which was first introduced in section 2.3). If we are using $R_t$ as the state variable, the general representation of the transition function would be written

$$R_{t+1} = R^M(R_t, x_t, W_{t+1}).$$

Our notation is exactly analogous to the notation for a general state variable $S_t$, but it opens the door to other modeling dimensions. For now, we illustrate the resource transition equation using some simple applications.

**Resource acquisition I - Purchasing resources for immediate use**
Let $R_t$ be the quantity of a single resource class we have available at the end of a time period, but before we have acquired new resources (for the following time period). The resource may be money available to spend on an election campaign, or the amount of oil, coal, grain or other commodities available to satisfy a market. Let $\hat{D}_t$ be the demand for the resource that occurs over time interval $t$, and let $x_t$ be the quantity of the resource that is acquired at time $t$ to be used during time interval $t+1$. The transition function would be written

$$R_{t+1} = \max\{0, R_t + x_t - \hat{D}_{t+1}\}.$$

**Resource acquisition II: purchasing futures**
Now assume that we are purchasing futures at time $t$ to be exercised at time $t' > t$. At the end of time period $t$, we would let $R_{tt'}$ be the number of futures we are holding that can be exercised during time period $t'$ (where $t' > t$). Now assume that we purchase $x_{tt'}$ additional futures to be used during time period $t'$. Our system dynamics would look like

$$R_{t+1,t'} = \begin{cases} \max\{0, (R_{tt} + R_{t,t+1}) + x_{t,t+1} - \hat{D}_{t+1}\}, & t' = t+1, \\ R_{tt'} + x_{tt'}, & t' \geq t+2. \end{cases}$$

In many problems, we can purchase resources on the spot market, which means we are allowed to see the actual demand before we make the decision. This decision would be represented by $x_{t+1,t+1}$, which means the amount purchased using the information that arrived during time interval $t+1$ to be used during time interval $t+1$ (of course, these decisions are usually the most expensive). In this case, the dynamics would be written

$$R_{t+1,t'} = \begin{cases} \max\{0, (R_{tt} + R_{t,t+1}) + x_{t,t+1} + x_{t+1,t+1} - \hat{D}_{t+1}\}, & t' = t+1, \\ R_{tt'} + x_{tt'}, & t' \geq t+2. \end{cases}$$

**Planning a path through college**
Consider a student trying to satisfy a set of course requirements (for example, number of science courses, language courses, departmentals, and so on). Let $R_{tc}$ be the number of courses taken that satisfy requirement $c$ at the end of semester $t$. Let $x_{tc}$ be the number of courses the student enrolled in at the end of semester $t$ for semester $t+1$ to satisfy requirement $c$. Finally let $\hat{F}_{tc}(x_{t-1})$ be the number of courses in which the student

received a failing grade during semester $t$ given $x_{t-1}$. This information depends on $x_{t-1}$ since a student cannot fail a course that she was not enrolled in. The system dynamics would look like

$$R_{t+1,c} = R_{t,c} + x_{t,c} - \hat{F}_{t+1,c}.$$

**Playing backgammon**

In backgammon, there are 24 points on which pieces may sit, plus a bar when you bump your opponent's pieces off the board, plus the state that a piece has been removed from the board (you win when you have removed all of your pieces). Let $i \in \{1, 2, \ldots, 26\}$ represent these positions. Let $R_{ti}$ be the number of your pieces on location $i$ at time $t$, where $R_{ti} < 0$ means that your opponent has $|R_{ti}|$ pieces on $i$. Let $x_{tii'}$ be the number of pieces moved from point $i$ to $i'$ at time $t$. If $\omega_t = W_t(\omega)$ is the outcome of the roll of the dice, then the allowable decisions can be written in the general form $x_t \in \mathcal{X}_t(\omega)$ where the feasible region $\mathcal{X}_t(\omega)$ captures the rules on how many pieces can be moved given $W_t(\omega)$. For example, if we only have two pieces on a point, $\mathcal{X}_t$ would restrict us from moving more than two pieces from this point. Let $\delta x_t$ be a column vector with element $i$ given by

$$\delta x_{ti} = \sum_{i'} x_{ti'i} - \sum_{i''} x_{tii''}.$$

Now let $\hat{y}_{t+1}$ be a variable similar to $x_t$ representing the moves of the opponent after we have finished our moves, and let $\delta\hat{y}_{t+1}$ be a column vector similar to $\delta x_t$. The transition equation would look like

$$R_{t+1} = R_t + \delta x_t + \delta\hat{y}_{t+1}.$$

**A portfolio problem**

Let $R_{tk}$ be the amount invested in asset $k \in \mathcal{K}$ where $\mathcal{K}$ may be individual stocks, mutual funds or asset classes such as bonds and money market instruments. Assume that each month we examine our portfolio and shift money from one asset to another. Let $x_{tkk'}$ be the amount we wish to move from asset $k$ to $k'$, where $x_{tkk}$ is the amount we hold in asset $k$. We assume the transition is made instantly (the issue of transaction costs are not relevant here). Now let $\hat{\rho}_{t+1,k}$ be the return for asset $k$ between $t$ and $t+1$. The transition equation would be given by

$$R_{t+1,k} = \hat{\rho}_{t+1,k}\left(\sum_{k' \in \mathcal{K}} x_{tk'k}\right).$$

We note that it is virtually always the case that the returns $\hat{\rho}_{tk}$ are correlated across the assets. When we use a sample realization $\hat{\rho}_{t+1}(\omega)$, we assume that these correlations are reflected in the sample realization.

### 5.7.3 Transition functions for complex resources*

When we are managing multiple, complex resources, each of which are described by an attribute vector $a$, it is useful to adopt special notation to describe the evolution of the system. Recall that the state variable $S_t$ might consist of both the resource state variable $R_t$ as well as other information. We need notation that specifically describes the evolution of $R_t$ separately from the other variables.

| t=40 pre-decision | t=40 post-decision | t=60 pre-decision |
|---|---|---|
| $\begin{pmatrix} St.Louis \\ 41.4 \end{pmatrix}$ | $\begin{pmatrix} LosAngeles \\ 65.0 \end{pmatrix}$ | $\begin{pmatrix} LosAngeles \\ 70.4 \end{pmatrix}$ |

**Figure 5.3** Pre- and post-decision attributes for a nomadic trucker.

We define the *attribute transition function* which describes how a specific entity with attribute vector $a$ is modified by a decision of type $d$. This is modeled using

$$a_{t+1} = a^M(a_t, d_t, W_{t+1}). \tag{5.16}$$

The function $a^M(\cdot)$ parallels the state transition function $S^M(\cdot)$, but it works at the level of a decision of type $d$ acting on a resource of type $a$. It is possible that there is random information affecting the outcome of the decision. For example, in section 5.3.3, we introduced a realistic version of our nomadic trucker where the attributes include dimensions such as the estimated time that the driver will arrive in a city (random delays can change this), and the maintenance status of the equipment (the driver may identify an equipment problem while moving).

As with our state variable, we let $a_t$ be the attribute just before we make a decision. Although we are acting on the resource with a decision of type $d$, we retain our notation for the post-decision state and let $a_t^x$ be the post-decision attribute vector. A simple example illustrates the pre- and post-decision attribute vectors for our nomadic trucker. Assume our driver has two attributes: location and the expected time at which he can be assigned to a new activity. Figure 5.3 shows that at time $t = 40$, we expect the driver to be available in St. Louis at time $t = 41.4$. At $t = 40$, we make the decision that as soon as the driver is available, we are going to assign him to a load going to Los Angeles, where we expect him (again at $t = 40$) to arrive at time $t = 65.0$. At time $t = 60$, he is still expected to be heading to Los Angeles, but we have received information that he has been delayed and now expect him to be available at time $t = 70.4$ (the delay is the new information).

As before, we can break down the effect of decisions and information using

$$\begin{aligned} a_t^x &= a^{M,x}(a_t, d_t), \\ a_{t+1} &= a^{M,W}(a_t^x, W_{t+1}). \end{aligned}$$

For algebraic purposes, it is also useful to define the indicator function

$$\begin{aligned} \delta_{a'}^x(a,d) &= \begin{cases} 1, & a_t^x = a' = a^{M,x}(a_t, d_t), \\ 0 & \text{otherwise.} \end{cases} \\ \Delta^x &= \text{Matrix with } \delta_{a'}^x(a,d) \text{ in row } a' \text{ and column } (a,d). \end{aligned}$$

The function $\delta^x(\cdot)$ (or matrix $\Delta^x$) gives the post-decision attribute vector resulting from a decision $d$ (in the case of $\Delta^x$, a set of decisions represented by $x_t$).

The attribute transition function is convenient since it describes many problems. Implicit in the notation is the behavior that the result of a decision acting on a resource of type $a$ does not depend on the status of any of the other resources. We can capture this behavior by defining a more general transition function for the resource vector, which we would write

$$R_{t+1} = R^M(R_t, x_t, W_{t+1}). \tag{5.17}$$

In this setting, we would represent the random information $W_{t+1}$ as exogenous changes to the resource vector, given by

$\hat{R}_{t+1,a}$ = The change in the number of resources with attribute vector $a$ due to information arriving during time interval $t+1$.

We can now write out the transition equation using

$$R_{t+1,a'} = \sum_{a \in \mathcal{A}} \sum_{d \in \mathcal{D}} \delta^x_{a'}(a,d) x_{tad} + \hat{R}_{t+1,a'}$$

or in matrix-vector form

$$R_{t+1} = \Delta^x R_t + \hat{R}_{t+1}.$$

### 5.7.4 Some special cases*

It is important to realize when special cases offer particular modeling challenges (or opportunities). Below we list a few we have encountered.

#### *Deterministic resource transition functions*

It is quite common in resource allocation problems that the attribute transition function is deterministic (the equipment never breaks down, there are no delays, resources do not randomly enter and leave the system). The uncertainty may arise not in the evolution of the resources that we are managing, but in the "other information" such as customer demands (for example, the loads that the driver might move) or our estimates of parameters (where we use our decisions to collect information on random variables with unknown distributions). If the attribute transition function is deterministic, then we would write $a^x_t = a^M(a_t, d_t)$ and we would have that $a_{t+1} = a^x_t$. In this case, we simply write

$$a_{t+1} = a^M(a_t, d_t).$$

This is an important special case since it arises quite often in practice. It does not mean the transition function is deterministic. For example, a driver moving over a network faces the decision, at node $i$, whether to go to node $j$. If he makes this decision, he will go to node $j$ deterministically, but the cost or time over the link from $i$ to $j$ may be random.

Another example arises when we are managing resources (people, equipment, blood) to serve demands that arrive randomly over time. The effect of a decision acting on a resource is deterministic. The only source of randomness is in the demands that arise in each time period. Let $R_t$ be the vector describing our resources, and let $\hat{D}_t$ be the demands that arose during time interval $t$. The state variable is $S_t = (R_t, \hat{D}_t)$ where $\hat{D}_t$ is purely exogenous. If $x_t$ is our decision vector which determines which resources are assigned to each demand, then the resource transition function $R_{t+1} = R^M(R_t, x_t)$ would be deterministic.

#### *Gaining and losing resources*

In addition to the attributes of the modified resource, we sometimes have to capture the fact that we may gain or lose resources in the process of completing a decision. We might define

$\rho_{t+1,a,d}$ = The multiplier giving the quantity of resources with attribute vector $a$ available after being acted on with decision $d$ at time $t$.

The multiplier may depend on the information available at time $t$ (in which case we would write it as $\rho_{tad}$), but is often random and depends on information that has not yet arrived (in which case we use $\rho_{t+1,a,d}$). Illustrations of gains and losses are given in the next set of examples.

---

■ **EXAMPLE 5.1**

A corporation is holding money in an index fund with a 180-day holding period (money moved out of this fund within the period incurs a four percent load) and would like to transfer them into a high yield junk bond fund. The attribute of the resource would be $a = (\text{Type, Age})$. There is a transaction cost (the cost of executing the trade) and a gain $\rho$, which is 1.0 for funds held more than 180 days, and 0.96 for funds held less than 180 days.

■ **EXAMPLE 5.2**

Transportation of liquefied natural gas - A company would like to purchase 500,000 tons of liquefied natural gas in southeast Asia for consumption in North America. Although in liquified form, the gas evaporates at a rate of 0.2 percent per day, implying $\rho = .998$.

---

## 5.8 THE CONTRIBUTION FUNCTION

We assume we have some measure of how well we are doing. This might be a cost that we wish to minimize, a contribution or reward if we are maximizing, or a more generic utility (which we typically maximize). We assume we are maximizing a contribution. In many problems, the contribution is a deterministic function of the state and action, in which case we would write

$C(S_t, x_t)$ = The contribution (cost if we are minimizing) earned by taking action $x_t$ while in state $S_t$ at time $t$.

We often write the contribution function as $C_t(S_t, x_t)$ to emphasize that it is being measured at time $t$ and therefore depends on information in the state variable $S_t$. Our contribution may be random. For example, we may invest $x$ dollars in an asset that earns a random return $\rho_{t+1}$ which we do not know until time $t+1$. We may think of the contribution as $\rho_{t+1}x_t$, which means the contribution is random. In this case, we typically will write

$\hat{C}_{t+1}(S_t, x_t, W_{t+1})$ = Contribution at time $t$ from being in state $S_t$, making decision $x_t$ which also depends on the information $W_{t+1}$.

We emphasize that the decision $x_t$ does not have access to the information $W_{t+1}$ (this is where our time indexing style eliminates any ambiguity about the information content of a variable). As a result, the decision $x_t$ has to work with the expected contribution, which we write

$$C_t(S_t, x_t) = \mathbb{E}\{\hat{C}_{t+1}(S_t, x_t, W_{t+1})|S_t\}.$$

The role that $W_{t+1}$ plays is problem-dependent, as illustrated in the examples below.

---

**EXAMPLE 5.1**

In asset acquisition problems, we order $x_t$ in time period $t$ to be used to satisfy demands $\hat{D}_{t+1}$ in the next time period. Our state variable is $S_t = R_t =$ the product on hand after demands in period $t$ have been satisfied. We pay a cost $c^p x_t$ in period $t$ and receive a revenue $p \min(R_t + x_t, \hat{D}_{t+1})$ in period $t+1$. Our total one-period contribution function is then

$$\hat{C}_{t,t+1}(R_t, x_t, \hat{D}_{t+1}) = p \min(R_t + x_t, \hat{D}_{t+1}) - c^p x_t.$$

The expected contribution is

$$C_t(S_t, x_t) = \mathbb{E}\{p \min(R_t + x_t, \hat{D}_{t+1}) - c^p x_t\}.$$

**EXAMPLE 5.2**

Now consider the same asset acquisition problem, but this time we place our orders in period $t$ to satisfy the known demand in period $t$. Our cost function contains both a fixed cost $c^f$ (which we pay for placing an order of any size) and a variable cost $c^p$. The cost function would look like

$$C_t(S_t, x_t) = \begin{cases} p \min(R_t + x_t, \hat{D}_t), & x_t = 0, \\ p \min(R_t + x_t, \hat{D}_t) - c^f - c^p x_t, & x_t > 0,. \end{cases}$$

Note that our contribution function no longer contains information from the next time period. If we did not incur a fixed cost $c^f$, then we would simply look at the demand $D_t$ and order the quantity needed to cover demand (as a result, there would never be any product left over). However, since we incur a fixed cost $c^f$ with each order, there is a benefit to ordering enough to cover the demand now and future demands. This benefit is captured through the value function.

---

There are many resource allocation problems where the contribution of a decision can be written using

$c_{tad}$ = The unit contribution of acting on a resource with attribute vector $a$ with decision $d$.

This contribution is incurred in period $t$ using information available in period $t$. In this case, our total contribution at time $t$ could be written

$$C_t(S_t, x_t) = \sum_{a \in \mathcal{A}} \sum_{d \in \mathcal{D}} c_{tad} x_{tad}.$$

In general, when we use a pre-decision state variable, it is best to think of $C_t(S_t, x_t)$ as an expectation of a function that may depend on future information. It is important to be aware that in some settings, the contribution function does not depend on future information.

It is surprisingly common for us to want to work with two contributions. The common view of a contribution function is that it contains revenues and costs that we want to maximize or minimize. In many operational problems, there can be a mixture of "hard dollars" and "soft dollars." The hard dollars are our quantifiable revenues and costs. But there are often other issues that are important in an operational setting, but which cannot always be easily quantified. For example, if we cannot cover all of the demand, we may wish to assess a penalty for not satisfying it. We can then manipulate this penalty to reduce the amount of unsatisfied demand. Examples of the use of soft-dollar bonuses and penalties abound in operational problems (see examples).

---

■ **EXAMPLE 5.1**

A trucking company has to pay the cost of a driver to move a load, but wants to avoid using inexperienced drivers for their high priority accounts (but has to accept the fact that it is sometimes necessary). An artificial penalty can be used to reduce the number of times this happens.

■ **EXAMPLE 5.2**

A charter jet company requires that in order for a pilot to land at night, he/she has to have landed a plane at night three times in the last 60 days. If the third time a pilot landed at night is at least 50 days ago, the company wants to encourage assignments of these pilots to flights with night landings so that they can maintain their status. A bonus can be assigned to encourage these assignments.

■ **EXAMPLE 5.3**

A student planning her schedule of courses has to face the possibility of failing a course, which may require taking either an extra course one semester or a summer course. She wants to plan out her course schedule as a dynamic program, but use a penalty to reduce the likelihood of having to take an additional course.

■ **EXAMPLE 5.4**

An investment banker wants to plan a strategy to maximize the value of an asset and minimize the likelihood of a very poor return. She is willing to accept lower overall returns in order to achieve this goal and can do it by incorporating an additional penalty when the asset is sold at a significant loss.

---

Given the presence of these so-called "soft dollars," it is useful to think of two contribution functions. We can let $C_t(S_t, x_t)$ be the hard dollars and $C_t^\pi(S_t, x_t)$ be the contribution function with the soft dollars included. The notation captures the fact that a set of soft bonuses and penalties represents a form of policy. So we can think of our policy as making decisions that maximize $C_t^\pi(S_t, x_t)$, but measure the value of the policy (in hard dollars), using $C_t(S_t, X^\pi(S_t))$.

## 5.9 THE OBJECTIVE FUNCTION

We are now ready to write out our objective function. Let $X_t^\pi(S_t)$ be a decision function (equivalent to a policy) that determines what decision we make given that we are in state $S_t$. Our optimization problem is to choose the best policy by choosing the best decision function from the family $(X_t^\pi(S_t))_{\pi\in\Pi}$. We wish to choose the best function that maximizes the total expected (discounted) contribution over a finite (or infinite) horizon. This would be written as

$$F_0^* = \max_{\pi\in\Pi} \mathbb{E}\left\{\sum_{t=0}^{T} \gamma^t C_t^\pi(S_t, X_t^\pi(S_t))|S_0\right\}, \tag{5.18}$$

where $\gamma$ discounts the money into time $t = 0$ values. We write the value of policy $\pi$ as

$$F_0^\pi = \mathbb{E}\left\{\sum_{t=0}^{T} \gamma^t C_t^\pi(S_t, X_t^\pi(S_t))|S_0\right\}.$$

In some communities, it is common to use an interest rate $r$, in which case the discount factor is

$$\gamma = \frac{1}{1+r}.$$

Important variants of this objective function are the infinite horizon problem ($T = \infty$) and the undiscounted finite horizon problem ($\gamma = 1$).

A separate problem class is the average reward, infinite horizon problem

$$F_0^\pi = \mathbb{E}\left\{\lim_{T\to\infty} \frac{1}{T}\sum_{t=0}^{T-1} C_t^\pi(S_t, X_t^\pi(S_t))|S_0\right\}. \tag{5.19}$$

Our optimization problem is to choose the best policy. In most practical applications, we can write the optimization problem as one of choosing the best policy, or

$$F_0^* = \max_{\pi\in\Pi} F_0^\pi. \tag{5.20}$$

It might be the case that a policy is characterized by a continuous parameter (the speed of a car, the temperature of a process, the price of an asset). In theory, we could have a problem where the optimal policy corresponds to a value of a parameter being equal to infinity. It is possible that $F_0^*$ exists, but that an optimal "policy" does not exist (because it requires finding a parameter equal to infinity). While this is more of a mathematical curiosity, we handle these situations by writing the optimization problem as

$$F_0^* = \sup_{\pi\in\Pi} F_0^\pi, \tag{5.21}$$

where "*sup*" is the supremum operator, which finds the smallest number greater than or equal to $F_0^\pi$ for any value of $\pi$. If we were minimizing, we would use "inf," which stands for "infimum," which is the largest value less than or equal to the value of any policy. It is common in more formal treatments to use "sup" instead of "max" or "inf" instead of "min" since these are more general. Our emphasis is on computation and approximation, where we consider only problems where a solution exists. For this reason, we use "max" and "min" throughout our presentation.

The expression (5.18) contains one important but subtle assumption that will prove to be critical later and which will limit the applicability of our techniques in some problem classes. Specifically, we assume the presence of what is known as *linear, additive utility*. That is, we have added up contributions for each time period. It does not matter if the contributions are discounted or if the contribution functions themselves are nonlinear. However, we will not be able to handle functions that look like

$$F^{\pi} = \mathbb{E}^{\pi}\left\{\left(\sum_{t\in\mathcal{T}}\gamma^t C_t(S_t, X_t^{\pi}(S_t))\right)^2\right\}. \tag{5.22}$$

The assumption of linear, additive utility means that the total contribution is a separable function of the contributions in each time period. While this works for many problems, it certainly does not work for all of them, as depicted in the examples below.

---

**■ EXAMPLE 5.1**

We may value a policy of managing a resource using a nonlinear function of the number of times the price of a resource dropped below a certain amount.

**■ EXAMPLE 5.2**

Assume we have to find the route through a network where the traveler is trying to arrive at a particular point in time. The value function is a nonlinear function of the total lateness, which means that the value function is not a separable function of the delay on each link.

**■ EXAMPLE 5.3**

Consider a mutual fund manager who has to decide how much to allocate between aggressive stocks, conservative stocks, bonds, and money market instruments. Let the allocation of assets among these alternatives represent a policy $\pi$. The mutual fund manager wants to maximize long term return, but needs to be sensitive to short term swings (the risk). He can absorb occasional downswings, but wants to avoid sustained downswings over several time periods. Thus, his value function must consider not only his return in a given time period, but also how his return looks over one-year, three-year and five-year periods.

---

In some cases these apparent instances of violations of linear, additive utility can be solved using a creatively defined state variable.

## 5.10 A MEASURE-THEORETIC VIEW OF INFORMATION**

For researchers interested in proving theorems or reading theoretical research articles, it is useful to have a more fundamental understanding of information.

When we work with random information processes and uncertainty, it is standard in the probability community to define a probability space, which consists of three elements. The first is the set of outcomes $\Omega$, which is generally assumed to represent all possible outcomes of the information process (actually, $\Omega$ can include outcomes that can never happen). If these outcomes are discrete, then all we would need is the probability of each outcome $p(\omega)$.

It is nice to have a terminology that allows for continuous quantities. We want to define the probabilities of our events, but if $\omega$ is continuous, we cannot talk about the probability of an outcome $\omega$. However we can talk about a set of outcomes $\mathcal{E}$ that represent some specific event (if our information is a price, the event $\mathcal{E}$ could be all the prices that constitute the event that the price is greater than some number). In this case, we can define the probability of an outcome $\mathcal{E}$ by integrating the density function $p(\omega)$ over all $\omega$ in the event $\mathcal{E}$.

Probabilists handle continuous outcomes by defining a set of events $\mathfrak{F}$, which is literally a "set of sets" because each element in $\mathfrak{F}$ is itself a set of outcomes in $\Omega$. This is the reason we resort to the script font $\mathfrak{F}$ as opposed to our calligraphic font for sets; it is easy to read $\mathcal{E}$ as "calligraphic E" and $\mathfrak{F}$ as "script F." The set $\mathfrak{F}$ has the property that if an event $\mathcal{E}$ is in $\mathfrak{F}$, then its complement $\Omega \setminus \mathcal{E}$ is in $\mathfrak{F}$, and the union of any two events $\mathcal{E}_X \cup \mathcal{E}_Y$ in $\mathfrak{F}$ is also in $\mathfrak{F}$. $\mathfrak{F}$ is called a "sigma-algebra" (which may be written "$\sigma$-algebra"), and is a countable union of outcomes in $\Omega$. An understanding of sigma-algebras is not important for computational work, but can be useful in certain types of proofs, as we see in the "why does it work" sections at the end of several chapters. Sigma-algebras are without question one of the more arcane devices used by the probability community, but once they are mastered, they are a powerful theoretical tool.

Finally, it is required that we specify a probability measure denoted $\mathcal{P}$, which gives the probability (or density) of an outcome $\omega$ which can then be used to compute the probability of an event in $\mathfrak{F}$.

We can now define a formal probability space for our exogenous information process as $(\Omega, \mathfrak{F}, \mathcal{P})$. If we wish to take an expectation of some quantity that depends on the information, say $Ef(W_t)$, then we would sum (or integrate) over the set $\omega$ multiplied by the probability (or density) $\mathcal{P}$.

It is important to emphasize that $\omega$ represents *all* the information that will become available, over all time periods. As a rule, we are solving a problem at time $t$, which means we do not have the information that will become available after time $t$. To handle this, we let $\mathfrak{F}_t$ be the sigma-algebra representing events that can be created using only the information up to time $t$. To illustrate, consider an information process $W_t$ consisting of a single 0 or 1 in each time period. $W_t$ may be the information that a customer purchases a jet aircraft, or the event that an expensive component in an electrical network fails. If we look over three time periods, there are eight possible outcomes, as shown in table 5.2.

Let $\mathcal{E}_{\{W_1\}}$ be the set of outcomes $\omega$ that satisfy some logical condition on $W_1$. If we are at time $t = 1$, we only see $W_1$. The event $W_1 = 0$ would be written

$$\mathcal{E}_{\{W_1=0\}} = \{\omega | W_1 = 0\} = \{1, 2, 3, 4\}.$$

The sigma-algebra $\mathfrak{F}_1$ would consist of the events

$$\{\mathcal{E}_{\{W_1=0\}}, \mathcal{E}_{\{W_1=1\}}, \mathcal{E}_{\{W_1 \in \{0,1\}\}}, \mathcal{E}_{\{W_1 \notin \{0,1\}\}}\}.$$

Now assume that we are at time $t = 2$ and have access to $W_1$ and $W_2$. With this information, we are able to divide our outcomes $\Omega$ into finer subsets. Our history $H_2$ consists of the elementary events $\mathcal{H}_2 = \{(0,0), (0,1), (1,0), (1,1)\}$. Let $h_2 = (0,1)$ be an element of

| Outcome | Time period | | |
|---|---|---|---|
| $\omega$ | 1 | 2 | 3 |
| 1 | 0 | 0 | 0 |
| 2 | 0 | 0 | 1 |
| 3 | 0 | 1 | 0 |
| 4 | 0 | 1 | 1 |
| 5 | 1 | 0 | 0 |
| 6 | 1 | 0 | 1 |
| 7 | 1 | 1 | 0 |
| 8 | 1 | 1 | 1 |

**Table 5.2** Set of demand outcomes

$H_2$. The event $\mathcal{E}_{\{h_2=(0,1)\}} = \{3,4\}$. At time $t = 1$, we could not tell the difference between outcomes 1, 2, 3, and 4; now that we are at time 2, we can differentiate between $\omega \in \{1,2\}$ and $\omega \in \{3,4\}$. The sigma-algebra $\mathfrak{F}_2$ consists of all the events $\mathcal{E}_{h_2}, h_2 \in \mathcal{H}_2$, along with all possible unions and complements.

Another event in $\mathfrak{F}_2$ is $\{\omega|(W_1, W_2) = (0,0)\} = \{1,2\}$. A third event in $\mathfrak{F}_2$ is the union of these two events, which consists of $\omega = \{1,2,3,4\}$ which, of course, is one of the events in $\mathfrak{F}_1$. In fact, every event in $\mathfrak{F}_1$ is an event in $\mathfrak{F}_2$, but not the other way around. The reason is that the additional information from the second time period allows us to divide $\Omega$ into finer set of subsets. Since $\mathfrak{F}_2$ consists of all unions (and complements), we can always take the union of events, which is the same as ignoring a piece of information. By contrast, we cannot divide $\mathfrak{F}_1$ into a finer subsets. The extra information in $\mathfrak{F}_2$ allows us to filter $\Omega$ into a finer set of subsets than was possible when we only had the information through the first time period. If we are in time period 3, $\mathfrak{F}$ will consist of each of the individual elements in $\Omega$ as well as all the unions needed to create the same events in $\mathfrak{F}_2$ and $\mathfrak{F}_1$.

From this example, we see that more information (that is, the ability to see more elements of $W_1, W_2, \ldots$) allows us to divide $\Omega$ into finer-grained subsets. For this reason, we can always write $\mathfrak{F}_{t-1} \subseteq \mathfrak{F}_t$. $\mathfrak{F}_t$ always consists of every event in $\mathfrak{F}_{t-1}$ in addition to other finer events. As a result of this property, $\mathfrak{F}_t$ is termed a *filtration*. It is because of this interpretation that the sigma-algebras are typically represented using the script letter $F$ (which literally stands for filtration) rather the more natural letter $H$ (which stands for history). The fancy font used to denote a sigma-algebra is used to designate that it is a set of sets (rather than just a set).

It is *always* assumed that information processes satisfy $\mathfrak{F}_{t-1} \subseteq \mathfrak{F}_t$. Interestingly, this is not always the case in practice. The property that information forms a filtration requires that we never "forget" anything. In real applications, this is not always true. Assume, for example, that we are doing forecasting using a moving average. This means that our forecast $f_t$ might be written as $f_t = (1/T)\sum_{t'=1}^{T} \hat{D}_{t-t'}$. Such a forecasting process "forgets" information that is older than $T$ time periods.

There are numerous textbooks on measure theory. For a nice introduction to measure-theoretic thinking (and in particular the value of measure-theoretic thinking), see Pollard (2002).

## 5.11 BIBLIOGRAPHIC NOTES

Section 5.2 - Figure 5.1 which describes the mapping from continuous to discrete time was outlined for me by Erhan Cinlar.

Section 5.3 - The multiattribute notation for multiple resource classes is based primarily on Powell et al. (2001).

Section 5.4 - The definition of states is amazingly confused in the stochastic control literature. The first recognition of the difference between the physical state and the state of knowledge appears to be in Bellman & Kalaba (1959) which used the term "hyperstate" to refer to the state of knowledge. The control literature has long used state to represent a sufficient statistic (see for example Kirk (1998)), representing the information needed to model the system forward in time. For an introduction to partially observable Markov decision processes, see White (1991). An excellent description of the modeling of Markov decision processes from an AI perspective is given in Boutilier et al. (1999), including a very nice discussion of factored representations. See also Guestrin et al. (2003) for an application of the concept of factored state spaces to a Markov decision process.

Section 5.5 - Our notation for decisions represents an effort to bring together the fields of dynamic programming and math programming. We believe this notation was first used in Powell et al. (2001). For a classical treatment of decisions from the perspective of Markov decision processes, see Puterman (1994). For examples of decisions from the perspective of the optimal control community, see Kirk (1998) and Bertsekas (2000). For examples of treatments of dynamic programming in economics, see Stokey & R. E. Lucas (1989) and Chow (1997).

Section 5.6 - Our representation of information follows classical styles in the probability literature (see, for example, Chung (1974)). Considerable attention has been given to the topic of supervisory control. A sample include Werbos (1992*b*).

## PROBLEMS

**5.1** A college student must plan what courses she takes over each of eight semesters. To graduate, she needs 34 total courses, while taking no more than five and no less than three courses in any semester. She also needs two language courses, one science course, eight departmental courses in her major and two math courses.

(a) Formulate the state variable for this problem in the most compact way possible.

(b) Give the transition function for our college student assuming that she successfully passes any course she takes. You will need to introduce variables representing her decisions.

(c) Give the transition function for our college student, but now allow for the random outcome that she may not pass every course.

**5.2** Assume that we have $N$ discrete resources to manage, where $R_a$ is the number of resources of type $a \in \mathcal{A}$ and $N = \sum_{a \in \mathcal{A}} R_a$. Let $\mathcal{R}$ be the set of possible values of the vector $R$. Show that

$$|\mathcal{R}| = \binom{N + |\mathcal{A}| - 1}{|\mathcal{A}| - 1},$$

where

$$\binom{X}{Y} = \frac{X!}{Y!(X-Y)!}$$

is the number of combinations of $X$ items taken $Y$ at a time.

**5.3** A broker is working in thinly traded stocks. He must make sure that he does not buy or sell in quantities that would move the price and he feels that if he works in quantities that are no more than 10 percent of the average sales volume, he should be safe. He tracks the average sales volume of a particular stock over time. Let $\hat{v}_t$ be the sales volume on day $t$, and assume that he estimates the average demand $f_t$ using $f_t = (1-\alpha)f_{t-1} + \alpha \hat{v}_t$. He then uses $f_t$ as his estimate of the sales volume for the next day. Assuming he started tracking demands on day $t = 1$, what information would constitute his state variable?

**5.4** How would your previous answer change if our broker used a 10-day moving average to estimate his demand? That is, he would use $f_t = 0.10 \sum_{i=1}^{10} \hat{v}_{t-i+1}$ as his estimate of the demand.

**5.5** The pharmaceutical industry spends millions managing a sales force to push the industry's latest and greatest drugs. Assume one of these salesmen must move between a set $\mathcal{I}$ of customers in his district. He decides which customer to visit next only after he completes a visit. For this exercise, assume that his decision does not depend on his prior history of visits (that is, he may return to a customer he has visited previously). Let $S_n$ be his state immediately after completing his $n^{th}$ visit that day.

(a) Assume that it takes exactly one time period to get from any customer to any other customer. Write out the definition of a state variable, and argue that his state is only his current location.

(b) Now assume that $\tau_{ij}$ is the (deterministic and integer) time required to move from location $i$ to location $j$. What is the state of our salesman at any time $t$? Be sure to consider both the possibility that he is at a location (having just finished with a customer) or between locations.

(c) Finally assume that the travel time $\tau_{ij}$ follows a discrete uniform distribution between $a_{ij}$ and $b_{ij}$ (where $a_{ij}$ and $b_{ij}$ are integers)?

**5.6** Consider a simple asset acquisition problem where $x_t$ is the quantity purchased at the end of time period $t$ to be used during time interval $t+1$. Let $D_t$ be the demand for the assets during time interval $t$. Let $R_t$ be the pre-decision state variable (the amount on hand before you have ordered $x_t$) and $R_t^x$ be the post-decision state variable.

(a) Write the transition function so that $R_{t+1}$ is a function of $R_t$, $x_t$, and $D_{t+1}$.

(b) Write the transition function so that $R_t^x$ is a function of $R_{t-1}^x$, $D_t$, and $x_t$.

(c) Write $R_t^x$ as a function of $R_t$, and write $R_{t+1}$ as a function of $R_t^x$.

**5.7** As a buyer for an orange juice products company, you are responsible for buying futures for frozen concentrate. Let $x_{tt'}$ be the number of futures you purchase in year $t$ that can be exercised during year $t'$.

(a) What is your state variable in year $t$?

(b) Write out the transition function.

**5.8** A classical inventory problem works as follows. Assume that our state variable $R_t$ is the amount of product on hand at the end of time period $t$ and that $D_t$ is a random variable giving the demand during time interval $(t-1, t)$ with distribution $p_d = \mathbb{P}(D_t = d)$. The demand in time interval $t$ must be satisfied with the product on hand at the beginning of the period. We can then order a quantity $x_t$ at the end of period $t$ that can be used to replenish the inventory in period $t+1$. Give the transition function that relates $R_{t+1}$ to $R_t$.

**5.9** Many problems involve the movement of resources over networks. The definition of the state of a single resource, however, can be complicated by different assumptions for the probability distribution for the time required to traverse a link. For each example below, give the state of the resource:

(a) You have a deterministic, static network, and you want to find the shortest path from an origin node $q$ to a destination node $r$. There is a known cost $c_{ij}$ for traversing each link $(i, j)$.

(b) Next assume that the cost $c_{ij}$ is a random variable with an unknown distribution. Each time you traverse a link $(i, j)$, you observe the cost $\hat{c}_{ij}$, which allows you to update your estimate $\bar{c}_{ij}$ of the mean of $c_{ij}$.

(c) Finally assume that when the traveler arrives at node $i$ he sees $\hat{c}_{ij}$ for each link $(i, j)$ out of node $i$.

(d) A taxicab is moving people in a set of cities $\mathcal{C}$. After dropping a passenger off at city $i$, the dispatcher may have to decide to reposition the cab from $i$ to $j$, $(i, j) \in \mathcal{C}$. The travel time from $i$ to $j$ is $\tau_{ij}$, which is a random variable with a discrete uniform distribution (that is, the probability that $\tau_{ij} = t$ is $1/T$, for $t = 1, 2, \ldots, T$). Assume that the travel time is known before the trip starts.

(e) Same as (d), but now the travel times are random with a geometric distribution (that is, the probability that $\tau_{ij} = t$ is $(1-\theta)\theta^{t-1}$, for $t = 1, 2, 3, \ldots$).

**5.10** In the figure below, a sailboat is making its way upwind from point A to point B. To do this, the sailboat must tack, whereby it sails generally at a 45 degree angle to the wind. The problem is that the angle of the wind tends to shift randomly over time. The skipper decides to check the angle of the wind each minute and must decide whether the boat should be on port or starboard tack. Note that the proper decision must consider the current location of the boat, which we may indicate by an $(x, y)$ coordinate.

(a) Formulate the problem as a dynamic program. Carefully define the state variable, decision variable, exogenous information and the contribution function.

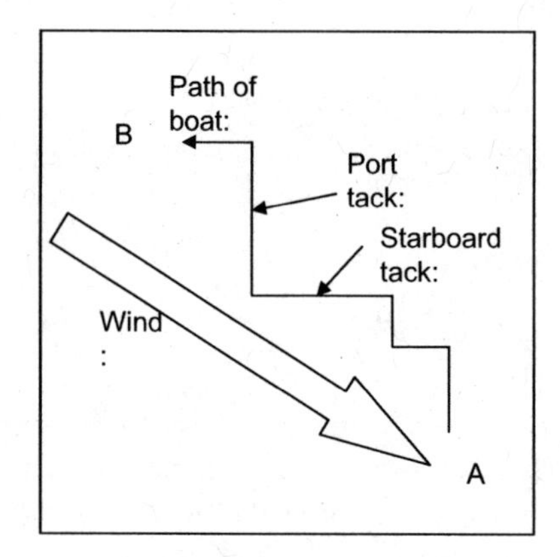


(b) Use $\delta$ to discretize any continuous variables (in practice, you might choose different levels of discretization for each variable, but we are going to keep it simple). In terms of $\delta$, give the size of the state space, the number of exogenous outcomes (in a single time period) and the action space. If you need an upper bound on a variable (e.g. wind speed), simply define an appropriate variable and express your answer in terms of this variable. All your answers should be expressed algebraically.

(c) Using a maximum wind speed of 30 miles per hour and $\delta = .1$, compute the size of your state, outcome and action spaces.

**5.11** Implement your model from exercise 5.10 as a Markov decision process, and solve it using the techniques of 3 (section 3.2). Choose a value of $\delta$ that makes your program computationally reasonable (run times under 10 minutes). Let $\bar{\delta}$ be the smallest value of $\delta$ that produces a run time (for your computer) of under 10 minutes, and compare your solution (in terms of the total contribution) for $\delta = \bar{\delta}^N$ for $N = 2, 4, 8, 16$. Evaluate the quality of the solution by simulating 1000 iterations using the value functions obtained using backward dynamic programming. Plot your average contribution function as a function of $\delta$.

**5.12** What is the difference between the *history* of a process, and the state of a process?

**5.13** As the purchasing manager for a major citrus juice company, you have the responsibility of maintaining sufficient reserves of oranges for sale or conversion to orange juice products. Let $x_{ti}$ be the amount of oranges that you decide to purchase from supplier $i$ in week $t$ to be used in week $t + 1$. Each week, you can purchase up to $\hat{q}_{ti}$ oranges (that is, $x_{ti} \leq \hat{q}_{ti}$) at a price $\hat{p}_{ti}$ from supplier $i \in \mathcal{I}$, where the price/quantity pairs $(\hat{p}_{ti}, \hat{q}_{ti})_{i \in \mathcal{I}}$ fluctuate from week to week. Let $s_0$ be your total initial inventory of oranges, and let $D_t$ be the number of oranges that the company needs for production during week $t$ (this is our demand). If we are unable to meet demand, the company must purchase additional oranges on the spot market at a spot price $\hat{p}_{ti}^{spot}$.

(a) What is the exogenous stochastic process for this system?

(b) What are the decisions you can make to influence the system?

(c) What would be the state variable for your problem?

(d) Write out the transition equations.

(e) What is the one-period contribution function?

(f) Propose a reasonable structure for a decision rule for this problem, and call it $X^{\pi}$. Your decision rule should be in the form of a function that determines how much to purchase in period $t$.

(g) Carefully and precisely, write out the objective function for this problem in terms of the exogenous stochastic process. Clearly identify what you are optimizing over.

(h) For your decision rule, what do we mean by the space of policies?

**5.14** Customers call in to a service center according to a (nonstationary) Poisson process. Let $\mathcal{E}$ be the set of events representing phone calls, where $t_e, e \in \mathcal{E}$ is the time that the call is made. Each customer makes a request that will require time $\tau_e$ to complete and will pay a reward $r_e$ to the service center. The calls are initially handled by a receptionist who determines $\tau_e$ and $r_e$. The service center does not have to handle all calls and obviously favors calls with a high ratio of reward per time unit required $(r_e/\tau_e)$. For this reason, the company adopts a policy that the call will be refused if $(r_e/\tau_e) < \gamma$. If the call is accepted, it is placed in a queue to wait for one of the available service representatives. Assume that the probability law driving the process is known, where we would like to find the right value of $\gamma$.

(a) This process is driven by an underlying exogenous stochastic process with element $\omega \in \Omega$. What is an instance of $\omega$?

(b) What are the decision epochs?

(c) What is the state variable for this system? What is the transition function?

(d) What is the action space for this system?

(e) Give the one-period reward function.

(f) Give a full statement of the objective function that defines the Markov decision process. Clearly define the probability space over which the expectation is defined, and what you are optimizing over.

**5.15** A major oil company is looking to build up its storage tank reserves, anticipating a surge in prices. It can acquire 20 million barrels of oil, and it would like to purchase this quantity over the next 10 weeks (starting in week 1). At the beginning of the week, the company contacts its usual sources, and each source $j \in \mathcal{J}$ is willing to provide $\hat{q}_{tj}$ million barrels at a price $\hat{p}_{tj}$. The price/quantity pairs $(\hat{p}_{tj}, \hat{q}_{tj})$ fluctuate from week to week. The company would like to purchase (in discrete units of millions of barrels) $x_{tj}$ million barrels (where $x_{tj}$ is discrete) from source $j$ in week $t \in \{1, 2, \ldots, 10\}$. Your goal is to acquire 20 million barrels while spending the least amount possible.

(a) What is the exogenous stochastic process for this system?

(b) What would be the state variable for your problem? Give an equation(s) for the system dynamics.

(c) Propose a structure for a decision rule for this problem and call it $X^\pi$.

(d) For your decision rule, what do we mean by the space of policies? Give examples of two different decision rules.

(e) Write out the objective function for this problem using an expectation over the exogenous stochastic process.

(f) You are given a budget of $300 million to purchase the oil, but you absolutely must end up with 20 million barrels at the end of the 10 weeks. If you exceed the initial budget of $300 million, you may get additional funds, but each additional $1 million will cost you $1.5 million. How does this affect your formulation of the problem?

**5.16** You own a mutual fund where at the end of each week $t$ you must decide whether to sell the asset or hold it for an additional week. Let $\hat{r}_t$ be the one-week return (e.g. $\hat{r}_t = 1.05$ means the asset gained five percent in the previous week), and let $p_t$ be the price of the asset if you were to sell it in week $t$ (so $p_{t+1} = p_t \hat{r}_{t+1}$). We assume that the returns $\hat{r}_t$ are independent and identically distributed. You are investing this asset for eventual use in your college education, which will occur in 100 periods. If you sell the asset at the end of time period $t$, then it will earn a money market rate $q$ for each time period until time period 100, at which point you need the cash to pay for college.

(a) What is the state space for our problem?

(b) What is the action space?

(c) What is the exogenous stochastic process that drives this system? Give a five time period example. What is the history of this process at time t?

(d) You adopt a policy that you will sell if the asset falls below a price $\bar{p}$ (which we are requiring to be independent of time). Given this policy, write out the objective function for the problem. Clearly identify exactly what you are optimizing over.

# CHAPTER 6

# STOCHASTIC APPROXIMATION METHODS

Stochastic approximation methods are the foundation of most approximate dynamic programming algorithms. They represent a technique whose simplicity is matched only by the depth and richness of the theory that supports them. Particularly attractive from a practical perspective is the ease with which they provide us with the means for solving, at least approximately, problems with considerable complexity. There is a price for this generality, but for many complex problems, we gladly pay it in return for a strategy that can provide insights into problems that would otherwise seem completely intractable.

This chapter provides a basic introduction to stochastic approximation theory by focusing on the problem of estimating the mean of a random variable with unknown distribution. This is especially important in dynamic programming because we use these methods to estimate the value of being in a state from random samples. Chapter 7 applies these methods to estimate a more general class of functional approximations for dynamic programming.

The basic problem can be stated as one of optimizing the expectation of a function, which we can state as

$$\max_{\theta \in \Theta} \mathbb{E}\{F(\theta, W)\}. \tag{6.1}$$

Here, $\theta$ is one or more parameters that we are trying to estimate. An element of $\theta$ may be $\theta_s = v(s)$ which is the value of being in state $s$. Alternatively, we may have an asset with attribute $a$, where $\theta_a$ is the value of the asset. $W$ may be a vector of random variables representing information that becomes known after we make the decision $\theta$. For example, we may make a decision based on an estimate of the value of a state, and then obtain a random observation of the value of the state which helps us improve the estimate.

*Approximate Dynamic Programming*. By Warren B. Powell

The general stochastic optimization problem in equation (6.1) also arises in many settings that have nothing to do with parameter estimation. For example, we may need to make a decision to allocate resources, denoted by $x$. After we allocate the resources, we then learn about the demands for these resources (captured by a random variable $D$). Because the demands become known after we decide on our allocation, we face the problem of solving $\min_x \mathbb{E}\{F(x, D)\}$.

We generally assume that we cannot compute $\mathbb{E}\{F(\theta, W)\}$, but we still have to optimize over $\theta$. In practice, we may not even know the probability distribution of $W$, but we assume that we do have access to observations of $W$, either from a physical process or by randomly generating realizations from a mathematical model (we describe how to do this later). This behavior is illustrated in the examples.

---

**EXAMPLE 6.1**

An information technology group uses planning tools to estimate the man-hours of programming time required to complete a project. Based on this estimate, the group then allocates programming resources to complete the project. After the project, the group can compute how many man-hours were actually needed to complete it, which can be used to estimate the distribution of the error between the initial estimate of what was required and what was actually required.

**EXAMPLE 6.2**

A venture capital firm has to estimate what it can expect for a return if it invests in an emerging technology company. Since these companies are privately held, the firm can only learn the return by agreeing to invest in the company.

**EXAMPLE 6.3**

An electric power company has to maintain a set of spare components in the event of failure. These components can require a year or two to build and are quite expensive, so it is necessary to have a reasonable number of spares with the right features and in the right location to respond to emergencies.

**EXAMPLE 6.4**

A business jet company is trying to estimate the demand for its services. Each week, it uses the actual demands to update its estimate of the average demand. The goal is to minimize the deviation between the estimate of the demand and the actual.

---

This chapter introduces the elegant and powerful theory of stochastic approximation methods (also known as stochastic gradient algorithms). These are properly viewed as the algorithm of last resort in stochastic optimization, but for many practical applications, they are the only option. As with other chapters in this volume, the core algorithms are separated from the proofs of convergence. However, readers are encouraged to delve into these proofs to gain an appreciation of the supporting theory.

## 6.1 A STOCHASTIC GRADIENT ALGORITHM

Our basic problem is to optimize a function that also depends on a random variable. This can be stated as

$$\min_{\theta \in \Theta} \mathbb{E}\left\{F(\theta, W)\right\}$$

where $\Theta$ is the set of allowable values for $\theta$ (many estimation problems are unconstrained). For example, assume that we are trying to find the mean $\mu$ of a random variable $R$. We wish to find a number $\theta$ that produces the smallest squared error between the estimate of the mean $\theta$ and that of a particular sample. This can be stated as

$$\min_{\theta} \mathbb{E}\left\{\frac{1}{2}(\theta - R)^2\right\}. \tag{6.2}$$

If we want to optimize (6.2), we could take the derivative and set it equal to zero (for a simple, unconstrained, continuously differentiable problem such as ours). However, let us assume that we cannot take the expectation easily. Instead, we can choose a sample observation of the random variable $R$ that is represented by $R(\omega)$. A sample of our function is now

$$F(\theta, \omega) = \frac{1}{2}(\theta - R(\omega))^2. \tag{6.3}$$

Let $\nabla F(\theta)$ be the gradient of $F(\theta) = \mathbb{E}F(\theta, R)$ with respect to $\theta$, and let $\nabla F(\theta, R(\omega))$ be the sample gradient, taken when $R = R(\omega)$. In our example, clearly

$$\nabla F(\theta, \omega) = (\theta - R(\omega)). \tag{6.4}$$

We call $\nabla F(\theta, \omega)$ a *stochastic gradient* because, obviously, it is a gradient and it is stochastic.

What can we do with our stochastic gradient? If we had an exact gradient, we could use the standard optimization sequence, given by

$$\bar{\theta}^n = \bar{\theta}^{n-1} - \alpha_{n-1}\nabla F(\theta^{n-1}, \omega^n), \tag{6.5}$$

where $\alpha_{n-1}$ is known as a stepsize because it tells us how far we should go in the direction of $\nabla F(\theta^{n-1}, \omega^n)$ (later we explain why the stepsize is indexed by $n-1$ instead of $n$). In some communities the stepsize is known as the *learning rate* or *smoothing factor*. $\theta^{n-1}$ is the estimate of $\theta$ computed from the previous iteration (using the sample realization $\omega^{n-1}$, while $\omega^n$ is the sample realization in iteration $n$ (the indexing tells us that $\theta^{n-1}$ was computed without $\omega^n$). When the function is deterministic, we would choose the stepsize by solving the one-dimensional optimization problem

$$\min_{\alpha} F\left(\bar{\theta}^{n-1} - \alpha\nabla F(\theta^{n-1}, \omega^n)\right). \tag{6.6}$$

If $\alpha^*$ is the stepsize that solves (6.6), we would then use this stepsize in (6.5) to find $\bar{\theta}^n$. However, we are using a stochastic gradient, and because of this, we cannot use sophisticated logic such as finding the best stepsize. Part of the problem is that a stochastic gradient can even point away from the optimal solution such that any positive stepsize actually makes the solution worse. For example, consider the problem of estimating the mean of a random variable $R$ where $\bar{\theta}^n$ is our estimate after $n$ iterations. Assume the mean is 10 and our current estimate of the mean is $\bar{\theta}^{n-1} = 7$. If we now observe $R^n = 3$ with $\alpha = .1$, our update would be

$$\begin{aligned} \bar{\theta}^n &= 7 - 0.1(7-3) \\ &= 6.6. \end{aligned}$$

Thus, our estimate has moved even further from the true value. This is the world we live in when we depend on stochastic gradients. We have no guarantee that the solution will improve from one iteration to the next.

**Remark:** Many authors will write equation (6.5) in the form

$$\bar{\theta}^{n+1} = \bar{\theta}^n - \alpha_n \nabla F(\theta^n, \omega^n). \tag{6.7}$$

With this style, we would say that $\theta^{n+1}$ is the estimate of $\theta$ to be used in iteration $n+1$ (although it was computed with the information from iteration $n$). We use the form in (6.5) because we will later allow the stepsizes to depend on the data, and the indexing tells us the information content of the stepsize (for theoretical reasons, it is important that the stepsize be computed using information up through $n-1$, hence our use of $\alpha_{n-1}$). We index $\bar{\theta}^n$ on the left-hand side of (6.5) using $n$ because the right-hand side has information from iteration $n$. It is often the case that time $t$ is also our iteration counter, and so it helps to be consistent with our time indexing notation. Finally, we use $\alpha_{n-1}$ rather than $\alpha_n$ in (6.5) to tell us that the stepsize used in iteration $n$ is not allowed to use information from iteration $n$ (in particular, it is not allowed to see $\omega^n$). This condition is easy to overlook, but is important for both theoretical reasons (we lose the ability to prove convergence without it - see section 6.8.3) as well as practical considerations (choosing the stepsize after you see the outcome of your experiment will generally produce stepsizes that overreact to the data).

If we put (6.4) and (6.5) together, we can write our updating algorithm in the following two forms:

$$\bar{\theta}^n = \bar{\theta}^{n-1} - \alpha_{n-1}\left(\bar{\theta}^{n-1} - R(\omega^n)\right) \tag{6.8}$$

$$\quad = (1-\alpha_{n-1})\bar{\theta}^{n-1} + \alpha_{n-1}R(\omega^n). \tag{6.9}$$

Equation (6.8) writes the updating in the standard form of a stochastic gradient algorithm. The form given in equation (6.9) arises in a number of settings. The demand forecasting community refers to it as exponential smoothing; the engineering systems community refers to it as a linear filter. We prefer to retain its derivation as a stochastic gradient algorithm and while we primarily use the form in (6.9), we still refer to $\alpha_{n-1}$ as a stepsize. For problems where we are trying to estimate the mean of a random variable, we will typically require that $0 \leq \alpha_{n-1} \leq 1$, because the units of the gradient and the units of the decision variable are the same.

There are many applications where the units of the gradient, and the units of the decision variable, are different. For example, consider the problem of ordering a quantity of product $x$ to satisfy a random demand $D$. We pay an overage cost $c^o$ for each unit that we order over the demand, and an underage cost $c^u$ for each unit of demand that is unsatisfied. The resulting problem requires that we solve

$$\min_x E\{c^o \max(x-D,0) + c^u \max(D-x,0)\}. \tag{6.10}$$

In this case, the derivative with respect to $x$ is either $c^o$ or $-c^u$ depending on whether $x > D$ or $x < D$. The units of $c^o$ and $c^u$ are in dollars, while $x$ has the units of our decision variable (e.g., how much product to order). In this situation, the stepsize has to be scaled so that the adjustments to the decision are not too large or too small.

Returning to our original problem of estimating the mean, we assume when running a stochastic gradient algorithm that $\bar{\theta}^0$ is an initial guess, and that $R(\omega^1)$ is our first observation. If our stepsize sequence uses an initial stepsize $\alpha_0 = 1$, then

$$\begin{aligned}\bar{\theta}^1 &= (1 - \alpha_0)\bar{\theta}^0 + \alpha_0 R(\omega^1) \\ &= R(\omega^1),\end{aligned}$$

which means we do not need the initial estimate for $\bar{\theta}^0$. Smaller initial stepsizes would only make sense if we had access to a reliable initial guess, and in this case, the stepsize should reflect the confidence in our original estimate (for example, we might be warm starting an algorithm from a previous iteration).

We can evaluate our performance using a mean squared statistical measure. If we have an initial estimate $\bar{\theta}^0$, we would use

$$MSE = \frac{1}{n}\sum_{m=1}^{n}(\bar{\theta}^{m-1} - R(\omega^m))^2. \tag{6.11}$$

However, it is often the case that the sequence of random variables $R(\omega^n)$ is nonstationary, which means they are coming from a distribution that is changing over time. (For example, in chapter 4 we would make random observations of the value of being in a state, which we referred to as $\hat{v}_t^n$, but these depended on a value function approximation for future events which was changing over time.) In this case, estimating the mean squared error is similar to our problem of estimating the mean of the random variable $R$, in which case we should use a standard stochastic gradient (smoothing) expression of the form

$$MSE^n = (1 - \beta_{n-1})MSE^{n-1} + \beta_{n-1}(\bar{\theta}^{n-1} - R(\omega^n))^2,$$

where $\beta_{n-1}$ is another stepsize sequence (which could be the same as $\alpha_{n-1}$).

## 6.2 DETERMINISTIC STEPSIZE RECIPES

One of the challenges in Monte Carlo methods is finding the stepsize $\alpha_n$. We refer to a method for choosing a stepsize as a *stepsize rule*, while other communities refer to them as *learning rate schedules*. A standard technique in deterministic problems (of the continuously differentiable variety) is to find the value of $\alpha_n$ so that $\bar{\theta}^n$ gives the smallest possible objective function value (among all possible values of $\alpha$). For a deterministic problem, this is generally not too hard. For a stochastic problem, it means calculating the objective function, which involves computing an expectation. For most applications, expectations are computationally intractable, which makes it impossible to find an optimal stepsize.

Throughout our presentation, we assume that we are using stepsizes to estimate a parameter $\theta$ which might be the value of being in a state $s$. In chapter 7, we use these techniques to estimate more general parameter vectors when we introduce the idea of using regression models to approximate a value function. We let $\bar{\theta}^{n-1}$ be the estimate of $\theta$ after $n-1$ iterations, and we let $\hat{\theta}^n$ be a random observation in iteration $n$ of the value of being in state $s$ ($\hat{\theta}^n$ might be a biased observation of the true value $\theta^n$). (In chapter 4, we used $\hat{v}^n$ as our random observation of being in a state, as shown in figure 4.1.)

Our updating equation looks like

$$\bar{\theta}^n = (1 - \alpha_{n-1})\bar{\theta}^{n-1} + \alpha_{n-1}\hat{\theta}^n. \tag{6.12}$$

Our iteration counter always starts at $n = 1$ (just as our first time interval starts with $t = 1$). The use of $\alpha_{n-1}$ in equation (6.12) means that we are computing $\alpha_{n-1}$ using information available at iteration $n - 1$ and before. Thus, we have an explicit assumption that we are not using $\hat{\theta}^n$ to compute the stepsize in iteration $n$. This is irrelevant when we use a deterministic stepsize sequence, but is critical in convergence proofs for stochastic stepsize formulas (introduced below). In most formulas, $\alpha_0$ is a parameter that has to be specified, although we will generally assume that $\alpha_0 = 1$, which means that we do not have to specify $\bar{\theta}_0$. The only reason to use $\alpha_0 < 1$ is when we have some *a priori* estimate of $\bar{\theta}_0$ which is better than $\hat{\theta}^1$.

*Stepsize rules are important! As you experiment with ADP, it is possible to find problems where provably convergent algorithms simply do not work, and the only reason is a poor choice of stepsizes. Inappropriate choices of stepsize rules have led many to conclude that "approximate dynamic programming does not work."*

There are two issues when designing a good stepsize rule. The first is the issue of whether the stepsize will produce a provably convergent algorithm. While this is primarily of theoretical interest, these conditions do provide important guidelines to follow to produce good behavior. The second issue is whether the rule produces the fastest rate of convergence. Both issues are very important in practice.

Below, we start with a general discussion of stepsize rules. Following this, we provide a number of examples of deterministic stepsize rules. These are formulas that depend only on the iteration counter $n$ (or more precisely, the number of times that we update a particular parameter). Section 6.3 then describes stochastic stepsize rules that adapt to the data.

The deterministic and stochastic rules presented in this section and section 6.3 are, for the most part, heuristically designed to achieve good rates of convergence, but are not supported by any theory that they will produce the best rate of convergence. Later (section 6.5) we provide a theory for choosing stepsizes that produce the fastest possible rate of convergence.

### 6.2.1 Properties for convergence

The theory for proving convergence of stochastic gradient algorithms was first developed in the early 1950's and has matured considerably since then (see section 6.8). However, all the proofs require three basic conditions

$$\alpha_{n-1} \geq 0, \quad n = 1, 2, \ldots, \tag{6.13}$$

$$\sum_{n=1}^{\infty} \alpha_{n-1} = \infty, \tag{6.14}$$

$$\sum_{n=1}^{\infty} (\alpha_{n-1})^2 < \infty. \tag{6.15}$$

Equation (6.13) obviously requires that the stepsizes be nonnegative. The most important requirement is (6.14), which states that the infinite sum of stepsizes must be infinite. If this condition did not hold, the algorithm might stall prematurely. Finally, condition 6.15 requires that the infinite sum of the squares of the stepsizes be finite. This condition, in effect, requires that the stepsize sequence converge "reasonably quickly." A good intuitive justification for this condition is that it guarantees that the *variance* of our estimate of the optimal solution goes to zero in the limit. Sections 6.8.2 and 6.8.3 illustrate two proof techniques that both lead to these requirements on the stepsize. Fifty years of research in this area has not been able to relax them.

Conditions (6.14) and (6.15) effectively require that the stepsizes decline according to an arithmetic sequence such as

$$\alpha_{n-1} = \frac{1}{n}. \tag{6.16}$$

This rule has an interesting property. Exercise 6.3 asks you to show that a stepsize of $1/n$ produces an estimate $\bar{\theta}^n$ that is simply an average of all previous observations, which is to say

$$\bar{\theta}^n = \frac{1}{n}\sum_{m=1}^{n}\hat{\theta}^m. \tag{6.17}$$

Of course, we have a nice name for equation (6.17): it is called a sample average. And we are all aware that in general (some modest technical conditions are required) as $n \rightarrow \infty$, $\bar{\theta}^n$ will converge (in some sense) to the mean of our random variable $R$. What is nice about equation (6.8) is that it is very easy to use (actually, much easier than equation (6.17)). Also, it lends itself nicely to adaptive estimation, where we may not know the sample size in advance.

The issue of the rate at which the stepsizes decrease is of considerable practical importance. Consider, for example, the stepsize sequence

$$\alpha_n = .5\alpha_{n-1},$$

which is a geometrically decreasing progression. This stepsize formula violates (6.14). More intuitively, the problem is that the stepsizes would decrease so quickly that it is likely that we would never reach the final solution.

Surprisingly, the "$1/n$" stepsize formula, which works in theory, tends not to work in practice because it drops to zero too quickly when applied to approximate dynamic programming applications. The reason is that we are usually updating the value function using biased estimates which are changing over time. For example, consider the updating expression we used for the post-decision state variable given in section 4.4.3, which we repeat here for convenience

$$\begin{aligned}\hat{v}_t^n &= C_t(S_t^n, x_t^n) + \bar{V}_t^{n-1}(S_t^{x,n}),\\ \bar{V}_{t-1}^n(S_{t-1}^{x,n}) &= (1-\alpha_{n-1})\bar{V}_{t-1}^{n-1}(S_{t-1}^{x,n}) + \alpha_{n-1}\hat{v}_t^n.\end{aligned}$$

$\hat{v}_t^n$ is our sample observation of an estimate of the value of being in state $S_t$, which we then smooth into the current approximation $\bar{V}_{t-1}^{n-1}(S_{t-1}^{x,n})$. If $\hat{v}_t^n$ were an unbiased estimate of the true value, then a stepsize of $1/n$ would be the best we could do (we show this later). However, $\hat{v}_t^n$ depends on $\bar{V}_t^{n-1}(S_t^x)$, which is an imperfect estimate of the value function for time $t$. What typically happens is that the value functions undergo a transient learning phase. Since we have not found the correct estimate for $\bar{V}_t^{n-1}(S_t^{x,n})$, the estimates $\hat{v}_t^n$ are biased, and the $1/n$ rule puts the highest weights on the early iterations when the estimates are the worst. The resulting behavior is illustrated in figure 6.1.

The remainder of this section presents a series of deterministic stepsize formulas designed to overcome this problem. These rules are the simplest to implement and are typically the best starting point when designing an ADP algorithm.

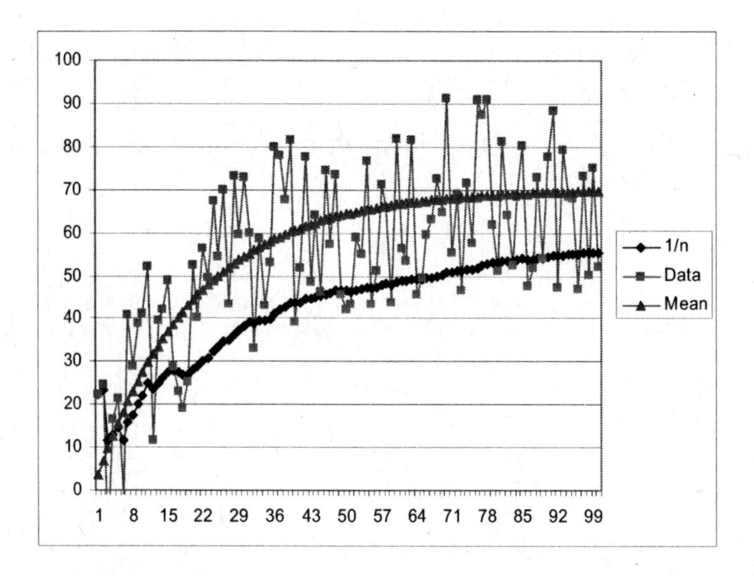


**Figure 6.1** Illustration of poor convergence of $1/n$ stepsize rule in the presence of transient data.

## Constant stepsizes

A constant stepsize rule is simply

$$\alpha_{n-1} = \begin{cases} 1 & \text{if } n = 1, \\ \bar{\alpha} & \text{otherwise,} \end{cases}$$

where $\bar{\alpha}$ is a stepsize that we have chosen. It is common to start with a stepsize of 1 so that we do not need an initial value $\bar{\theta}^0$ for our statistic.

Constant stepsizes are popular when we are estimating not one but many parameters (for large scale applications, these can easily number in the thousands or millions). In these cases, no single rule is going to be right for all of the parameters and there is enough noise that any reasonable stepsize rule will work well. Constant stepsizes are easy to code (no memory requirements) and, in particular, easy to tune (there is only one parameter). Perhaps the biggest point in their favor is that we simply may not know the rate of convergence, which means that we run the risk with a declining stepsize rule of allowing the stepsize to decline too quickly, producing a behavior we refer to as "apparent convergence."

In dynamic programming, we are typically trying to estimate the value of being in a state using observations that are not only random, but which are also changing systematically as we try to find the best policy. As a general rule, as the noise in the observations of the values increases, the best stepsize decreases. But if the values are increasing rapidly, we want a larger stepsize. Choosing the best stepsize requires striking a balance between stabilizing the noise and responding to the changing mean. Figure 6.2 illustrates observations that are coming from a process with relatively low noise but where the mean is changing quickly (6.2a), and observations that are very noisy but where the mean is not changing at all (6.2b). For the first, the ideal stepsize is relatively large, while for the second, the best stepsize is quite small.

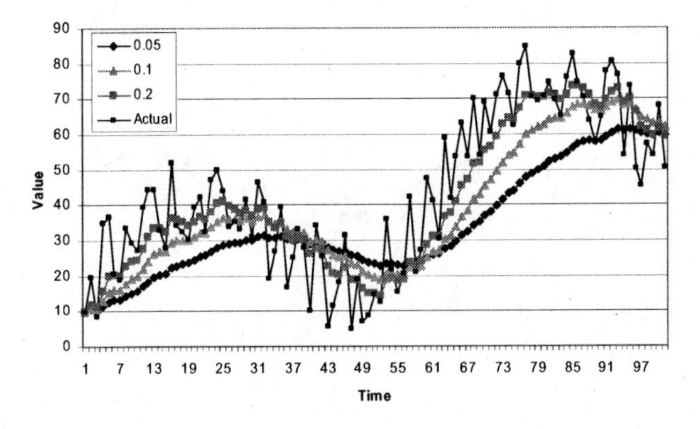


6.2a: Low-noise

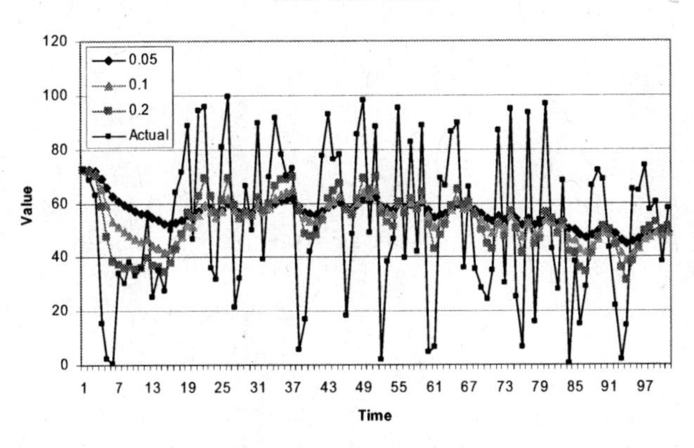


6.2b: High-noise

**Figure 6.2** Illustration of the effects of smoothing using constant stepsizes. Case (a) represents a low-noise dataset, with an underlying nonstationary structure; case (b) is a high-noise dataset from a stationary process.

### Generalized harmonic stepsizes

A generalization of the $1/n$ rule is the generalized harmonic sequence given by

$$\alpha_{n-1} = \frac{a}{a+n-1}. \tag{6.18}$$

This rule satisfies the conditions for convergence, but produces larger stepsizes for $a > 1$ than the $1/n$ rule. Increasing $a$ slows the rate at which the stepsize drops to zero, as illustrated in figure 6.3. In practice, it seems that despite theoretical convergence proofs to the contrary, the stepsize $1/n$ can decrease to zero far too quickly, resulting in "apparent convergence" when in fact the solution is far from the best that can be obtained.

### Polynomial learning rates

An extension of the basic harmonic sequence is the stepsize

$$\alpha_{n-1} = \frac{1}{(n)^{\beta}}, \tag{6.19}$$

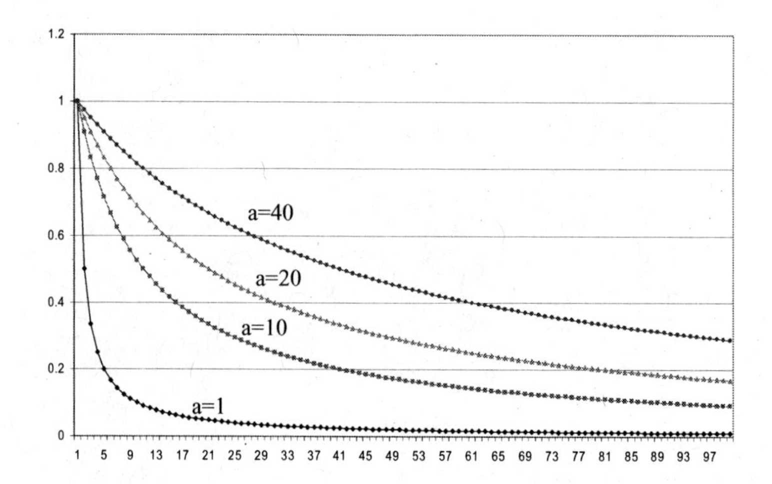


**Figure 6.3** Stepsizes for $a/(a+n)$ while varying $a$.

where $\beta \in (\frac{1}{2}, 1]$. Smaller values of $\beta$ slow the rate at which the stepsizes decline, which improves the responsiveness in the presence of initial transient conditions. The best value of $\beta$ depends on the degree to which the initial data is transient, and as such is a parameter that needs to be tuned.

## McClain's formula

McClain's formula (McClain (1974)) is an elegant way of obtaining $1/n$ behavior initially but approaching a specified constant in the limit. The formula is given by

$$\alpha_n = \frac{\alpha_{n-1}}{1 + \alpha_{n-1} - \bar{\alpha}}. \tag{6.20}$$

where $\bar{\alpha}$ is a specified parameter. Note that steps generated by this model satisfy the following properties

$$\begin{aligned} \alpha_n > \alpha_{n+1} > \bar{\alpha} \quad &\text{if} \quad \alpha > \bar{\alpha}, \\ \alpha_n < \alpha_{n+1} < \bar{\alpha} \quad &\text{if} \quad \alpha < \bar{\alpha}. \end{aligned}$$

McClain's rule, illustrated in figure 6.4, combines the features of the "$1/n$" rule which is ideal for stationary data, and constant stepsizes for nonstationary data. If we set $\bar{\alpha} = 0$, then it is easy to verify that McClain's rule produces $\alpha_{n-1} = 1/n$. In the limit, $\alpha_n \rightarrow \bar{\alpha}$. The value of the rule is that the $1/n$ averaging generally works quite well in the very first iterations (this is a major weakness of constant stepsize rules), but avoids going to zero. The rule can be effective when you are not sure how many iterations are required to start converging, and it can also work well in nonstationary environments.

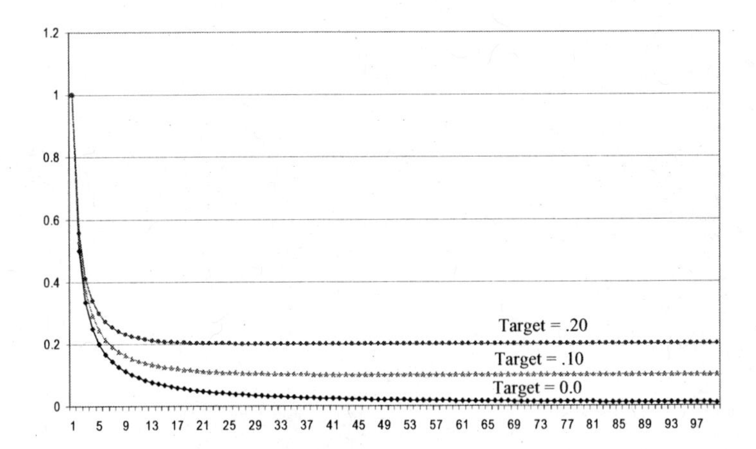


**Figure 6.4** The McClain stepsize rule with varying targets.

## Search-then-converge learning rule

The search-then-converge (STC) stepsize rule (Darken & Moody (1992)) is a variation on the harmonic stepsize rule that produces delayed learning. It was originally proposed as

$$\alpha_n = \alpha_0 \frac{\left(1 + \frac{\beta}{\alpha_0} \frac{n}{\tau}\right)}{\left(1 + \frac{\beta}{\alpha_0} \frac{n}{\tau} + \frac{n^2}{\tau}\right)}. \tag{6.21}$$

where $\alpha_0$, $\beta$ and $\tau$ are parameters to be determined. A more compact and slightly more general version of this formula is

$$\alpha_{n-1} = \alpha_0 \frac{\left(\frac{b}{n} + a\right)}{\left(\frac{b}{n} + a + n^\beta\right)}. \tag{6.22}$$

If $\beta = 1$, then this formula is similar to the STC rule. In addition, if $b = 0$, then it is the same as the $a/(a+n)$ rule. The addition of the term $b/n$ to the numerator and the denominator can be viewed as a kind of $a/(a+n)$ rule where $a$ is very large but declines with $n$. The effect of the $b/n$ term, then, is to keep the stepsize larger for a longer period of time, as illustrated in figure 6.5. This can help algorithms that have to go through an extended learning phase when the values being estimated are relatively unstable. The relative magnitude of $b$ depends on the number of iterations which are expected to be run, which can range from several dozen to several million.

This class of stepsize rules is termed "search-then-converge" because they provide for a period of high stepsizes (while searching is taking place) after which the stepsize declines (to achieve convergence). The degree of delayed learning is controlled by the parameter $b$, which can be viewed as playing the same role as the parameter $a$ but which declines as the algorithm progresses.

The exponent $\beta$ in the denominator has the effect of increasing the stepsize in later iterations (see figure 6.6). With this parameter, it is possible to accelerate the reduction of the stepsize in the early iterations (by using a smaller $a$) but then slow the descent in later iterations (to sustain the learning process). This may be useful for problems where there is an extended transient phase requiring a larger stepsize for a larger number of iterations.

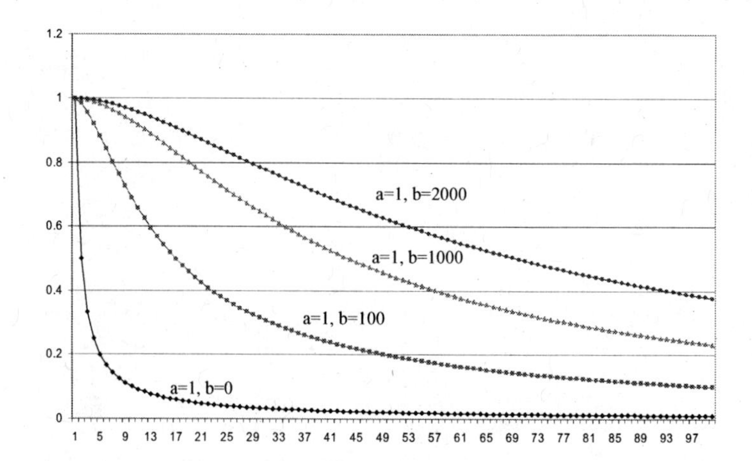


**Figure 6.5** Stepsizes for $(b/n + a)/(b/n + a + n)$ while varying $b$.

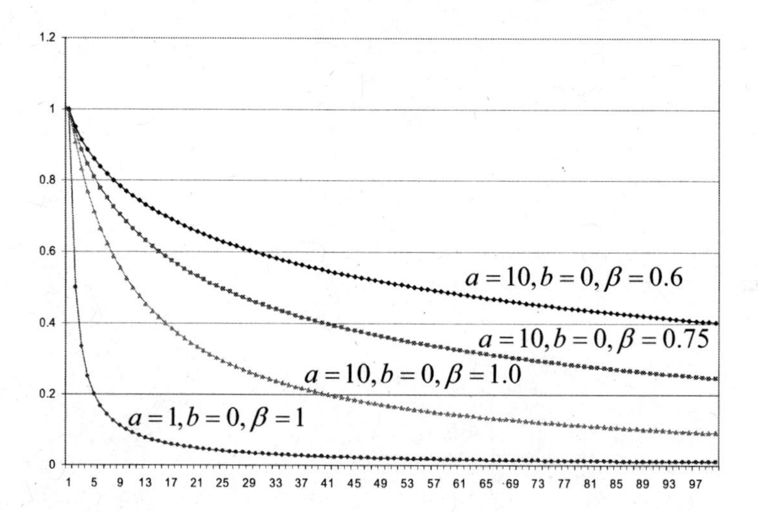


**Figure 6.6** Stepsizes for $(b/n + a)/(b/n + a + n^{\beta})$ while varying $\beta$.

## 6.3 STOCHASTIC STEPSIZES

There is considerable appeal to the idea that the stepsize should depend on the actual trajectory of the algorithm. For example, if we are consistently observing that our estimate $\bar{\theta}^{n-1}$ is smaller (or larger) than the observations $\hat{\theta}^n$, then it suggests that we are trending upward (or downward). When this happens, we typically would like to use a larger stepsize to increase the speed at which we reach a good estimate. When the stepsizes depend on the observations $\hat{\theta}^n$, then we say that we are using a *stochastic stepsize*.

In this section, we first review the case for stochastic stepsizes, then present the revised theoretical conditions for convergence, and finally outline a series of recipes that have been suggested in the literature (including some that have not).

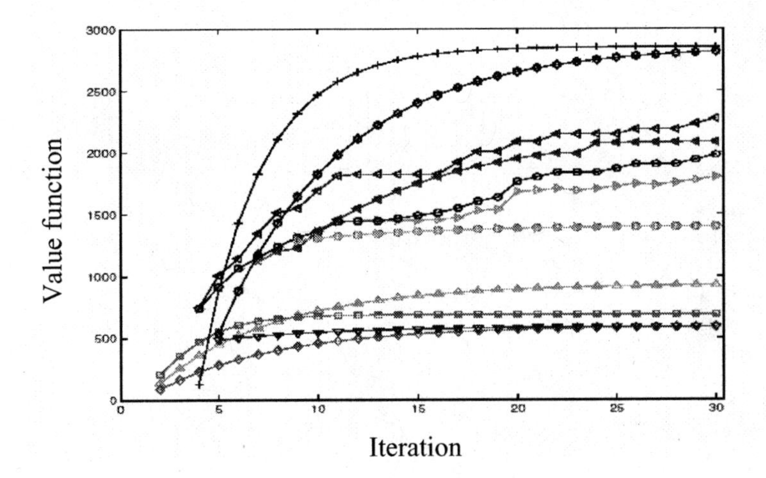


**Figure 6.7** Different parameters can undergo significantly different initial rates.

### 6.3.1 The case for stochastic stepsizes

Assume that our estimates are consistently under or consistently over the actual observations. This can easily happen during early iterations due to either a poor initial starting point or the use of biased estimates (which is common in dynamic programming) during the early iterations. For large problems, it is possible that we have to estimate thousands of parameters. It seems unlikely that all the parameters will approach their true value at the same rate. Figure 6.7 shows the change in estimates of the value of being in different states, illustrating the wide variation in learning rates that can occur within the same dynamic program.

Stochastic stepsizes try to adjust to the data in a way that keeps the stepsize larger while the parameter being estimated is still changing quickly. Balancing noise against the change in the underlying signal, particularly when both of these are unknown, is a difficult challenge.

### 6.3.2 Convergence conditions

When the stepsize depends on the history of the process, the stepsize itself becomes a random variable. This change requires some subtle modifications to our requirements for convergence (equations (6.14) and (6.15)). For technical reasons, our convergence criteria change to

$$\alpha_n \geq 0, \text{ almost surely,} \tag{6.23}$$

$$\sum_{n=0}^{\infty} \alpha_n = \infty \text{ almost surely,} \tag{6.24}$$

$$\mathbb{E}\left\{\sum_{n=0}^{\infty}(\alpha_n)^2\right\} < \infty. \tag{6.25}$$

The condition "almost surely" (universally abbreviated "a.s.") means that equation (6.24) holds for every sample path $\omega$, and not just on average. More precisely, we mean every sample path $\omega$ that might actually happen (we exclude sample paths where the probability that the sample path would happen is zero).

For the reasons behind these conditions, go to our "Why does it work" section (section 6.8). It is important to emphasize, however, that these conditions are completely unverifiable and are purely for theoretical reasons. The real issue with stochastic stepsizes is whether they contribute to the rate of convergence.

### 6.3.3 Recipes for stochastic stepsizes

To present our stochastic stepsize formulas, we need to define a few quantities. Recall that our basic updating expression is given by

$$\bar{\theta}^n = (1 - \alpha_{n-1})\bar{\theta}^{n-1} + \alpha_{n-1}\hat{\theta}^n.$$

$\bar{\theta}^{n-1}$ is our estimate of the next observation, given by $\hat{\theta}^n$. The difference between the estimate and the actual can be treated as the error, given by

$$\varepsilon^n = \bar{\theta}^{n-1} - \hat{\theta}^n.$$

We may wish to smooth the error in the estimate, which we designate by the function

$$S(\varepsilon^n) = (1 - \beta)S(\varepsilon^{n-1}) + \beta\varepsilon^n.$$

Some formulas depend on tracking changes in the sign of the error. This can be done using the indicator function

$$1_{\{X\}} = \begin{cases} 1 & \text{if the logical condition } X \text{ is true,} \\ 0 & \text{otherwise.} \end{cases}$$

Thus, $1_{\varepsilon^n \varepsilon^{n-1} < 0}$ indicates if the sign of the error has changed in the last iteration.

Following is a series of formulas that adjust the stepsize based on the observed errors in the estimates.

#### *Kesten's rule*

Kesten's rule (Kesten (1958)) was one of the earliest stepsize rules which took advantage of a simple principle. If we are far from the optimal, the errors tend to all have the same sign. As we get close, the errors tend to alternate. Exploiting this simple observation, Kesten proposed the following simple rule:

$$\alpha_{n-1} = \frac{a}{a + K^n - 1}, \tag{6.26}$$

where $a$ is a parameter to be calibrated. $K^n$ counts the number of times that the sign of the error has changed, where we use

$$K^n = \begin{cases} n & \text{if } n = 1, 2, \\ K^{n-1} + 1_{\{\varepsilon^n \varepsilon^{n-1} < 0\}} & \text{if } n > 2. \end{cases} \tag{6.27}$$

Kesten's rule is particularly well suited to initialization problems. It slows the reduction in the stepsize as long as the error exhibits the same sign (and indication that the algorithm is still climbing into the correct region). However, the stepsize declines monotonically. This is typically fine for most dynamic programming applications, but can encounter problems in situations with delayed learning.

### Mirozahmedov's rule

Mirozahmedov & Uryasev (1983) formulates an adaptive stepsize rule that increases or decreases the stepsize in response to whether the inner product of the successive errors is positive or negative, along similar lines as in Kesten's rule.

$$\alpha_n = \alpha_{n-1} \exp\left[\left(a\hat{\varepsilon}_t^n \hat{\varepsilon}^{n-1} - \delta\right) \alpha_{n-1}\right] \tag{6.28}$$

where $a$ and $\delta$ are some fixed constants. A variation of this rule where $\delta$ is zero is proposed by Ruszczyński & Syski (1986).

### Gaivoronski's rule

Gaivoronski (1988) proposes a stepsize which is computed as a function of the ratio of the progress to the path of the algorithm. The progress is measured in terms of the difference in the values of the smoothed estimate between a certain number of iterations. The path is measured as the sum of absolute values of the differences between successive estimates for the same number of iterations.

$$\alpha_n = \begin{cases} \gamma_1 \alpha_{n-1} & \text{if } \Phi^{n-1} \leq \gamma_2, \\ \alpha_{n-1} & \text{otherwise.} \end{cases}$$

$\Phi_n$ is computed using

$$\Phi_n = \frac{\left|\bar{\theta}^{n-k} - \bar{\theta}^n\right|}{\sum_{i=n-k}^{n-1} \left|\bar{\theta}^i - \bar{\theta}^{i+1}\right|},$$

where $\gamma_1$ and $\gamma_2$ are constants.

### Stochastic gradient adaptive stepsize rule

This class of rules uses stochastic gradient logic to update the stepsize. We first compute

$$\psi^n = (1 - \alpha_{n-1})\psi^{n-1} + \varepsilon^n. \tag{6.29}$$

The stepsize is then given by

$$\alpha_n = \left[\alpha_{n-1} + \nu\psi^{n-1}\varepsilon^n\right]_{\alpha_-}^{\alpha_+}, \tag{6.30}$$

where $\alpha_+$ and $\alpha_-$ are, respectively, upper and lower limits on the stepsize. $[\cdot]_{\alpha_-}^{\alpha_+}$ represents a projection back into the interval $[\alpha_-, \alpha_+]$, and $\nu$ is a scaling factor. $\psi^{n-1}\varepsilon^n$ is a stochastic gradient that indicates how we should change the stepsize to improve the error. Since the stochastic gradient has units that are the square of the units of the error, while the stepsize is unitless, $\nu$ has to perform an important scaling function. The equation $\alpha_{n-1} + \nu\psi^{n-1}\varepsilon^n$ can easily produce stepsizes that are larger than 1 or smaller than 0, so it is customary to specify an allowable interval (which is generally smaller than (0,1)). This rule has provable convergence, but in practice, $\nu$, $\alpha_+$ and $\alpha_-$ all have to be tuned.

The remaining formulas are drawn from the forecasting literature. These problems are characterized by nonstationary series which can exhibit shifts in the mean. For these problems, it can be useful to have a stepsize rule that moves upward when it detects what appears to be a structural change in the signal.

### *Trigg's formula*

Trigg's formula (Trigg & Leach (1967) is given by

$$\alpha_n = \frac{|S(\varepsilon^n)|}{S(|\varepsilon^n|)}. \tag{6.31}$$

The formula takes advantage of the simple property that smoothing on the absolute value of the errors is greater than or equal to the absolute value of the smoothed errors. If there is a series of errors with the same sign, that can be taken as an indication that there is a significant difference between the true mean and our estimate of the mean, which means we would like larger stepsizes.

This is the first of the adaptive stepsize formulas that uses the ratio of absolute value of the smoothed error over the smoothed absolute values of errors. Although appealing, experiments with Trigg's formula indicated that it was too responsive to what were nothing more than random sequences of errors with the same sign. The overresponsiveness of Trigg's adaptive formula has produced variants that dampen this behavior.

### *Damped Trigg formula*

Trigg's formula has been found to react too quickly to sequences of positive or negative errors. This can be fixed by smoothing on the noise, giving us

$$\alpha_n = (1 - \alpha_{n-1}^{\text{mcclain}})\alpha_{n-1} + \alpha_{n-1}^{\text{mcclain}}\nu\left(\frac{|\varepsilon^n + \varepsilon^{n-1}|}{|\varepsilon^n| + |\varepsilon^{n-1}|}\right), \tag{6.32}$$

where $\alpha_n^{\text{mcclain}}$ is as in (6.20) and $\nu$ is a constant in the interval $(0, 1]$ to be calibrated. A variation on this model is the following

$$\alpha_n = \left|0.5 - \left(\frac{|\varepsilon^n + \varepsilon^{n-1}|}{|\varepsilon^n| + |\varepsilon^{n-1}|}\right)\right|. \tag{6.33}$$

The damped version of Trigg's formula was designed to reduce the tendency of Trigg's formula to jump around.

The next two stepsize rules are variants of the McClain and Trigg formulas.

### *Adaptive McClain formula (Godfrey's rule)*

Trigg's formula can be too volatile. McClain's formula is deterministic and always decreases, which can be inappropriate in some settings. A way of combining the two yields

$$\alpha_n = \frac{\alpha_{n-1}}{1 + \alpha_{n-1} - \alpha_{n-1}^{\text{trigg}}}. \tag{6.34}$$

where $\alpha_{n-1}^{\text{trigg}}$ is the adaptive target step at iteration $n$ given by Trigg's formula. Godfrey's rule (invented by Greg Godfrey) is like Trigg's formula with a shock absorber. McClain's formula moves toward the target with a rate comparable to an arithmetic sequence. By using Trigg's formula as the target, changes in the Trigg stepsize are damped by McClain's formula. When Trigg's formula changes, Godfrey's rule moves toward the new target with a $1/n$ rate.

### *Adaptive $1/n$ (Belgacem's rule)*

In this formula, we use a $1/n$ stepsize, but we reset the iteration counter when certain conditions are satisfied.

$$\alpha_{n-1} = \frac{1}{K^n}, \tag{6.35}$$

where

$$K^n = \begin{cases} 1 & \text{if } n = 1, \\ \left\{ \begin{array}{ll} K^{n-1} + 1 & \text{if } \alpha_{n-1}^{\text{trigg}} \leq \bar{\alpha}, \\ 1 & \text{otherwise.} \end{array} \right\} & \text{if } n > 1. \end{cases} \tag{6.36}$$

Belgacem's rule (developed by Belgacem Bouzaiene-Ayari) uses a $1/n$ rule but resets the counter when it detects what appears to be a change in the underlying signal by using Trigg's formula as a trigger. $\bar{\alpha}$ is a tunable parameter that controls how quickly the counter $K^n$ is reset.

### 6.3.4 Experimental notes

Throughout our presentation, we represent the stepsize at iteration $n$ using $\alpha_{n-1}$. For discrete, lookup-table representations of value functions (as we are doing here), the stepsize should reflect how many times we have visited a specific state. If $n(S)$ is the number of times we have visited state $S$, then the stepsize for updating $\bar{V}(S)$ should be $\alpha_{n(S)}$. For notational simplicity, we suppress this capability, but it can have a significant impact on the empirical rate of convergence.

A word of caution is offered when testing out stepsize rules. It is quite easy to test out these ideas in a controlled way in a simple spreadsheet on randomly generated data, but there is a big gap between showing a stepsize that works well in a spreadsheet and one that works well in specific applications. Stochastic stepsize rules work best in the presence of transient data where the degree of noise is not too large compared to the change in the signal (the mean). As the variance of the data increases, stochastic stepsize rules begin to suffer and simpler (deterministic) rules tend to work better.

## 6.4 COMPUTING BIAS AND VARIANCE

In section 6.5 we develop a theory for finding stepsizes that produce the fastest rate of convergence. But before we present these results, it helps to present some results on the bias and variance of our estimators which are true for any stepsize rule, and which play an important role in our development of an optimal stepsize rule.

As before, we assume we are updating our estimate of $\theta$ using

$$\bar{\theta}^n = (1 - \alpha_{n-1})\bar{\theta}^{n-1} + \alpha_{n-1}\hat{\theta}^n,$$

where $\hat{\theta}^n$ is an unbiased observation (that is, $\mathbb{E}\hat{\theta}^n = \theta^n$) that is assumed to be independent of $\bar{\theta}^{n-1}$. We emphasize that the parameter we are trying to estimate, $\theta^n$, varies with $n$ just as expected value functions vary with $n$ (recall the behavior of the value function when using value iteration from chapter 3). We are interested in estimating the variance of $\bar{\theta}^n$ and its bias, which is defined by $\bar{\theta}^{n-1} - \theta^n$.

We start by computing the variance of $\bar{\theta}^n$. We assume that our observations of $\theta$ can be represented using

$$\hat{\theta}^n = \theta^n + \varepsilon^n,$$

where $\mathbb{E}\varepsilon^n = 0$ and $Var[\varepsilon^n] = \sigma^2$. Previously, $\varepsilon^n$ was the error between our previous estimate and our latest observation. Here, we treat this as an exogenous measurement error. With this model, we can compute the variance of $\bar{\theta}^n$ using

$$Var[\bar{\theta}^n] = \lambda^n \sigma^2, \tag{6.37}$$

where $\lambda^n$ can be computed from the simple recursion

$$\lambda^n = \begin{cases} (\alpha_{n-1})^2, & n = 1, \\ (1-\alpha_{n-1})^2\lambda^{n-1} + (\alpha_{n-1})^2, & n > 1. \end{cases} \tag{6.38}$$

To see this, we start with $n = 1$. For a given (deterministic) initial estimate $\bar{\theta}^0$, we first observe that the variance of $\bar{\theta}^1$ is given by

$$\begin{aligned} Var[\bar{\theta}^1] &= Var[(1-\alpha_0)\bar{\theta}^0 + \alpha_0\hat{\theta}^1] \\ &= (\alpha_0)^2 Var[\hat{\theta}^1] \\ &= (\alpha_0)^2\sigma^2. \end{aligned}$$

For general $\bar{\theta}^n$, we use a proof by induction. Assume that $Var[\bar{\theta}^{n-1}] = \lambda^{n-1}\sigma^2$. Then, since $\bar{\theta}^{n-1}$ and $\hat{\theta}^n$ are independent, we find

$$\begin{aligned} Var[\bar{\theta}^n] &= Var\left[(1-\alpha_{n-1})\bar{\theta}^{n-1} + \alpha_{n-1}\hat{\theta}^n\right] \\ &= (1-\alpha_{n-1})^2 Var\left[\bar{\theta}^{n-1}\right] + (\alpha_{n-1})^2 Var[\hat{\theta}^n] \\ &= (1-\alpha_{n-1})^2\lambda^{n-1}\sigma^2 + (\alpha_{n-1})^2\sigma^2 \qquad (6.39) \\ &= \lambda^n\sigma^2. \qquad (6.40) \end{aligned}$$

Equation (6.39) is true by assumption (in our induction proof), while equation (6.40) establishes the recursion in equation (6.38). This gives us the variance, assuming of course that $\sigma^2$ is known.

The bias of our estimate is the difference between our current estimate and the true value, given by

$$\beta^n = \mathbb{E}[\bar{\theta}^{n-1}] - \theta^n.$$

We note that $\bar{\theta}^{n-1}$ is our estimate of $\theta^n$ computed using the information up through iteration $n-1$. Of course, our formula for the bias assumes that $\theta^n$ is known. These two results for the variance and bias are called the *parameters-known* formulas.

We next require the mean-squared error, which can be computed using

$$\mathbb{E}\left[\left(\bar{\theta}^{n-1} - \theta^n\right)^2\right] = \lambda^{n-1}\sigma^2 + (\beta^n)^2. \tag{6.41}$$

See exercise 6.4 to prove this. This formula gives the variance around the known mean, $\theta^n$. For our purposes, it is also useful to have the variance around the observations $\hat{\theta}^n$. Let

$$\nu^n = \mathbb{E}\left[\left(\bar{\theta}^{n-1} - \hat{\theta}^n\right)^2\right]$$

be the mean squared error (including noise and bias) between the current estimate $\bar{\theta}^{n-1}$ and the observation $\hat{\theta}^n$. It is possible to show that (see exercise 6.5)

$$\nu^n = (1 + \lambda^{n-1})\sigma^2 + (\beta^n)^2, \tag{6.42}$$

where $\lambda^n$ is computed using (6.38).

In practice, we do not know $\sigma^2$, and we certainly do not know $\theta^n$. As a result, we have to estimate both parameters from our data. We begin by providing an estimate of the bias using

$$\bar{\beta}^n = (1 - \eta_{n-1})\bar{\beta}^{n-1} + \eta_{n-1}(\bar{\theta}^{n-1} - \hat{\theta}^n),$$

where $\eta_{n-1}$ is a (typically simple) stepsize rule used for estimating the bias and variance. As a general rule, we should pick a stepsize for $\eta_{n-1}$ which produces larger stepsizes than $\alpha_{n-1}$ because we are more interested in tracking the true signal than producing an estimate with a low variance. We have found that a constant stepsize such as .10 works quite well on a wide range of problems, but if precise convergence is needed, it is necessary to use a rule where the stepsize goes to zero such as the harmonic stepsize rule (equation (6.18)).

To estimate the variance, we begin by finding an estimate of the total variation $\nu^n$. Let $\bar{\nu}^n$ be the estimate of the total variance which we might compute using

$$\bar{\nu}^n = (1 - \eta_{n-1})\bar{\nu}^{n-1} + \eta_{n-1}(\bar{\theta}^{n-1} - \hat{\theta}^n)^2$$

Using $\bar{\nu}^n$ as our estimate of the total variance, we can compute an estimate of $\sigma^2$ using

$$(\bar{\sigma}^n)^2 = \frac{\bar{\nu}^n - (\bar{\beta}^n)^2}{1 + \lambda^{n-1}}.$$

We can use $(\bar{\sigma}^n)^2$ in equation (6.37) to obtain an estimate of the variance of $\bar{\theta}^n$.

If we are doing true averaging (as would occur if we use a stepsize of $1/n$), we can get a more precise estimate of the variance for small samples by using the recursive form of the small sample formula for the variance

$$(\hat{\sigma}^2)^n = \frac{n-2}{n-1}(\hat{\sigma}^2)^{n-1} + \frac{1}{n}(\bar{\theta}^{n-1} - \hat{\theta}^n)^2. \tag{6.43}$$

$(\hat{\sigma}^2)^n$ is an estimate of the variance of $\hat{\theta}^n$. The variance of our estimate $\bar{\theta}^{(n)}$ is computed using

$$(\bar{\sigma}^2)^n = \frac{1}{n}(\hat{\sigma}^2)^n.$$

## 6.5 OPTIMAL STEPSIZES

Given the variety of stepsize formulas we can choose from, it seems natural to ask whether there is an optimal stepsize rule. Before we can answer such a question, we have to define exactly what we mean by it. Assume that we are trying to estimate a parameter (such as a value of being in a state or the slope of a value function) that we denote by $\theta^n$ that may be changing over time. At iteration $n$, our estimate of $\theta^n$, $\bar{\theta}^n$, is a random variable that depends on our stepsize rule. To express this dependence, let $\alpha$ represent a stepsize rule, and let $\bar{\theta}^n(\alpha)$ be the estimate of the parameter $\theta$ after iteration $n$ using stepsize rule $\alpha$. We would like to choose a stepsize rule to minimize

$$\min_{\alpha} \mathbb{E}(\bar{\theta}^n(\alpha) - \theta^n)^2. \tag{6.44}$$

Here, the expectation is over the entire history of the algorithm and requires (in principle) knowing the true value of the parameter being estimated. If we could solve this problem (which requires knowing certain parameters about the underlying distributions), we would obtain a deterministic stepsize rule. In practice, we do not generally know these parameters which need to be estimated from data, producing a stochastic stepsize rule.

There are other objective functions we could use. For example, instead of minimizing the distance to an unknown parameter sequence $\theta^n$, we could solve the minimization problem

$$\min_{\alpha} \mathbb{E}\left\{(\bar{\theta}^n(\alpha) - \hat{\theta}^{n+1})^2\right\}, \tag{6.45}$$

where we are trying to minimize the deviation between our prediction, obtained at iteration $n$, and the actual observation at $n+1$. Here, we are again proposing an unconditional expectation, which means that $\bar{\theta}^n(\alpha)$ is a random variable within the expectation. Alternatively, we could condition on our history up to iteration $n$

$$\min_{\alpha} \mathbb{E}^n \left\{ (\bar{\theta}^n(\alpha) - \hat{\theta}^{n+1})^2 \right\} \tag{6.46}$$

where $\mathbb{E}^n$ means that we are taking the expectation given what we know at iteration $n$ (which means that $\bar{\theta}^n(\alpha)$ is a constant). (For readers familiar with the material in section 5.10, we would write the expectation as $\mathbb{E}\left\{(\bar{\theta}^n(\alpha) - \hat{\theta}^{n+1})^2 | \mathfrak{F}^n\right\}$, where $\mathfrak{F}^n$ is the sigma-algebra generated by the history of the process up through iteration $n$.) In this formulation $\bar{\theta}^n(\alpha)$ is now deterministic at iteration $n$ (because we are conditioning on the history up through iteration $n$), whereas in (6.45), $\bar{\theta}^n(\alpha)$ is random (since we are not conditioning on the history). The difference between these two objective functions is subtle but significant.

We begin our discussion of optimal stepsizes in section 6.5.1 by addressing the case of estimating a constant parameter which we observe with noise. Section 6.5.2 considers the case where we are estimating a parameter that is changing over time, but where the changes have mean zero. Finally, section 6.5.3 addresses the case where the mean may be drifting up or down with nonzero mean, a situation that we typically face when approximating a value function.

### 6.5.1 Optimal stepsizes for stationary data

Assume that we observe $\hat{\theta}^n$ at iteration $n$ and that the observations $\hat{\theta}^n$ can be described by

$$\hat{\theta}^n = \theta + \varepsilon^n$$

where $\theta$ is an unknown constant and $\varepsilon^n$ is a stationary sequence of independent and identically distributed random deviations with mean 0 and variance $\sigma^2$. We can approach the problem of estimating $\theta$ from two perspectives: choosing the best stepsize and choosing the best linear combination of the estimates. That is, we may choose to write our estimate $\bar{\theta}^n$ after $n$ observations in the form

$$\bar{\theta}^n = \sum_{m=1}^{n} a_m^n \hat{\theta}_m.$$

For our discussion, we will fix $n$ and work to determine the coefficients $a_m$ (recognizing that they can depend on the iteration). We would like our statistic to have two properties: It should be unbiased, and it should have minimum variance (that is, it should solve (6.44)). To be unbiased, it should satisfy

$$\begin{aligned}
\mathbb{E}\left[\sum_{m=1}^{n} a_m \hat{\theta}_m\right] &= \sum_{m=1}^{n} a_m \mathbb{E}\hat{\theta}_m \\
&= \sum_{m=1}^{n} a_m \theta \\
&= \theta,
\end{aligned}$$

which implies that we must satisfy

$$\sum_{m=1}^{n} a_m = 1.$$

The variance of our estimator is given by:

$$Var(\bar{\theta}^n) = Var\left[\sum_{m=1}^{n} a_m \hat{\theta}_m\right].$$

We use our assumption that the random deviations are independent, which allows us to write

$$\begin{aligned} Var(\bar{\theta}^n) &= \sum_{m=1}^{n} Var[a_m \hat{\theta}_m] \\ &= \sum_{m=1}^{n} a_m^2 Var[\hat{\theta}_m] \\ &= \sigma^2 \sum_{m=1}^{n} a_m^2. \end{aligned} \tag{6.47}$$

Now we face the problem of finding $a_1, \ldots, a_n$ to minimize (6.47) subject to the requirement that $\sum_m a_m = 1$. This problem is easily solved using the Lagrange multiplier method. We start with the nonlinear programming problem

$$\min_{\{a_m\}} \sum_{m=1}^{n} a_m^2$$

subject to

$$\sum_{m=1}^{n} a_m = 1, \tag{6.48}$$

$$a_m \geq 0. \tag{6.49}$$

We relax constraint (6.48) and add it to the objective function

$$\min_{\{a_m\}} L(a, \lambda) = \sum_{m=1}^{n} a_m^2 - \lambda \left( \sum_{m=1}^{n} a_m - 1 \right)$$

subject to (6.49). We are now going to try to solve $L(a, \lambda)$ (known as the "Lagrangian") and hope that the coefficients $a$ are all nonnegative. If this is true, we can take derivatives and set them equal to zero

$$\frac{\partial L(a, \lambda)}{\partial a_m} = 2a_m - \lambda. \tag{6.50}$$

The optimal solution $(a^*, \lambda^*)$ would then satisfy

$$\frac{\partial L(a, \lambda)}{\partial a_m} = 0.$$

This means that at optimality

$$a_m = \lambda/2,$$

which tells us that the coefficients $a_m$ are all equal. Combining this result with the requirement that they sum to one gives the expected result:

$$a_m = \frac{1}{n}.$$

In other words, our best estimate is a sample average. From this (somewhat obvious) result, we can obtain the optimal stepsize, since we already know that $\alpha_{n-1} = 1/n$ is the same as using a sample average.

This result tells us that if the underlying data is stationary, and we have no prior information about the sample mean, then the best stepsize rule is the basic $1/n$ rule. Using any other rule requires that there be some violation in our basic assumptions. In practice, the most common violation is that the observations are not stationary because they are derived from a process where we are searching for the best solution.

### 6.5.2 Optimal stepsizes for nonstationary data - I

Assume now that our parameter evolves over time (iterations) according to the process

$$\theta^n = \theta^{n-1} + \xi^n, \tag{6.51}$$

where $\mathbb{E}\xi^n = 0$ is a zero mean drift term with variance $(\sigma^\xi)^2$. As before, we measure $\theta^n$ with an error according to

$$\hat{\theta}^n = \theta^n + \varepsilon^n.$$

We want to choose a stepsize so that we minimize the mean squared error. This problem can be solved using the Kalman filter. The Kalman filter is a powerful recursive regression technique, but we adapt it here for the problem of estimating a single parameter. Typical applications of the Kalman filter assume that the variance of $\xi^n$, given by $(\sigma^\xi)^2$, and the variance of the measurement error, $\varepsilon^n$, given by $\sigma^2$, are known. In this case, the Kalman filter would compute a stepsize (generally referred to as the gain) using

$$\alpha_n = \frac{(\sigma^\xi)^2}{\nu^n + \sigma^2}, \tag{6.52}$$

where $\nu^n$ is computed recursively using

$$\nu^n = (1 - \alpha_{n-1})\nu^{n-1} + (\sigma^\xi)^2. \tag{6.53}$$

Remember that $\alpha_0 = 1$, so we do not need a value of $\nu^0$. For our application, we do not know the variances so these have to be estimated from data. We first estimate the bias using

$$\bar{\beta}^n = (1 - \eta_{n-1})\,\bar{\beta}^{n-1} + \eta_{n-1}\left(\bar{\theta}^{n-1} - \hat{\theta}^n\right), \tag{6.54}$$

where $\eta_{n-1}$ is a simple stepsize rule such as the harmonic stepsize rule or McClain's formula. We then estimate the total error sum of squares using

$$\bar{\nu}^n = (1 - \eta_{n-1})\,\bar{\nu}^{n-1} + \eta_{n-1}\left(\bar{\theta}^{n-1} - \hat{\theta}^n\right)^2. \tag{6.55}$$

Finally, we estimate the variance of the error using

$$(\bar{\sigma}^n)^2 = \frac{\bar{\nu}^n - (\bar{\beta}^n)^2}{1 + \bar{\lambda}^{n-1}}, \tag{6.56}$$

where $\bar{\lambda}^{n-1}$ is computed using (6.38). We use $(\bar{\sigma}^n)^2$ as our estimate of $\sigma^2$. We then propose to use $\left(\bar{\beta}^n\right)^2$ as our estimate of $(\sigma^\xi)^2$. This is purely an approximation, but experimental work suggests that it performs quite well, and it is relatively easy to implement.

### 6.5.3 Optimal stepsizes for nonstationary data - II

In dynamic programming, we are trying to estimate the value of being in a state (call it $v$) by $\bar{v}$ which is estimated from a sequence of random observations $\hat{v}$. The problem we encounter is that $\hat{v}$ might depend on a value function approximation which is steadily increasing, which means that the observations $\hat{v}$ are nonstationary. Furthermore, unlike the assumption made by the Kalman filter that the mean of $\hat{v}$ is varying in a zero-mean way, our observations of $\hat{v}$ might be steadily increasing. This would be the same as assuming that $\mathbb{E}\xi = \mu > 0$ in the section above. In this section, we derive the Kalman filter learning rate for biased estimates.

Our challenge is to devise a stepsize that strikes a balance between minimizing error (which prefers a smaller stepsize) and responding to the nonstationary data (which works better with a large stepsize). We return to our basic model

$$\hat{\theta}^n = \theta^n + \varepsilon^n,$$

where $\theta^n$ varies over time, but it might be steadily increasing or decreasing. This would be similar to the model in the previous section (equation (6.51)) but where $\xi^n$ has a nonzero mean. As before we assume that $\{\varepsilon^n\}_{n=1,2,\ldots}$ are independent and identically distributed with mean value of zero and variance, $\sigma^2$. We perform the usual stochastic gradient update to obtain our estimates of the mean

$$\bar{\theta}^n(\alpha_{n-1}) = (1-\alpha_{n-1})\,\bar{\theta}^{n-1} + \alpha_{n-1}\hat{\theta}^n. \tag{6.57}$$

We wish to find $\alpha_{n-1}$ that solves,

$$\min_{\alpha_{n-1}} F(\alpha_{n-1}) = \mathbb{E}\left[\left(\bar{\theta}^n(\alpha_{n-1}) - \theta^n\right)^2\right]. \tag{6.58}$$

It is important to realize that we are trying to choose $\alpha_{n-1}$ to minimize the *unconditional* expectation of the error between $\bar{\theta}^n$ and the true value $\theta^n$. For this reason, our stepsize rule will be deterministic, since we are not allowing it to depend on the information obtained up through iteration $n$.

We assume that the observation at iteration $n$ is unbiased, which is to say

$$\mathbb{E}\left[\hat{\theta}^n\right] = \theta^n. \tag{6.59}$$

But the smoothed estimate is biased because we are using simple smoothing on nonstationary data. We denote this bias as

$$\begin{aligned}\beta^{n-1} &= \mathbb{E}\left[\bar{\theta}^{n-1} - \theta^n\right] \\ &= \mathbb{E}\left[\bar{\theta}^{n-1}\right] - \theta^n.\end{aligned} \tag{6.60}$$

We note that $\beta^{n-1}$ is the bias computed after iteration $n-1$ (that is, after we have computed $\bar{\theta}^{n-1}$). $\beta^{n-1}$ is the bias when we use $\bar{\theta}^{n-1}$ as an estimate of $\theta^n$.

The variance of the observation $\hat{\theta}^n$ is computed as follows:

$$\begin{aligned}Var\left[\hat{\theta}^n\right] &= \mathbb{E}\left[\left(\hat{\theta}^n - \theta^n\right)^2\right] \\ &= \mathbb{E}\left[(\varepsilon^n)^2\right] \\ &= \sigma^2.\end{aligned} \tag{6.61}$$

We now have what we need to derive an optimal stepsize for nonstationary data with a mean that is steadily increasing (or decreasing). We refer to this as the *bias-adjusted Kalman filter* stepsize rule (or BAKF), in recognition of its close relationship to the Kalman filter learning rate. We state the formula in the following theorem:

**Theorem 6.5.1** *The optimal stepsizes* $(\alpha_m)_{m=0}^n$ *that minimize the objective function in equation (6.58) can be computed using the expression*

$$\alpha_{n-1} = 1 - \frac{\sigma^2}{(1+\lambda^{n-1})\sigma^2 + (\beta^n)^2}, \tag{6.62}$$

*where* $\lambda$ *is computed recursively (see equation* (6.38)*) using*

$$\lambda^n = \begin{cases} (\alpha_{n-1})^2, & n = 1 \\ (1-\alpha_{n-1})^2\lambda^{n-1} + (\alpha_{n-1})^2, & n > 1. \end{cases} \tag{6.63}$$

**Proof:** We present the proof of this result because it brings out some properties of the solution that we exploit later when we handle the case where the variance and bias are unknown. Let $F(\alpha_{n-1})$ denote the objective function from the problem stated in (6.58).

$$\begin{aligned} F(\alpha_{n-1}) &= \mathbb{E}\left[\left(\bar{\theta}^n(\alpha_{n-1}) - \theta^n\right)^2\right] & (6.64) \\ &= \mathbb{E}\left[\left((1-\alpha_{n-1})\,\bar{\theta}^{n-1} + \alpha_{n-1}\hat{\theta}^n - \theta^n\right)^2\right] & (6.65) \\ &= \mathbb{E}\left[\left((1-\alpha_{n-1})\left(\bar{\theta}^{n-1} - \theta^n\right) + \alpha_{n-1}\left(\hat{\theta}^n - \theta^n\right)\right)^2\right] & (6.66) \\ &= (1-\alpha_{n-1})^2\,\mathbb{E}\left[\left(\bar{\theta}^{n-1} - \theta^n\right)^2\right] + (\alpha_{n-1})^2\,\mathbb{E}\left[\left(\hat{\theta}^n - \theta^n\right)^2\right] \\ &\quad +2\alpha_{n-1}(1-\alpha_{n-1})\underbrace{\mathbb{E}\left[\left(\bar{\theta}^{n-1} - \theta^n\right)\left(\hat{\theta}^n - \theta^n\right)\right]}_{I}. & (6.67) \end{aligned}$$

Equation (6.64) is true by definition, while (6.65) is true by definition of the updating equation for $\bar{\theta}^n$. We obtain (6.66) by adding and subtracting $\alpha_{n-1}\theta^n$. To obtain (6.67), we expand the quadratic term and then use the fact that the stepsize rule, $\alpha_{n-1}$, is deterministic, which allows us to pull it outside the expectations. Then, the expected value of the cross-product term, *I*, vanishes under the assumption of independence of the observations and the objective function reduces to the following form

$$F(\alpha_{n-1}) = (1-\alpha_{n-1})^2\,\mathbb{E}\left[\left(\bar{\theta}^{n-1} - \theta^n\right)^2\right] + (\alpha_{n-1})^2\,\mathbb{E}\left[\left(\hat{\theta}^n - \theta^n\right)^2\right]. \tag{6.68}$$

In order to find the optimal stepsize, $\alpha_{n-1}^*$, that minimizes this function, we obtain the first-order optimality condition by setting $\frac{\partial F(\alpha_{n-1})}{\partial \alpha_{n-1}} = 0$, which gives us

$$-2\left(1-\alpha_{n-1}^*\right)\mathbb{E}\left[\left(\bar{\theta}^{n-1} - \theta^n\right)^2\right] + 2\alpha_{n-1}^*\mathbb{E}\left[\left(\hat{\theta}^n - \theta^n\right)^2\right] = 0. \tag{6.69}$$

Solving this for $\alpha_{n-1}^*$ gives us the following result

$$\alpha_{n-1}^* = \frac{\mathbb{E}\left[\left(\bar{\theta}^{n-1} - \theta^n\right)^2\right]}{\mathbb{E}\left[\left(\bar{\theta}^{n-1} - \theta^n\right)^2\right] + \mathbb{E}\left[\left(\hat{\theta}^n - \theta^n\right)^2\right]}. \tag{6.70}$$

Using $\mathbb{E}\left[\left(\hat{\theta}^n - \theta^n\right)^2\right] = \sigma^2$ and $\mathbb{E}\left[\left(\bar{\theta}^{n-1} - \theta^n\right)^2\right] = \lambda^{n-1}\sigma^2 + (\beta^n)^2$ (from (6.41)) in (6.70) gives us

$$\begin{aligned} \alpha_{n-1} &= \frac{\lambda^{n-1}\sigma^2 + (\beta^n)^2}{\lambda^{n-1}\sigma^2 + (\beta^n)^2 + \sigma^2} \\ &= 1 - \frac{\sigma^2}{(1+\lambda^{n-1})\,\sigma^2 + (\beta^n)^2}, \end{aligned}$$

which is our desired result (equation (6.62)). □

A brief remark is in order. We have taken considerable care to make sure that $\alpha_{n-1}$, used to update $\bar{\theta}^{n-1}$ to obtain $\bar{\theta}^n$, is a function of information available up through iteration $n-1$. Yet, the expression for $\alpha_{n-1}$ in equation (6.62) includes $\beta^n$. At this point in our derivation, however, $\beta^n$ is deterministic, which means that it is known (in theory at least) at iteration $n-1$. Below, we have to address the problem that we do not actually know $\beta^n$ and have to estimate it from data.

Before we turn to the problem of estimating the bias and variance, there are two interesting corollaries that we can easily establish: the optimal stepsize if the data comes from a stationary series, and the optimal stepsize if there is no noise. Earlier we proved this case by solving a linear regression problem, and noted that the best estimate was a sample average, which implied a particular stepsize. Here, we directly find the optimal stepsize as a corollary of our earlier result.

**Corollary 6.5.1** *For a sequence with a static mean, the optimal stepsizes are given by*

$$\alpha_{n-1} = \frac{1}{n} \quad \forall\, n = 1, 2, \ldots. \tag{6.71}$$

**Proof:** In this case, the mean $\theta^n = \theta$ is a constant. Therefore, the estimates of the mean are unbiased, which means $\beta^n = 0 \quad \forall t = 2, \ldots,$. This allows us to write the optimal stepsize as

$$\alpha_{n-1} = \frac{\lambda^{n-1}}{1+\lambda^{n-1}}. \tag{6.72}$$

Substituting (6.72) into (6.63) gives us

$$\alpha_n = \frac{\alpha_{n-1}}{1+\alpha_{n-1}}. \tag{6.73}$$

If $\alpha_0 = 1$, it is easy to verify (6.71). □

For the case where there is no noise ($\sigma^2 = 0$), we have the following:

**Corollary 6.5.2** *For a sequence with zero noise, the optimal stepsizes are given by*

$$\alpha_{n-1} = 1 \quad \forall\, n = 1, 2, \ldots. \tag{6.74}$$

The corollary is proved by simply setting $\sigma^2 = 0$ in equation (6.62).

As a final result, we obtain

**Corollary 6.5.3** *In general,*

$$\alpha_{n-1} \geq \frac{1}{n} \quad \forall\, n = 1, 2, \ldots.$$

**Proof:** We leave this more interesting proof as an exercise to the reader (see exercise 6.14).

Corollary 6.5.3 is significant since it establishes one of the conditions needed for convergence of a stochastic approximation method. An open theoretical question, as of this writing, is whether the BAKF stepsize rule also satisfies the requirement that $\sum_{n=1}^{\infty} \alpha_n = \infty$.

The problem with using the stepsize formula in equation (6.62) is that it assumes that the variance $\sigma^2$ and the bias $(\beta^n)^2$ are known. This can be problematic in real instances, especially the assumption of knowing the bias, since computing this basically requires knowing the real function. If we have this information, we do not need this algorithm.

As an alternative, we can try to estimate these quantities from data. Let

$$\begin{aligned}
(\bar{\sigma}^2)^n &= \text{Estimate of the variance of the error after iteration } n, \\
\bar{\beta}^n &= \text{Estimate of the bias after iteration } n, \\
\bar{\nu}^n &= \text{Estimate of the variance of the bias after iteration } n.
\end{aligned}$$

To make these estimates, we need to smooth new observations with our current best estimate, something that requires the use of a stepsize formula. We could attempt to find an optimal stepsize for this purpose, but it is likely that a reasonably chosen deterministic formula will work fine. One possibility is McClain's formula (equation (6.20)):

$$\eta_n = \frac{\eta_{n-1}}{1 + \eta_{n-1} - \bar{\eta}}.$$

A limit point such as $\bar{\eta} \in (0.05, 0.10)$ appears to work well across a broad range of functional behaviors. The property of this stepsize that $\eta_n \rightarrow \bar{\eta}$ can be a strength, but it does mean that the algorithm will not tend to converge in the limit, which requires a stepsize that goes to zero. If this is needed, we suggest a harmonic stepsize rule:

$$\eta_{n-1} = \frac{a}{a + n - 1},$$

where $a$ in the range between 5 and 10 seems to work quite well for many dynamic programming applications.

Care needs to be used in the early iterations. For example, if we let $\alpha_0 = 1$, then we do not need an initial estimate for $\bar{\theta}^0$ (a trick we have used throughout). However, since the formulas depend on an estimate of the variance, we still have problems in the second iteration. For this reason, we recommend forcing $\eta_1$ to equal 1 (in addition to using $\eta_0 = 1$). We also recommend using $\alpha_n = 1/(n+1)$ for the first few iterations, since the estimates of $(\bar{\sigma}^2)^n$, $\bar{\beta}^n$ and $\bar{\nu}^n$ are likely to be very unreliable in the very beginning.

Figure 6.8 summarizes the entire algorithm. Note that the estimates have been constructed so that $\alpha_n$ is a function of information available up through iteration $n$.

Figure 6.9 illustrates the behavior of the bias-adjusted Kalman filter stepsize rule for two signals: very low noise (figure 6.9a) and with higher noise (figure 6.9b). For both cases, the signal starts small and rises toward an upper limit of 1.0 (on average). In both figures, we also show the stepsize $1/n$. For the low-noise case, the stepsize stays quite large. For the high noise case, the stepsize roughly tracks $1/n$ (note that it never goes below $1/n$).

## 6.6 SOME EXPERIMENTAL COMPARISONS OF STEPSIZE FORMULAS

George & Powell (2006) reports on a series of experimental comparisons of stepsize formulas. The methods were tested on a series of functions which increase monotonically

---

**Step 0.** Initialization:

**Step 0a.** Set the baseline to its initial value, $\bar{\theta}_0$.

**Step 0b.** Initialize the parameters - $\bar{\beta}_0, \bar{\nu}_0$ and $\bar{\lambda}_0$.

**Step 0c.** Set initial stepsizes $\alpha_0 = \eta_0 = 1$, and specify the stepsize rule for $\eta$.

**Step 0d.** Set the iteration counter, $n = 1$.

**Step 1.** Obtain the new observation, $\hat{\theta}^n$.

**Step 2.** Smooth the baseline estimate.

$$\bar{\theta}^n = (1 - \alpha_{n-1})\bar{\theta}^{n-1} + \alpha_{n-1}\hat{\theta}^n$$

**Step 3.** Update the following parameters:

$$\begin{aligned}
\varepsilon^n &= \bar{\theta}^{n-1} - \hat{\theta}^n, \\
\bar{\beta}^n &= (1 - \eta_{n-1})\,\bar{\beta}^{n-1} + \eta_{n-1}\varepsilon^n, \\
\bar{\nu}^n &= (1 - \eta_{n-1})\bar{\nu}^{n-1} + \eta_{n-1}(\varepsilon^n)^2, \\
(\bar{\sigma}^2)^n &= \frac{\bar{\nu}^n - (\bar{\beta}^n)^2}{1 + \bar{\lambda}^{n-1}}.
\end{aligned}$$

**Step 4.** Evaluate the stepsizes for the next iteration.

$$\begin{aligned}
\alpha_n &= \begin{cases} 1/(n+1) & n = 1, 2, \\ 1 - \frac{(\bar{\sigma}^2)^n}{\bar{\nu}^n}, & n > 2, \end{cases} \\
\eta_n &= \frac{a}{a + n - 1}. \quad \text{Note that this gives us } \eta_1 = 1.
\end{aligned}$$

**Step 5.** Compute the coefficient for the variance of the smoothed estimate of the baseline.

$$\bar{\lambda}^n = (1 - \alpha_{n-1})^2\bar{\lambda}^{n-1} + (\alpha_{n-1})^2.$$

**Step 6.** If $n < N$, then $n = n + 1$ and go to Step 1, else stop.

---

**Figure 6.8** The bias-adjusted Kalman filter stepsize rule.

at different rates, after which they level off. Two classes of functions were used, but they all increased monotonically to a limiting value. The first class, which we denote by $f^I(n)$, increases at a geometrically decreasing rate right from the beginning (similar to the behavior we see with value iteration). The argument $n$ is referred to as the "iteration" since we view these functions as reflecting the change in a value function over the iterations. The second class, denoted $f^{II}(n)$, remains constant initially and then undergoes a delayed increase. For each class of functions, there are five variations with different overall rates of increase.

Each function was measured with an error term with an assumed variance. Three levels of variance were used: a) low noise, measured relative to the change in the function, b) medium noise, where the level of noise is comparable to the structural change in the function, and c) high noise, where the noise is large relative to the change in the function (the function almost appears flat).

The idea of an adaptive stepsize is most dramatically illustrated for the functions that start off constant and then rise quickly at a later iteration. Standard stepsize formulas that decline monotonically over time run the risk of dropping to small values. When the function starts to increase, the stepsizes have dropped to such small values that they cannot respond quickly. Figure 6.10 compares the search-then-converge formula to the biased-adjusted Kalman filter stepsize rule (the algorithm given in figure 6.8) to the search-then-converge

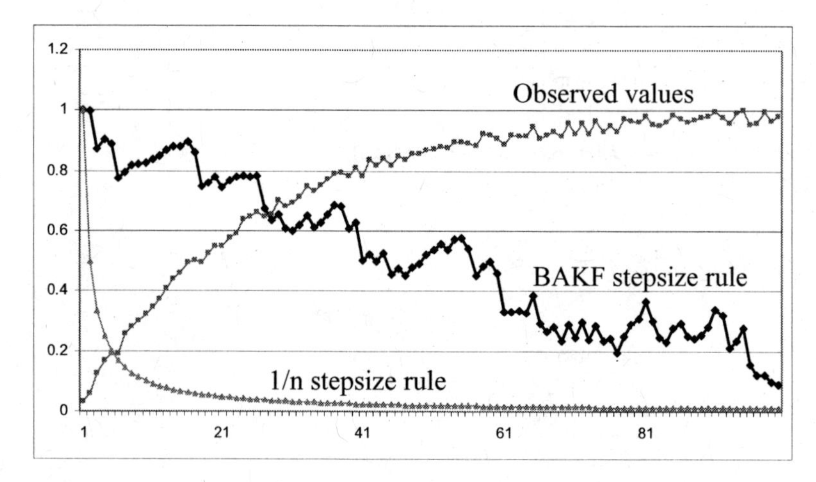


6.9a Bias-adjusted Kalman filter for a signal with low noise.

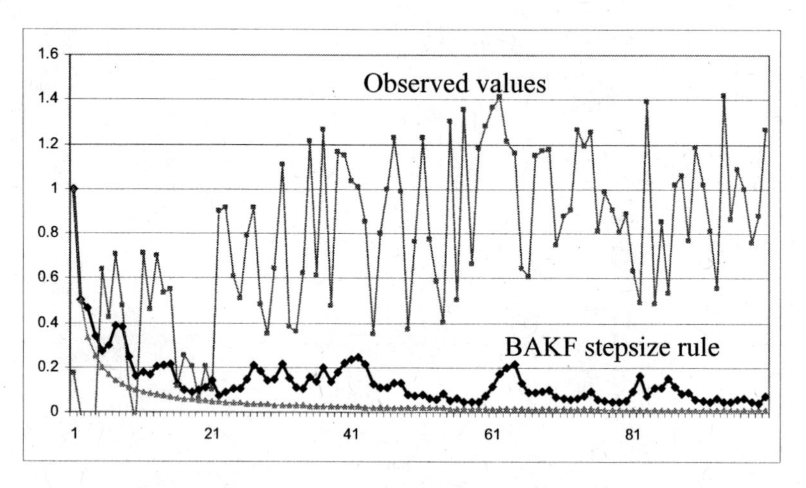


6.9b Bias-adjusted Kalman filter for a signal with higher noise.

**Figure 6.9** The BAKF stepsize rule for low-noise (a) and high-noise (b). Each figure shows the signal, the BAKF stepsizes and the stepsizes produced by the $1/n$ stepsize rule.

rule and Trigg's rule. This behavior can actually occur in dynamic programs which require a number of steps before receiving a reward. Examples include most games (backgammon, checkers, tic-tac-toe) where it is necessary to play an entire game before we learn if we won (we receive a reward) or lost.

Many dynamic programs exhibit the behavior of our type I functions where the value function rises steadily and then levels off. The problem is that the rate of increase can vary widely. It helps to use a larger stepsize for parameters that are increasing quickly.

Table 6.1 shows the results of a series of experiments comparing different stepsize rules for the class I functions ($f^I(n)$). The experiments were run using three noise levels and were measured at three different points $n$ along the curve. At $n = 25$, the curve is still rising quickly; $n = 50$ corresponds to the "elbow" of the curve; and at $n = 75$ the curve is

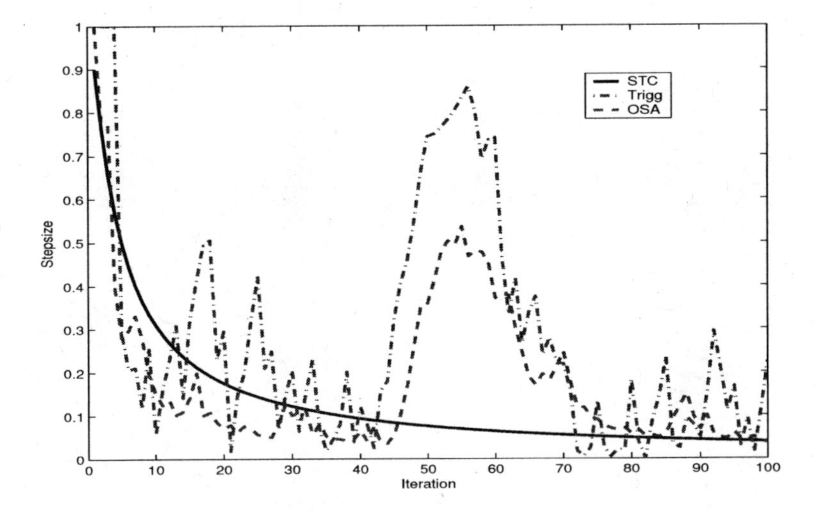


**Figure 6.10** Comparison of stepsize formulas

stabilizing. The stepsizes considered were $1/n$, $1/n^\beta$ (with $\beta = .85$), STC (search-then-converge) using parameters optimized for this problem class, McClain (with target 0.10), Kesten's rule, the stochastic gradient adaptive stepsize rule (SGASR), and the bias-adjusted Kalman filter (BAKF) (equation (6.62)). The table gives the average mean squared error and the standard deviation of the average. Table 6.2 provides the same statistics for the second class of functions which exhibit delayed learning. For these functions $n = 25$ occurs when the function is in its initial stable period; $n = 50$ is the point where the curve is rising the most; and $n = 75$ corresponds to where the curve is again stable.

These results show that the adaptive stepsize formulas (in particular the bias-adjusted Kalman filter) works best with lower levels of noise. If the noise level is quite high, stochastic formulas can be thrown off. For these cases, the family of arithmetically declining stepsizes ($1/n, 1/n^\beta$, STC, and McClain) all work reasonably well. The risk of using $1/n$ or even $1/n^\beta$ is that they may decline too quickly for functions which are still changing after many of iterations. McClain avoids this by imposing a target. For problems where we are running more than a few dozen iterations, McClain is similar to using a constant stepsize.

The real value of an adaptive stepsize rule in dynamic programming is its ability to respond to different rates of convergence for different parameters (such as the value of being in a specific state). Some parameters may converge much more slowly than others and will benefit from a larger stepsize. Since we cannot tune a stepsize rule for each parameter, an adaptive stepsize rule may work best. Table 6.3 shows the percent error for a steady-state nomadic trucker application (which can be solved optimally) using three values for the discount factor $\gamma$. Since updates of the value function depend on the value function approximation itself, convergence tends to be slow (especially for the higher discount factors). Here the bias-adjusted Kalman filter works particularly well.

There does not appear to be a universally accepted stepsize rule. Stochastic stepsize rules offer tremendous appeal. For example, if we are estimating the value of being in each state, it may be the case that each of these values moves according to its own process. Some may be relatively stationary, while others move quickly before stabilizing. The value of an adaptive stepsize rule appears to depend on the nature of the data and our ability to tune a

| Variance | $n$ | $1/n$ | $1/n^\beta$ | STC | McClain | Kesten | SGASR | BAKF |
|---|---|---|---|---|---|---|---|---|
| | 25 | 5.721 | 3.004 | 0.494 | 1.760 | 0.368 | 0.855 | **0.365** |
| | *std. dev.* | *0.031* | *0.024* | *0.014* | *0.021* | *0.014* | *0.025* | *0.015* |
| $\sigma^2 = 1$ | 50 | 5.697 | 2.374 | 0.322 | 0.502 | 0.206 | 0.688 | **0.173** |
| | *std. dev.* | *0.021* | *0.015* | *0.008* | *0.010* | *0.008* | *0.021* | *0.008* |
| | 75 | 4.993 | 1.713 | 0.200 | 0.169 | 0.135 | 0.578 | **0.126** |
| | *std. dev.* | *0.016* | *0.011* | *0.005* | *0.005* | *0.006* | *0.018* | *0.006* |
| | 25 | 6.216 | 3.527 | **1.631** | 2.404 | 2.766 | 3.031 | 1.942 |
| | *std. dev.* | *0.102* | *0.080* | *0.068* | *0.073* | *0.107* | *0.129* | *0.078* |
| $\sigma^2 = 10$ | 50 | 5.911 | 2.599 | **0.871** | 0.968 | 1.571 | 1.610 | 1.118 |
| | *std. dev.* | *0.070* | *0.050* | *0.037* | *0.038* | *0.064* | *0.075* | *0.047* |
| | 75 | 5.127 | 1.870 | **0.588** | 0.655 | 1.146 | 1.436 | 1.039 |
| | *std. dev.* | *0.053* | *0.035* | *0.025* | *0.028* | *0.049* | *0.070* | *0.045* |
| | 25 | 10.088 | **7.905** | 12.484 | 8.066 | 25.958 | 72.842 | 13.420 |
| | *std. dev.* | *0.358* | *0.317* | *0.530* | *0.341* | *1.003* | *2.490* | *0.596* |
| $\sigma^2 = 100$ | 50 | 8.049 | **5.078** | 6.675 | 5.971 | 15.548 | 73.830 | 9.563 |
| | *std. dev.* | *0.239* | *0.194* | *0.289* | *0.258* | *0.655* | *2.523* | *0.427* |
| | 75 | 6.569 | **3.510** | 4.277 | 5.241 | 10.690 | 71.213 | 9.625 |
| | *std. dev.* | *0.182* | *0.137* | *0.193* | *0.236* | *0.460* | *2.477* | *0.421* |

**Table 6.1** A comparison of stepsize rules for *class I* functions, including search-then-converge (STC), McClain, Kesten, stochastic gradient adaptive stepsize rule (SGASR), and the bias-adjusted Kalman filter (BAKF). The best result for each problem is shown in bold (from George & Powell (2006)).

deterministic rule. It may be the case that if we can properly tune a single, deterministic stepsize rule, then this will work the best. The challenge is that this tuning process can be time-consuming. Furthermore, it is not always obvious when a particular stepsize rule is not working.

Despite the apparent benefits of adaptive stepsize rules, deterministic rules, such as variants of the STC rule, remain quite popular. One issue is that not only do the adaptive rules introduce a series of additional computations, but also the logic introduces additional statistics that have to be computed and stored for each parameter. In some applications, there can be tens of thousands of such parameters, which introduces significant computational and memory overhead. Just as important, while adaptive formulas can do a better job of estimating a value function, this does not always translate into a better policy.

## 6.7 CONVERGENCE

A practical issue that arises with all stochastic approximation algorithms is that we simply do not have reliable, implementable stopping rules. Proofs of convergence in the limit are an important theoretical property, but they provide no guidelines or guarantees in practice. A good illustration of the issue is given in figure 6.11. Figure 6.11a shows the objective function for a dynamic program over 100 iterations (in this application, a single iteration required approximately 20 minutes of CPU time). The figure shows the objective function for an ADP algorithm which was run 100 iterations, at which point it appeared to be

| Variance | $n$ | $1/n$ | $1/n^\beta$ | STC | McClain | Kesten | SGASR | BAKF |
|---|---|---|---|---|---|---|---|---|
| $\sigma^2 = 1$ | 25 | 0.416 | 0.300 | 0.227 | 0.231 | 0.285 | 0.830 | **0.209** |
| | *std. dev.* | *0.008* | *0.007* | *0.009* | *0.007* | *0.011* | *0.024* | *0.008* |
| | 50 | 13.437 | 10.711 | 3.718 | 6.232 | 2.685 | **1.274** | 1.813 |
| | *std. dev.* | *0.033* | *0.032* | *0.040* | *0.036* | *0.042* | *0.045* | *0.061* |
| | 75 | 30.403 | 15.818 | 0.466 | 1.186 | 0.252 | 0.605 | **0.230** |
| | *std. dev.* | *0.041* | *0.033* | *0.011* | *0.016* | *0.009* | *0.019* | *0.009* |
| $\sigma^2 = 10$ | 25 | 0.784 | **0.715** | 1.920 | 0.770 | 2.542 | 2.655 | 1.319 |
| | *std. dev.* | *0.031* | *0.030* | *0.076* | *0.033* | *0.097* | *0.123* | *0.057* |
| | 50 | 13.655 | 10.962 | 4.689 | 6.754 | **4.109** | 4.870 | 4.561 |
| | *std. dev.* | *0.104* | *0.102* | *0.130* | *0.116* | *0.140* | *0.164* | *0.155* |
| | 75 | 30.559 | 15.989 | **1.133** | 1.677 | 1.273 | 1.917 | 1.354 |
| | *std. dev.* | *0.127* | *0.103* | *0.047* | *0.055* | *0.055* | *0.089* | *0.056* |
| $\sigma^2 = 100$ | 25 | **4.500** | 5.002 | 19.399 | 6.359 | 25.887 | 71.974 | 12.529 |
| | *std. dev.* | *0.199* | *0.218* | *0.765* | *0.271* | *0.991* | *2.484* | *0.567* |
| | 50 | 15.682 | 13.347 | 14.210 | **11.607** | 18.093 | 72.855 | 13.726 |
| | *std. dev.* | *0.346* | *0.345* | *0.600* | *0.438* | *0.751* | *2.493* | *0.561* |
| | 75 | 32.069 | 17.709 | 7.602 | **6.407** | 11.119 | 73.286 | 10.615 |
| | *std. dev.* | *0.409* | *0.338* | *0.333* | *0.278* | *0.481* | *2.504* | *0.461* |

**Table 6.2** A comparison of stepsize rules for *class II* functions which undergo delayed learning (from George & Powell (2006))

| $\gamma$ | $n$ | $1/n$ | | $1/n^\beta$ | | STC | | SGASR | | BAKF | |
|---|---|---|---|---|---|---|---|---|---|---|---|
| 0.80 | 2 | 43.55 | *0.14* | 52.33 | *0.12* | 41.93 | *0.14* | 50.63 | *0.14* | **41.38** | *0.16* |
| | 5 | 17.39 | *0.12* | 25.53 | *0.11* | 12.04 | *0.13* | 19.05 | *0.14* | **11.76** | *0.15* |
| | 10 | 10.54 | *0.08* | 14.91 | *0.08* | **4.29** | *0.04* | 7.70 | *0.07* | 4.86 | *0.05* |
| 0.90 | 2 | 51.27 | *0.14* | 64.23 | *0.09* | 48.39 | *0.13* | 60.78 | *0.11* | **46.15** | *0.12* |
| | 5 | 29.22 | *0.12* | 42.12 | *0.11* | 19.20 | *0.12* | 27.61 | *0.12* | **15.26** | *0.14* |
| | 10 | 22.72 | *0.11* | 31.36 | *0.08* | 8.55 | *0.07* | 9.84 | *0.10* | **7.45** | *0.09* |
| 0.95 | 2 | 62.64 | *0.19* | 76.92 | *0.08* | 58.75 | *0.14* | 72.76 | *0.12* | **54.93** | *0.15* |
| | 5 | 45.26 | *0.21* | 60.95 | *0.09* | 33.21 | *0.15* | 41.64 | *0.15* | **24.38** | *0.14* |
| | 10 | 39.39 | *0.20* | 51.83 | *0.08* | 21.46 | *0.11* | 18.78 | *0.11* | **14.93** | *0.10* |

**Table 6.3** Percentage error in the estimates from the optimal values, averaged over all the resource states, as a function of the average number of observations per state, for different discount factors $\gamma$. The figures in italics denote the standard deviations of the values to the left (from George & Powell (2006)).

flattening out (evidence of convergence). Figure 6.11b is the objective function for the same algorithm run for 400 iterations. A solid line that shows the best objective function after 100 iterations is shown at the same level on the graph where the algorithm was run for 400 iterations. As we see, the algorithm was nowhere near convergence after 100 iterations.

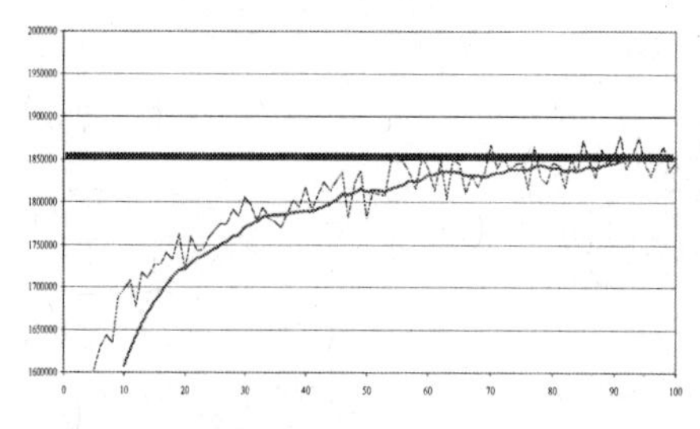

6.11a: Objective function over 100 iterations.

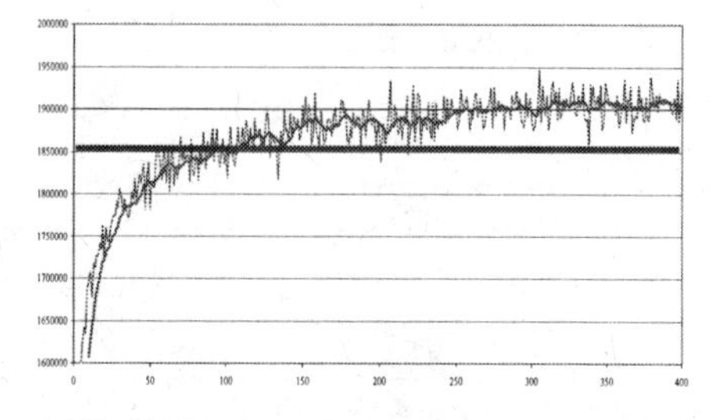

6.11b: Objective function over 400 iterations.

**Figure 6.11** The objective function, plotted over 100 iterations (a), displays "apparent convergence." The same algorithm, continued over 400 iterations (b), shows significant improvement.

We refer to this behavior as "apparent convergence," and it is particularly problematic on large-scale problems where run times are long. Typically, the number of iterations needed before the algorithm "converges" requires a level of subjective judgment. When the run times are long, wishful thinking can interfere with this process.

Complicating the analysis of convergence in approximate dynamic programming is the behavior in some problems to go through periods of stability which are simply a precursor to breaking through to new plateaus. During periods of exploration, an ADP algorithm might discover a strategy that opens up new opportunities, moving the performance of the algorithm to an entirely new level.

Special care has to be made in the choice of stepsize rule. In any algorithm using a declining stepsize, it is possible to show a stabilizing objective function simply because the stepsize is decreasing. This problem is exacerbated when using algorithms based on value iteration, where updates to the value of being in a state depend on estimates of the values of future states, which can be biased. We recommend that initial testing of an ADP algorithm start with inflated stepsizes. After getting a sense for the number of iterations needed for the algorithm to stabilize, decrease the stepsize (keeping in mind that the number of iterations required to convergence may increase) to find the right tradeoff between noise and rate of convergence.

## 6.8 WHY DOES IT WORK?**

Stochastic approximation methods have a rich history starting with the seminal paper Robbins & Monro (1951) and followed by Blum (1954*a*) and Dvoretzky (1956). The serious reader should see Kushner & Yin (1997) for a modern treatment of the subject. Wasan (1969) is also a useful reference for fundamental results on stochastic convergence theory. A separate line of investigation was undertaken by researchers in eastern European community focusing on constrained stochastic optimization problems (Gaivoronski (1988), Ermoliev (1983), Ermoliev (1988), Ruszczyński (1980), Ruszczyński (1987)). This work is critical to our fundamental understanding of Monte Carlo-based stochastic learning methods.

The theory behind these proofs is fairly deep and requires some mathematical maturity. For pedagogical reasons, we start in section 6.8.1 with some probabilistic preliminaries, after which section 6.8.2 presents one of the original proofs, which is relatively more accessible and which provides the basis for the universal requirements that stepsizes must satisfy for theoretical proofs. Section 6.8.3 provides a more modern proof based on the theory of martingales.

### 6.8.1 Some probabilistic preliminaries

The goal in this section is to prove that these algorithms work. But what does this mean? The solution $\bar{\theta}^n$ at iteration $n$ is a random variable. Its value depends on the sequence of sample realizations of the random variables over iterations 1 to $n$. If $\omega = (\omega^1, \omega^2, \ldots, \omega^n, \ldots)$ represents the sample path that we are following, we can ask what is happening to the limit $\lim_{n\to\infty} \bar{\theta}^n(\omega)$. If the limit is $\theta^*$, does $\theta^*$ depend on the sample path $\omega$?

In the proofs below, we show that the algorithms converge *almost surely*. What this means is that

$$\lim_{n\to\infty} \bar{\theta}^n(\omega) = \theta^*$$

for all $\omega \in \Omega$ that can occur with positive measure. This is the same as saying that we reach $\theta^*$ with probability 1. Here, $\theta^*$ is a deterministic quantity that does not depend on the sample path. Because of the restriction $p(\omega) > 0$, we accept that in theory, there could exist a sample outcome that can never occur that would produce a path that converges to some other point. As a result, we say that the convergence is "almost sure," which is universally abbreviated as "*a.s.*" Almost sure convergence establishes the core theoretical property that the algorithm will eventually settle in on a single point. This is an important property for an algorithm, but it says nothing about the rate of convergence (an important issue in approximate dynamic programming).

Let $x \in \Re^n$. At each iteration $n$, we sample some random variables to compute the function (and its gradient). The sample realizations are denoted by $\omega^n$. We let $\omega = (\omega^1, \omega^2, \ldots,)$ be a realization of all the random variables over all iterations. Let $\Omega$ be the set of all possible realizations of $\omega$, and let $\mathfrak{F}$ be the $\sigma$-algebra on $\Omega$ (that is to say, the set of all possible events that can be defined using $\Omega$). We need the concept of the history up through iteration $n$. Let

$H^n$ = A random variable giving the history of all random variables up through iteration $n$.

A sample realization of $H^n$ would be

$$\begin{aligned} h^n &= H^n(\omega) \\ &= (\omega^1, \omega^2, \ldots, \omega^n). \end{aligned}$$

We could then let $\Omega^n$ be the set of all outcomes of the history (that is, $h^n \in H^n$) and let $\mathcal{H}^n$ be the $\sigma$-algebra on $\Omega^n$ (which is the set of all events, including their complements and unions, defined using the outcomes in $\Omega^n$). Although we could do this, this is not the convention followed in the probability community. Instead, we define a sequence of $\sigma$-algebras $\mathfrak{F}^1, \mathfrak{F}^2, \ldots, \mathfrak{F}^n$ as the sequence of $\sigma$-algebras on $\Omega$ that can be generated as we have access to the information through the first $1, 2, \ldots, n$ iterations, respectively. What does this mean? Consider two outcomes $\omega \neq \omega'$ for which $H^n(\omega) = H^n(\omega')$. If this is the case, then any event in $\mathfrak{F}^n$ that includes $\omega$ must also include $\omega'$. If we say that a function is $\mathfrak{F}^n$-measurable, then this means that it must be defined in terms of the events in $\mathfrak{F}^n$, which is in turn equivalent to saying that we cannot be using any information from iterations $n+1, n+2, \ldots$.

We would say, then, that we have a standard probability space $(\Omega, \mathfrak{F}, \mathcal{P})$ where $\omega \in \Omega$ represents an elementary outcome, $\mathfrak{F}$ is the $\sigma$-algebra on $\mathfrak{F}$ and $\mathcal{P}$ is a probability measure on $\Omega$. Since our information is revealed iteration by iteration, we would also then say that we have an increasing set of $\sigma$-algebras $\mathfrak{F}^1 \subseteq \mathfrak{F}^2 \subseteq \ldots \subseteq \mathfrak{F}^n$ (which is the same as saying that $\mathcal{F}^n$ is a filtration).

### 6.8.2 An older proof

Enough with probabilistic preliminaries. Let $F(\theta, \omega)$ be a $\mathfrak{F}$-measurable function. We wish to solve the unconstrained problem

$$\max_{\theta} \mathbb{E}\left\{F(\theta, \omega)\right\} \tag{6.75}$$

with $\theta^*$ being the optimal solution. Let $g(\theta, \omega)$ be a stochastic ascent vector that satisfies

$$g(\theta, \omega)^T \nabla F(\theta, \omega) \geq 0. \tag{6.76}$$

For many problems, the most natural ascent vector is the gradient itself

$$g(\theta, \omega) \quad = \quad \nabla F(\theta, \omega) \tag{6.77}$$

which clearly satisfies (6.76).

We assume that $F(\theta) = \mathbb{E}\left\{F(\theta, \omega)\right\}$ is continuously differentiable and convex, with bounded first and second derivatives so that for finite $M$

$$-M \leq g(\theta, \omega)^T \nabla^2 F(\theta) g(\theta, \omega) \leq M. \tag{6.78}$$

A stochastic gradient algorithm (sometimes called a stochastic approximation method) is given by

$$\bar{\theta}^n \quad = \quad \bar{\theta}^{n-1} + \alpha_{n-1} g(\bar{\theta}^{n-1}, \omega^n). \tag{6.79}$$

We first prove our result using the proof technique of Blum (1954*a*) that generalized the original stochastic approximation procedure proposed by Robbins & Monro (1951) to multidimensional problems. This approach does not depend on more advanced concepts such as martingales and, as a result, is accessible to a broader audience. This proof helps the reader understand the basis for the conditions $\sum_{n=0}^{\infty} \alpha_n = \infty$ and $\sum_{n=0}^{\infty} (\alpha_n)^2 < \infty$ that are required of all stochastic approximation algorithms.

We make the following (standard) assumptions on stepsizes

$$\alpha_n \geq 0 \quad n \geq 0, \tag{6.80}$$

$$\sum_{n=0}^{\infty} \alpha_n = \infty, \tag{6.81}$$

$$\sum_{n=0}^{\infty} (\alpha_n)^2 < \infty. \tag{6.82}$$

We want to show that under suitable assumptions, the sequence generated by (6.79) converges to an optimal solution. That is, we want to show that

$$\lim_{n \to \infty} \bar{\theta}^n \quad = \quad \theta^* \ a.s. \tag{6.83}$$

We now use Taylor's theorem (remember Taylor's theorem from freshman calculus?), which says that for any continuously differentiable convex function $F(\theta)$, there exists a parameter $0 \leq \gamma \leq 1$ that satisfies

$$F(\theta) = F(\bar{\theta}^0) + \nabla F(\bar{\theta}^0 + \gamma(\theta - \bar{\theta}^0))(\theta - \bar{\theta}^0). \tag{6.84}$$

This is the first-order version of Taylor's theorem. The second-order version takes the form

$$F(\theta) = F(\bar{\theta}^0) + \nabla F(\bar{\theta}^0)(\theta - \bar{\theta}^0) + \frac{1}{2}(\theta - \bar{\theta}^0)^T \nabla^2 F(\bar{\theta}^0 + \gamma(\theta - \bar{\theta}^0))(\theta - \bar{\theta}^0) \tag{6.85}$$

for some $0 \leq \gamma \leq 1$. We use the second-order version. Replace $\bar{\theta}^0$ with $\bar{\theta}^{n-1}$, and replace $\theta$ with $\bar{\theta}^n$. Also, we can simplify our notation by using

$$g^n = g(\bar{\theta}^{n-1}, \omega^n). \tag{6.86}$$

This means that

$$\begin{aligned}
\theta - \bar{\theta}^0 &= \bar{\theta}^n - \bar{\theta}^{n-1} \\
&= (\bar{\theta}^{n-1} + \alpha_{n-1} g^n) - \bar{\theta}^{n-1} \\
&= \alpha_{n-1} g^n.
\end{aligned}$$

From our stochastic gradient algorithm (6.79), we may write

$$\begin{aligned}
F(\bar{\theta}^n, \omega^n) &= F(\bar{\theta}^{n-1} + \alpha_{n-1} g^n, \omega^n) \\
&= F(\bar{\theta}^{n-1}, \omega^n) + \nabla F(\bar{\theta}^{n-1}, \omega^n)(\alpha_{n-1} g^n) \\
&\quad + \frac{1}{2}(\alpha_{n-1} g^n)^T \nabla^2 F(\bar{\theta}^{n-1} + \gamma(\alpha_{n-1} g^n))(\alpha_{n-1} g^n).
\end{aligned} \tag{6.87}$$

It is now time to use a *standard mathematician's trick*. We sum both sides of (6.87) to get

$$\begin{aligned}
\sum_{n=1}^{N} F(\bar{\theta}^n, \omega^n) &= \sum_{n=1}^{N} F(\bar{\theta}^{n-1}, \omega^n) + \sum_{n=1}^{N} \nabla F(\bar{\theta}^{n-1}, \omega^n)(\alpha_{n-1} g^n) \\
&\quad + \frac{1}{2} \sum_{n=1}^{N} (\alpha_{n-1} g^n)^T \nabla^2 F\left(\bar{\theta}^{n-1} + \theta(\alpha_{n-1} g^n)\right)(\alpha_{n-1} g^n).
\end{aligned} \tag{6.88}$$

Note that the terms $F(\bar{\theta}^n), n = 2, 3, \ldots, N$ appear on both sides of (6.88). We can cancel these. We then use our lower bound on the quadratic term (6.78) to write

$$F(\bar{\theta}^N, \omega^N) \geq F(\bar{\theta}^0, \omega^1) + \sum_{n=1}^{N} \nabla F(\bar{\theta}^{n-1}, \omega^n)(\alpha_{n-1} g^n) + \frac{1}{2} \sum_{n=1}^{N} (\alpha_{n-1})^2(-M). \tag{6.89}$$

We now want to take the limit of both sides of (6.89) as $N \rightarrow \infty$. In doing so, we want to show that everything must be bounded. We know that $F(\bar{\theta}^N)$ is bounded (*almost surely*) because we assumed that the original function was bounded. We next use the assumption (6.15) that the infinite sum of the squares of the stepsizes is also bounded to conclude that the rightmost term in (6.89) is bounded. Finally, we use (6.76) to claim that all the terms in the remaining summation ($\sum_{n=1}^{N} \nabla F(\bar{\theta}^n)(\alpha_{n-1} g^n)$) are positive. That means that this term is also bounded (from both above and below).

What do we get with all this boundedness? Well, if

$$\sum_{n=1}^{\infty} \alpha_{n-1} \nabla F(\bar{\theta}^n, \omega^n) g^n < \infty \quad a.s. \tag{6.90}$$

and (from (6.14))

$$\sum_{n=1}^{\infty} \alpha_{n-1} = \infty. \tag{6.91}$$

We can conclude that

$$\sum_{n=1}^{\infty} \nabla F(\bar{\theta}^{n-1}, \omega^n) g^n < \infty. \tag{6.92}$$

Since all the terms in (6.92) are positive, they must go to zero. (Remember, everything here is true *almost surely*; after a while, it gets a little boring to keep saying *almost surely* every time. It is a little like reading Chinese fortune cookies and adding the automatic phrase "under the sheets" at the end of every fortune.)

We are basically done except for some relatively difficult (albeit important if you are ever going to do your own proofs) technical points to really prove convergence. At this point, we would use technical conditions on the properties of our ascent vector $g^n$ to argue that if $\nabla F(\bar{\theta}^n, \omega^n) g^n \to 0$ then $\nabla F(\bar{\theta}^n, \omega^n) \to 0$, (it is okay if $g^n$ goes to zero as $F(\bar{\theta}^n, \omega^n)$ goes to zero, but it cannot go to zero too quickly).

This proof was first proposed in the early 1950's by Robbins and Monro and became the basis of a large area of investigation under the heading of stochastic approximation methods. A separate community, growing out of the Soviet literature in the 1960's, addressed these problems under the name of stochastic gradient (or stochastic quasi-gradient) methods. More modern proofs are based on the use of martingale processes, which do not start with Taylor's formula and do not (always) need the continuity conditions that this approach needs.

Our presentation does, however, help to present several key ideas that are present in most proofs of this type. First, concepts of almost sure convergence are virtually standard. Second, it is common to set up equations such as (6.87) and then take a finite sum as in (6.88) using the alternating terms in the sum to cancel all but the first and last elements of the sequence of some function (in our case, $F(\bar{\theta}^{n-1}, \omega^n)$). We then establish the boundedness of this expression as $N \to \infty$, which will require the assumption that $\sum_{n=1}^{\infty} (\alpha_{n-1})^2 < \infty$. Then, the assumption $\sum_{n=1}^{\infty} \alpha_{n-1} = \infty$ is used to show that if the remaining sum is bounded, then its terms must go to zero.

More modern proofs will use functions other than $F(\bar{\theta})$. Popular is the introduction of so-called Lyapunov functions, which are artificial functions that provide a measure of optimality. These functions are constructed for the purpose of the proof and play no role in the algorithm itself. For example, we might let $T^n = ||\bar{\theta}^n - \theta^*||$ be the distance between our current solution $\bar{\theta}^n$ and the optimal solution. We will then try to show that $T^n$ is suitably reduced to prove convergence. Since we do not know $\theta^*$, this is not a function we can actually measure, but it can be a useful device for proving that the algorithm actually converges.

It is important to realize that stochastic gradient algorithms of all forms do not guarantee an improvement in the objective function from one iteration to the next. First, a sample gradient $g^n$ may represent an appropriate ascent vector for a sample of the function

$F(\bar{\theta}^n, \omega^n)$ but not for its expectation. In other words, randomness means that we may go in the wrong direction at any point in time. Second, our use of a nonoptimizing stepsize, such as $\alpha_{n-1} = 1/n$, means that even with a good ascent vector, we may step too far and actually end up with a lower value.

### 6.8.3 A more modern proof

Since the original work by Robbins and Monro, more powerful proof techniques have evolved. Below we illustrate a basic martingale proof of convergence. The concepts are somewhat more advanced, but the proof is more elegant and requires milder conditions. A significant generalization is that we no longer require that our function be differentiable (which our first proof required). For large classes of resource allocation problems, this is a significant improvement.

First, just what is a martingale? Let $\omega_1, \omega_2, \ldots, \omega_t$ be a set of exogenous random outcomes, and let $h_t = H_t(\omega) = (\omega_1, \omega_2, \ldots, \omega_t)$ represent the history of the process up to time $t$. We also let $\mathfrak{F}_t$ be the $\sigma$-algebra on $\Omega$ generated by $H_t$. Further, let $U_t$ be a function that depends on $h_t$ (we would say that $U_t$ is a $\mathfrak{F}_t$-measurable function), and bounded ($\mathbb{E}|U_t| < \infty, \ \forall t \geq 0$). This means that if we know $h_t$, then we know $U_t$ deterministically (needless to say, if we only know $h_t$, then $U_{t+1}$ is still a random variable). We further assume that our function satisfies

$$\mathbb{E}[U_{t+1}|\mathfrak{F}_t] = U_t.$$

If this is the case, then we say that $U_t$ is a *martingale*. Alternatively, if

$$\mathbb{E}[U_{t+1}|\mathfrak{F}_t] \leq U_t \qquad (6.93)$$

then we say that $U_t$ is a *supermartingale*. If $U_t$ is a supermartingale, then it has the property that it drifts downward, usually to some limit point $U^*$. What is important is that it only drifts downward in expectation. That is, it could easily be the case that $U_{t+1} > U_t$ for specific outcomes. This captures the behavior of stochastic approximation algorithms. Properly designed, they provide solutions that improve on average, but where from one iteration to another the results can actually get worse.

Finally, assume that $U_t \geq 0$. If this is the case, we have a sequence $U_t$ that drifts downward but which cannot go below zero. Not surprisingly, we obtain the following key result:

**Theorem 6.8.1** *Let $U_t$ be a positive supermartingale. Then, $U_t$ converges to a finite random variable $U^*$ a.s.*

So what does this mean for us? We assume that we are still solving a problem of the form

$$\max_\theta \mathbb{E}\{F(\theta, \omega)\}, \qquad (6.94)$$

where we assume that $F(\theta, \omega)$ is continuous and concave (but we do not require differentiability). Let $\bar{\theta}^n$ be our estimate of $\theta$ at iteration $n$ (remember that $\bar{\theta}^n$ is a random variable). Instead of watching the evolution of a process of time, we are studying the behavior of an algorithm over iterations. Let $F^n = \mathbb{E}F(\bar{\theta}^n)$ be our objective function at iteration $n$ and let $F^*$ be the optimal solution. If we are maximizing, we know that $F^n \leq F^*$. If we let $U^n = F^* - F^n$, then we know that $U^n \geq 0$ (this assumes that we can find the true expectation, rather than some approximation of it). A stochastic algorithm will not guarantee that $F^n \geq F^{n-1}$, but if we have a good algorithm, then we may be able to show that $U^n$ is a supermartingale, which at least tells us that in the limit, $U^n$ will approach some limit $\bar{U}$. With additional work, we might be able to show that $\bar{U} = 0$, which means that we have found the optimal solution.

A common strategy is to define $U^n$ as the distance between $\bar{\theta}^n$ and the optimal solution, which is to say

$$U^n = (\bar{\theta}^n - \theta^*)^2. \tag{6.95}$$

Of course, we do not know $\theta^*$, so we cannot actually compute $U^n$, but that is not really a problem for us (we are just trying to prove convergence). Note that we immediately get $U^n \geq 0$ (without an expectation). If we can show that $U^n$ is a supermartingale, then we get the result that $U^n$ converges to a random variable $U^*$ (which means the algorithm converges). Showing that $U^* = 0$ means that our algorithm will (eventually) produce the optimal solution.

We are solving this problem using a stochastic gradient algorithm

$$\bar{\theta}^n = \bar{\theta}^{n-1} - \alpha_{n-1} g^n, \tag{6.96}$$

where $g^n$ is our stochastic gradient. If $F$ is differentiable, we would write

$$g^n = \nabla_\theta F(\bar{\theta}^{n-1}, \omega^n).$$

But in general, $F$ may be nondifferentiable, in which case we may have multiple gradients at a point $\bar{\theta}^{n-1}$ (for a single sample realization). In this case, we write

$$g^n \in \partial_\theta F(\bar{\theta}^{n-1}, \omega^n),$$

where $\partial_\theta F(\bar{\theta}^{n-1}, \omega^n)$ refers to the set of subgradients at $\bar{\theta}^{n-1}$. We assume our problem is unconstrained, so $\nabla_\theta F(\bar{\theta}^*, \omega^n) = 0$ if $F$ is differentiable. If it is nondifferentiable, we would assume that $0 \in \partial_\theta F(\bar{\theta}^*, \omega^n)$.

Throughout our presentation, we assume that $\theta$ (and hence $g^n$) is a scalar (exercise 6.13 provides an opportunity to redo this section using vector notation). In contrast with the previous section, we are now going to allow our stepsizes to be stochastic. For this reason, we need to slightly revise our original assumptions about stepsizes (equations (6.80) to (6.82)) by assuming

$$\alpha_n \geq 0 \text{ a.s.}, \tag{6.97}$$

$$\sum_{n=0}^{\infty} \alpha_n = \infty \text{ a.s.}, \tag{6.98}$$

$$\mathbb{E}\left[\sum_{n=0}^{\infty} (\alpha_n)^2\right] < \infty. \tag{6.99}$$

The requirement that $\alpha_n$ be nonnegative "almost surely" (a.s.) recognizes that $\alpha_n$ is a random variable. We can write $\alpha_n(\omega)$ as a sample realization of the stepsize (that is, this is the stepsize at iteration $n$ if we are following sample path $\omega$). When we require that $\alpha_n \geq 0$ "almost surely" we mean that $\alpha_n(\omega) \geq 0$ for all $\omega$ where the probability (more precisely, probability measure) of $\omega$, $p(\omega)$, is greater than zero (said differently, this means that the probability that $\mathbb{P}[\alpha_n \geq 0] = 1$). The same reasoning applies to the sum of the stepsizes given in equation (6.98). As the proof unfolds, we will see the reason for needing the conditions (and why they are stated as they are).

We next need to assume some properties of the stochastic gradient $g^n$. Specifically, we need to assume the following:

**Assumption 1** - $\mathbb{E}[g^{n+1}(\bar{\theta}^n - \theta^*)|\mathfrak{F}^n] \geq 0,$

**Assumption 2** - $|g^n| \leq B_g,$

**Assumption 3** - For any $\theta$ where $|\theta - \theta^*| > \delta,\ \delta > 0$, there exists $\epsilon > 0$ such that $\mathbb{E}[g^{n+1}|\mathfrak{F}^n] > \epsilon$.

Assumption 1 assumes that on average, the gradient $g^n$ points toward the optimal solution $\theta^*$. This is easy to prove for deterministic, differentiable functions. While this may be harder to establish for stochastic problems or problems where $F(\theta)$ is nondifferentiable, we do not have to assume that $F(\theta)$ is differentiable. Nor do we assume that a particular gradient $g^{n+1}$ moves toward the optimal solution (for a particular sample realization, it is entirely possible that we are going to move away from the optimal solution). Assumption 2 assumes that the gradient is bounded. Assumption 3 requires that the expected gradient cannot vanish at a nonoptimal value of $\theta$. This assumption will be satisfied for any concave function.

To show that $U^n$ is a supermartingale, we start with

$$\begin{aligned} U^{n+1} - U^n &= (\bar{\theta}^{n+1} - \theta^*)^2 - (\bar{\theta}^n - \theta^*)^2 \\ &= \left((\bar{\theta}^n - \alpha_n g^{n+1}) - \theta^*\right)^2 - (\bar{\theta}^n - \theta^*)^2 \\ &= \left((\bar{\theta}^n - \theta^*)^2 - 2\alpha_n g^{n+1}(\bar{\theta}^n - \theta^*) + (\alpha_n g^{n+1})^2\right) - (\bar{\theta}^n - \theta^*)^2 \\ &= (\alpha_n g^{n+1})^2 - 2\alpha_n g^{n+1}(\bar{\theta}^n - \theta^*). \end{aligned} \tag{6.100}$$

Taking conditional expectations on both sides gives

$$\mathbb{E}[U^{n+1}|\mathfrak{F}^n] - \mathbb{E}[U^n|\mathfrak{F}^n] = \mathbb{E}[(\alpha_n g^{n+1})^2|\mathfrak{F}^n] - 2\mathbb{E}[\alpha_n g^{n+1}(\bar{\theta}^n - \theta^*)|\mathfrak{F}^n]. \tag{6.101}$$

We note that

$$\mathbb{E}[\alpha_n g^{n+1}(\bar{\theta}^n - \theta^*)|\mathfrak{F}^n] = \alpha_n \mathbb{E}[g^{n+1}(\bar{\theta}^n - \theta^*)|\mathfrak{F}^n] \tag{6.102}$$

$$\geq 0. \tag{6.103}$$

Equation (6.102) is subtle but important, as it explains a critical piece of notation in this book. Keep in mind that we may be using a stochastic stepsize formula, which means that $\alpha_n$ is a random variable. We assume that $\alpha_n$ is $\mathfrak{F}^n$-measurable, which means that we are not allowed to use information from iteration $n+1$ to compute it. This is why we use $\alpha_{n-1}$ in updating equations such as equation (6.5) instead of $\alpha_n$. When we condition on $\mathfrak{F}^n$ in equation (6.102), $\alpha_n$ is deterministic, allowing us to take it outside the expectation. This allows us to write the conditional expectation of the product of $\alpha_n$ and $g^{n+1}$ as the product of the expectations. Equation (6.103) comes from Assumption 1 and the nonnegativity of the stepsizes.

Recognizing that $\mathbb{E}[U^n|\mathfrak{F}^n] = U^n$ (given $\mathfrak{F}^n$), we may rewrite (6.101) as

$$\begin{aligned} \mathbb{E}[U^{n+1}|\mathfrak{F}^n] &= U^n + \mathbb{E}[(\alpha_n g^{n+1})^2|\mathfrak{F}^n] - 2\mathbb{E}[\alpha_n g^{n+1}(\bar{\theta}^n - \theta^*)|\mathfrak{F}^n] \\ &\leq U^n + \mathbb{E}[(\alpha_n g^{n+1})^2|\mathfrak{F}^n]. \end{aligned} \tag{6.104}$$

Because of the positive term on the right-hand side of (6.104), we cannot directly get the result that $U^n$ is a supermartingale. But hope is not lost. We appeal to a neat little trick that works as follows. Let

$$W^n = U^n + \sum_{m=n}^{\infty} (\alpha_m g^{m+1})^2. \quad (6.105)$$

We are going to show that $W^n$ is a supermartingale. From its definition, we obtain

$$W^n = W^{n+1} + U^n - U^{n+1} + (\alpha_n g^{n+1})^2. \quad (6.106)$$

Taking conditional expectations of both sides gives

$$W^n = \mathbb{E}\left[W^{n+1}|\mathfrak{F}^n\right] + U^n - \mathbb{E}\left[U^{n+1}|\mathfrak{F}^n\right] + \mathbb{E}\left[(\alpha_n g^{n+1})^2|\mathfrak{F}^n\right]$$

which is the same as

$$\mathbb{E}[W^{n+1}|\mathfrak{F}^n] = W^n - \underbrace{\left(U^n + \mathbb{E}\left[(\alpha_n g^{n+1})^2|\mathfrak{F}^n\right] - \mathbb{E}[U^{n+1}|\mathfrak{F}^n]\right)}_{I}.$$

We see from equation (6.104) that $I \geq 0$. Removing this term gives us the inequality

$$\mathbb{E}[W^{n+1}|\mathfrak{F}^n] \leq W^n. \quad (6.107)$$

This means that $W^n$ is a supermartingale. It turns out that this is all we really need because $\lim_{n\to\infty} W^n = \lim_{n\to\infty} U^n$. This means that

$$\lim_{n\to\infty} U^n \to U^* \quad a.s. \quad (6.108)$$

Now that we have the basic convergence of our algorithm, we have to ask: but what is it converging to? For this result, we return to equation (6.100) and sum it over the values $n = 0$ up to some number $N$, giving us

$$\sum_{n=0}^{N}(U^{n+1} - U^n) = \sum_{n=0}^{N}(\alpha_n g^{n+1})^2 - 2\sum_{n=0}^{N} \alpha_n g^{n+1}(\bar{\theta}^n - \theta^*). \quad (6.109)$$

The left-hand side of (6.109) is an alternating sum (sometimes referred to as a telescoping sum), which means that every element cancels out except the first and the last, giving us

$$U^{N+1} - U^0 = \sum_{n=0}^{N}(\alpha_n g^{n+1})^2 - 2\sum_{n=0}^{N} \alpha_n g^{n+1}(\bar{\theta}^n - \theta^*).$$

Taking expectations of both sides gives

$$\mathbb{E}[U^{N+1} - U^0] = \mathbb{E}\left[\sum_{n=0}^{N}(\alpha_n g^{n+1})^2\right] - 2\mathbb{E}\left[\sum_{n=0}^{N} \alpha_n g^{n+1}(\bar{\theta}^n - \theta^*)\right]. \quad (6.110)$$

We want to take the limit of both sides as $N$ goes to infinity. To do this, we have to appeal to the *Dominated Convergence Theorem* (DCT), which tells us that

$$\lim_{N\to\infty} \int_x f^n(x)dx = \int_x \left(\lim_{N\to\infty} f^n(x)\right) dx$$

if $|f^n(x)| \leq g(x)$ for some function $g(x)$ where

$$\int_x g(x) < \infty.$$

For our application, the integral represents the expectation (we would use a summation instead of the integral if $x$ were discrete), which means that the DCT gives us the conditions needed to exchange the limit and the expectation. Above, we showed that $\mathbb{E}[U^{n+1}|\mathfrak{F}^n]$ is bounded (from (6.104) and the boundedness of $U^0$ and the gradient). This means that the right-hand side of (6.110) is also bounded for all $n$. The DCT then allows us to take the limit as $N$ goes to infinity inside the expectations, giving us

$$U^* - U^0 = \mathbb{E}\left[\sum_{n=0}^{\infty}(\alpha_n g^{n+1})^2\right] - 2\mathbb{E}\left[\sum_{n=0}^{\infty}\alpha_n g^{n+1}(\bar{\theta}^n - \theta^*)\right].$$

We can rewrite the first term on the right-hand side as

$$\mathbb{E}\left[\sum_{n=0}^{\infty}(\alpha_n g^{n+1})^2\right] \leq \mathbb{E}\left[\sum_{n=0}^{\infty}(\alpha_n)^2(B)^2\right] \tag{6.111}$$

$$= B^2\,\mathbb{E}\left[\sum_{n=0}^{\infty}(\alpha_n)^2\right] \tag{6.112}$$

$$< \infty. \tag{6.113}$$

Equation (6.111) comes from Assumption 2 which requires that $|g^n|$ be bounded by $B$, which immediately gives us Equation (6.112). The requirement that $\mathbb{E}\sum_{n=0}^{\infty}(\alpha_n)^2 < \infty$ (equation (6.82)) gives us (6.113), which means that the first summation on the right-hand side of (6.110) is bounded. Since the left-hand side of (6.110) is bounded, we can conclude that the second term on the right-hand side of (6.110) is also bounded.

Now let

$$\begin{aligned}\beta^n &= \mathbb{E}\left[g^{n+1}(\bar{\theta}^n - \theta^*)\right] \\ &= \mathbb{E}\left[\mathbb{E}\left[g^{n+1}(\bar{\theta}^n - \theta^*)|\mathfrak{F}^n\right]\right] \\ &\geq 0,\end{aligned}$$

since $\mathbb{E}[g^{n+1}(\bar{\theta}^n - \theta^*)|\mathfrak{F}^n] \geq 0$ from Assumption 1. This means that

$$\sum_{n=0}^{\infty}\alpha_n\beta^n < \infty \text{ a.s.} \tag{6.114}$$

But, we have required that $\sum_{n=0}^{\infty}\alpha_n = \infty$ a.s. (equation (6.98)). Since $\alpha_n \geq 0$ and $\beta^n \geq 0$ (a.s.), we conclude that

$$\lim_{n\to\infty}\beta^n \to 0 \text{ a.s.} \tag{6.115}$$

If $\beta^n \to 0$, then $\mathbb{E}[g^{n+1}(\bar{\theta}^n - \theta^*)] \to 0$, which allows us to conclude that $\mathbb{E}[g^{n+1}(\bar{\theta}^n - \theta^*)|\mathfrak{F}^n] \to 0$ (the expectation of a nonnegative random variable cannot be zero unless the random variable is always zero). But what does this tell us about the behavior of $\bar{\theta}^n$? Knowing that $\beta^n \to 0$ does not necessarily imply that $g^{n+1} \to 0$ or $\bar{\theta}^n \to \theta^*$. There are three scenarios:

1) $\bar{\theta}^n \rightarrow \theta^*$ for all $n$, and of course all sample paths $\omega$. If this were the case, we are done.

2) $\bar{\theta}^{n_k} \rightarrow \theta^*$ for a subsequence $n_1, n_2, \ldots, n_k, \ldots$. For example, it might be that the sequence $\bar{\theta}^1, \bar{\theta}^3, \bar{\theta}^5, \ldots \rightarrow \theta^*$, while $\mathbb{E}[g^2|\mathfrak{F}^1], \mathbb{E}[g^4|\mathfrak{F}^3], \ldots, \rightarrow 0$. This would mean that for the subsequence $n_k$, $U^{n_k} \rightarrow 0$. But we already know that $U^n \rightarrow U^*$ where $U^*$ is the unique limit point, which means that $U^* = 0$. But if this is the case, then this is the limit point for every sequence of $\bar{\theta}^n$.

3) There is no subsequence $\bar{\theta}^{n_k}$ which has $\bar{\theta}^*$ as its limit point. This means that $\mathbb{E}[g^{n+1}|\mathfrak{F}^n] \rightarrow 0$. However, assumption 3 tells us that the expected gradient cannot vanish at a nonoptimal value of $\theta$. This means that this case cannot happen.

This completes the proof. □

## 6.9 BIBLIOGRAPHIC NOTES

Section 6.1 - The theoretical foundation for estimating value functions from Monte Carlo estimates has its roots in stochastic approximation theory, originated by Robbins & Monro (1951), with important early contributions made by Kiefer & Wolfowitz (1952), Blum (1954*b*) and Dvoretzky (1956). For thorough theoretical treatments of stochastic approximation theory, see Wasan (1969), Kushner & Clark (1978) and Kushner & Yin (1997). Very readable treatments of stochastic optimization can be found in Pflug (1996) and Spall (2003).

Section 6.2 - A number of different communities have studied the problem of "stepsizes," including the business forecasting community (Brown (1959), Holt et al. (1960), Brown (1963), Giffin (1971), Trigg (1964), Gardner (1983)), artificial intelligence (Jaakkola et al. (1994), Darken & Moody (1991), Darken et al. (1992), Sutton & Singh (1994)), stochastic programming (Kesten (1958) , Mirozahmedov & Uryasev (1983) , Pflug (1988), Ruszczyński & Syski (1986)) and signal processing (Goodwin & Sin (1984), Benveniste et al. (1990), Mathews & Xie (1993), Stengel (1994), Douglas & Mathews (1995)). The neural network community refers to "learning rate schedules"; see Haykin (1999).

Section 6.4 - This section is based on George & Powell (2006).

Section 6.5 - This section is based on the review in George & Powell (2006), along with the development of the optimal stepsize rule. Our proof that the optimal stepsize is $1/n$ for stationary data is based on Kmenta (1997).

Section 6.6 - These experiments are drawn from George & Powell (2006).

Section 6.8.2 - This proof is based on Blum (1954*a*), which generalized the original paper by Robbins & Monro (1951).

Section 6.8.3 - The proof in section 6.8.3 uses standard techniques drawn from several sources, notably Wasan (1969), Chong (1991), Kushner & Yin (1997) and, for this author, Cheung & Powell (2000).

## PROBLEMS

**6.1** We are going to solve a classic stochastic optimization problem known as the newsvendor problem. Assume we have to order $x$ assets after which we try to satisfy a random demand $D$ for these assets, where $D$ is randomly distributed between 100 and 200. If $x > D$, we have ordered too much and we pay $5(x - D)$. If $x < D$, we have an underage, and we have to pay $20(D - x)$.

(a) Write down the objective function in the form $\min_x \mathbb{E}f(x, D)$.

(b) Derive the stochastic gradient for this function.

(c) Find the optimal solution analytically [Hint: take the expectation of the stochastic gradient, set it equal to zero and solve for the quantity $\mathbb{P}(D \leq x^*)$. From this, find $x^*$.]

(d) Since the gradient is in units of dollars while $x$ is in units of the quantity of the asset being ordered, we encounter a scaling problem. Choose as a stepsize $\alpha_{n-1} = \alpha_0/n$ where $\alpha_0$ is a parameter that has to be chosen. Use $x^0 = 100$ as an initial solution. Plot $x^n$ for 1000 iterations for $\alpha_0 = 1, 5, 10, 20$. Which value of $\alpha_0$ seems to produce the best behavior?

(e) Repeat the algorithm (1000 iterations) 10 times. Let $\omega = (1, \ldots, 10)$ represent the 10 sample paths for the algorithm, and let $x^n(\omega)$ be the solution at iteration $n$ for sample path $\omega$. Let *Var*$(x^n)$ be the variance of the random variable $x^n$ where

$$\bar{V}(x^n) = \frac{1}{10}\sum_{\omega=1}^{10}(x^n(\omega) - x^*)^2$$

Plot the standard deviation as a function of $n$ for $1 \leq n \leq 1000$.

**6.2** A customer is required by her phone company to pay for a minimum number of minutes per month for her cell phone. She pays 12 cents per minute of guaranteed minutes, and 30 cents per minute that she goes over her minimum. Let $x$ be the number of minutes she commits to each month, and let $M$ be the random variable representing the number of minutes she uses each month, where $M$ is normally distributed with mean 300 minutes and a standard deviation of 60 minutes.

(a) Write down the objective function in the form $\min_x \mathbb{E}f(x, M)$.

(b) Derive the stochastic gradient for this function.

(c) Let $x^0 = 0$ and choose as a stepsize $\alpha_{n-1} = 10/n$. Use 100 iterations to determine the optimum number of minutes the customer should commit to each month.

**6.3** Show that if we use a stepsize rule $\alpha_{n-1} = 1/n$, then $\bar{\theta}^n$ is a simple average of $\hat{\theta}^1, \hat{\theta}^2, \ldots, \hat{\theta}^n$ (thus proving equation 6.17).

**6.4** Show that $\mathbb{E}\left[\left(\bar{\theta}^{n-1} - \theta^n\right)^2\right] = \lambda^{n-1}\sigma^2 + (\beta^n)^2$. [Hint: Add and subtract $\mathbb{E}\bar{\theta}^{n-1}$ inside the expectation and expand.]

**6.5** Show that $\mathbb{E}\left[\left(\bar{\theta}^{n-1} - \hat{\theta}^n\right)^2\right] = (1 + \lambda^{n-1})\sigma^2 + (\beta^n)^2$ (which proves equation 6.42). [Hint: See previous exercise.]

**6.6** Derive the small sample form of the recursive equation for the variance given in (6.43). Recall that if

$$\bar{\theta}^n = \frac{1}{n}\sum_{m=1}^{n}\hat{\theta}^m$$

then an estimate of the variance of $\hat{\theta}$ is

$$Var[\hat{\theta}] = \frac{1}{n-1}\sum_{m=1}^{n}(\hat{\theta}^m - \bar{\theta}^n).$$

**6.7** We are going to again try to use approximate dynamic programming to estimate a discounted sum of random variables (we first saw this in chapter 4):

$$F^T = \mathbb{E}\sum_{t=0}^{T}\gamma^t R_t,$$

where $R_t$ is a random variable that is uniformly distributed between 0 and 100 (you can use this information to randomly generate outcomes, but otherwise you cannot use this information). This time we are going to use a discount factor of $\gamma = .95$. We assume that $R_t$ is independent of prior history. We can think of this as a single state Markov decision process with no decisions.

(a) Using the fact that $\mathbb{E}R_t = 50$, give the exact value for $F^{100}$.

(b) Propose an approximate dynamic programming algorithm to estimate $F^T$. Give the value function updating equation, using a stepsize $\alpha_t = 1/t$.

(c) Perform 100 iterations of the approximate dynamic programming algorithm to produce an estimate of $F^{100}$. How does this compare to the true value?

(d) Compare the performance of the following stepsize rules: Kesten's rule, the stochastic gradient adaptive stepsize rule (use $\nu = .001$), $1/n^\beta$ with $\beta = .85$, the Kalman filter rule, and the optimal stepsize rule. For each one, find both the estimate of the sum and the variance of the estimate.

**6.8** Consider a random variable given by $R = 10U$ (which would be uniformly distributed between 0 and 10). We wish to use a stochastic gradient algorithm to estimate the mean of $R$ using the iteration $\bar{\theta}^n = \bar{\theta}^{n-1} - \alpha_{n-1}(R^n - \bar{\theta}^{n-1})$, where $R^n$ is a Monte Carlo sample of $R$ in the $n^{th}$ iteration. For each of the stepsize rules below, use equation (6.11) to measure the performance of the stepsize rule to determine which works best, and compute an estimate of the bias and variance at each iteration. If the stepsize rule requires choosing a parameter, justify the choice you make (you may have to perform some test runs).

(a) $\alpha_{n-1} = 1/n$.

(b) Fixed stepsizes of $\alpha_n = .05, .10$ and $.20$.

(c) The stochastic gradient adaptive stepsize rule (equations 6.29)-(6.30)).

(d) The Kalman filter (equations (6.52)-(6.56)).

(e) The optimal stepsize rule (algorithm 6.8).

**6.9** Repeat exercise 6.8 using

$$R^n = 10(1 - e^{-0.1n}) + 6(U - 0.5).$$

**6.10** Repeat exercise 6.8 using

$$R^n = \left(10/(1 + e^{-0.1(50-n)})\right) + 6(U - 0.5).$$

**6.11** Let $U$ be a uniform $[0, 1]$ random variable, and let

$$\mu^n = 1 - \exp(-\theta_1 n).$$

Now let $\hat{R}^n = \mu^n + \theta_2(U^n - .5)$. We wish to try to estimate $\mu^n$ using

$$\bar{R}^n = (1 - \alpha_{n-1})\bar{R}^{n-1} + \alpha_{n-1}\hat{R}^n.$$

In the exercises below, estimate the mean (using $\bar{R}^n$) and compute the standard deviation of $\bar{R}^n$ for $n = 1, 2, \ldots, 100$, for each of the following stepsize rules:

- $\alpha_{n-1} = 0.10$.
- $\alpha_{n-1} = a/(a + n - 1)$ for $a = 1, 10$.
- Kesten's rule.
- Godfrey's rule.
- The bias-adjusted Kalman filter stepsize rule.

For each of the parameter settings below, compare the rules based on the average error (1) over all 100 iterations and (2) in terms of the standard deviation of $\bar{R}^{100}$.

(a) $\theta_1 = 0, \theta_2 = 10$.

(b) $\theta_1 = 0.05, \theta_2 = 0$.

(c) $\theta_1 = 0.05, \theta_2 = 0.2$.

(d) $\theta_1 = 0.05, \theta_2 = 0.5$.

(e) Now pick the single stepsize that works the best on all four of the above exercises.

**6.12** An oil company covers the annual demand for oil using a combination of futures and oil purchased on the spot market. Orders are placed at the end of year $t - 1$ for futures that can be exercised to cover demands in year $t$. If too little oil is purchased this way, the company can cover the remaining demand using the spot market. If too much oil is purchased with futures, then the excess is sold at 70 percent of the spot market price (it is not held to the following year – oil is too valuable and too expensive to store).

To write down the problem, model the exogenous information using

$$\begin{aligned}
\hat{D}_t &= \text{Demand for oil during year } t, \\
\hat{p}^s_t &= \text{Spot price paid for oil purchased in year } t, \\
\hat{p}^f_{t,t+1} &= \text{Futures price paid in year } t \text{ for oil to be used in year } t+1.
\end{aligned}$$

The demand (in millions of barrels) is normally distributed with mean 600 and standard deviation of 50. The decision variables are given by

$$\bar{\theta}^f_{t,t+1} = \text{Number of futures to be purchased at the end of year } t \text{ to be used in year } t+1.$$
$$\bar{\theta}^s_t = \text{Spot purchases made in year } t.$$

(a) Set up the objective function to minimize the expected total amount paid for oil to cover demand in a year $t+1$ as a function of $\bar{\theta}^f_t$. List the variables in your expression that are not known when you have to make a decision at time $t$.

(b) Give an expression for the stochastic gradient of your objective function. That is, what is the derivative of your function for a particular sample realization of demands and prices (in year $t+1$)?

(c) Generate 100 years of random spot and futures prices as follows:

$$\hat{p}^f_t = 0.80 + 0.10U^f_t,$$
$$\hat{p}^s_{t,t+1} = \hat{p}^f_t + 0.20 + 0.10U^s_t,$$

where $U^f_t$ and $U^s_t$ are random variables uniformly distributed between 0 and 1. Run 100 iterations of a stochastic gradient algorithm to determine the number of futures to be purchased at the end of each year. Use $\bar{\theta}^f_0 = 30$ as your initial order quantity, and use as your stepsize $\alpha_t = 20/t$. Compare your solution after 100 years to your solution after 10 years. Do you think you have a good solution after 10 years of iterating?

**6.13** The proof in section 6.8.3 was performed assuming that $\theta$ is a scalar. Repeat the proof assuming that $\theta$ is a vector. You will need to make adjustments such as replacing Assumption 2 with $\|g^n\| < B$. You will also need to use the triangle inequality which states that $\|a+b\| \leq \|a\| + \|b\|$.

**6.14** Prove corollary 6.5.3.

## CHAPTER 7

# APPROXIMATING VALUE FUNCTIONS

The algorithms in chapter 4 assumed a standard lookup-table representation for a value function. That is to say, if we visited a discrete state $S_t^n$ we would then observe an estimate of the value of being in state $S_t^n$, which we typically referred to as $\hat{v}^n$, after which we updated the value of being in state $S_t^n$ using

$$\bar{V}_t^n(S_t^n) = (1 - \alpha_{n-1})\bar{V}_t^{n-1}(S_t^n) + \alpha_{n-1}\hat{v}_t^n.$$

The problem with discrete lookup-table representations is that they suffer from the first curse of dimensionality: multidimensional state variables. Unless $S_t^n$ is a scalar (and even this can cause problems for continuous variables), estimating a value for each state will often pose serious (and possibly intractable) statistical problems.

The statistical problem is compounded by the fact that we need to do more than just estimate the value of states we have visited. We also need to estimate values of states we *might* visit in order to solve decision problems such as

$$x_t^n = \arg\max_{x_t \in \mathcal{X}_t^n} \left( C(S_t^n, x_t) + \gamma \bar{V}_t^{n-1}(S^{M,x}(S_t^n, x_t)) \right).$$

Thus, while it is certainly useful to update our estimate of the value of being in state $S_t^n$, as we can see we need to be able to reasonably approximate $\bar{V}_t(S_t^x)$ for $S_t^x = S^{M,x}(S_t^n, x_t)$ for all $x_t \in \mathcal{X}_t^n$. It becomes quickly apparent that while we never have to loop over every state (as we did in chapter 3), we still have the problem of statistically estimating the value of being in a large number of states.

Ultimately, the challenge of estimating value functions draws on the entire field of statistics. Approximate dynamic programming introduces some unique challenges to the problem of statistically estimating value functions, but in the end, it all boils down to statistical estimation.

This chapter introduces two types of general purpose techniques designed for working with complex state spaces. The first depends on the use of aggregation methods, and assumes that we have access to one or more functions which can aggregate the attribute space into smaller sets. The second is general regression methods where we assume that analytical functions, with a relatively small number of parameters, can be formed to capture the important properties of the attribute vector.

## 7.1 APPROXIMATION USING AGGREGATION

We are going to start by using the context of managing a single complex entity with attribute $a_t$. Our entire presentation could be made using the context of an abstract system in state $S_t$, but we feel this setting will help to develop the intuition behind the methods we are going to present. We assume that the attribute vector has no special structure. The examples provide some illustrations.

---

**EXAMPLE 7.1**

A manufacturing operation has machines that can be set up to paint parts in different colors. The state of the machine is the color that it has been prepared to use for the next few jobs.

**EXAMPLE 7.2**

A team of medical technicians is being scheduled to visit different hospitals to train the staff of each hospital in the use of advanced equipment. Each technician is characterized by (a) his/her training, (b) their current location, (c) where they live, and (d) how many hours they have been working that day.

**EXAMPLE 7.3**

A student planning her college career has to plan which courses to take each semester. At the end of each semester, her status is described by the courses she has taken, or more compactly, by the number of courses she has completed toward each type of requirement (number of science courses, number of language courses, number of departmental courses, and so on).

---

For each problem, we assume that we are going to act on an entity with attribute $a_t$ with a decision of type $d_t$. We make a decision by solving a problem of the form

$$\max_{d_t \in \mathcal{D}} \left( C(a_t, d_t) + \gamma \bar{v}(a^{M,x}(a_t, d_t)) \right),$$

where, as before, $a^{M,x}(a_t, d_t)$ is the attribute of the resource after being acted on by decision $d_t$. We assume that the attribute vector consists of $I$ dimensions, where $I$ is not too large (large problems may have a few dozen or a hundred attributes, but we do not anticipate problems with thousands of attributes). In the classical language of dynamic programming, our challenge is to estimate the value $v(a)$ of being in state $a$. However, we assume that the attribute space $\mathcal{A}$ is too large to enumerate, which means we are going to have a difficult time estimating the value of being in any particular state (in a statistically reliable way). For some problems, the state space can be enumerated, but it is large enough that we will encounter statistical problems getting accurate estimates of the value of being in each state.

### 7.1.1 ADP and aggregation

For many years, researchers addressed the "curse of dimensionality" by using aggregation to reduce the size of the state space. The idea was to aggregate the original problem, solve it exactly (using the techniques of chapter 3), and then disaggregate it back to obtain an approximate solution to the original problem. Not only would the value function be defined over this smaller state space, so would the one-step transition matrix. In fact, simplifying the transition matrix (since its size is given by the square of the number of states) was even more important than simplifying the states over which the value function was defined.

Perhaps one of the most powerful features of approximate dynamic programming is that even if we use aggregation, we do not have to simplify the state of the system. The transition function $S_{t+1} = S^M(S_t, x_t, W_{t+1})$ always uses the original, disaggregate (possibly continuous) state vector. Aggregation is only used to approximate the value function. For example, in our nomadic trucker problem it is necessary to capture location, domicile, fleet type, equipment type, number of hours he has driven that day, how many hours he has driven on each of the past seven days, and the number of days he has been driving since he was last at home. All of this information is needed to simulate the driver forward in time. But we might estimate the value function using only the location of the driver, his domicile and fleet type. We are not trying to simplify how we represent the entity; rather, we only want to simplify how we approximate the value function. If our nomadic trucker is described by attribute vector $a_t$, the transition function $a_{t+1} = a^M(a_t, d_t, W_{t+1})$ may represent the attribute vector at a high level of detail (some values may be continuous). But the decision problem

$$\max_{a_t \in \mathcal{A}} \left( C(a_t, d_t) + \gamma \mathbb{E} \bar{V}_{t+1}(G(a_{t+1})) \right) \tag{7.1}$$

uses a value function $\bar{V}_{t+1}(G(a_{t+1}))$, where $G(\cdot)$ is an aggregation function that maps the original (and very detailed) state $a$ into something much simpler. The aggregation function $G$ may ignore an attribute, discretize it, or use any of a variety of ways to reduce the number of possible values of an attribute. This also reduces the number of parameters we have to estimate. In what follows, we drop the explicit reference of the aggregation function $G$ and simply use $\bar{V}_{t+1}(a_{t+1})$. The aggregation is implicit in the value function approximation.

Some examples of aggregation include:

**Spatial -** A transportation company is interested in estimating the value of truck drivers at a particular location. Locations may be calculated at the level of a five-digit zip code (there are about 55,000 in the United States), three-digit zip code (about 1,000), or the state level (48 contiguous states).

**Temporal -** A bank may be interested in estimating the value of holding an asset at a point in time. Time may be measured by the day, week, month, or quarter.

**Continuous parameters -** An attribute of an aircraft may be its fuel level; an attribute of a traveling salesman may be how long he has been away from home; an attribute of a water reservoir may be the depth of the water; the state of the cash reserve of a mutual fund is the amount of cash on hand at the end of the day. These are examples of systems with at least one attribute that is at least approximately continuous. The variables may all be discretized into intervals.

**Hierarchical classification -** A portfolio problem may need to estimate the value of investing money in the stock of a particular company. It may be useful to aggregate companies by industry segment (for example, a particular company might be in the chemical industry, and it might be further aggregated based on whether it is viewed as a domestic or multinational company). Similarly, problems of managing large inventories of parts (for cars, for example) may benefit by organizing parts into part families (transmission parts, engine parts, dashboard parts).

The examples below provide additional illustrations.

---

■ **EXAMPLE 7.1**

The state of a jet aircraft may be characterized by multiple attributes which include spatial and temporal dimensions (location and flying time since the last maintenance check), as well other attributes. A continuous parameter could be the fuel level, an attribute that lends itself to hierarchical aggregation might be the specific type of aircraft. We can reduce the number of states (attributes) of this resource by aggregating each dimension into a smaller number of potential outcomes.

■ **EXAMPLE 7.2**

The state of a portfolio might consist of the number of bonds which are characterized by the source of the bond (a company, a municipality or the federal government), the maturity (six months, 12 months, 24 months), when it was purchased, and its rating by bond agencies. Companies can be aggregated up by industry segment. Bonds can be further aggregated by their bond rating.

■ **EXAMPLE 7.3**

Blood stored in blood banks can be characterized by type, the source (which might indicate risks for diseases), age (it can be stored for up to 42 days), and the current location where it is being stored. A national blood management agency might want to aggregate the attributes by ignoring the source (ignoring an attribute is a form of aggregation), discretizing the age from days into weeks, and aggregating locations into more aggregate regions.

■ **EXAMPLE 7.4**

The value of an asset is determined by its current price, which is continuous. We can estimate the asset using a price discretized to the nearest dollar.

---

| Aggregation level | Location | Fleet type | Domicile | Size of state space |
|---|---|---|---|---|
| 0 | Sub-region | Fleet | Region | $400 \times 5 \times 100 = 200,000$ |
| 1 | Region | Fleet | Region | $100 \times 5 \times 100 = 50,000$ |
| 2 | Region | Fleet | Zone | $100 \times 5 \times 10 = 5,000$ |
| 3 | Region | Fleet | - | $100 \times 5 \times 1 = 500$ |
| 4 | Zone | - | - | $10 \times 1 \times 1 = 10$ |

**Table 7.1** Examples of aggregations on the attribute space for the nomadic trucker problem. '-' indicates that the particular attribute is ignored.

There are many applications where aggregation is naturally hierarchical. For example, in our nomadic trucker problem we might want to estimate the value of a truck based on three attributes: location, home domicile, and fleet type. The first two represent geographical locations, which can be represented (for this example) at three levels of aggregation: 400 sub-regions, 100 regions, and 10 zones. Table 7.1 illustrates five levels of aggregation that might be used. In this example, each higher level can be represented as an aggregation of the previous level.

Aggregation is also useful for continuous variables. Assume that our state variable is the amount of cash we have on hand, a number that might be as large as $10 million dollars. We might discretize our state space in units of $1 million, $100 thousand, $10 thousand, $1,000, $100, and $10. This discretization produces a natural hierarchy since 10 segments at one level of aggregation naturally group into one segment at the next level of aggregation.

Hierarchical aggregation is often the simplest to work with, but in most cases there is no reason to assume that the structure is hierarchical. In fact, we may even use overlapping aggregations (sometimes known as "soft" aggregation), where the same attribute $a$ aggregates into multiple elements in $\mathcal{A}^g$. For example, assume that $a$ represents an $(x, y)$ coordinate in a continuous space which has been discretized into the set of points $(x_i, y_i)_{i \in \mathcal{I}}$. Further assume that we have a distance metric $\rho((x, y), (x_i, y_i))$ that measures the distance from any point $(x, y)$ to every aggregated point $(x_i, y_i)$ , $i \in \mathcal{I}$. We might use an observation at the point $(x, y)$ to update estimates at each $(x_i, y_i)$ with a weight that declines with $\rho((x, y), (x_i, y_i))$.

### 7.1.2 Modeling aggregation

We present aggregation in the context of managing a single entity with attribute $a_t$ since the concepts are easier to visualize in this setting. We begin by defining a family of aggregation functions

$$G^g : \mathcal{A} \rightarrow \mathcal{A}^{(g)}.$$

$\mathcal{A}^{(g)}$ represents the $g^{th}$ level of aggregation of the attribute space $\mathcal{A}$. Let

$$a^{(g)} = G^g(a), \text{ the } g^{th} \text{ level aggregation of the attribute vector } a.$$

$$\mathcal{G} = \text{The set of indices corresponding to the levels of aggregation.}$$

In this section, we assume we have a single aggregation function $G$ that maps the disaggregate attribute $a \in \mathcal{A} = \mathcal{A}^{(0)}$ into an aggregated space $\mathcal{A}^{(g)}$. In section 7.1.3, we let $g \in \mathcal{G} = \{0, 1, 2, \ldots\}$ and we work with all levels of aggregation at the same time.

To begin our study of aggregation, we first need to characterize how we sample different states (at the disaggregate level). For this discussion, we assume we have two exogenous processes: At iteration $n$, the first process chooses an attribute to sample (which we denote by $\hat{a}^n$), and the second produces an observation of the value of being in state $\hat{a}^n \in \mathcal{A}$, which we denote by $\hat{v}^n$ (or $\hat{v}^n_a$). There are different ways of choosing an attribute $\hat{a}$; we could choose it at random from the attribute space, or by finding $d^n$ by solving (7.1) and then choosing an attribute $\hat{a} = a^M(a_t, d_t, W_{t+1})$.

We need to characterize the errors that arise in our estimate of the value function. Let

$$\nu^{(g)}_a = \text{The true value of being in state } a \text{ at aggregation level } g.$$

Here, $a$ is the original, disaggregated attribute vector. We let $\nu_a = \nu^{(0)}_a$ be the true (expected) value of being in state $a$. $\nu^{(g)}_a$ is the expected value of being in aggregated state $G(a)$. We can think of $\nu^{(g)}_a$ as an average over all the values of attributes $a$ such that $G(a) = \bar{a}$. In fact, there are different ways to weight all these disaggregate values, and we have to specify the weighting in a specific application. We might weight by how many times we actually visit a state (easy and practical, but it means the aggregate measure depends on how you are visiting states), or we might weight all attributes equally (this takes away the dependence on the policy, but we might include states we would never visit).

We need a model of how we sample and measure values. Assume that at iteration $n$, we sample a (disaggregated) state $\hat{a}^n$, and we then observe the value of being in this state with the noisy observation

$$\hat{v}^n = \nu_{\hat{a}^n} + \varepsilon^n.$$

We let $\omega$ be a sample realization of the sequence of attributes (states) and values, given by

$$\omega = (\hat{a}^1, \hat{v}^1, \hat{a}^2, \hat{v}^2, \ldots).$$

Let

$\bar{v}^{(g,n)}_a$ = The estimate of the value associated with the attribute vector $a$ at the $g^{th}$ level of aggregation after $n$ observations.

Throughout our discussion, a bar over a variable means it was computed from sample observations. A hat means the variable was an exogenous observation. If there is nothing (such as $\nu$ or $\beta$), then it means this is the true value (which is not known to us).

For $g = 0$, $a = \hat{a}^n$. For $g > 0$, the subscript $a$ in $\bar{v}^{(g,n)}_a$ refers to $G^g(\hat{a}^n)$, or the $g^{th}$ level of aggregation of $\hat{a}^n$. Given an observation $(\hat{a}^n, \hat{v}^n)$, we would update our estimate of being in state $a = \hat{a}^n$ using

$$\bar{v}^{(g,n)}_a = (1 - \alpha^{(g)}_{a,n-1})\bar{v}^{(g,n-1)}_a + \alpha^{(g)}_{a,n-1}\hat{v}^n.$$

Here, we have written the stepsize $\alpha^{(g)}_{a,n-1}$ to explicitly represent the dependence on the attribute and level of aggregation. Implicit is that this is also a function of the number of times that we have updated $\bar{v}^{(g,n)}_a$ by iteration $n$, rather than a function of $n$ itself.

To illustrate, imagine that our nomadic trucker has attributes $a = (\text{Loc}, \text{Equip}, \text{Home}, \text{DOThrs}, \text{Days})$, where "Loc" is location, "Equip" denotes the type of trailer (long, short, refrigerated), "Home" is the location of where he lives, "DOThrs" is a vector giving the number of hours the driver has worked on each of the last eight days, and "Days" is the number of days the driver has been away from home. We are going to estimate the value $\bar{V}(a)$ for different levels of aggregation of $a$, where we aggregate purely by ignoring certain dimensions of $a$. We start with our original disaggregate observation $\hat{v}(a)$, which we are going to write as

$$\hat{v}\begin{pmatrix} \text{Loc} \\ \text{Equip} \\ \text{Home} \\ \text{DOThrs} \\ \text{Days} \end{pmatrix} = \max_d \left(C(a,d) + \bar{V}(a^M(a,d))\right).$$

We now wish to use this estimate of the value of a driver with attribute $a$ to produce value functions at different levels of aggregation. We can do this by simply smoothing this disaggregate estimate in with estimates at different levels of aggregation, as in

$$\bar{v}^{(1,n)}\begin{pmatrix} \text{Loc} \\ \text{Equip} \\ \text{Home} \end{pmatrix} = (1-\alpha^{(1)}_{a,n-1})\bar{v}^{(1,n-1)}\begin{pmatrix} \text{Loc} \\ \text{Equip} \\ \text{Home} \end{pmatrix} + \alpha^{(1)}_{a,n-1}\hat{v}\begin{pmatrix} \text{Loc} \\ \text{Equip} \\ \text{Home} \\ \text{DOThrs} \\ \text{Days} \end{pmatrix},$$

$$\bar{v}^{(2,n)}\begin{pmatrix} \text{Loc} \\ \text{Equip} \end{pmatrix} = (1-\alpha^{(2)}_{a,n-1})\bar{v}^{(2,n-1)}\begin{pmatrix} \text{Loc} \\ \text{Equip} \end{pmatrix} + \alpha^{(2)}_{a,n-1}\hat{v}\begin{pmatrix} \text{Loc} \\ \text{Equip} \\ \text{Home} \\ \text{DOThrs} \\ \text{Days} \end{pmatrix},$$

$$\bar{v}^{(3,n)}\begin{pmatrix} \text{Loc} \end{pmatrix} = (1-\alpha^{(3)}_{a,n-1})\bar{v}^{(3,n-1)}\begin{pmatrix} \text{Loc} \end{pmatrix} + \alpha^{(3)}_{a,n-1}\hat{v}\begin{pmatrix} \text{Loc} \\ \text{Equip} \\ \text{Home} \\ \text{DOThrs} \\ \text{Days} \end{pmatrix}.$$

In the first equation, we are smoothing the value of a driver based on a five-dimensional attribute vector, given by $\hat{v}(a)$, in with an approximation indexed by a three-dimensional attribute vector. The second equation does the same using value function approximation indexed by a two-dimensional attribute vector, while the third equation does the same with a one-dimensional attribute vector. It is very important to keep in mind that the stepsize must reflect the number of times an attribute has been updated.

We can estimate the variance of $\bar{v}^{(g,n)}_a$ using the techniques described in section 6.4. Let

$(s^2_a)^{(g,n)}$ = The estimate of the variance of observations made of attribute $a$, using data from aggregation level $g$, after $n$ observations.

$(s^2_a)^{(g,n)}$ is the estimate of the variance of the observations $\hat{v}$ when we observe an attribute $\hat{a}^n$ which aggregates to attribute $a$ (that is, $G^g(\hat{a}^n) = a$). We are really interested in the variance of our estimate of the mean, $\bar{v}^{(g,n)}_a$. In section 6.4, we showed that

$$\begin{aligned} (\bar{\sigma}^2_a)^{(g,n)} &= Var[\bar{v}^{(g,n)}_a] \\ &= \lambda^{(g,n)}_a (s^2_a)^{(g,n)}, \end{aligned} \tag{7.2}$$

where $(s_a^2)^{(g,n)}$ is an estimate of the variance of the observations $\hat{v}_t^n$ at the $g^{th}$ level of aggregation (computed below), and $\lambda_a^{(g,n)}$ can be computed from the recursion

$$\lambda_a^{(g,n)} = \begin{cases} (\alpha_{a,n-1}^{(g)})^2, & n = 1, \\ (1-\alpha_{a,n-1}^{(g)})^2 \lambda_a^{(g,n-1)} + (\alpha_{a,n-1}^{(g)})^2, & n > 1. \end{cases}$$

Note that if the stepsize $\alpha_{a,n-1}^{(g)}$ goes to zero, then $\lambda_a^{(g,n)}$ will also go to zero, as will $(\bar{\sigma}_a^2)^{(g,n)}$. We now need to compute $(s_a^2)^{(g,n)}$ which is the estimate of the variance of observations $\hat{v}^n$ for attributes $\hat{a}^n$ for which $G^g(\hat{a}^n) = a$ (the observations of attributes that aggregate up to $a$). Let $\bar{\nu}_a^{(g,n)}$ be the total variation, given by

$$\bar{\nu}_a^{(g,n)} = (1-\eta_{n-1})\bar{\nu}_a^{(g,n-1)} + \eta_{n-1}(\bar{v}_a^{(g,n-1)} - \hat{v}_a^n)^2,$$

where $\eta_{n-1}$ is a stepsize formula such as McClain's (section 6.2). We refer to $\bar{\nu}_a^{(g,n)}$ as the total variation because it captures deviations that arise both due to measurement noise (the randomness when we compute $\hat{v}_a^n$) and bias (since $\bar{v}_a^{(g,n-1)}$ is a biased estimate of the mean of $\hat{v}_a^n$).

We finally need an estimate of the bias. There are two types of bias. In section 6.4 we measured the transient bias that arose from the use of smoothing applied to a nonstationary time series using

$$\bar{\beta}_a^{(g,n)} = (1-\eta_{n-1})\bar{\beta}_a^{(g,n-1)} + \eta_{n-1}(\hat{v}^n - \bar{v}_a^{(g,n-1)}). \tag{7.3}$$

This bias arises because the observations $\hat{v}^n$ may be steadily increasing (or decreasing) with the iterations (the usual evolution of the value function that we first saw with value iteration). When we smooth on past observations, we obtain an estimate $\bar{v}_a^{(g,n-1)}$ that tends to underestimate (overestimate if $\hat{v}^n$ tends to decrease) the true mean of $\hat{v}^n$.

The second form of bias is the aggregation bias given by the difference between the estimate at an aggregate level and the disaggregate level. We compute the aggregation bias using

$$\bar{\mu}_a^{(g,n)} = \bar{v}_a^{(g,n)} - \bar{v}_a^{(0,n)}. \tag{7.4}$$

Using the same reasoning presented in section 6.4, we can separate out the effect of bias to obtain an estimate of the variance of the error using

$$(s_a^2)^{(g,n)} = \frac{\bar{\nu}_a^{(g,n)} - (\bar{\beta}_a^{(g,n)})^2}{1+\lambda^{n-1}}. \tag{7.5}$$

In the next section, we put the estimate of aggregation bias, $\bar{\mu}_a^{(g,n)}$, to work.

The relationships are illustrated in figure 7.1, which shows a simple function defined over a single, continuous attribute (for example, the price of an asset). If we select a particular attribute $a$, we find we have only two observations for that attribute, versus seven for that section of the function. If we use an aggregate approximation, we would produce a single number over that range of the function, creating a bias between the true function and the aggregated estimate. As the illustration shows, the size of the bias depends on the shape of the function in that region.

One method for choosing the best level of aggregation is to choose the level that minimizes $(\bar{\sigma}_a^2)^{(g,n)} + (\bar{\mu}_a^{(g,n)})^2$, which captures both bias and variance. In the next section, we use the bias and variance to develop a method that uses estimates at all levels of aggregation at the same time.

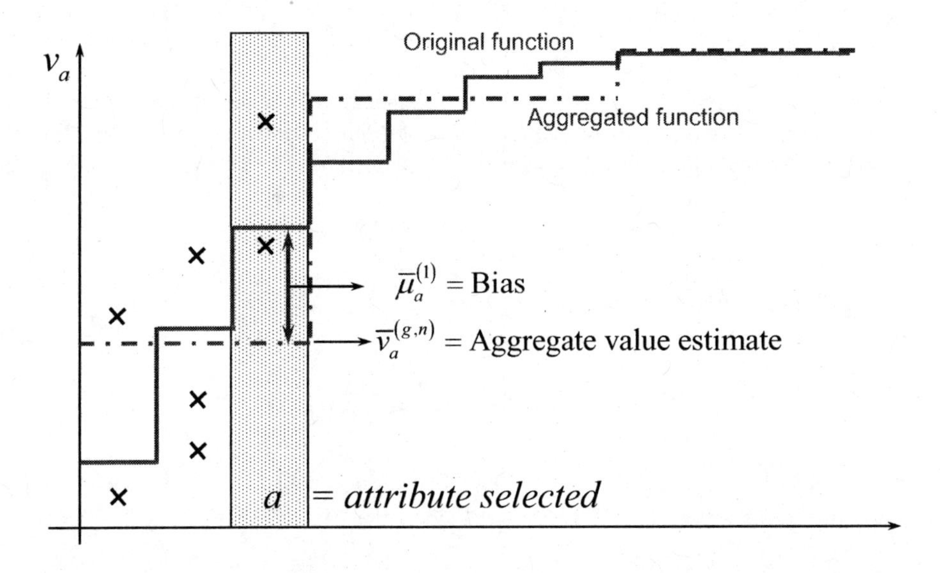


**Figure 7.1** Illustration of a disaggregate function, an aggregated approximation and a set of samples. For a particular attribute $a$, we show the estimate and the bias.

### 7.1.3 Combining multiple levels of aggregation

Rather than try to pick the best level of aggregation, it is intuitively appealing to use a weighted sum of estimates at different levels of aggregation. The simplest strategy is to use

$$\bar{v}_a^n = \sum_{g \in \mathcal{G}} w^{(g)} \bar{v}_a^{(g)}, \tag{7.6}$$

where $w^{(g)}$ is the weight applied to the $g^{th}$ level of aggregation. We would expect the weights to be positive and sum to one, but we can also view these simply as coefficients in a regression function. In such a setting, we would normally write the regression as

$$\bar{V}(a|\theta) = \theta_0 + \sum_{g \in \mathcal{G}} \theta_g \bar{v}_a^{(g)}$$

(see section 7.2.2 for a discussion of general regression methods.) The problem with this strategy is that the weight does not depend on the attribute $a$. Intuitively, it makes sense to put a higher weight on attributes $a$ which have more observations, or where the estimated variance is lower. This behavior is lost if the weight does not depend on $a$.

In practice, we will generally observe certain attributes much more frequently than others, suggesting that the weights should depend on $a$. To accomplish this, we need to use

$$\bar{v}_a^n = \sum_{g \in \mathcal{G}} w_a^{(g)} \bar{v}_a^{(g,n)}.$$

Now the weight depends on the attribute, allowing us to put a higher weight on the disaggregate estimates when we have a lot of observations. This is clearly the most natural, but when the attribute space is large, we face the challenge of computing thousands (perhaps hundreds of thousands) of weights. If we are going to go this route, we need a fairly simple method to compute the weights.

We can view the estimates $(\bar{v}^{(g,n)})_{g\in\mathcal{G}}$ as different ways of estimating the same quantity. There is an extensive statistics literature on this problem. For example, it is well known that the weights that minimize the variance of $\bar{v}_a^n$ in equation (7.6) are given by

$$w_a^{(g)} \propto \left((\bar{\sigma}_a^2)^{(g,n)}\right)^{-1}.$$

Since the weights should sum to one, we obtain

$$w_a^{(g)} = \left(\frac{1}{(\bar{\sigma}_a^2)^{(g,n)}}\right)\left(\sum_{g\in\mathcal{G}}\frac{1}{(\bar{\sigma}_a^2)^{(g,n)}}\right)^{-1}. \tag{7.7}$$

These weights work if the estimates are unbiased, which is clearly not the case. This is easily fixed by using the total variation (variance plus the square of the bias), producing the weights

$$w_a^{(g,n)} = \frac{1}{\left((\bar{\sigma}_a^2)^{(g,n)} + \left(\bar{\mu}_a^{(g,n)}\right)^2\right)}\left(\sum_{g'\in\mathcal{G}}\frac{1}{\left((\bar{\sigma}_a^2)^{(g',n)} + \left(\bar{\mu}_a^{(g',n)}\right)^2\right)}\right)^{-1}. \tag{7.8}$$

These are computed for each level of aggregation $g \in \mathcal{G}$. Furthermore, we compute a different set of weights for each attribute $a$. $(\bar{\sigma}_a^2)^{(g,n)}$ and $\bar{\mu}_a^{(g,n)}$ are easily computed recursively using equations (7.2) and (7.4), which makes the approach well suited to large scale applications. Note that if the stepsize used to smooth $\hat{v}^n$ goes to zero, then the variance $(\bar{\sigma}_a^2)^{(g,n)}$ will also go to zero as $n \to \infty$. However, the bias $\bar{\beta}_a^{(g,n)}$ will in general not go to zero.

Figure 7.2 shows the average weight put on each level of aggregation (when averaged over all the attributes $a$) for a particular application. The behavior illustrates the intuitive property that the weights on the aggregate level are highest when there are only a few observations, with a shift to the more disaggregate level as the algorithm progresses. This is a very important behavior when approximating value functions. It is simply not possible to produce good value function approximations with only a few data points, so it is important to use simple functions (with only a few parameters).

The weights computed using (7.2) minimize the variance in the estimate $\bar{v}_a^{(g)}$ if the estimates at different levels of aggregation are independent, but this is simply not going to be the case. $\bar{v}_a^{(0)}$ (an estimate of $a$ at the most disaggregate level) and $\bar{v}_a^{(1)}$ will be correlated since $\bar{v}_a^{(1)}$ is estimated using some of the same observations used to estimate $\bar{v}_a^{(0)}$. So it is fair to ask if the weights produce accurate estimates.

To get a handle on this question, consider the scalar function in figure 7.3a. At the disaggregate level, the function is defined for 10 discrete values. This range is then divided into three larger intervals, and an aggregated function is created by estimating the function over each of these three larger ranges. Instead of using the weights computed using (7.8), we can fit a regression of the form

$$\hat{v}^n = \theta_0\bar{v}_a^{(0,n)} + \theta_1\bar{v}_a^{(1,n)}. \tag{7.9}$$

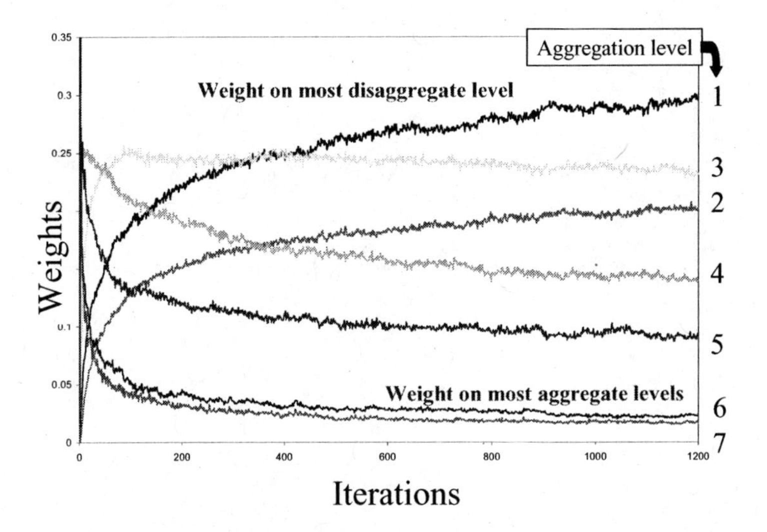


**Figure 7.2** Average weight (across all attributes) for each level of aggregation using equation (7.8).

The parameters $\theta_0$ and $\theta_1$ can be fit using regression techniques. Note that while we would expect $\theta_0 + \theta_1$ to be approximately 1, there is no formal guarantee of this. If we use only two levels of aggregation, we can find $(\theta_0, \theta_1)$ using linear regression and can compare these weights to those computed using equation (7.8) where we assume independence.

For this example, the weights are shown in figure 7.3b. The figure illustrates that the weight on the disaggregate level is highest when the function has the greatest slope, which produces the highest biases. When we compute the optimal weights (which captures the correlation), the weight on the disaggregate level for the portion of the curve that is flat is zero, as we would expect. Note that when we assume independence, the weight on the disaggregate level (when the slope is zero) is no longer zero. Clearly a weight of zero is best because it means that we are aggregating all the points over the interval into a single estimate, which is going to be better than trying to produce three individual estimates.

One would expect that using the optimal weights, which captures the correlations between estimates at different levels of aggregation, would also produce better estimates of the function itself. This does not appear to be the case. We compared the errors between the estimated function and the actual function using both methods for computing weights, using three levels of noise around the function. The results are shown in figure 7.4, which indicates that there is virtually no difference in the accuracy of the estimates produced by the two methods. This observation has held up under a number of experiments.

### 7.1.4 State aliasing and aggregation

An issue that arises when we use aggregation is that two different states (call them $S_1$ and $S_2$) may have the same behavior (as a result of aggregation), despite the fact that the states are different, and perhaps should exhibit different behaviors. When this happens, we refer to $S_1$ as an alias of $S_2$ (and vice versa). We refer to this behavior as *aliasing*.

In approximate dynamic programming, we do not (as a rule) aggregate the state variable as we step forward through time. In most applications, transition functions can handle state variables at a high level of detail. However, we may aggregate the state for the purpose

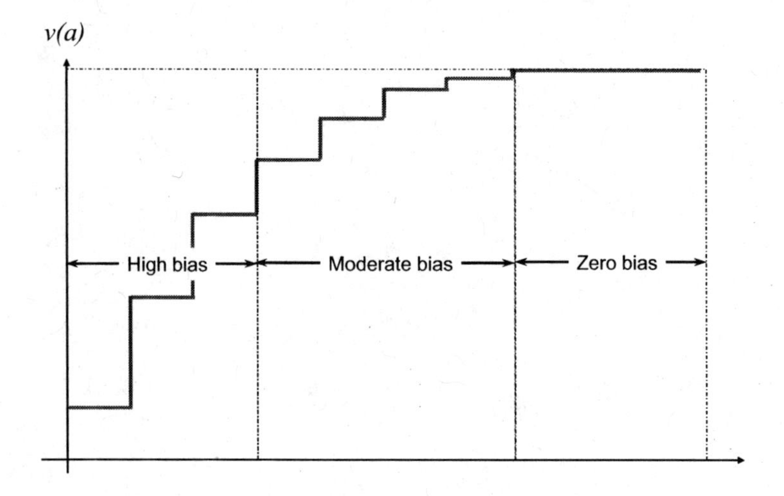


7.3a: Scalar, nonlinear function

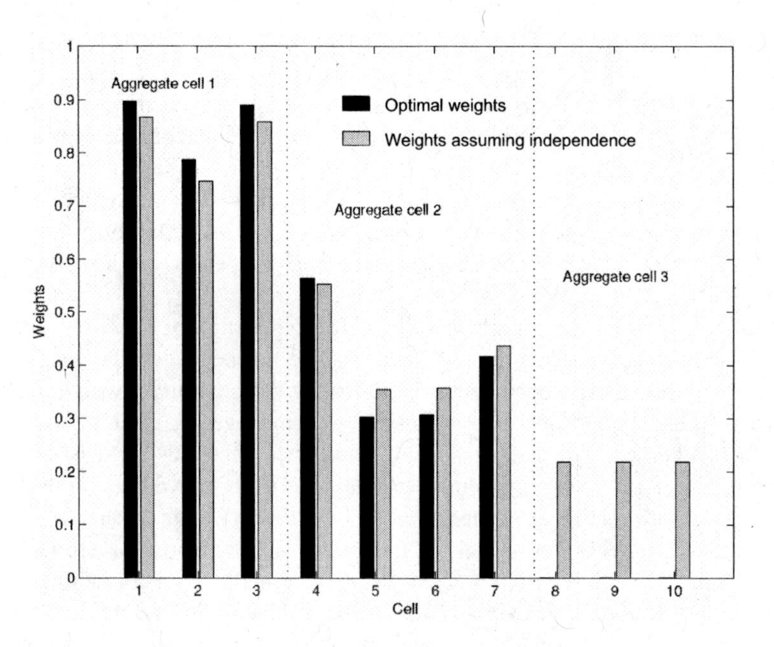


7.3b: Weight given to disaggregate level

**Figure 7.3** The weight given to the disaggregate level for a two-level problem at each of 10 points, with and without the independence assumption (from George et al. (2005)).

of computing the value function. In this case, if we are in state $S_i$ and take action $x_i$, we might expect to end up in state $S_i'$ where $S_1' \neq S_2'$. But we might have $\bar{V}(S_1') = \bar{V}(S_2')$ as a result of aggregation. In this case, it might happen that we make the same decision even though the difference in the state $S_1$ and $S_2$ might call for different decisions. Welcome to approximate dynamic programming.

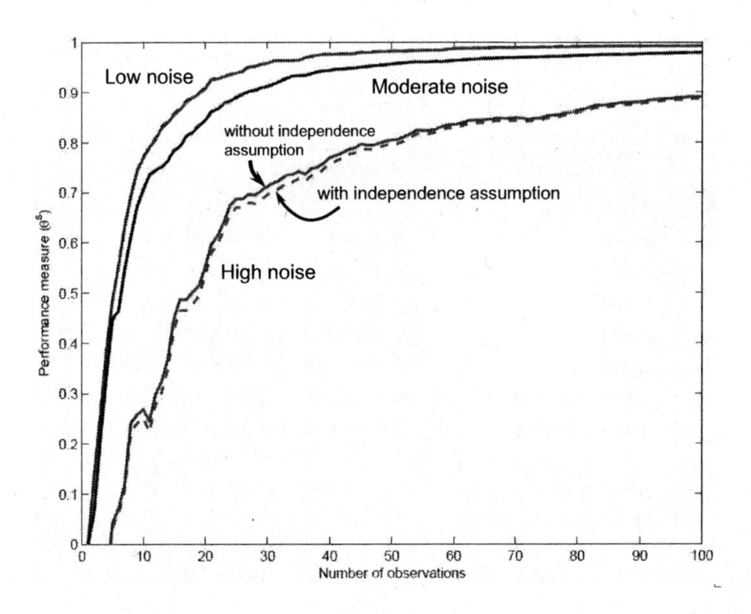


**Figure 7.4** The effect of ignoring the correlation between estimates at different levels of aggregation

## 7.2 APPROXIMATION METHODS USING REGRESSION MODELS

Up to now, we have focused on lookup-table representations of value functions, where if we are in a (discrete) state $s$, we compute an approximation $\bar{v}(s)$ that is an estimate of the value of being in state $s$. Using aggregation (even mixtures of estimates at different levels of aggregation) is still a form of look-up table (we are just using a simpler lookup-table). An advantage of this strategy is that it avoids the need to exploit specialized structure in the state variable (other than the definition of levels of aggregation). A disadvantage is that it does not allow you to take advantage of structure in the state variable.

There has been considerable interest in estimating value functions using regression methods. A classical presentation of linear regression poses the problem of estimating a parameter vector $\theta$ to fit a model that predicts a variable $y$ using a set of observations $(x_i)_{i \in \mathcal{I}}$, where we assume a model of the form

$$y = \theta_0 + \sum_{i=1}^{I} \theta_i x_i + \varepsilon. \tag{7.10}$$

In the language of approximate dynamic programming, the independent variables $x_i$ are created using *basis functions* which reduce potentially large state variables into a small number of *features*, a term widely used in the artificial intelligence community. If we are playing the game of tic-tac-toe, we might want to know if our opponent has captured the center square, or we might want to know the number of corner squares we own. If we are managing a fleet of taxis, we might define for each region of the city a feature that gives the number of taxis in the region, plus 0.5 times the number of taxis in each of the neighboring regions.

Using this language, instead of an independent variable $x_i$, we would have a basis function $\phi_f(S)$, where $f \in \mathcal{F}$ is a *feature*. $\phi_f(S)$ might be an indicator variable (e.g.,

1 if we have an 'X' in the center square of our tic-tac-toe board), a discrete number (the number of X's in the corners of our tic-tac-toe board), or a continuous quantity (the price of an asset, the amount of oil in our inventories, the amount of $AB-$ blood on hand at the hospital). Some problems might have fewer than 10 features; others may have dozens; and some may have hundreds of thousands. In general, however, we would write our value function in the form

$$\bar{V}(S|\theta) = \sum_{f \in \mathcal{F}} \theta_f \phi_f(S).$$

In a time dependent model, the parameter vector $\theta$ and basis functions $\phi$ may also be indexed by time (but not necessarily).

In the remainder of this section, we provide a brief review of linear regression, followed by some examples of regression models. We close with a more advanced presentation that provides insights into the geometry of basis functions (including a better understanding of why they are called "basis functions").

### 7.2.1 Linear regression review

Let $y^n$ be the $n^{th}$ observation of our dependent variable (what we are trying to predict) based on the observation $(x_1^n, x_2^n, \ldots, x_I^n)$ of our independent (or explanatory) variables (the $x_i$ are equivalent to the basis functions we used earlier). Our goal is to estimate a parameter vector $\theta$ that solves

$$\min_\theta \sum_{m=1}^{n} \left( y^m - (\theta_0 + \sum_{i=1}^{I} \theta_i x_i^m) \right)^2. \tag{7.11}$$

This is the standard linear regression problem. Let $\bar{\theta}^n$ be the optimal solution for this problem. Throughout this section, we assume that the underlying process from which the observations $y^n$ are drawn is stationary (an assumption that is almost never satisfied in approximate dynamic programming).

If we define $x_0 = 1$, we let

$$x^n = \begin{pmatrix} x_0^n \\ x_1^n \\ \vdots \\ x_I^n \end{pmatrix}$$

be an $I+1$-dimensional column vector of observations. Throughout this section, and unlike the rest of the book, we use traditional vector operations, where $x^T x$ is an inner product (producing a scalar) while $x x^T$ is an outer product, producing a matrix of cross terms.

Letting $\theta$ be the column vector of parameters, we can write our model as

$$y = \theta^T x + \varepsilon.$$

We assume that the errors $(\varepsilon^1, \ldots, \varepsilon^n)$ are independent and identically distributed. We do not know the parameter vector $\theta$, so we replace it with an estimate $\bar{\theta}$ which gives us the predictive formula

$$\bar{y}^n = (\bar{\theta})^T x^n,$$

where $\bar{y}^n$ is our predictor of $y^{n+1}$. Our prediction error is

$$\hat{\varepsilon}^n = y^n - (\bar{\theta})^T x^n.$$

Our goal is to choose $\theta$ to minimize the mean squared error

$$\min_{\theta} \sum_{m=1}^{n} (y^m - \theta^T x^m)^2. \tag{7.12}$$

It is well known that this can be solved very simply. Let $X^n$ be the $n$ by $I+1$ matrix

$$X^n = \begin{pmatrix} x_0^1 & x_1^1 & & x_I^1 \\ x_0^2 & x_1^2 & & x_I^2 \\ \vdots & \vdots & \cdots & \vdots \\ x_0^n & x_1^n & & x_I^n \end{pmatrix}.$$

Next, denote the vector of observations of the dependent variable as

$$Y^n = \begin{pmatrix} y^1 \\ y^2 \\ \vdots \\ y^n \end{pmatrix}.$$

The optimal parameter vector $\bar{\theta}$ (after $n$ observations) is given by

$$\bar{\theta} \quad = \quad [(X^n)^T X^n]^{-1} (X^n)^T Y^n. \tag{7.13}$$

Solving a static optimization problem such as (7.12), which produces the elegant equations for the optimal parameter vector in (7.13), is the most common approach taken by the statistics community. It has little direct application in approximate dynamic programming since our problems tend to be recursive in nature, reflecting the fact that at each iteration we obtain new observations, which require updates to the parameter vector. In addition, our observations tend to be notoriously nonstationary. Later, we show how to overcome this problem using the methods of recursive statistics.

### 7.2.2 Illustrations using regression models

There are many problems where we can exploit structure in the state variable, allowing us to propose functions characterized by a small number of parameters which have to be estimated statistically. Section 7.1.3 represented one version where we had a parameter for each (possibly aggregated) state. The only structure we assumed was implicit in the ability to specify a series of one or more aggregation functions.

The remainder of this section illustrates the use of regression models in specific applications. The examples use a specific method for estimating the parameter vector $\theta$ that will typically prove to be somewhat clumsy in practice. Section 7.3 describes methods for estimating $\theta$ recursively.

#### *Pricing an American option*

Consider the problem of determining the value of an American-style put option which gives us the right to sell an asset (or contract) at a specified price at any of a set of discrete time periods. For example, we might be able to exercise the option on the last day of the month over the next 12 months.

Assume we have an option that allows us to sell an asset at \$1.20 at any of four time periods. We assume a discount factor of 0.95 to capture the time value of money. If we wait until time period 4, we must exercise the option, receiving zero if the price is over \$1.20. At intermediate periods, however, we may choose to hold the option even if the price is below \$1.20 (of course, exercising it if the price is above \$1.20 does not make sense). Our problem is to determine whether to hold or exercise the option at the intermediate points.

From history, we have found 10 samples of price trajectories which are shown in table 7.2. If we wait until time period 4, our payoff is shown in table 7.3, which is zero if the

| Stock prices | | | | |
|---|---|---|---|---|
| | Time period | | | |
| Outcome | 1 | 2 | 3 | 4 |
| 1 | 1.21 | 1.08 | 1.17 | 1.15 |
| 2 | 1.09 | 1.12 | 1.17 | 1.13 |
| 3 | 1.15 | 1.08 | 1.22 | 1.35 |
| 4 | 1.17 | 1.12 | 1.18 | 1.15 |
| 5 | 1.08 | 1.15 | 1.10 | 1.27 |
| 6 | 1.12 | 1.22 | 1.23 | 1.17 |
| 7 | 1.16 | 1.14 | 1.13 | 1.19 |
| 8 | 1.22 | 1.18 | 1.21 | 1.28 |
| 9 | 1.08 | 1.11 | 1.09 | 1.10 |
| 10 | 1.15 | 1.14 | 1.18 | 1.22 |

**Table 7.2** Ten sample realizations of prices over four time periods

price is above 1.20, and $1.20 - p_4$ for prices below \$1.20.

At time $t = 3$, we have access to the price history $(p_1, p_2, p_3)$. Since we may not be able to assume that the prices are independent or even Markovian (where $p_3$ depends only on $p_2$), the entire price history represents our state variable, along with an indicator that tells us if we are still holding the asset. We wish to predict the value of holding the option at time $t = 4$. Let $V_4(a_4)$ be the value of the option if we are holding it at time 4, given the state (which includes the price $p_4$) at time 4. Now let the conditional expectation at time 3 be

$$\bar{V}_3(a_3) = \mathbb{E}\{V_4(a_4)|a_3\}.$$

Our goal is to approximate $\bar{V}_3(a_3)$ using information we know at time 3. We propose a linear regression of the form

$$Y = \theta_0 + \theta_1 X_1 + \theta_2 X_2 + \theta_3 X_3,$$

| Option value at $t = 4$ | | | | |
|---|---|---|---|---|
| | Time period | | | |
| Outcome | 1 | 2 | 3 | 4 |
| 1 | - | - | - | 0.05 |
| 2 | - | - | - | 0.07 |
| 3 | - | - | - | 0.00 |
| 4 | - | - | - | 0.05 |
| 5 | - | - | - | 0.00 |
| 6 | - | - | - | 0.03 |
| 7 | - | - | - | 0.01 |
| 8 | - | - | - | 0.00 |
| 9 | - | - | - | 0.10 |
| 10 | - | - | - | 0.00 |

**Table 7.3** The payout at time 4 if we are still holding the option

where

$$\begin{aligned} Y &= V_4, \\ X_1 &= p_2, \\ X_2 &= p_3, \\ X_3 &= (p_3)^2. \end{aligned}$$

The variables $X_1, X_2$ and $X_3$ are our basis functions. Keep in mind that it is important that our explanatory variables $X_i$ must be a function of information we have at time $t = 3$, whereas we are trying to predict what will happen at time $t = 4$ (the payoff). We would then set up the data matrix given in table 7.4.

We may now run a regression on this data to determine the parameters $(\theta_i)_{i=0}^3$. It makes sense to consider only the paths which produce a positive value in the fourth time period, since these represent the sample paths where we are most likely to still be holding the asset at the end. The linear regression is only an approximation, and it is best to fit the approximation in the region of prices which are the most interesting (we could use the same reasoning to include some "near misses"). We only use the value function to estimate the value of holding the asset, so it is this part of the function we wish to estimate. For our illustration, however, we use all 10 observations, which produces the equation

$$\bar{V}_3 \approx 0.0056 - 0.1234 p_2 + 0.6011 p_3 - 0.3903 (p_3)^2.$$

$\bar{V}_3$ is an approximation of the expected value of the price we would receive if we hold the option until time period 4. We can now use this approximation to help us decide what to do at time $t = 3$. Table 7.5 compares the value of exercising the option at time 3 against holding the option until time 4, computed as $\gamma \bar{V}_3(a_3)$. Taking the larger of the two payouts,

| Regression data | | | | |
|---|---|---|---|---|
| | Independent variables | | | Dependent variable |
| Outcome | $X_1$ | $X_2$ | $X_3$ | $Y$ |
| 1 | 1.08 | 1.17 | 1.3689 | 0.05 |
| 2 | 1.12 | 1.17 | 1.3689 | 0.07 |
| 3 | 1.08 | 1.22 | 1.4884 | 0.00 |
| 4 | 1.12 | 1.18 | 1.3924 | 0.05 |
| 5 | 1.15 | 1.10 | 1.2100 | 0.00 |
| 6 | 1.22 | 1.23 | 1.5129 | 0.03 |
| 7 | 1.44 | 1.13 | 1.2769 | 0.01 |
| 8 | 1.18 | 1.21 | 1.4641 | 0.00 |
| 9 | 1.11 | 1.09 | 1.1881 | 0.10 |
| 10 | 1.14 | 1.18 | 1.3924 | 0.00 |

**Table 7.4** The data table for our regression at time 3

we find, for example, that we would hold the option given samples 1-4, 6, 8, and 10, but would sell given samples 5, 7, and 9.

We can repeat the exercise to estimate $\bar{V}_2(a_t)$. This time, our dependent variable "$Y$" can be calculated two different ways. The simplest is to take the larger of the two columns from table 7.5 (marked in bold). So, for sample path 1, we would have $Y_1 = \max\{.03, 0.03947\} = 0.03947$. This means that our observed value is actually based on our approximate value function $\bar{V}_3(a_3)$.

An alternative way of computing the observed value of holding the option in time 3 is to use the approximate value function to determine the decision, but then use the actual price we receive when we eventually exercise the option. Using this method, we receive 0.05 for the first sample path because we decide to hold the asset at time 3 (based on our approximate value function) after which the price of the option turns out to be worth 0.05. Discounted, this is worth 0.0475. For sample path 2, the option proves to be worth 0.07 which discounts back to 0.0665 (we decided to hold at time 3, and the option was worth 0.07 at time 4). For sample path 5 the option is worth 0.10 because we decided to exercise at time 3.

Regardless of which way we compute the value of the problem at time 3, the remainder of the procedure is the same. We have to construct the independent variables "$Y$" and regress them against our observations of the value of the option at time 3 using the price history $(p_1, p_2)$. Our only change in methodology would occur at time 1 where we would have to use a different model (because we do not have a price at time 0).

### *Playing "lose tic-tac-toe"*

The game of "lose tic-tac-toe" is the same as the familiar game of tic-tac-toe, with the exception that now you are trying to make the other person get three in a row. This nice

| Rewards | | |
|---|---|---|
| | Decision | |
| Outcome | Exercise | Hold |
| 1 | 0.03 | $0.04155 \times .95 =$ **0.03947** |
| 2 | 0.03 | $0.03662 \times .95 =$ **0.03479** |
| 3 | 0.00 | $0.02397 \times .95 =$ **0.02372** |
| 4 | 0.02 | $0.03346 \times .95 =$ **0.03178** |
| 5 | **0.10** | $0.05285 \times .95 = 0.05021$ |
| 6 | 0.00 | $0.00414 \times .95 =$ **0.00394** |
| 7 | **0.07** | $0.00899 \times .95 = 0.00854$ |
| 8 | 0.00 | $0.01610 \times .95 =$ **0.01530** |
| 9 | **0.11** | $0.06032 \times .95 = 0.05731$ |
| 10 | 0.02 | $0.03099 \times .95 =$ **0.02944** |

**Table 7.5** The payout if we exercise at time 3, and the expected value of holding based on our approximation. The best decision is indicated in bold.

twist on the popular children's game provides the setting for our next use of regression methods in approximate dynamic programming.

Unlike our exercise in pricing options, representing a tic-tac-toe board requires capturing a discrete state. Assume the cells in the board are numbered left to right, top to bottom as shown in figure 7.5a. Now consider the board in figure 7.5b. We can represent the state of the board after the $t^{th}$ play using

$$a_{ti} = \begin{cases} 1 & \text{if cell } i \text{ contains an "X,"} \\ 0 & \text{if cell } i \text{ is blank,} \\ -1 & \text{if cell } i \text{ contains an "O,"} \end{cases}$$

$$a_t = (a_{ti})_{i=1}^9.$$

We see that this simple problem has up to $3^9 = 19,683$ states. While many of these states will never be visited, the number of possibilities is still quite large, and seems to overstate the complexity of the game.

We quickly realize that what is important about a game board is not the status of every cell as we have represented it. For example, rotating the board does not change a thing, but it does represent a different state. Also, we tend to focus on strategies (early in the game when it is more interesting) such as winning the center of the board or a corner. We might start defining variables (basis functions) such as

$$\phi_1(a_t) = \text{1 if there is an "X" in the center of the board, 0 otherwise,}$$

$$\phi_2(a_t) = \text{The number of corner cells with an "X,"}$$

$$\phi_3(a_t) = \text{The number of instances of adjacent cells with an "X" (horizontally, vertically, or diagonally).}$$

There are, of course, numerous such functions we can devise, but it is unlikely that we could come up with more than a few dozen (if that) which appeared to be useful. It is important to realize that we do not need a value function to tell us to make obvious moves.

Once we form our basis functions, our value function approximation is given by

$$\bar{V}_t(a_t) = \sum_{f \in \mathcal{F}} \theta_{tf} \phi_f(a_t).$$

We note that we have indexed the parameters by time (the number of plays) since this might play a role in determining the value of the feature being measured by a basis function, but it is reasonable to try fitting a model where $\theta_{tf} = \theta_f$. We estimate the parameters $\theta$ by playing the game (and following some policy) after which we see if we won or lost. We let $Y^n = 1$ if we won the $n^{th}$ game, 0 otherwise. This also means that the value function is trying to approximate the probability of winning if we are in a particular state.

We may play the game by using our value functions to help determine a policy. Another strategy, however, is simply to allow two people (ideally, experts) to play the game and use this to collect observations of states and game outcomes. This is an example of supervisory learning. If we lack a "supervisor" then we have to depend on simple strategies combined with the use of slowly learned value function approximations. In this case, we also have to recognize that in the early iterations, we are not going to have enough information to reliably estimate the coefficients for a large number of basis functions.

### 7.2.3 A geometric view of basis functions*

For readers comfortable with linear algebra, we can obtain an elegant perspective on the geometry of basis functions. In section 7.2.1, we found the parameter vector $\theta$ for a regression model by minimizing the expected square of the errors between our model and a set of observations. Assume now that we have a "true" value function $V(s)$ which gives the value of being in state $s$, and let $p(s)$ be the probability of visiting state $s$. We wish to find the approximate value function that best fits $V(s)$ using a given set of basis functions $(\phi_f(s))_{f \in \mathcal{F}}$. If we minimize the expected square of the errors between our approximate model and the true value function, we would want to solve

$$\min_\theta F(\theta) = \sum_{s \in \mathcal{S}} p(s) \left( V(s) - \sum_{f \in \mathcal{F}} \theta_f \phi_f(s) \right)^2, \tag{7.14}$$

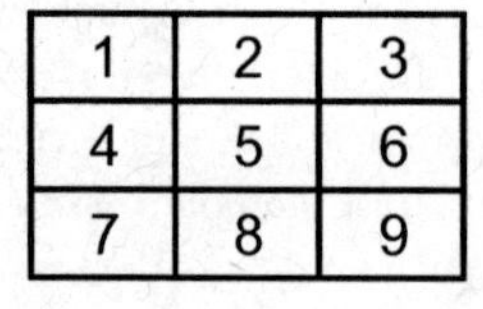


7.5a

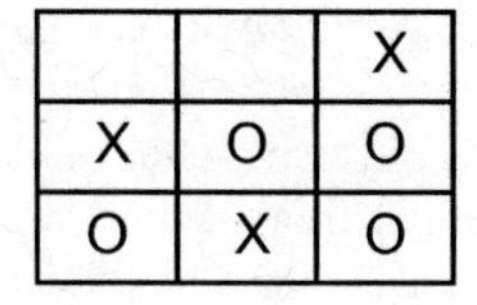


7.5b

**Figure 7.5** Some tic-tac-toe boards. (7.5a) Our indexing scheme. (7.5b) Sample board.

where we have weighted the error for state $s$ by the probability of actually being in state $s$. Our parameter vector $\theta$ is unconstrained, so we can find the optimal value by taking the derivative and setting this equal to zero. Differentiating with respect to $\theta_{f'}$ gives

$$\frac{\partial F(\theta)}{\partial \theta_{f'}} = -2\sum_{s\in\mathcal{S}} p(s)\left(V(s) - \sum_{f\in\mathcal{F}} \theta_f \phi_f(s)\right)\phi_{f'}(s).$$

Setting the derivative equal to zero and rearranging gives

$$\sum_{s\in\mathcal{S}} p(s)V(s)\phi_{f'}(s) = \sum_{s\in\mathcal{S}} p(s)\sum_{f\in\mathcal{F}} \theta_f\phi_f(s)\phi_{f'}(s). \tag{7.15}$$

At this point, it is much more elegant to revert to matrix notation. Define an $|\mathcal{S}| \times |\mathcal{S}|$ diagonal matrix $D$ where the diagonal elements are the state probabilities $p(s)$, as follows

$$D = \begin{pmatrix} p(1) & 0 & & 0 \\ 0 & p(2) & & 0 \\ \vdots & 0 & \cdots & \vdots \\ 0 & \vdots & & p(|\mathcal{S}|) \end{pmatrix}.$$

Let $V$ be the column vector giving the value of being in each state

$$V = \begin{pmatrix} V(1) \\ V(2) \\ \vdots \\ V(|\mathcal{S}|) \end{pmatrix}.$$

Finally, let $\Phi$ be an $|\mathcal{S}| \times |\mathcal{F}|$ matrix of the basis functions given by

$$\Phi = \begin{pmatrix} \phi_1(1) & \phi_2(1) & & \phi_{|\mathcal{F}|}(1) \\ \phi_1(2) & \phi_2(2) & & \phi_{|\mathcal{F}|}(2) \\ \vdots & \vdots & \cdots & \vdots \\ \phi_1(|\mathcal{S}|) & \phi_2(|\mathcal{S}|) & & \phi_{|\mathcal{F}|}(|\mathcal{S}|) \end{pmatrix}.$$

Recognizing that equation (7.15) is for a particular feature $f'$, with some care it is possible to see that equation (7.15) for all features is given by the matrix equation

$$\Phi^T DV = \Phi^T D\Phi\theta. \tag{7.16}$$

It helps to keep in mind that $\Phi$ is an $|\mathcal{S}| \times |\mathcal{F}|$ matrix, $D$ is an $|\mathcal{S}| \times |\mathcal{S}|$ diagonal matrix, $V$ is an $|\mathcal{S}| \times 1$ column vector, and $\theta$ is an $|\mathcal{F}| \times 1$ column vector. The reader should carefully verify that (7.16) is the same as (7.15).

Now, pre-multiply both sides of (7.16) by $(\Phi^T D\Phi)^{-1}$. This gives us the optimal value of $\theta$ as

$$\theta = (\Phi^T D\Phi)^{-1}\Phi^T DV. \tag{7.17}$$

This equation is closely analogous to the normal equations of linear regression, given by equation (7.13), with the only difference being the introduction of the scaling matrix $D$ which captures the probability that we are going to visit a state.

Now, pre-multiply both sides of (7.17) by $\Phi$, which gives

$$\Phi\theta = \bar{V} = \Phi(\Phi^T D\Phi)^{-1}\Phi^T DV.$$

$\Phi\theta$ is, of course, our approximation of the value function, which we have denoted by $\bar{V}$. This, however, is the best possible value function given the set of functions $\phi = (\phi_f)_{f\in\mathcal{F}}$. If the vector $\phi$ formed a complete basis over the space formed by the value function $V(s)$ and the state space $\mathcal{S}$, then we would obtain $\Phi\theta = \bar{V} = V$. Since this is generally not the case, we can view $\bar{V}$ as the nearest point projection (where "nearest" is defined as a weighted measure using the state probabilities $p(s)$) onto the space formed by the basis functions. In fact, we can form a projection operator $\Pi$ defined by

$$\Pi = \Phi(\Phi^T D\Phi)^{-1}\Phi^T D$$

so that $\bar{V} = \Pi V$ is the value function closest to $V$ that can be produced by the set of basis functions.

This discussion brings out the geometric view of basis functions (and at the same time, the reason why we use the term "basis function"). There is an extensive literature on basis functions that has evolved in the approximation literature.

### 7.2.4 Approximating continuous functions

There is an extensive literature on the use of approximation methods, primarily for continuous functions which may not have a particular structure such as convexity. These problems, which arise in many applications in engineering and economics, require the use of approximation methods that can adapt to a wide range of functions. Interpolation techniques, orthogonal polynomials, Fourier approximations and splines are just some of the most popular techniques.

A particularly powerful approximation strategy is the use of locally polynomial approximations. Also known as kernel-based methods, this strategy involves estimating relatively simple approximations that only apply to a small part of the function. Thus, a function may be both convex and concave, but locally it might be described by a quadratic function. Instead of using a single quadratic function, we can use a family of functions, each centered around a particular part of the state space.

The use of such strategies in approximate dynamic programming is relatively recent. The bibliographic remarks contain additional references.

## 7.3 RECURSIVE METHODS FOR REGRESSION MODELS

Estimating regression models to approximate a value function involves all the same tools and statistical issues that students would encounter in any course on regression. The only difference in dynamic programming is that our data are usually generated internal to the algorithm, which means that we have the opportunity to update our regression model after every iteration. This is both an opportunity and a challenge. Traditional methods for estimating the parameters of a regression model either involve solving a system of linear equations or solving a nonlinear programming problem to find the best parameters. Both methods are generally too slow in the context of dynamic programming. The remainder of this section describes some simple updating methods that have been used in the context of approximate dynamic programming.

### 7.3.1 Parameter estimation using a stochastic gradient algorithm

In our original representation, we effectively had a basis function for each state $a$ and the parameters were the value of being in each state, given by $v(a)$, where $\bar{v}_a^n$ is our estimate of $v(a)$ after the $n^{th}$ iteration. Our updating step was given by

$$\bar{v}_a^n = \bar{v}_a^{n-1} - \alpha_{n-1}\left[\bar{v}_a^{n-1} - \hat{v}_a^n\right].$$

This update is a step in the algorithm required to solve

$$\min_v \mathbb{E}\frac{1}{2}(v - \hat{v})^2,$$

where $\hat{v}$ is a sample estimate of $V(a)$. When we parameterize the value function, we create a function that we can represent using $\bar{V}_a(\theta)$. We want to find $\theta$ that solves

$$\min_\theta \mathbb{E}\frac{1}{2}(\bar{V}_a(\theta) - \hat{v})^2.$$

Applying our standard stochastic gradient algorithm, we obtain the updating step

$$\bar{\theta}^n = \bar{\theta}^{n-1} - \alpha_{n-1}(\bar{V}_a(\bar{\theta}^{n-1}) - \hat{v}(\omega^n))\nabla_\theta \bar{V}_a(\theta^n). \tag{7.18}$$

Since $\bar{V}_a(\theta^n) = \sum_{f\in\mathcal{F}} \theta_f \phi_f(a)$, the gradient with respect to $\theta$ is given by

$$\nabla_\theta \bar{V}_a(\theta^n) = \begin{pmatrix} \frac{\partial \bar{V}_a(\theta^n)}{\partial \theta_1} \\ \frac{\partial \bar{V}_a(\theta^n)}{\partial \theta_2} \\ \vdots \\ \frac{\partial \bar{V}_a(\theta^n)}{\partial \theta_F} \end{pmatrix} = \begin{pmatrix} \phi_1(a^n) \\ \phi_2(a^n) \\ \vdots \\ \phi_F(a^n) \end{pmatrix} = \Phi(a^n).$$

Thus, the updating equation (7.18) is given by

$$\begin{aligned} \bar{\theta}^n &= \bar{\theta}^{n-1} - \alpha_{n-1}(\bar{V}_a(\bar{\theta}^{n-1}) - \hat{v}(\omega^n))\Phi(a^n) \\ &= \bar{\theta}^{n-1} - \alpha_{n-1}(\bar{V}_a(\bar{\theta}^{n-1}) - \hat{v}(\omega^n)) \begin{pmatrix} \phi_1(a^n) \\ \phi_2(a^n) \\ \vdots \\ \phi_F(a^n) \end{pmatrix}. \end{aligned} \tag{7.19}$$

Using a stochastic gradient algorithm requires that we have some starting estimate $\bar{\theta}^0$ for the parameter vector, although $\theta^0 = 0$ is a common choice.

An important practical problem is the scaling of the stepsize. Unlike our previous work, we now have what is known as a scaling problem. The units of $\bar{\theta}^{n-1}$ and the units of $(\bar{V}_a(\bar{\theta}^{n-1}) - \hat{v}(\omega^n))\Phi(a^n)$ may be completely different. What we have learned about stepsizes still applies, except that we may need an initial stepsize that is quite different than 1.0 (our common starting point). Our experimental work has suggested that the following policy works well: When you choose a stepsize formula, scale the first value of the stepsize so that the change in $\bar{\theta}^n$ in the early iterations of the algorithm is approximately 20 to 50 percent (you will typically need to observe several iterations). You want to see individual elements of $\bar{\theta}^n$ moving consistently in the same direction during the early iterations. If the stepsize is too large, the values can swing wildly, and the algorithm may not converge at all. If the changes are too small, the algorithm may simply stall out. It is very tempting to run the algorithm for a period of time and then conclude that it appears to have converged (presumably to a good solution). While it is important to see the individual elements moving in the same direction (consistently increasing or decreasing) in the early iterations, it is also important to see oscillatory behavior toward the end.

### 7.3.2 Recursive least squares for stationary data

Equation (7.13) for estimating the parameters of a regression equation is far too expensive to be useful in dynamic programming applications. Even for a relatively small number of parameters (which may not be that small), the matrix inverse is going to be too slow for most applications. Fortunately, it is possible to compute these formulas recursively. The updating equation for $\theta$ is

$$\bar{\theta}^n = \bar{\theta}^{n-1} - H^n x^n \hat{\varepsilon}^n, \tag{7.20}$$

where $H^n$ is a matrix computed using

$$H^n = \frac{1}{\gamma^n} B^{n-1}. \tag{7.21}$$

$B^{n-1}$ is an $I+1$ by $I+1$ matrix which is updated recursively using

$$B^n = B^{n-1} - \frac{1}{\gamma^n}(B^{n-1}x^n(x^n)^T B^{n-1}). \tag{7.22}$$

$\gamma^n$ is a scalar computed using

$$\gamma^n \quad = \quad 1 + (x^n)^T B^{n-1} x^n. \tag{7.23}$$

The derivation of equations (7.20)-(7.23) is given in section 7.6.1. Equation (7.20) has the feel of a stochastic gradient algorithm, but it has one significant difference. Instead of using a typical stepsize, we have the $H^n$ which serves as a scaling matrix.

It is possible in any regression problem that the matrix $(X^n)^T X^n$ (in equation (7.13)) is non-invertible. If this is the case, then our recursive formulas are not going to overcome this problem. When this happens, we will observe $\gamma^n = 0$. Alternatively, the matrix may be invertible, but unstable, which occurs when $\gamma^n$ is very small (say, $\gamma^n < \epsilon$ for some small $\epsilon$). When this occurs, the problem can be circumvented by using

$$\bar{\gamma}^n = \gamma^n + \delta,$$

where $\delta$ is a suitably chosen small perturbation that is large enough to avoid instabilities. Some experimentation is likely to be necessary, since the right value depends on the scale of the parameters being estimated.

The only missing step in our algorithm is initializing $B^n$. One strategy is to collect a sample of $m$ observations where $m$ is large enough to compute $B^m$ using full inversion. Once we have $B^m$, we can then proceed to update it using the formula above. A second strategy is to use $B^0 = \epsilon I$, where $I$ is the identity matrix and $\epsilon$ is a "small constant." This strategy is not guaranteed to give the exact values, but should work well if the number of observations is relatively large.

In our dynamic programming applications, the observations $y^n$ will represent estimates of the value of being in a state, and our independent variables will be either the states of our system (if we are estimating the value of being in each state) or the basis functions, in which case we are estimating the coefficients of the basis functions. The equations assume implicitly that the estimates come from a stationary series.

There are many problems where the number of basis functions can be extremely large. In these cases, even the efficient recursive expressions in this section cannot avoid the fact that we are still updating a matrix where the number of rows and columns is the number of states (or basis functions). If we are only estimating a few dozen or a few hundred parameters, this can be fine. If the number of parameters extends into the thousands, even this strategy would probably bog down. It is very important to work out the approximate dimensionality of the matrices before using these methods.

### 7.3.3 Recursive least squares for nonstationary data

It is generally the case in approximate dynamic programming that our observations $y^n$ (typically, updates to an estimate of a value function) come from a nonstationary process. Instead of minimizing total errors (as we do in equation (7.11)) it makes sense to minimize a geometrically weighted sum of errors

$$\min_{\theta} \sum_{m=1}^{n} \lambda^{n-m} \left( y^m - (\theta_0 + \sum_{i=1}^{I} \theta_i x_i^m) \right)^2, \tag{7.24}$$

where $\lambda$ is a discount factor that we use to discount older observations. If we repeat the derivation in section 7.3.2, the only changes we have to make are in the expression for $\gamma^n$, which is now given by

$$\gamma^n = \lambda + (x^n)^T B^{n-1} x^n, \tag{7.25}$$

and the updating formula for $B^n$, which is now given by

$$B^n = \frac{1}{\lambda} \left( B^{n-1} - \frac{1}{\gamma^n} (B^{n-1} x^n (x^n)^T B^{n-1}) \right).$$

$\lambda$ works in a way similar to a stepsize, although in the opposite direction. Setting $\lambda = 1$ means we are putting an equal weight on all observations, while smaller values of $\lambda$ puts more weight on more recent observations.

We could use this logic and view $\lambda$ as a tunable parameter. Of course, a constant goal in the design of approximate dynamic programming algorithms is to avoid the need to tune yet another parameter. Ideally, we would like to take advantage of the theory we developed for automatically adapting stepsizes (in sections 6.2 and 6.5) to automatically adjust $\lambda$. For the special case where our regression model is just a constant (in which case $x^n = 1$), we can develop a simple relationship between $\alpha_n$ and the discount factor (which we now compute at each iteration, so we write it as $\lambda_n$). Let $G^n = (H^n)^{-1}$, which means that our updating equation is now given by

$$\bar{\theta}^n = \bar{\theta}^{n-1} - (G^n)^{-1} x^n \hat{\varepsilon}^n.$$

The matrix $G^n$ is updated recursively using

$$G^n = \lambda_n G^{n-1} + x^n (x^n)^T, \tag{7.26}$$

with $G^0 = 0$. For the case where $x^n = 1$ (in which case $G^n$ is also a scalar), $(G^n)^{-1}x^n = (G^n)^{-1}$ plays the role of our stepsize, so we would like to write $\alpha_n = G^n$. Assume that $\alpha_{n-1} = \left(G^{n-1}\right)^{-1}$. Equation (7.26) implies that

$$\begin{aligned}\alpha_n &= (\lambda_n G^{n-1} + 1)^{-1} \\ &= \left(\frac{\lambda_n}{\alpha_{n-1}} + 1\right)^{-1}.\end{aligned}$$

Solving for $\lambda_n$ gives

$$\lambda_n = \alpha_{n-1}\left(\frac{1-\alpha_n}{\alpha_n}\right). \tag{7.27}$$

Note that if $\lambda_n = 1$, then we want to put equal weight on all the observations (which would be optimal if we have stationary data). We know that in this setting, the best stepsize is $\alpha_n = 1/n$. Substituting this stepsize into equation (7.27) verifies this identity.

The value of equation (7.27) is that it allows us to use our knowledge of stepsizes from chapter 6. We note that this result applies only for the case where our regression model is just a constant, but we have used this logic for general regression models with excellent results. Of particular value is the ability to use the optimal stepsize rule in section 6.5.3 for the types of transient observations that we encounter in approximate dynamic programming.

### 7.3.4 Recursive estimation using multiple observations

The previous methods assume that we get one observation and use it to update the parameters. Another strategy is to sample several paths and solve a classical least-squares problem for estimating the parameters. In the simplest implementation, we would choose a set of realizations $\hat{\Omega}^n$ (rather than a single sample $\omega^n$) and follow all of them, producing a set of estimates $(\hat{v}^n(\omega))_{\omega \in \hat{\Omega}^n}$ that we can use to update the value function.

If we have a set of observations, we then face the classical problem of finding a vector of parameters $\hat{\theta}^n$ that best match all of these value function estimates. Thus, we want to solve

$$\hat{\theta}^n = \arg\min_{\hat{\theta}} \frac{1}{|\hat{\Omega}^n|} \sum_{\omega \in \hat{\Omega}^n} (\bar{V}(\hat{\theta}) - \hat{v}^n(\omega))^2.$$

This is the standard parameter estimation problem faced in the statistical estimation community. If $\bar{V}(\theta)$ is linear in $\theta$, then we can use the usual formulas for linear regression. If the function is more general, we would typically resort to nonlinear programming algorithms to solve the problem. In either case, $\hat{\theta}^n$ is still an update that needs to be smoothed in with the previous estimate $\bar{\theta}^{n-1}$, which we would do using

$$\bar{\theta}^n = (1 - \alpha_{n-1})\bar{\theta}^{n-1} + \alpha_{n-1}\hat{\theta}^n. \tag{7.28}$$

One advantage of this strategy is that in contrast with the updates that depend on the gradient of the value function, updates of the form given in equation (7.28) do not encounter a scaling problem, and therefore we return to our more familiar territory where $0 < \alpha_n \leq 1$. Of course, as the sample size $\hat{\Omega}$ increases, the stepsize should also be increased because there is more information in $\hat{\theta}^n$. Using stepsizes based on the Kalman filter (sections 6.5.2 and 6.5.3) will automatically adjust to the amount of noise in the estimate.

The usefulness of this particular strategy will be very problem-dependent. In many applications, the computational burden of producing multiple estimates $\hat{v}^n(\omega), \omega \in \hat{\Omega}^n$ before producing a parameter update will simply be too costly.

### 7.3.5 Recursive time-series estimation

Up to now, we have used regression models that depend on explanatory variables (which we sometimes refer to as basis functions) that depend on the state of the system. The goal has been to represent the value of being in a state $S_t$ using a small number of parameters by using the structure of the state variable.

Now consider a somewhat different use of linear regression. Let $\hat{v}^n$ be an estimate of the value of being in some state $S_t = s$ (which we suppress to keep the notation simple). By now we are well aware that the random observations $\hat{v}^n$ tend to grow (or shrink) over time. If they are growing, then the standard updating equation

$$\bar{v}^n = (1 - \alpha_{n-1})\bar{v}^{n-1} + \alpha_{n-1}\hat{v}^n \tag{7.29}$$

will generally produce an estimate $\bar{v}^n$ which is less than $\hat{v}^{n+1}$. We can minimize this by estimating a regression model of the form

$$\hat{v}^{n+1} = \theta_0\hat{v}^n + \theta_1\bar{v}^{n-1} + \cdots + \theta_F\bar{v}^{n-F} + \hat{\varepsilon}^{n+1}. \tag{7.30}$$

After iteration $n$, our estimate of the value of being in state $s$ is

$$\bar{v}^n = \bar{\theta}_0^n\hat{v}^n + \bar{\theta}_1^n\bar{v}^{n-1} + \cdots + \bar{\theta}_F^n\bar{v}^{n-F}. \tag{7.31}$$

Contrast this update with our simple exponential smoothing (such as equation (7.29)). We see that exponential smoothing is simply a form of linear regression where the parameter vector $\theta$ is determined by the stepsize.

We can update our parameter vector $\bar{\theta}^n$ using the techniques described earlier in this section. We need to keep in mind that we are estimating the value of being in a state $s$ using an $\tau + 1$-dimensional parameter vector. There are applications where we are trying to estimate the value of being in hundreds of thousands of states. For such large applications, we have to decide if we are going to have a different parameter vector $\theta$ for each state $s$, or one for the entire system.

There are many variations on this basic idea which draw on the entire field of time series analysis. For example, we can use our recursive formulas to estimate a more general time-series model. At iteration $n$, let the elements of our basis function be given by

$$\phi^n = (\phi_f(S^n))_{f \in \mathcal{F}},$$

which is the observed value of each function given the state vector $S^n$. If we wished to include a constant term, we would define a basis function $\phi_0 = 1$. Let $F = |\mathcal{F}|$ be the number of basis functions used to explain the value function, so $\phi^n$ is an $F$-dimensional column vector.

We can combine the information in our basis functions (which depend on the state at time $t$) with the history of observations of the updated values, which we represent using the column vector

$$\tilde{v}^n = (\hat{v}^n, \bar{v}^{n-1}, \ldots, \bar{v}^{n-\tau})^T$$

just as we did in sections 7.3.1 to 7.3.4 above. Taken together, $\phi^n$ and $\tilde{v}^n$ is our population of potential explanatory variables that we can use to help us predict $\hat{v}^{n+1}$. $\phi^n$ is an $F$-dimensional column vector, while $\tilde{v}^n$ is a $\tau + 1$-dimensional column vector. We can formulate a model using

$$\hat{v}^{n+1} = (\theta^\phi)^T \phi^n + (\theta^v)^T \tilde{v}^n + \hat{\varepsilon}^{n+1},$$

where $\theta^\phi$ and $\theta^v$ are, respectively, $F$ and $\tau + 1$-dimensional parameter vectors. This gives us a prediction of the value of being in state $S^n$ using

$$\bar{v}^n = (\bar{\theta}^{\phi,n})^T \phi^n + (\bar{\theta}^{v,n})^T \tilde{v}^n.$$

Again, we update our estimate of the parameter vector $\bar{\theta}^n = (\bar{\theta}^{\phi,n}, \bar{\theta}^{v,n})$ using the same techniques we presented above.

It is important to be creative. For example, we could use a time series model based on differences, as in

$$\begin{aligned}\hat{v}^{n+1} &= \theta_0 \hat{v}^n + \theta_1(\bar{v}^{n-1} - \hat{v}^n) + \theta_2(\bar{v}^{n-2} - \bar{v}^{n-1}) + \cdots \\ &\quad + \theta_F(\bar{v}^{n-F} - \bar{v}^{n-F+1}) + \hat{\varepsilon}^{n+1}.\end{aligned}$$

Alternatively, we can use an extensive library of techniques developed under the umbrella of time-series modeling.

The signal processing community has developed a broad range of techniques for estimating models in a dynamic environment. This community refers to models such as (7.30) and (7.32) as *linear filters*. Using these techniques to improve the process of estimating value functions remains an active area of research. Our sense is that if the variance of $\hat{v}^n$ is large relative to the rate at which the mean is changing, then we are unlikely to derive much value from a more sophisticated model which considers the history of prior estimates of the value function. However, there are problems where the mean of $\bar{v}^n$ is either changing quickly, or changes for many iterations (due to a slow learning process). For these problems, a time-series model which captures the behavior of the estimates of the value function over time may improve the convergence of the algorithm.

### 7.3.6 Mixed strategies

For the vast majority of problems, it is simply not possible to estimate the value of being in each state. As a result, it is almost essential that we be able to identify structure in the problem that allows us to create functional approximations. However, there are problems where the state variable may include "easy" components and "hard" components. We may find that we have good intuition about the effect of the easy components which allows us to develop an approximation, but we struggle with the hard components.

Consider the example of modeling a financial portfolio. Let $R_{ta}$ be the amount of money we have invested in asset class $a$, where the asset classes could be individual stocks, or broader classes such as domestic equities, international equities, corporate bonds, and so on. The resource state of this system is given by $R_t = (R_{ta})_{a \in \mathcal{A}}$, where $\mathcal{A}$ is our set of asset classes. We feel that the value of a particular allocation (captured by $R_t$) is affected by a set of financial statistics $f_t = (f_{ti})_{i \in \mathcal{I}}$. For example, $f_{t1}$ might be the value of the Euro (in dollars), and $f_{t2}$ might be the S&P 500 index (we are going to assume that the set $\mathcal{I}$ is fairly small). Our state variable would be given by

$$S_t = (R_t, f_t).$$

Now assume that we feel comfortable approximating the effect of the resource variable $R_t$ using an approximation such as

$$\bar{V}_t^R(R_t|\theta) = \sum_{a\in\mathcal{A}} \left(\theta_{0a}(R_{ta})^{\theta_{1a}}\right).$$

The problem is that we feel that the value of being in state $S_t$ is a function of both $R_t$ and the financial variables $f_t$. However, we may not have good intuition about how the financial statistics $f_t$ interact with the value of the portfolio (but we are sure that it does interact). We can create a family of value function approximations with the form

$$\bar{V}_t(S_t) = \bar{V}_t^R(R_t|\theta, f_t).$$

By this we mean that we are going to estimate an approximation $\bar{V}_t^R(R_t|\theta)$ for each value of the vector $f_t$. This requires estimating a vector $\theta$ for each value of $f_t$.

This will work if the set of values of $f_t$ is not too large. But do not be timid. We have solved problems of this sort where $f_t$ takes on 10,000 different possibilities. We refer to this approach as a mixed strategy since we are effectively using a lookup-table format for the state $f_t$, while using a continuous functional approximation for $R_t$. It works if we are willing to live with a separable approximation for the resource variable $R_t$, but where we are not comfortable assuming any sort of relationship between $f_t$ and $R_t$. We have found that strategies such as this can work quite effectively even for fairly coarse discretizations of $f_t$.

## 7.4 NEURAL NETWORKS

Neural networks represent an unusually powerful and general class of approximation strategies that have been widely used in approximate dynamic programming, primarily in classical engineering applications. There are a number of excellent textbooks on the topic, so our presentation is designed only to introduce the basic idea and encourage readers to experiment with this technology if simpler models are not effective.

Up to now, we have considered approximation functions of the form

$$\bar{V}(S) = \sum_{f\in\mathcal{F}} \theta_f \phi_f(S),$$

where $\mathcal{F}$ is our set of features, and $(\phi_f(S))_{f\in\mathcal{F}}$ are the basis functions which extract what are felt to be the important characteristics of the state variable which explain the value of being in a state. We have seen that when we use an approximation that is linear in the parameters, we can estimate the parameters $\theta$ recursively using standard methods from linear regression. For example, if $R_a$ is the number of resources with attribute $a$, our approximation might look like

$$\bar{V}(R|\theta) = \sum_{a\in\mathcal{A}} \left(\theta_{1a}R_a + \theta_{2a}R_a^2\right).$$

Now assume that we feel that the best function might not be quadratic in $R_a$, but we are not sure of the precise form. We might want to estimate a function of the form

$$\bar{V}(R|\theta) = \sum_{a\in\mathcal{A}} \left(\theta_{1a}R_a + \theta_{2a}R_a^{\theta_3}\right).$$

Now we have a function that is nonlinear in the parameter vector $(\theta_1, \theta_2, \theta_3)$, where $\theta_1$ and $\theta_2$ are vectors and $\theta_3$ is a scalar. If we have a training dataset of state-value observations, $(\hat{v}_n, R_n)_{n=1}^N$, we can find $\theta$ by solving

$$\min_\theta \sum_{n=1}^N \left(\hat{v}_n - \bar{V}(R_n|\theta)\right)^2,$$

which generally requires the use of nonlinear programming algorithms. One challenge is that nonlinear optimization problems do not lend themselves to the simple recursive updating equations that we obtained for linear (in the parameters) functions. But more problematic is that we have to experiment with various functional forms to find the one that fits best.

Neural networks offer a much more flexible set of architectures, and at the same time can be updated recursively. The technology has matured to the point that there are a number of commercial packages available which implement the algorithms. However, applying the technology to specific dynamic programming problems can be a nontrivial challenge. In addition, it is not possible to know in advance which problem classes will benefit most from the additional generality in contrast with the simpler strategies that we have covered in this chapter.

Neural networks are, ultimately, a form of statistical model which, for our application, predicts the value of being in a state as a function of the state, using a series of observations of states and values. The simplest neural network is nothing more than a linear regression model. If we are in post-decision state $S_t^x$, we are going to make the random transition $S_{t+1} = S^{M,W}(S_t^x, W_{t+1}(\omega))$ and then observe a random value $\hat{v}_{t+1}$ from being in state $S_{t+1}$. We would like to estimate a statistical function $f_t(S_t^x)$ (the same as $\bar{V}_t(S_t^x)$) that predicts $\hat{v}_{t+1}$. To write this in the traditional notation of regression, let $X_t = S_t^x$ where $X_t = (X_{t1}, X_{t2}, \ldots, X_{tI})$ (we assume that all the components $X_{ti}$ are numerical). If we use a linear model, we might write

$$f_t(X_t) = \theta_0 + \sum_{i=1}^I \theta_i X_{ti}.$$

In the language of neural networks, we have $I$ inputs (we have $I + 1$ parameters since we also include a constant term), which we wish to use to estimate a single output $\hat{v}_{t+1}$ (a random observations of being in a state). The relationships are illustrated in figure 7.6 where we show the $I$ inputs which are then "flowed" along the links to produce $f_t(X_t)$. After this, we then learn the sample realization $\hat{v}_{t+1}$ that we were trying to predict, which allows us to compute the error $\epsilon_{t+1} = \hat{v}_{t+1} - f_t(X_t)$. We would like to find a vector $\theta$ that solves

$$\min_\theta \mathbb{E}\frac{1}{2}(f_t(X_t) - \hat{v}_{t+1})^2.$$

Let $F(\theta) = \mathbb{E}\big(0.5(f_t(X_t) - \hat{v}_{t+1})^2\big)$, and let $F(\theta, \hat{v}_{t+1}(\omega)) = 0.5(f_t(X_t) - \hat{v}_{t+1}(\omega))^2$ where $\hat{v}_{t+1}(\omega)$ is a sample realization of the random variable $\hat{v}_{t+1}(\omega)$. As before, we can solve this iteratively using a stochastic gradient algorithm

$$\theta_{t+1} = \theta_t - \alpha_t \nabla_\theta F(\theta_t, \hat{v}_{t+1}(\omega)),$$

where $\nabla_\theta F(\theta_t, \hat{v}_{t+1}(\omega)) = \epsilon_{t+1}$.

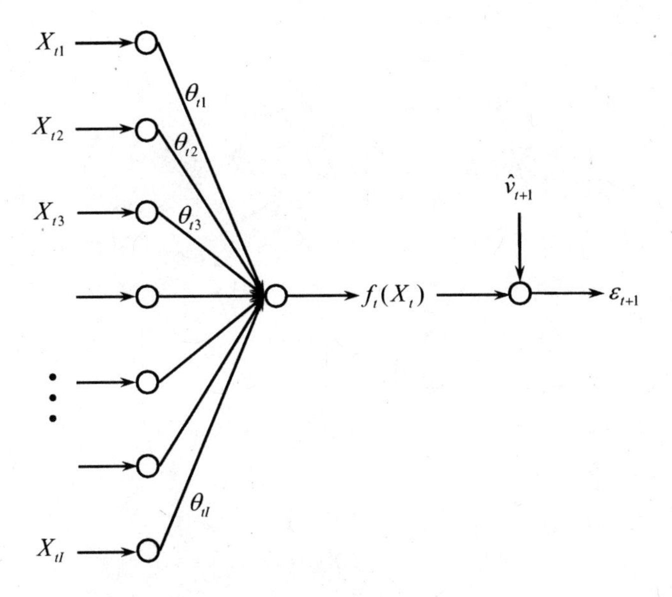


**Figure 7.6** Neural networks with a single layer.

We illustrated our linear model by assuming that the inputs were the individual dimensions of the state variable which we denoted $X_{ti}$. We may not feel that this is the best way to represent the state of the system (imagine representing the states of a Connect-4 game board). We may feel it is more effective (and certainly more compact) if we have access to a set of basis functions $\phi_f(S_t),\ f \in \mathcal{F}$, where $\phi_f(S_t)$ captures a relevant feature of our system. In this case, we would be using our standard basis function representation, where each basis function provides one of the inputs to our neural network.

This was a simple illustration, but it shows that if we have a linear model, we get the same basic class of algorithms that we have already used. A richer model, given in figure 7.7, illustrates a more classical neural network. Here, the "input signal" $X_t$ (this can be the state variable or the set of basis functions) is communicated through several layers. Let $X_t^{(1)} = X_t$ be the input to the first layer (recall that $X_{ti}$ might be the $i^{th}$ dimension of the state variable itself, or a basis function). Let $\mathcal{I}^{(1)}$ be the set of inputs to the first layer (for example, the set of basis functions).

Here, the first linear layer produces $J$ outputs given by

$$Y_{tj}^{(2)} = \sum_{i \in \mathcal{I}^{(1)}} \theta_{ij}^{(1)} X_{ti}^{(1)}, \quad j \in \mathcal{I}^{(2)}.$$

$Y_{tj}^{(2)}$ becomes the input to a nonlinear *perceptron* node which is characterized by a nonlinear function that may dampen or magnify the input. A typical functional form for a perceptron node is the logistics function given by

$$\sigma(y) = \frac{1}{1 + e^{-ay}},$$

where $a$ is a scaling coefficient. The function $\sigma(y)$ is illustrated in figure 7.8. $\sigma(x)$ introduces nonlinear behavior into the communication of the "signal" $X_t$.

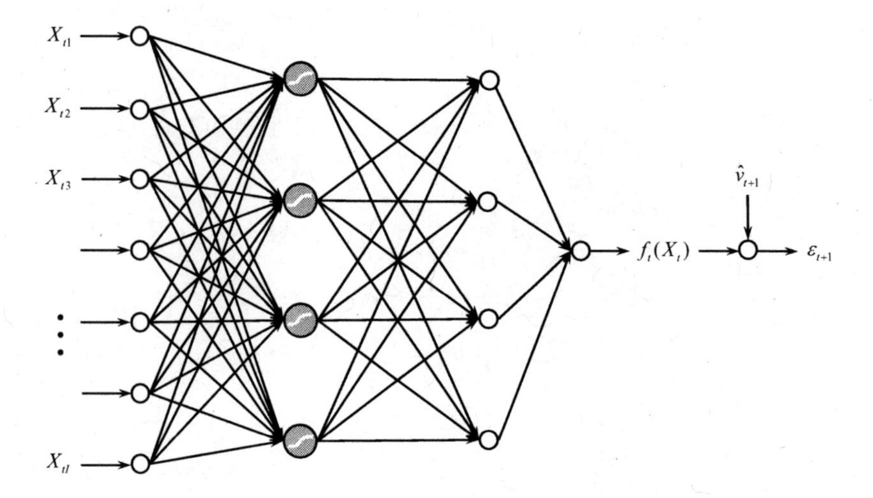


**Figure 7.7** A three-layer neural network.

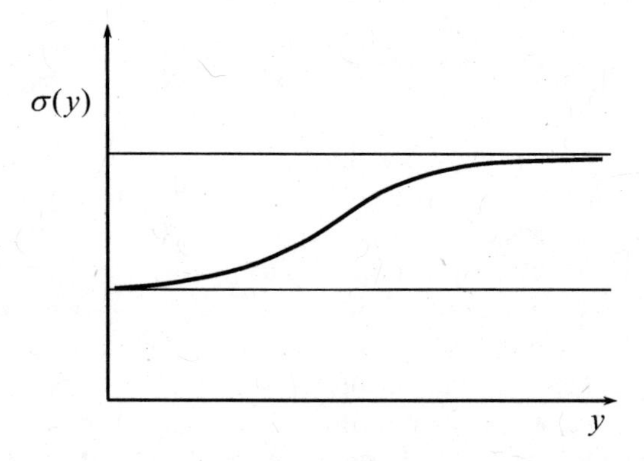


**Figure 7.8** Illustrative logistics function for introducing nonlinear behavior into neural networks.

We next calculate

$$X_{ti}^{(2)} = \sigma(Y_{ti}^{(2)}), \quad i \in \mathcal{I}^{(2)}$$

and use $X_{ti}^{(2)}$ as the input to the second linear layer. We then compute

$$Y_{tj}^{(3)} = \sum_{i \in \mathcal{I}^{(2)}} \theta_{ij}^{(2)} X_{ti}^{(2)}, \quad j \in \mathcal{I}^{(3)}.$$

Finally, we compute the single output using

$$f_t = \sum_{i \in \mathcal{I}^{(3)}} \theta_i^{(3)} X_{ti}^{(3)}.$$

As before, $f_t$ is our estimate of the value of the input $X_t$. This is effectively our value function approximation $\bar{V}_t(S_t^x)$ which we update using the observation $\hat{v}_{t+1}$. We update the parameter vector $\theta = (\theta^{(1)}, \theta^{(2)}, \theta^{(3)})$ using the same stochastic gradient algorithms we used for a single layer network. The only difference is that the derivatives have to capture the fact that changing $\theta^{(1)}$, for example, impacts the "flows" through the rest of the network. The derivatives are slightly more difficult to compute, but the logic is basically the same.

Our presentation above assumes that there is a single output, which is to say that we are trying to match a scalar quantity $\hat{v}_{t+1}$, the observed value of being in a state. In some settings, $\hat{v}_{t+1}$ might be a vector. For example, in chapter 11 (see, in particular, section 11.1), we describe problems where $\hat{v}_{t+1}$ is the gradient of a value function, which of course would be multidimensional. In this case, $f_t$ would also be a vector which would be estimated using

$$f_{tj} = \sum_{i \in \mathcal{I}^{(3)}} \theta_{ij}^{(3)} X_{ti}^{(3)}, \quad j \in \mathcal{I}^{(4)},$$

where $|\mathcal{I}^{(4)}|$ is the dimensionality of the output layer (that is, the dimensionality of $\hat{v}_{t+1}$).

This presentation should be viewed as nothing more than a very simple illustration of an extremely rich field. The advantage of neural networks is that they offer a much richer class of nonlinear functions ("nonlinear architectures" in the language of neural networks) which can be trained in an iterative way, consistent with the needs of approximate dynamic programming. This said, neural networks are no panacea (a statement that can be made about almost anything). As with our simple linear models, the updating mechanisms struggle with scaling (the units of $X_t$ and the units of $\hat{v}_{t+1}$ may be completely different) which has to be handled by the parameter vectors $\theta^{(\ell)}$. Neural networks typically introduce significantly more parameters which can also introduce problems with stability. All the challenges with the choice of stepsizes (section 6.2) still apply. There are significant opportunities for taking advantage of problem structure, which can be reflected both in the design of the inputs (as with the choice of basis functions) as well as the density of the internal networks (these do not have to be dense).

## 7.5 VALUE FUNCTION APPROXIMATION FOR BATCH PROCESSES

A classic problem in dynamic programming arises when customers arrive to a shuttle bus which periodically departs, serving a group of customers all at the same time. We refer to these as batch processes, and they arrive in two forms: positive drift, where customers arrive according to an exogenous process and are served in batches, and negative drift, where an exogenous demand depletes an inventory which then has to be served in batches. Some examples of positive and negative drift problems are given in the examples.

---

**■ EXAMPLE 7.1**

A software company continually updates its software product. From time to time, it ships a new release to its customer base. Most of the costs of shipping an update, which include preparing announcements, posting software on the internet, printing new CD's, and preparing manuals summarizing the new features, are relatively independent of how many changes have been made to the software.

**EXAMPLE 7.2**

E-Z Pass is an automated toll collection system. Users provide a credit card, and the system deducts $25 to provide an initial account balance. This balance is reduced each time the traveler passes through one of the automated toll booths. When the balance goes below a minimum level, another $25 is charged to the credit card to restore the balance.

**EXAMPLE 7.3**

Shipments accumulate at a freight dock where they are loaded onto an outbound truck. Periodically, the truck fills and it is dispatched, but sometimes it is dispatched before it is full to avoid excessive service penalties for holding the shipments.

**EXAMPLE 7.4**

An oil company monitors its total oil reserves, which are constantly drawn down. Periodically, it acquires new reserves either through exploration or by purchasing known reserves owned by another company.

---

We use the context of a shuttle serving customers in batches to illustrate the power of using value function approximations. If all the customers are the same, this is a fairly simple dynamic program that is easy to solve with the techniques of chapter 3, although they are even easier to solve (approximately) using continuous value function approximation. More significantly is that we can, with modest additional effort, provide high quality solutions to problems where there are multiple customer types. Normally, this dimension would explode the size of the state space (a classic example of the curse of dimensionality), but with value function approximations, we can handle this complication with relative ease.

### 7.5.1 A basic model

A standard model of a batch service process

$K$ = The capacity of a batch if it is dispatched.

$R_t$ = The number of customers waiting at time $t$ before a batch has been dispatched.

$A_t$ = A random variable giving the number of customers arriving during time interval $t$.

$x_t$ = 1 if we dispatch the vehicle, 0 otherwise.

$y_t$ = The number of customers departing in a batch at time $t$.

$c^f$ = The cost of dispatching a vehicle.

$c^h$ = The cost per customer left waiting (per time period).

$y_t$ is given by

$$y_t = x_t \min\{K, R_t\}.$$

The transition function is given by

$$\begin{aligned} R_t^x &= R_t - y_t. \\ R_{t+1} &= R_t^x + A_{t+1}. \end{aligned}$$

The cost function (we are minimizing here) is given by

$$C_t(R_t, y_t) = c^f x_t + c^h R_t^x.$$

The value function is governed by Bellman's equation. Using the expectation form (around the pre-decision state variable), this is given by

$$V_t(R_t) = \max_{x_t \in \{0,1\}} \left( C_t(R_t, x_t) + \mathbb{E}\left\{V_{t+1}(R_{t+1})\right\} \right). \tag{7.32}$$

The value function is illustrated in figure 7.9. It turns out it has a few nice properties which

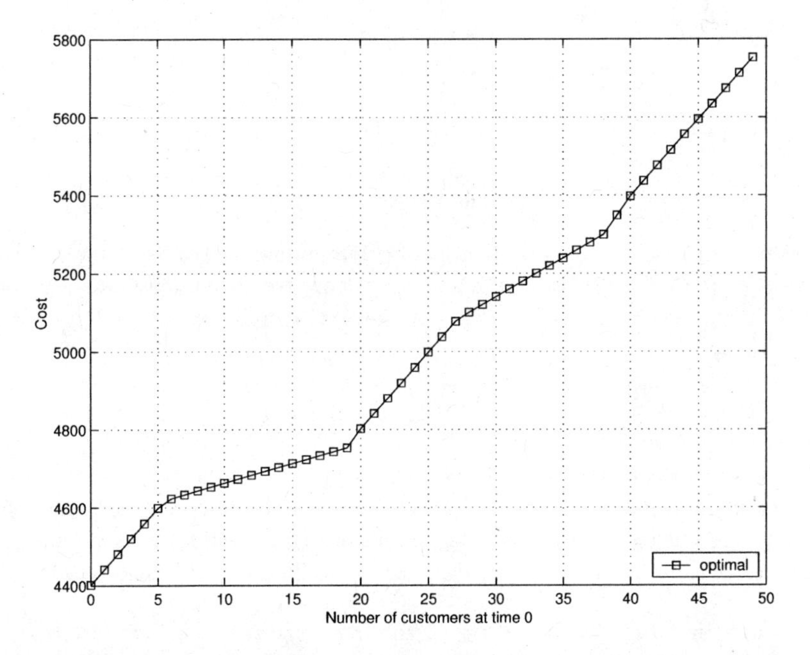


**Figure 7.9** Shape of the value function for the positive-drift batch replenishment problem, from Papadaki & Powell (2003).

can be proven. The first is that it increases monotonically (rather, it never decreases). The second is that it is concave over the range $R \in (nK, (n+1)(K) - 1)$ for $n = 0, 1, \ldots$. The third is that the function is $K$-convex, which means that for $R^+ \geq R^-$ it satisfies

$$V(R^+ + K) - V(R^- + K) \geq V(R^+) - V(R^-). \tag{7.33}$$

For example, if we have a vehicle that can take 20 customers, $K$-convexity means that $V(35) - V(25) \geq V(15) - V(5)$. The value function is convex when measured on a lattice $K$ units apart.

We can exploit these properties in the design of an approximation strategy.

### 7.5.2 Approximating the value function

While finding an optimal policy is nice, we are more interested in obtaining good quality solutions using methods that are scalable to more complex problems. The most important property of the function is that it rises monotonically, suggesting that a linear approximation is likely to work reasonably well. This means that our value function approximation will look like

$$\bar{V}_t(R_t) = \bar{v}_t R_t. \tag{7.34}$$

If we replace the equation in (7.32) and use the linear approximation, we obtain

$$\widetilde{V}_t(R_t) = \max_{x_t \in \{0,1\}} \left( C_t(R_r, x_t) + \bar{v}_t R_t^x \right). \tag{7.35}$$

Solving this is quite easy, since we only have to try two values of $x_t$. We can get an estimate of the slope of the function using

$$\hat{v}_t^n = \widetilde{V}_t(R_t + 1) - \widetilde{V}_t(R_t),$$

which we then smooth to obtain our estimate of $\bar{v}_{t-1}^n$

$$\bar{v}_{t-1}^n = (1 - \alpha_{n-1})\bar{v}_{t-1}^{n-1} + \alpha_{n-1}\hat{v}_t^n.$$

Another strategy is to recognize that the most important part of the curve corresponds to values $0 \leq R \leq K$, over which the function is concave. We can use the techniques of chapter 11 to produce a concave approximation which would more accurately capture the function in this region.

We can compare these approximations to the optimal solution for the scalar case, since this is one of the few problems where we can obtain an optimal solution. The results are shown in figure 7.10 for both the linear and nonlinear (concave) approximations. Note that the linear case works better with fewer iterations. It is easier to estimate a single slope rather than an entire piecewise linear function. As we run more iterations, the nonlinear function works better. For some large scale problems, it may be impossible to run hundreds of iterations (or even dozens). For such applications, a linear approximation is best.

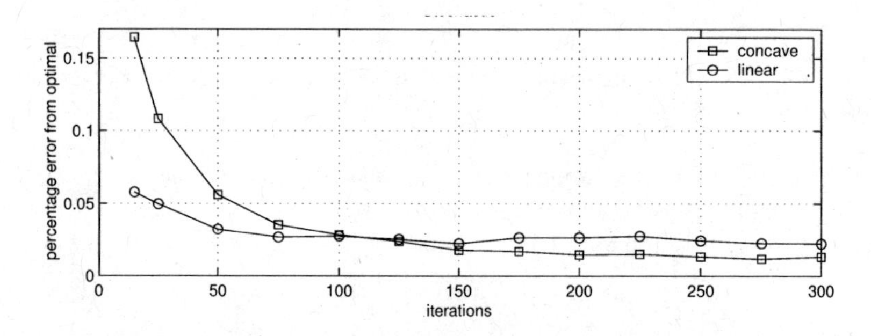


**Figure 7.10** Percent error produced by linear and nonlinear approximations as a function of the training iterations, from Papadaki & Powell (2003).

### 7.5.3 Solving a multiclass problem using linear approximations

Now assume that we have different types of customers arriving to our queue. Using our standard notation, let $R_{ta}$ be the quantity of customers of type $a$ at time $t$ (we continue to use a post-decision state variable). We are going to assume that our attribute is not too big (dozens, hundreds, but not thousands).

For this problem, we still have a scalar decision variable $x_t$ that indicates whether we are dispatching the vehicle or not. However now we have a nontrivial problem of determining how many customers of each type to put on the vehicle. Let

$R_{ta}$ = The number of customers of type $a$ at the end of period $t$,

$A_{ta}$ = The number of arrivals of customers of type $a$ during time interval $t$,

$y_{ta}$ = The number of customers of type $a$ that we put on the vehicle.

We let $R_t$, $A_t$ and $y_t$ be the corresponding vectors over all the customer types. We require that $\sum_{a\in\mathcal{A}} y_{ta} \leq K$. The transition equation is now given by

$$R_{t+1,a} = R_{t,a} - y_{ta} + A_{t+1,a}.$$

For this problem, nonlinear value functions become computationally much more difficult. Our linear value function now looks like

$$\bar{V}_t(R_t) = \sum_{a\in\mathcal{A}} \bar{v}_{ta} R_{ta},$$

which means that we are solving, at each point in time

$$\widetilde{V}_t(R_t) = \max_{x_t\in\{0,1\}} \Big(C_t(R_t, x_t) + \sum_{a\in\mathcal{A}} \bar{v}_{ta} R_{ta}(R_t, x_t)\Big). \tag{7.36}$$

This is characterized by the parameters $(\bar{v}_{ta})_{a\in\mathcal{A}}$, which means we have to estimate one parameter per customer type. We can estimate this parameter by computing a numerical derivative for each customer type. Let $e_a$ be a $|\mathcal{A}|$-dimensional vector of zeroes with a 1 in the $a^{th}$ element. Then compute

$$\hat{v}_{ta} = \widetilde{V}_{t-1}(R_t + e_a) - \widetilde{V}_{t-1}(R_t). \tag{7.37}$$

Equation (7.37) is a sample estimate of the slope, which we then smooth to obtain an updated estimate of the slope

$$\bar{v}^n_{ta} = (1 - \alpha_{n-1})\bar{v}^{n-1}_{ta} + \alpha_{n-1}\hat{v}^n_{ta}. \tag{7.38}$$

Computing $\widetilde{V}_t(R_t + e_a)$ for each product type $a$ can be computationally burdensome in applications with large numbers of product types. It can be approximated by assuming that the decision variable $x_t$ does not change when we increment the resource vector.

Solving equation (7.36) requires that we determine both $x_t$ and $y_t$. We only have two values of $x_t$ to evaluate, so this is not too hard, but how do we determine the vector $y_t$? We need to know something about how the customers differ. Assume that our customers are differentiated by their "value" where the holding cost is a function of value. Given this and given the linear value function approximation, it is not surprising that it is optimal to put as many of the most valuable customers into the vehicle as we can, and then move to the second most valuable, and so on, until we fill up the vehicle. This is how we determine the vector $y_t$.

Given this method of determining $y_t$, finding the best value of $x_t$ is not very difficult, since we simply have to compute the cost for $x_t = 0$ and $x_t = 1$. The next question is, How well does it work? We saw that it worked quite well for the case of a single customer type. With multiple customer types, we are no longer able to find the optimal solution. The

| | Method | linear | linear | linear | linear | linear | linear | linear | linear |
|---|---|---|---|---|---|---|---|---|---|
| | | scalar | scalar | scalar | scalar | mult. | mult. | mult. | mult. |
| | Iterations | (25) | (50) | (100) | (200) | (25) | (50) | (100) | (200) |
| $c^h > c^r/K$ | periodic | 0.602 | 0.597 | 0.591 | 0.592 | 0.633 | 0.626 | 0.619 | 0.619 |
| | aperiodic | 0.655 | 0.642 | 0.639 | 0.635 | 0.668 | 0.660 | 0.654 | 0.650 |
| $c^h \simeq c^r/K$ | periodic | 0.822 | 0.815 | 0.809 | 0.809 | 0.850 | 0.839 | 0.835 | 0.835 |
| | aperiodic | 0.891 | 0.873 | 0.863 | 0.863 | 0.909 | 0.893 | 0.883 | 0.881 |
| $c^h < c^r/K$ | periodic | 0.966 | 0.962 | 0.957 | 0.956 | 0.977 | 0.968 | 0.965 | 0.964 |
| | aperiodic | 0.976 | 0.964 | 0.960 | 0.959 | 0.985 | 0.976 | 0.971 | 0.969 |
| Average | | 0.819 | 0.809 | 0.803 | 0.802 | 0.837 | 0.827 | 0.821 | 0.820 |

**Table 7.6** Costs returned by the value function approximation as a fraction of the costs returned by a dispatch-when-full policy

classic "curse of dimensionality" catches up with us, since we are not able to enumerate all possibilities of the resource vector $R_t$.

As an alternative, we can compare against a sensible dispatch policy. Assume that we are going to dispatch the vehicle whenever it is full, but we will never hold it more than a maximum time $\tau$. Further assume that we are going to test a number of values of $\tau$ and find the one that minimizes total costs. We call this an optimized "dispatch-when-full" (DWF) policy (a bit misleading, since the limit on the holding time means that we may be dispatching the vehicle before it is full).

When testing the policy, it is important to control the relationship of the holding cost $c^h$ to the *average* dispatch cost per customer, $c^r/K$. For $c^h < c^r/K$, the best strategy tends to be to hold the vehicle until it is full. If $c^h > c^r/K$, the best strategy will often be to limit how long the vehicle is held. When $c^h \simeq c^r/K$, the strategy gets more complicated.

A series of simulations (reported in Papadaki & Powell (2003)) were run on datasets with two types of customer arrival patterns: periodic, where the arrivals varied according to a fixed cycle (a period of high arrival rates followed by a period of low arrival rates). The results are given in table 7.6 for both the single and multiproduct problems, where the results are expressed as a fraction of the costs returned by our optimized DWF policy. The results show that the linear approximation always outperforms the DWF policy, even for the case $c^h < c^r/K$ where DWF should be nearly optimal. The multiproduct results are very close to the single product results. The computational complexity of the value function approximation for each forward pass is almost the same as DWF. If it is possible to estimate the value functions off-line, then there is very little additional burden for using an ADP policy.

## 7.6 WHY DOES IT WORK?**

### 7.6.1 Derivation of the recursive estimation equations

Here we derive the recursive estimation equations given by equations (7.20)-(7.23). To begin, we note that the matrix $(X^n)^T X^n$ is an $I+1$ by $I+1$ matrix where the element for row $i$, column $j$ is given by

$$[(X^n)^T X^n]_{i,j} = \sum_{m=1}^{n} x_i^m x_j^m.$$

This term can be computed recursively using

$$[(X^n)^T X^n]_{i,j} = \sum_{m=1}^{n-1} (x_i^m x_j^m) + x_i^n x_j^n.$$

In matrix form, this can be written

$$[(X^n)^T X^n] = [(X^{n-1})^T X^{n-1}] + x^n (x^n)^T.$$

Keeping in mind that $x^n$ is a column vector, $x^n(x^n)^T$ is an $I+1$ by $I+1$ matrix formed by the cross products of the elements of $x^n$. We now use the Sherman-Morrison formula (see section 7.6.2 for a derivation) for updating the inverse of a matrix

$$[A + uu^T]^{-1} = A^{-1} - \frac{A^{-1}uu^T A^{-1}}{1 + u^T A^{-1} u}, \tag{7.39}$$

where $A$ is an invertible $n \times n$ matrix, and $u$ is an $n$-dimensional column vector. Applying this formula to our problem, we obtain

$$\begin{aligned} [(X^n)^T X^n]^{-1} &= [(X^{n-1})^T X^{n-1} + x^n (x^n)^T]^{-1} \\ &= [(X^{n-1})^T X^{n-1}]^{-1} \\ &\quad - \frac{[(X^{n-1})^T X^{n-1}]^{-1} x^n (x^n)^T [(X^{n-1})^T X^{n-1}]^{-1}}{1 + (x^n)^T [(X^{n-1})^T X^{n-1}]^{-1} x^n}. \end{aligned} \tag{7.40}$$

The term $(X^n)^T Y^n$ can also be updated recursively using

$$(X^n)^T Y^n = (X^{n-1})^T Y^{n-1} + x^n (y^n). \tag{7.41}$$

To simplify the notation, let

$$\begin{aligned} B^n &= [(X^n)^T X^n]^{-1}, \\ \gamma^n &= 1 + (x^n)^T [(X^{n-1})^T X^{n-1}]^{-1} x^n. \end{aligned}$$

This simplifies our inverse updating equation (7.40) to

$$B^n = B^{n-1} - \frac{1}{\gamma^n}(B^{n-1} x^n (x^n)^T B^{n-1}).$$

Recall that

$$\bar{\theta}^n = [(X^n)^T X^n]^{-1} (X^n)^T Y^n. \tag{7.42}$$

Combining (7.42) with (7.40) and (7.41) gives

$$\begin{aligned}
\bar{\theta}^n &= [(X^n)^T X^n]^{-1}(X^n)^T Y^n \\
&= \left(B^{n-1} - \frac{1}{\gamma^n}(B^{n-1}x^n(x^n)^T B^{n-1})\right)((X^{n-1})^T Y^{n-1} + x^n y^n), \\
&= B^{n-1}(X^{n-1})^T Y^{n-1} \\
&\quad - \frac{1}{\gamma^n}B^{n-1}x^n(x^n)^T B^{n-1}\left[(X^{n-1})^T Y^{n-1} + x^n y^n\right] + B^{n-1}x^n y^n.
\end{aligned}$$

We can start to simplify by using $\bar{\theta}^{n-1} = B^{n-1}(X^{n-1})^T Y^{n-1}$. We are also going to bring the term $x^n B^{n-1}$ inside the square brackets. Finally, we are going to bring the last term $B^{n-1}x^n y^n$ inside the brackets by taking the coefficient $B^{n-1}x^n$ outside the brackets and multiplying the remaining $y^n$ by the scalar $\gamma^n = 1 + (x^n)^T B^{n-1}x^n$, giving us

$$\begin{aligned}
\bar{\theta}^n &= \bar{\theta}^{n-1} - \frac{1}{\gamma^n}B^{n-1}x^n\left[(x^n)^T(B^{n-1}(X^{n-1})^T Y^{n-1})\right. \\
&\quad \left. + (x^n)^T B^{n-1}x^n y^n - (1 + (x^n)^T B^{n-1}x^n)y^n\right].
\end{aligned}$$

Again, we use $\bar{\theta}^{n-1} = B^{n-1}(X^{n-1})^T Y^{n-1}$ and observe that there are two terms $(x^n)^T B^{n-1}x^n y^n$ that cancel, leaving

$$\bar{\theta}^n = \bar{\theta}^{n-1} - \frac{1}{\gamma^n}B^{n-1}x^n\left((x^n)^T\bar{\theta}^{n-1} - y^n\right).$$

We note that $(\bar{\theta}^{n-1})^T x^n$ is our prediction of $y^n$ using the parameter vector from iteration $n-1$ and the explanatory variables $x^n$. $y^n$ is, of course, the actual observation, so our error is given by

$$\hat{\varepsilon}^n = y^n - (\bar{\theta}^{n-1})^T x^n.$$

Let

$$H^n = -\frac{1}{\gamma^n}B^{n-1}.$$

We can now write our updating equation using

$$\bar{\theta}^n = \bar{\theta}^{n-1} - H^n x^n \hat{\varepsilon}^n. \tag{7.43}$$

### 7.6.2 The Sherman-Morrison updating formula

The Sherman-Morrison matrix updating formula (also known as the Woodbury formula or the Sherman-Morrison-Woodbury formula) assumes that we have a matrix $A$ and that we are going to update it with the outer product of the column vector $u$ to produce the matrix $B$, given by

$$B = A + uu^T. \tag{7.44}$$

Pre-multiply by $B^{-1}$ and post-multiply by $A^{-1}$, giving

$$A^{-1} = B^{-1} + B^{-1}uu^T A^{-1}. \tag{7.45}$$

Post-multiply by $u$

$$\begin{aligned} A^{-1}u &= B^{-1}u + B^{-1}uu^T A^{-1}u \\ &= B^{-1}u\left(1 + u^T A^{-1}u\right). \end{aligned}$$

Note that $u^T A^{-1} u$ is a scalar. Divide through by $\left(1 + u^T A^{-1}u\right)$

$$\frac{A^{-1}u}{(1 + u^T A^{-1}u)} = B^{-1}u.$$

Now post-multiply by $u^T A^{-1}$

$$\frac{A^{-1}uu^T A^{-1}}{(1 + u^T A^{-1}u)} = B^{-1}uu^T A^{-1}. \tag{7.46}$$

Equation (7.45) gives us

$$B^{-1}uu^T A^{-1} = A^{-1} - B^{-1}. \tag{7.47}$$

Substituting (7.47) into (7.46) gives

$$\frac{A^{-1}uu^T A^{-1}}{(1 + u^T A^{-1}u)} = A^{-1} - B^{-1}. \tag{7.48}$$

Solving for $B^{-1}$ gives us

$$\begin{aligned} B^{-1} &= [A + uu^T]^{-1} \\ &= A^{-1} - \frac{A^{-1}uu^T A^{-1}}{(1 + u^T A^{-1}u)}, \end{aligned}$$

which is the desired formula.

## 7.7 BIBLIOGRAPHIC NOTES

The use of regression models to approximate value functions dates to the origins of dynamic programming.

Sections 7.1.2-7.1.3 - Aggregation has been a widely used technique in dynamic programming as a method to overcome the curse of dimensionality. Early work focused on picking a fixed level of aggregation (Whitt (1978), Bean et al. (1987)), or using adaptive techniques that change the level of aggregation as the sampling process progresses (Bertsekas & Castanon (1989), Mendelssohn (1982), Bertsekas & Tsitsiklis (1996)), but which still use a fixed level of aggregation at any given time. Much of the literature on aggregation has focused on deriving error bounds (Zipkin (1980*b*), Zipkin (1980*a*)). A recent discussion of aggregation in dynamic programming can be found in Lambert et al. (2002). For a good discussion of aggregation as a general technique in modeling, see Rogers et al. (1991). The material in section 7.1.3 is based on George et al. (2005) and George & Powell (2006). LeBlanc & Tibshirani (1996) and Yang (2001) provide excellent discussions of mixing estimates from different sources. For a discussion of soft state aggregation, see Singh et al. (1995).

Section 7.2 - Basis functions have their roots in the modeling of physical processes. A good introduction to the field from this setting is Heuberger et al. (2005). Schweitzer & Seidmann (1985) describes generalized polynomial approximations for Markov decision processes for use in value iteration, policy iteration and the linear programming method. Menache et al. (2005) discusses basis function adaptations in the context of reinforcement learning. For a very nice discussion of the use of basis functions in approximate dynamic programming, see Tsitsiklis & Van Roy (1996) and Van Roy (2001). Tsitsiklis & Van Roy (1997) proves convergence of iterative stochastic algorithms for fitting the parameters of a regression model when the policy is held fixed.

Section 7.2.2 - The first use of approximate dynamic programming for evaluating an American call option is given in Longstaff & Schwartz (2001), but the topic has been studied for decades (see Taylor (1967)). Tsitsiklis & Van Roy (2001) also provide an alternative ADP algorithm for American call options. Clement et al. (2002) provides formal convergence results for regression models used to price American options.

Section 7.2.3 - This presentation is based on Tsitsiklis & Van Roy (1997).

Section 7.2.4 - An excellent introduction to continuous approximation techniques is given in Judd (1998) in the context of economic systems and computational dynamic programming. Ormoneit & Sen (2002) and Ormoneit & Glynn (2002) discuss the use of kernel-based regression methods in an approximate dynamic programming setting, providing convergence proofs for specific algorithmic strategies. For a thorough introduction to locally polynomial regression methods, see Fan (1996). An excellent discussion of a broad range of statistical learning methods can be found in Hastie et al. (2001).

Section 7.3 - Ljung & Soderstrom (1983) and Young (1984) provide nice treatments of recursive statistics. Lagoudakis et al. (2002) and Bradtke & Barto (1996) present least squares methods in the context of reinforcement learning (artificial intelligence). Choi & Van Roy (2006) uses the Kalman filter to perform scaling for stochastic gradient updates, avoiding the scaling problems inherent in stochastic gradient updates such as equation (7.19). Nediç & Bertsekas (2003) describes the use of least squares equation with a linear (in the parameters) value function approximation using policy iteration and proves convergence for TD($\lambda$) with general $\lambda$. Bertsekas et al. (2004) presents a scaled method for estimating linear value function approximations within a temporal differencing algorithm. Section 7.3.5 is based on Soderstrom et al. (1978).

Section 7.4 - Bertsekas & Tsitsiklis (1996) provides an excellent discussion of neural networks in the context of approximate dynamic programming. Haykin (1999) presents a much more in-depth presentation of neural networks, including a chapter on approximate dynamic programming using neural networks.

Section 7.5 - The study of batch processes is a classical problem in operations research and dynamic programming (see, for example, Arrow et al. (1951), Whitin (1953), Dvoretzky et al. (1952), and Bellman et al. (1955) as well as Porteus (1990)), but the study of this problem using approximate dynamic programming is surprisingly recent. Adelman (2004) describes an application of approximate dynamic programming to a problem class known as the inventory routing problem, or more broadly as the joint replenishment problem, where a vehicle replenishes multiple inventories as part of a tour. The presentation in this section is based on Papadaki & Powell (2002) and Papadaki & Powell (2003).

Section 7.6.2 - The Sherman-Morrison updating formulas are given in a number of references, such as Ljung & Soderstrom (1983) and Golub & Loan (1996).

## PROBLEMS

**7.1** In a spreadsheet, create a $4 \times 4$ grid where the cells are numbered 1, 2, ..., 16 starting with the upper left-hand corner and moving left to right, as shown below. We are

| 1 | 2 | 3 | 4 |
|---|---|---|---|
| 5 | 6 | 7 | 8 |
| 9 | 10 | 11 | 12 |
| 13 | 14 | 15 | 16 |

going to treat each number in the cell as the mean of the observations drawn from that cell. Now assume that if we observe a cell, we observe the mean plus a random variable that is uniformly distributed between $-1$ and $+1$. Next define a series of aggregation where aggregation 0 is the disaggregate level, aggregation 1 divides the grid into four $2 \times 2$ cells, and aggregation 2 aggregates everything into a single cell. After $n$ iterations, let $\bar{v}_a^{(g,n)}$ be the estimate of cell "$a$" at the $n^{th}$ level of aggregation, and let

$$\bar{v}_a^n = \sum_{g \in \mathcal{G}} w_a^{(g)} \bar{v}_a^{(g,n)}$$

be your best estimate of cell $a$ using a weighted aggregation scheme. Compute an overall error measure using

$$(\bar{\sigma}^2)^n = \sum_{a \in \mathcal{A}} (\bar{v}_a^n - \nu_a)^2,$$

where $\nu_a$ is the true value (taken from your grid) of being in cell $a$. Also let $w^{(g,n)}$ be the average weight after $n$ iterations given to the aggregation level $g$ when averaged over all cells at that level of aggregation (for example, there is only one cell for $w^{(2,n)}$). Perform 1000 iterations where at each iteration you randomly sample a cell and measure it with noise. Update your estimates at each level of aggregation, and compute the variance of your estimate with and without the bias correction.

(a) Plot $w^{(g,n)}$ for each of the three levels of aggregation at each iteration. Do the weights behave as you would expect? Explain.

(b) For each level of aggregation, set the weight given to that level equal to one (in other words, we are using a single level of aggregation) and plot the overall error as a function of the number of iterations.

(c) Add to your plot the average error when you use a weighted average, where the weights are determined by equation (7.7) without the bias correction.

(d) Finally add to your plot the average error when you used a weighted average, but now determine the weights by equation (7.8), which uses the bias correction.

(e) Repeat the above assuming that the noise is uniformly distributed between $-5$ and $+5$.

**7.2** Show that

$$\sigma_a^2 = (\sigma_a^2)^{(g)} + (\beta_a^{(g)})^2 \tag{7.49}$$

which breaks down the total variation in an estimate at a level of aggregation is the sum of the variation of the observation error plus the bias squared.

**7.3** Show that the matrix $H^n$ in the recursive updating formula from equation (7.43)

$$\bar{\theta}^n = \bar{\theta}^{n-1} - H^n x^n \hat{\varepsilon}^n$$

reduces to $H^n = 1/n$ for the case of a single parameter (which means we are using $Y$ =constant, with no independent variables).

**7.4** An airline has to decide when to bring an aircraft in for a major engine overhaul. Let $s_t$ represent the state of the engine in terms of engine wear, and let $d_t$ be a nonnegative amount by which the engine deteriorates during period $t$. At the beginning of period $t$, the airline may decide to continue operating the aircraft ($z_t = 0$) or to repair the aircraft ($z_t = 1$) at a cost of $c^R$, which has the effect of returning the aircraft to $s_{t+1} = 0$ . If the airline does not repair the aircraft, the cost of operation is $c^o(s_t)$, which is a nondecreasing, convex function in $s_t$.

(a) Define what is meant by a control limit policy in dynamic programming, and show that this is an instance of a monotone policy.

(b) Formulate the one-period reward function $C_t(s_t, z_t)$, and show that it is submodular.

(c) Show that the decision rule is monotone in $s_t$. (Outline the steps in the proof, and then fill in the details.)

(d) Assume that a control limit policy exists for this problem, and let $\gamma$ be the control limit. Now, we may write $C_t(s_t, z_t)$ as a function of one variable: the state $s$. Using our control limit structure, we can write the decision $z_t$ as the decision rule $z^\pi(s_t)$. Illustrate the shape of $C_t(s_t, z^\pi(s))$ by plotting it over the range $0 \leq s \leq 3\gamma$ (in theory, we may be given an aircraft with $s > \gamma$ initially).

**7.5** A dispatcher controls a finite capacity shuttle that works as follows: In each time period, a random number $A_t$ arrives. After the arrivals occur, the dispatcher must decide whether to call the shuttle to remove up to $M$ customers. The cost of dispatching the shuttle is $c$, which is independent of the number of customers on the shuttle. Each time period that a customer waits costs $h$. If we let $z = 1$ if the shuttle departs and 0 otherwise, then our one-period reward function is given by

$$c_t(s, z) = cz + h[s - Mz]^+,$$

where $M$ is the capacity of the shuttle. Show that $c_t(s, a)$ is submodular where we would like to minimize $r$. Note that we are representing the state of the system after the customers arrive.

**7.6** Assume that a control limit policy exists for our shuttle problem in exercise 2 that allows us to write the optimal dispatch rule as a function of $s$, as in $z^{\pi}(s)$. We may write $r(s, z)$ as a function of one variable, the state $s$.

(a) Illustrate the shape of $r(s, z(s))$ by plotting it over the range $0 < s < 3M$ (since we are allowing there to be more customers than can fill one vehicle, assume that we are allowed to send $z = 0, 1, 2, \ldots$ vehicles in a single time period).

(b) Let $c = 10$, $h = 2$, and $M = 5$, and assume that $A_t = 1$ with probability 0.6 and is 0 with probability 0.4. Set up and solve a system of linear equations for the optimal value function for this problem in steady state.

**7.7** A general aging and replenishment problem arises as follows: Let $s_t$ be the "age" of our process at time $t$. At time $t$, we may choose between a decision $d = C$ to continue the process, incurring a cost $g(s_t)$ or a decision $d = R$ to replenish the process, which incurs a cost $K + g(0)$. Assume that $g(s_t)$ is convex and increasing. The state of the system evolves according to

$$s_{t+1} = \begin{cases} s_t + D_t & \text{if } d = C, \\ 0 & \text{if } d = R, \end{cases}$$

where $D_t$ is a nonnegative random variable giving the degree of deterioration from one epoch to another (also called the "drift").

(a) Prove that the structure of this policy is monotone. Clearly state the conditions necessary for your proof.

(b) How does your answer to part (1) change if the random variable $D_t$ is allowed to take on negative outcomes? Give the weakest possible conditions on the distribution of required to ensure the existence of a monotone policy.

(c) Now assume that the action is to reduce the state variable by an amount $q \leq s_t$ at a cost of $cq$ (instead of $K$ ). Further assume that $g(s) = as^2$ . Show that this policy is also monotone. Say as much as you can about the specific structure of this policy.

# CHAPTER 8

# ADP FOR FINITE HORIZON PROBLEMS

In chapter 4 we provided a very general introduction to approximate dynamic programming. Here, we continue this presentation but provide a much more in-depth presentation of different ADP algorithms that have been developed over the years. In this chapter, we focus purely on finite horizon problems. Since most authors present this material in the context of infinite horizon problems, a word of explanation is in order.

Our justification for starting with a finite horizon framework is based on pedagogical, practical, and theoretical reasons. Pedagogically, finite horizon problems require a more careful modeling of the dynamics of the problem; stationary models allow us to simplify the modeling by ignoring time indices, but this hides the modeling of the time dependencies of decisions and information. By starting with a finite horizon model, we are forced to learn how to model the evolution of information and decisions.

More practically, finite horizon models are the natural framework for a vast array of operational problems where the data is nonstationary and/or where the important behavior falls in the initial transient phase of the problem. Even when a problem is stationary, the decision of what to do *now* depends on value function approximations that often depend on the initial starting state. If we were able to compute an exact value function, we would be able to use this value function for any starting state. For more complex problems where we have to depend on approximations, the best approximation may quite easily depend on the initial state.

The theoretical justification is that certain convergence proofs depend on our ability to obtain unbiased sample estimates of the value of being in a state by following a path into the future. With finite horizon models, we only have to follow the path to the end of

*Approximate Dynamic Programming*. By Warren B. Powell

the horizon. With infinite horizon problems, authors either assume the path is infinite or depend on the presence of zero-cost, absorbing states.

## 8.1 STRATEGIES FOR FINITE HORIZON PROBLEMS

In this section, we sketch the basic strategies for using forward dynamic programming methods to approximate policies and value functions. The techniques are most easily illustrated using a post-decision state variable, after which we describe the modifications needed to handle a pre-decision state variable. We then describe a strategy known as *Q-learning* that is very popular in the research literature.

### 8.1.1 Single-pass procedures

There are two simple strategies for estimating the value function for finite horizon problems. The first uses a single-pass procedure. Here, we step forward in time using an approximation of the value function from the previous iteration. Assume we are in state $S_t^n$. We would find $\hat{v}_t^n$ using

$$\hat{v}_t^n = \max_{x_t \in \mathcal{X}_t^n} \left( C_t(S_t^n, x_t) + \gamma \bar{V}_t^{n-1}(S^{M,x}(S_t^n, x_t)) \right).$$

The next step would be to update the value function approximation. Before, we updated the value of being in state $S_t^n$ using

$$\bar{V}_{t-1}^n(S_{t-1}^{x,n}) = (1 - \alpha_{n-1})\bar{V}_{t-1}^{n-1}(S_{t-1}^{x,n}) + \alpha_{n-1}\hat{v}_t^n.$$

This is fine if we are using a lookup-table representation of a value function. However, by now we realize that there is a rich array of methods for approximating a value function (hierarchical aggregation, linear regression, neural networks), each of which involve their own updating methods. For this reason, we are going to represent the process of updating the value function using

$$\bar{V}_{t-1}^n \leftarrow U^V(\bar{V}_{t-1}^{n-1}, S_{t-1}^{x,n}, \hat{v}_t^n).$$

Although we are updating the value function around state $S_{t-1}^{x,n}$, we may be updating the entire value function.

We refer to this basic strategy as a *single-pass algorithm*, which is illustrated in figure 8.1. Recall (from chapter 4) that choosing a sample path may mean nothing more than choosing a random number seed in our random number generator, or it may involve choosing a sample of realizations from history. In this implementation we are approximating the value function around the post-decision state, which is why $\hat{v}_t^n$ is used to update $\bar{V}_{t-1}^{n-1}$. These updates occur as we step forward in time, so that by the time we reach the end of our horizon, we are done.

Figure 8.1 is known as a single-pass procedure because all the calculations are finished at the end of each forward pass. The updates of the value function take place as the algorithm progresses forward in time. The algorithm is fairly easy to implement, but may not provide the fastest convergence.

### 8.1.2 Double-pass procedures

As an alternative, we can use a *double-pass* procedure, which is illustrated in figure 8.2. In this version we step forward through time creating a trajectory of states, actions, and

---

**Step 0.** Initialization:

**Step 0a.** Initialize $\bar{V}_t^0, \ t \in \mathcal{T}$.

**Step 0b.** Set $n = 1$.

**Step 0c.** Initialize $S_0^1$.

**Step 1.** Choose a sample path $\omega^n$.

**Step 2.** Do for $t = 0, 1, 2, \ldots, T$:

**Step 2a.** Solve:

$$\hat{v}_t^n = \max_{x_t \in \mathcal{X}_t^n} \left( C_t(S_t^n, x_t) + \gamma \bar{V}_t^{n-1}(S^{M,x}(S_t^n, x_t)) \right) \tag{8.1}$$

and let $x_t^n$ be the value of $x_t$ that solves (8.1).

**Step 2b.** If $t > 0$, update the value function:

$$\bar{V}_{t-1}^n \leftarrow U^V(\bar{V}_{t-1}^{n-1}, S_{t-1}^{x,n}, \hat{v}_t^n).$$

**Step 2c.** Update the states:

$$\begin{aligned} S_t^{x,n} &= S^{M,x}(S_t^n, x_t^n), \\ S_{t+1}^n &= S^{M,W}(S_t^{x,n}, W_{t+1}(\omega^n)). \end{aligned}$$

**Step 3.** Increment $n$. If $n \leq N$ go to Step 1.

**Step 4.** Return the value functions $(\bar{V}_t^N)_{t=1}^T$.

---

**Figure 8.1** Single-pass version of the approximate dynamic programming algorithm.

outcomes. We then step backwards through time, updating the value of being in a state using information from the same trajectory in the future.

The idea of stepping backward through time to produce an estimate of the value of being in a state was first introduced in the control theory community under the name of *backpropagation through time* (BTT). The result of our backward pass is $\hat{v}_t^n$, which is the contribution from the sample path $\omega^n$ and a particular policy. Our policy is, quite literally, the set of decisions produced by the value function approximation $\bar{V}^{n-1}$. Unlike our forward-pass algorithm (where $\hat{v}_t^n$ depends on the approximation $\bar{V}_t^{n-1}(S_t^x)$), $\hat{v}_t^n$ is a valid, unbiased estimate of the value of being in state $S_t^n$ at time $t$ and following the policy produced by $\bar{V}^{n-1}$. Of course, since $\bar{V}^{n-1}$ is a biased estimate of the value of being in each state, our policy is not an optimal policy. But as a rule, it tends to produce better estimates than the forward-pass algorithm.

These two strategies are easily illustrated using our simple asset selling problem. For this illustration, we are going to slightly simplify the model we provided earlier, where we assumed that the change in price, $\hat{p}_t$, was the exogenous information. If we use this model, we have to retain the price $p_t$ in our state variable (even the post-decision state variable). For our illustration, we are going to assume that the exogenous information is the price itself, so that $p_t = \hat{p}_t$. We further assume that $\hat{p}_t$ is independent of all previous prices (a pretty strong assumption). For this model, the pre-decision state is $S_t = (R_t, p_t)$ while the post-decision state variable is simply $S_t^x = R_t^x = R_t - x_t$ which indicates whether we are holding the asset or not. Further, $S_{t+1} = S_t^x$ since the resource transition function is deterministic.

---

**Step 0.** Initialization:

**Step 0a.** Initialize $\bar{V}_t^0, \ t \in \mathcal{T}$.

**Step 0b.** Initialize $S_0^1$.

**Step 0c.** Set $n = 1$.

**Step 1.** Choose a sample path $\omega^n$

**Step 2:** Do for $t = 0, 1, 2, \ldots, T$:

**Step 2a:** Find

$$x_t^n = \arg\max_{x_t \in \mathcal{X}_t^n} \left( C_t(S_t^n, x_t) + \gamma \bar{V}_t^{n-1}(S^{M,x}(S_t^n, x_t)) \right)$$

**Step 2b:** Update the states

$$S_t^{x,n} = S^{M,x}(S_t^n, x_t^n),$$
$$S_{t+1}^n = S^{M,W}(S_t^{x,n}, W_{t+1}(\omega^n)).$$

**Step 3:** Do for $t = T, T-1, \ldots, 1$:

**Step 3a:** Compute $\hat{v}_t^n$ using the decision $x_t^n$ from the forward pass:

$$\hat{v}_t^n = C_t(S_t^n, x_t^n) + \gamma \hat{v}_{t+1}^n.$$

**Step 3b.** Update the value function approximations:

$$\bar{V}_{t-1}^n \leftarrow U^V(\bar{V}_{t-1}^{n-1}, S_{t-1}^{x,n}, \hat{v}_t^n).$$

**Step 4.** Increment $n$. If $n \leq N$ go to Step 1.

**Step 5.** Return the value functions $(\bar{V}_t^n)_{t=1}^T$.

---

**Figure 8.2** Double-pass version of the approximate dynamic programming algorithm.

With this model, a single-pass algorithm (figure 8.1) is performed by stepping forward through time, $t = 1, 2, \ldots, T$. At time $t$, we first sample $\hat{p}_t$ and we find

$$\hat{v}_t^n = \max_{x_t \in \{0,1\}} \left( \hat{p}_t^n x_t + (1 - x_t)(-c_t + \bar{v}_t^{n-1}) \right). \tag{8.2}$$

Assume that the holding cost $c_t = 2$ for all time periods.

Table 8.1 illustrates three iterations of a single-pass algorithm for a three-period problem. We initialize $\bar{v}_t^0 = 0$ for $t = 0, 1, 2, 3$. Our first decision is $x_1$ after we see $\hat{p}_1$. The first column shows the iteration counter, while the second shows the stepsize $\alpha_{n-1} = 1/n$. For the first iteration, we always choose to sell because $\bar{v}_t^0 = 0$, which means that $\hat{v}_t^1 = \hat{p}_t^1$. Since our stepsize is 1.0, this produces $\bar{v}_{t-1}^1 = \hat{p}_t^1$ for each time period.

In the second iteration, our first decision problem is

$$\begin{aligned} \hat{v}_1^2 &= \max\{\hat{p}_1^2, -c_1 + \bar{v}_1^1\} \\ &= \max\{24, -2 + 34\} \\ &= 32, \end{aligned}$$

which means $x_1^2 = 0$ (since we are holding). We then use $\hat{v}_1^2$ to update $\bar{v}_0^2$ using

$$\begin{aligned} \bar{v}_0^2 &= (1 - \alpha_1)\bar{v}_0^1 + \alpha_1 \hat{v}_1^1 \\ &= (0.5)30.0 + (0.5)32.0 \\ &= 31.0 \end{aligned}$$

| | | $t=0$ | $t=1$ | | | | $t=2$ | | | | $t=3$ | | | |
|---|---|---|---|---|---|---|---|---|---|---|---|---|---|---|
| Iteration | $\alpha_{n-1}$ | $\bar{v}_0$ | $\hat{v}_1$ | $\hat{p}_1$ | $x_1$ | $\bar{v}_1$ | $\hat{v}_2$ | $\hat{p}_2$ | $x_2$ | $\bar{v}_2$ | $\hat{v}_3$ | $\hat{p}_3$ | $x_3$ | $\bar{v}_3$ |
| 0 | | 0 | | | | 0 | | | | 0 | | | | 0 |
| 1 | 1 | 30 | 30 | 30 | 1 | 34 | 34 | 34 | 1 | 31 | 31 | 31 | 1 | 0 |
| 2 | 0.50 | 31 | 32 | 24 | 0 | 31.5 | 29 | 21 | 0 | 29.5 | 30 | 30 | 1 | 0 |
| 3 | 0.3 | 32.3 | 35 | 35 | 1 | 30.2 | 27.5 | 24 | 0 | 30.7 | 33 | 33 | 1 | 0 |

**Table 8.1** Illustration of a single-pass algorithm

| | | $t=0$ | $t=1$ | | | | $t=2$ | | | | $t=3$ | | | |
|---|---|---|---|---|---|---|---|---|---|---|---|---|---|---|
| Iteration | Pass | $\bar{v}_0$ | $\hat{v}_1$ | $\hat{p}_1$ | $x_1$ | $\bar{v}_1$ | $\hat{v}_2$ | $\hat{p}_2$ | $x_2$ | $\bar{v}_2$ | $\hat{v}_3$ | $\hat{p}_3$ | $x_3$ | $\bar{v}_3$ |
| 0 | | 0 | | | | 0 | | | | 0 | | | | 0 |
| 1 | Forward | $\rightarrow$ | $\rightarrow$ | 30 | 1 | $\rightarrow$ | $\rightarrow$ | 34 | 1 | $\rightarrow$ | $\rightarrow$ | 31 | 1 | |
| 1 | Back | 30 | 30 | $\leftarrow$ | $\leftarrow$ | 34 | 34 | $\leftarrow$ | $\leftarrow$ | 31 | 31 | $\leftarrow$ | $\leftarrow$ | 0 |
| 2 | Forward | $\rightarrow$ | $\rightarrow$ | 24 | 0 | $\rightarrow$ | $\rightarrow$ | 21 | 0 | $\rightarrow$ | $\rightarrow$ | 27 | 1 | |
| 2 | Back | 26.5 | 23 | $\leftarrow$ | $\leftarrow$ | 29.5 | 25 | $\leftarrow$ | $\leftarrow$ | 29 | 27 | $\leftarrow$ | $\leftarrow$ | 0 |

**Table 8.2** Illustration of a double-pass algorithm

Repeating this logic, we hold again for $t = 2$ but we always sell at $t = 3$ since this is the last time period. In the third pass, we again sell in the first time period, but hold for the second time period.

It is important to realize that this problem is quite simple, and as a result we are not quite following the algorithm in figure 8.1. If we sell, we are no longer holding the asset and the forward pass should stop (more precisely, we should continue to simulate the process given that we have sold the asset). Instead, even if we sell the asset, we step forward in time and continue to evaluate the state that we are holding the asset (the value of the state where we are not holding the asset is, of course, zero). Normally, we evaluate only the states that we transition to (see step 2b), but for this problem, we are actually visiting all the states (since there is, in fact, only one state that we really need to evaluate).

Now consider a double-pass algorithm. Table 8.2 illustrates the forward pass, followed by the backward pass. Each line of the table only shows the numbers determined during the forward or backward pass. In the first pass, we always sell (since the value of the future is zero), which means that at each time period the value of holding the asset is the price in that period.

In the second pass, it is optimal to hold for two periods until we sell in the last period. The value $\hat{v}_t^2$ for each time period is the contribution of the rest of the trajectory which, in this case, is the price we receive in the last time period. So, since $x_1 = x_2 = 0$ followed by $x_3 = 1$, the value of holding the asset at time 3 is the \$27 price we receive for selling in that time period. The value of holding the asset at time $t = 2$ is the holding cost of -2 plus $\hat{v}_3^2$, giving $\hat{v}_2^2 = -2 + \hat{v}_3^2 = -2 + 27 = 25$. Similarly, holding the asset at time 1 means $\hat{v}_1^2 = -2 + \hat{v}_2^2 = -2 + 25 = 23$. The smoothing of $\hat{v}_t^n$ with $\bar{v}_{t-1}^{n-1}$ to produce $\bar{v}_{t-1}^n$ is the same as for the single pass algorithm.

The value of implementing the double-pass algorithm depends on the problem. For example, imagine that our asset is an expensive piece of replacement equipment for a jet aircraft. We hold the part in inventory until it is needed, which could literally be years for certain parts: This means there could be hundreds of time periods (if each time period is a day) where we are holding the part. Estimating the value of the part now (which would determine whether we order the part to hold in inventory) using a single-pass algorithm could produce extremely slow convergence. A double-pass algorithm would work dramatically better. But if the part is used frequently, staying in inventory for only a few days, then the single-pass algorithm will work fine.

### 8.1.3 Value iteration using a pre-decision state variable

Our presentation up to now has focused on using a post-decision state variable, which gives us a much simpler process of finding decisions. If we use a pre-decision state variable, we make decisions by solving

$$\hat{v}_t^n = \max_{x_t} \left( C(S_t^n, x_t) + \gamma \mathbb{E}\{V_{t+1}(S_{t+1})\} \right), \tag{8.3}$$

where $S_{t+1} = S^M(S_t^n, x_t, W_{t+1})$, and $S_t^n$ is the state that we are in at time $t$, iteration $n$. One reason to use the pre-decision state variable is that for some problems, computing the expectation is easy. If this is not the case, then we have to approximate the expectation. For example, we might use

$$\hat{v}_t^n = \max_{x_t \in \mathcal{X}_t^n} \left( C(S_t^n, x_t) + \gamma \sum_{\hat{\omega} \in \hat{\Omega}^n} p^n(\hat{\omega}) \bar{V}_{t+1}^{n-1}(S^M(S_t^n, x_t, W_{t+1}(\hat{\omega}))) \right). \tag{8.4}$$

Either way, we update the value of being in state $S_t^n$ using

$$\bar{V}_t^n(S_t^n) = (1 - \alpha_{n-1}) \bar{V}_t^{n-1}(S_t^n) + \alpha_{n-1} \hat{v}_t^n.$$

Keep in mind that if we can compute an expectation (or if we approximate it using a large sample $\hat{\Omega}$), then the stepsize should be much larger than when we are using a single sample realization (as we did with the post-decision formulation).

An outline of the overall algorithm is given by figure 8.3.

## 8.2 $Q$-LEARNING

Return for the moment to the classical way of making decisions using dynamic programming. Normally we would want to solve

$$x_t^n = \arg\max_{x_t \in \mathcal{X}_t^n} \left\{ C_t(S_t^n, x_t) + \gamma \mathbb{E} \bar{V}_{t+1}^{n-1}\left(S_{t+1}(S_t^n, x_t, W_{t+1})\right) \right\}. \tag{8.5}$$

Solving equation (8.5) can be problematic for two different reasons. The first is that we may not know the underlying distribution of the exogenous information process. If we do not know the probability of an outcome, then we cannot compute the expectation. These are problems where we do not have a model of the information process. The second reason is that while we may know the probability distribution, the expectation

---

**Step 0.** Initialization:

**Step 0a.** Initialize $\bar{V}_t^0, \ t \in \mathcal{T}$.

**Step 0b.** Set $n = 1$.

**Step 0c.** Initialize $S^0$.

**Step 1.** Sample $\omega^n$.

**Step 2.** Do for $t = 0, 1, \ldots, T$:

**Step 2a:** Choose $\hat{\Omega}^n \subseteq \Omega$ and solve:

$$\hat{v}_t^n = \max_{x_t \in \mathcal{X}_t^n} \left( C_t(S_t^{n-1}, x_t) + \gamma \sum_{\hat{\omega} \in \hat{\Omega}^n} p^n(\hat{\omega}) \bar{V}_{t+1}^{n-1}(S^M(S_t^{n-1}, x_t, W_{t+1}(\hat{\omega}))) \right)$$

and let $x_t^n$ be the value of $x_t$ that solves the maximization problem.

**Step 2b:** Compute:

$$S_{t+1}^n = S^M(S_t^n, x_t^n, W_{t+1}(\omega^n)).$$

**Step 2c.** Update the value function:

$$\bar{V}_t^n \leftarrow U^V(\bar{V}_t^{n-1}, S_t^n, \hat{v}_t^n)$$

**Step 3.** Increment $n$. If $n \leq N$, go to Step 1.

**Step 4.** Return the value functions $(\bar{V}_t^n)_{t=1}^T$.

---

**Figure 8.3** Approximate dynamic programming using a pre-decision state variable.

may be computationally intractable. This typically arises when the information process is characterized by a vector of random variables.

In the previous section, we circumvented this problem by approximating the expectation by using a subset of outcomes (see equation (8.4)), but this can be computationally clumsy for many problems. One thought is to solve the problem for a single sample realization

$$x_t^n = \arg\max_{x_t \in \mathcal{X}_t^n} \left( C_t(S_t^n, x_t) + \gamma \bar{V}_{t+1}^{n-1}(S_{t+1}(S_t^n, x_t, W_{t+1}(\omega^n))) \right). \tag{8.6}$$

The problem is that this means we are choosing $x_t$ for a particular realization of the future information $W_{t+1}(\omega^n)$. This problem is probably solvable, but it is not likely to be a reasonable approximation (we can always do much better if we make a decision now knowing what is going to happen in the future). But what if we choose the decision $x_t^n$ first, then observe $W_{t+1}(\omega^n)$ (so we are not using this information when we choose our action) and then compute the cost? Let the resulting cost be represented using

$$\hat{q}_{t+1}^n = C_t(S_t^n, x_t^n) + \gamma \bar{V}_{t+1}^{n-1}(S^M(S_t^n, x_t^n, W_{t+1}(\omega^n))).$$

We could now smooth these values to obtain

$$\bar{Q}_t^n(S_t, x_t) = (1 - \alpha_{n-1}) \bar{Q}_t^{n-1}(S_t^n, x_t^n) + \alpha_{n-1} \hat{q}_{t+1}^n.$$

We use $\bar{Q}_t^n(S_t, x_t)$ as an approximation of

$$Q_t(S_t, x_t) = \mathbb{E}\left\{ C_t(S_t, x_t) + \gamma V_{t+1}(S^M(S_t, x_t, W_{t+1})) | S_t \right\}.$$

The functions $Q_t(S_t, x_t)$ are known as *Q-factors* and they capture the value of being in a state and taking a particular action. If we think of $(S, x)$ as an augmented state (see section

---

**Step 0.** Initialization:

**Step 0a.** Initialize an approximation for the value function $\bar{Q}_t^0(S_t, x_t)$ for all states $S_t$ and decisions $x_t \in \mathcal{X}_t, t \in \mathcal{T}$.

**Step 0b.** Set $n = 1$.

**Step 0c.** Initialize $S_0^1$.

**Step 1.** Choose a sample path $\omega^n$.

**Step 2.** Do for $t = 0, 1, \ldots, T$:

**Step 2a:** Find the decision using the current $Q$-factors:

$$x_t^n = \arg\max_{x_t \in \mathcal{X}_t^n} \bar{Q}_t^{n-1}(S_t^n, x_t).$$

**Step 2b.** Compute:

$$\hat{q}_{t+1}^n = C_t(S_t^n, x_t^n) + \gamma \bar{V}_{t+1}^{n-1}(S^M(S_t^n, x_t^n, W_{t+1}(\omega^n))).$$

**Step 2c.** Update $\bar{Q}_t^{n-1}$ and $\bar{V}_t^{n-1}$ using:

$$\begin{aligned} \bar{Q}_t^n(S_t^n, x_t^n) &= (1 - \alpha_{n-1})\bar{Q}_t^{n-1}(S_t^n, x_t^n) + \alpha_{n-1}\hat{q}_{t+1}^n, \\ \bar{V}_t^n(S_t^n) &= \max_{x_t} \bar{Q}_t^n(S_t^n, x_t^n). \end{aligned}$$

**Step 2d.** Find the next state:

$$S_{t+1}^n = S^M(S_t^n, x_t^n, W_{t+1}(\omega^n)).$$

**Step 3.** Increment $n$. If $n \leq N$ go to Step 1.

**Step 4.** Return the Q-factors $(\bar{Q}_t^n)_{t=1}^T$.

---

**Figure 8.4** A $Q$-learning algorithm.

5.4.3), then a $Q$-factor can be viewed as the value of being in state $(S, x)$. We can now choose an action by solving

$$x_t^n = \arg\max_{x_t \in \mathcal{X}_t^n} \bar{Q}_t^{n-1}(S_t^n, x_t). \tag{8.7}$$

This strategy is known as $Q$-learning. The complete algorithm is summarized in figure 8.4.

$Q$-learning shares certain similarities with dynamic programming using a post-decision value function. In particular, both avoid the need to directly compute or approximate the expectation. However, $Q$-learning accomplishes this goal by creating an artificial post-decision state given by the state/action pair $(S, x)$. We then have to learn the value of being in $(S, x)$, rather than the value of being in state $S$ alone (which is already very hard for most problems).

$Q$-factors can be written as a function of the value function around the post-decision state variable. The post-decision state variable requires finding $\bar{V}_t^{n-1}(S_t^x)$ with $S_t^x = S^{M,W}(S_{t-1}, W_t(\omega))$, and then finding an action by solving

$$\begin{aligned} \hat{v}_t^n &= \max_{x_t \in \mathcal{X}_t^n} \left( C_t(S_t^n, x_t) + \gamma \bar{V}_t^{n-1}(S^{M,x}(S_t^x, x_t)) \right) \\ &= \max_{x_t \in \mathcal{X}_t^n} \hat{Q}_t^{x,n}, \end{aligned}$$

where

$$\hat{Q}_t^{x,n} = C_t(S_t^n, x_t) + \gamma \bar{V}_t^{n-1}(S^{M,x}(S_t^x, x_t)) \tag{8.8}$$

is a form of $Q$-factor computed using the post-decision value function. Unlike the original $Q$-factor, which was computed using $W_{t+1}(\omega)$, this is computed using $W_t(\omega)$. Most significantly, $Q$-learning is typically presented in settings where states and actions are discrete, which means we have one $Q$-factor for each state-action pair. By contrast, if we use the post-decision value function we may use general regression models which allows us to avoid computing a value for each state-action pair. Computationally, the difference is dramatic.

Many authors describe $Q$-learning as a technique for "model-free" dynamic programming (in particular, problems where we do not know the probability law for the exogenous information process). The reason is that we might assume that an exogenous process is generating the random outcomes. We feel this is misleading, since virtually any ADP algorithm that depends on Monte Carlo samples can be presented as a "model-free" application. In our view, the real value of $Q$-learning is the generality of $(S, x)$ as a post-decision state variable. The value of using the post-decision state $S_t^x$ is that it depends on taking advantage of the structure of $S_t^x$. However, $S_t^x$ will *never* be more complicated than $S_t$ (the post-decision state variable will never contain more information than the pre-decision state), which means that $S_t^x$ is *always* simpler than a post-decision state created by pairing the state with the action.

There are many applications where the complexity of the pre-decision and post-decision states are the same (consider modeling a board game), but there are other applications where the post-decision state is much simpler than the pre-decision state. A good example arises with our nomadic trucker problem. Here, $S_t = (R_t, D_t)$ where $R_{ta} = 1$ if the trucker has attribute $a$, and $D_t$ is a vector where $D_{tb}$ is the number of loads with attribute $b \in \mathcal{B}$. Thus, the dimensionality of $R_t$ is $|\mathcal{A}|$ and the dimensionality of $D_t$ is $|\mathcal{B}|$, which means that $|\mathcal{S}| = |\mathcal{A}| \times |\mathcal{B}|$. $|\mathcal{A}|$ and $|\mathcal{B}|$ may be on the order of hundreds to tens of thousands (or more). If we use $Q$-learning, we have to estimate $Q(S, x)$ for all $x \in \mathcal{X}$, which means we have to estimate $|\mathcal{A}| \times |\mathcal{B}| \times |\mathcal{X}|$ parameters. With a post-decision state vector, we only have to estimate $V_t(R_t^x)$, where the dimension of $R_t^x$ is $|\mathcal{A}|$.

## 8.3 TEMPORAL DIFFERENCE LEARNING

The previous section introduced two variations of approximate dynamic programming, one using a pure forward pass, with the second using a forward with a backward pass. These variants can be viewed as special cases of a general class of methods known as *temporal difference learning* (often abbreviated as "TD learning"). The method is widely used in approximate dynamic programming and a rich theory has evolved around it.

### 8.3.1 The basic idea

This approach is most easily described using the problem of managing a single resource that initially has attribute $a_t$ and we make decision $d_t^\pi$ (using policy $\pi$), after which we observe the information $W_{t+1}$ which produces an entity with attribute $a_{t+1} = a^M(a_t, d_t, W_{t+1})$. The contribution from this transition is given by $C_t(a_t, d_t)$. Imagine, now, that we continue this until the end of our horizon $T$. For simplicity, we are going to drop discounting. In this case, the contribution along this path would be

$$\hat{v}_{ta}^n = C_t(a_t, d_t^\pi) + C_{t+1}(a_{t+1}^n, d_{t+1}^\pi) + \ldots + C_T(a_T^n, d_T^\pi).$$

This is the contribution from following the path produced by a combination of the information from outcome $\omega^n$ (this determines $W_{t+1}, W_{t+2}, \ldots, W_T$) and policy $\pi$. This is an estimate of the value of a resource that starts with attribute $a_t$, for which we follow the decisions produced by policy $\pi$ until the end of our horizon.

Reverting back to our vector notation (where we use state $S_t$ and decision $x_t$), our path cost is given by

$$\hat{v}_t^n = C_t(S_t, x_t^\pi) + C_{t+1}(S_{t+1}^n, x_{t+1}^\pi) + \cdots + C_T(S_T^n, x_T^\pi). \tag{8.9}$$

$\hat{v}^n$ is an unbiased sample estimate of the value of being in state $S_t$ and following policy $\pi$ over sample path $\omega^n$. We use our standard stochastic gradient algorithm to estimate the value of being in state $S_t$ using

$$\bar{V}_t^n(S_t) = \bar{V}_t^{n-1}(S_t) - \alpha_n \left[\bar{V}_t^{n-1}(S_t) - \hat{v}_t^n\right]. \tag{8.10}$$

We can obtain a richer class of algorithms by breaking down our path cost in (8.9) by using

$$\begin{aligned}\hat{v}_t^n &= \sum_{\tau=t}^{T} C_\tau(S_\tau, x_\tau) \\ &\quad \underbrace{-\left\{\sum_{\tau=t}^{T}(\bar{V}_\tau^{n-1}(S_\tau) - \bar{V}_{\tau+1}^{n-1}(S_{\tau+1}))\right\} + (\bar{V}_t^{n-1}(S_t) - \bar{V}_{T+1}^{n-1}(S_{T+1}))}_{=0}.\end{aligned}$$

We now use the fact that $\bar{V}_{T+1}^{n-1}(S_{T+1}) = 0$ (this is where our finite horizon model is useful). Rearranging gives

$$\hat{v}_t^n = \bar{V}_t^{n-1}(S_t) + \sum_{\tau=t}^{T}\left(C_\tau(S_\tau, x_\tau) + \bar{V}_{\tau+1}^{n-1}(S_{\tau+1}) - \bar{V}_\tau^{n-1}(S_\tau)\right).$$

Let

$$D_\tau = C_\tau(S_\tau, x_\tau) + \bar{V}_{\tau+1}^{n-1}(S_{\tau+1}) - \bar{V}_\tau^{n-1}(S_\tau). \tag{8.11}$$

The terms $D_\tau$ are called *temporal differences*. If we were using a standard single-pass algorithm, then at time $t$, $\hat{v}_t^n = C_t(S_t, x_t) + \bar{V}_{t+1}^{n-1}(S_{t+1})$ would be our sample observation of being in state $S_t$, while $\bar{V}_t^{n-1}(S_t)$ is our current estimate of the value of being in state $S_t$. This means that the temporal difference at time $t$, $D_t = \hat{v}_t^n - \bar{V}_t^{n-1}(S_t)$, is the error in our estimate of the value of being in state $S_t$.

Using (8.11), we can write $\hat{v}_t^n$ in the more compact form

$$\hat{v}_t^n = \bar{V}_t^{n-1}(S_t) + \sum_{\tau=t}^{T} D_\tau. \tag{8.12}$$

Substituting (8.12) into (8.10) gives

$$\begin{aligned}\bar{V}_t^n(S_t) &= \bar{V}_t^{n-1}(S_t) - \alpha_{n-1}\left[\bar{V}_t^{n-1}(S_t) - \left(\bar{V}_t^{n-1}(S_t) + \sum_{\tau=t}^{T} D_\tau\right)\right] \\ &= \bar{V}_t^{n-1}(S_t) + \alpha_{n-1}\sum_{\tau=t}^{T-1} D_\tau. \end{aligned} \tag{8.13}$$

The temporal differences $D_\tau$ are the errors in our estimates of the value of being in state $S_\tau$. We learned in chapter 6 that these errors are stochastic gradients for the problem of minimizing estimation error. The name reflects the historical development within the field of approximate dynamic programming. We can think of each term in (8.13) as a correction to the estimate of the value function. It makes sense that updates farther along the path should not be given as much weight as those earlier in the path. As a result, it is common to introduce an artificial discount factor $\lambda$, producing updates of the form

$$\bar{V}_t^n(S_t) = \bar{V}_t^{n-1}(S_t) + \alpha_{n-1} \sum_{\tau=t}^{T} \lambda^{\tau-t} D_\tau. \tag{8.14}$$

We derived this formula without a discount factor. We leave as an exercise to the reader to show that if we have a discount factor $\gamma$, then the temporal-difference update becomes

$$\bar{V}_t^n(S_t) = \bar{V}_t^{n-1}(S_t) + \alpha_{n-1} \sum_{\tau=t}^{T} (\gamma\lambda)^{\tau-t} D_\tau. \tag{8.15}$$

Equation (8.15) shows that the discount factor $\gamma$, which is typically viewed as capturing the time value of money, and the artificial discount $\lambda$, which is a purely algorithmic device, have exactly the same effect. Not surprisingly, modelers in operations research have often used a discount factor $\gamma$ set to a much smaller number than would be required to capture the time-value of money. Artificial discounting allows us to look into the future, but then discount the results when we feel that the results are not perfectly accurate.

Updates of the form given in equation (8.14) produce an algorithm that is known as TD($\lambda$) (or, temporal difference learning with discount $\lambda$). We could let $\lambda = \gamma$, but in practice, it is common to use $\lambda < \gamma$ to produce a form of heuristic discounting that accounts for the fact that we are typically following a suboptimal policy. If $\lambda = 0$, then we get the one-period look-ahead update equivalent to the single-pass algorithm in figure 8.1. If $\lambda = 1$, then we obtain the double-pass algorithm in figure 8.2.

The updating formula in equation (8.14) requires that we step all the way to the end of the horizon before updating our estimates of the value. There is, however, another way of implementing the updates. The temporal differences $D_\tau$ are computed as the algorithm steps forward in time. As a result, our updating formula can be implemented recursively. Assume we are at time $t'$ in our simulation. We would simply execute

$$\bar{V}_t^n(S_t) := \bar{V}_t^n(S_t) + \alpha_{n-1}\lambda^{t'-t} D_{t'} \quad \text{for all } t \le t'. \tag{8.16}$$

Here, our notation ":=" means that we take the current value of $\bar{V}_t^n(S_t)$, add $\alpha_{n-1}\lambda^{t'-t}D_{t'}$ to it to obtain an updated value of $\bar{V}_t^n(S_t)$. When we reach time $t' = T$, our value functions would have undergone a complete update. We note that at time $t'$, we need to update the value function for every $t \le t'$.

### 8.3.2 Variations

There are a number of variations of temporal difference learning. We have presented the algorithm in the context of a finite horizon problem where we update the value of being in a state by always following policy $\pi$ until we reach the end of our horizon. This is equivalent to our double-pass procedure in figure 8.2. The only difference is that the double-pass algorithm implicitly assumes that our policy is determined by our approximate value function (which is generally the case) whereas it could, in fact, be some rule (our policy) that simply specifies the action given the state.

Another strategy is to look out to a time period $T = t + \tau$ that is not necessarily the end of our horizon. We would then normally use an approximation of the value function $V_{T+1}(S_{T+1})$ in equations (8.12) and (8.13). If we use this idea but only look one period into the future, we are effectively using our single-pass algorithm (figure 8.1).

## 8.4 POLICY ITERATION

We can create an approximate dynamic programming version of policy iteration, as shown in figure 8.5. In it, we sweep forward in time using a set of value functions $V_t^{\pi,n-1}$ to make decisions. We then add up the contributions along this path until the end of the horizon, which is stored as $\hat{v}^m$. We repeat this exercise using the same value functions $V_t^{\pi,n-1}$ but using a different sample realization $\omega^m$ for $m = 1, 2, \ldots, M$. The cost of each trajectory, $\hat{v}^m$, is then smoothed (typically averaged) into an estimate $\bar{V}_t^{n,m}(S_t^n)$. After $M$ repetitions of this process, we obtain $\bar{V}_t^{n,M}(S_t^n)$ which we then use to produce a new policy.

As $M \rightarrow \infty$, this algorithm starts to look like real policy iteration. The problem is that we are not controlling what states we are visiting. As a result, we cannot obtain any guarantees that the value functions are converging to the optimal solution.

If we compare this version of policy iteration to our first introduction to policy iteration (section 3.5), we notice one key difference. When we first did policy iteration, we found the value of a policy (which we called $v^\pi$) and used this value to update our policy, which is to say we found an action $X^\pi(s)$ for each state $s$. In this chapter, we assume that we are doing approximate dynamic programming because our state space is too large to enumerate. This means that we cannot explicitly compute an action $x$ for each state $s$. But we can solve the function in equation (8.17), which is to say that we are going to directly use the value function approximation. Technically, this is the same as determining an action for each state, but how we manage the calculations is quite different.

Policy iteration comes in different flavors, reflecting the diversity of ways in which we can express a policy. Examples include the following:

(a) Value function approximations - In figure 8.5, a policy is represented by equation (8.17), where we use the value function approximation $\bar{V}_t^{\pi,n-1}(S_t^x)$ to determine the action we take now. Our "policy" is determined by the value function approximation.

(b) Lookup-table - A policy can be expressed in the form $X_t^\pi(S_t)$ which returns a decision $x_t$ if we are in state $S_t$. These are typically used for problems where the set of states and actions are small. We could revise our algorithm in figure 8.5 by defining

$$X_t^\pi(S_t) = \arg\max_{x_t \in \mathcal{X}_t} \left( C_t(S_t, x_t) + \gamma V_t^{\pi,n-1}(S^x) \right)$$

for all $S_t$. Then, instead of solving the optimization problem in equation (8.17), we would just use $x_t^n = X_t^\pi(S_t^n)$.

(c) Regression function - Imagine that our decision is a quantity (the temperature of a process, the price we should sell our oil, how many customers should be admitted to the hospital). Perhaps we think we can write our decision function in the form

$$X^\pi(S) = \theta_0 + \theta_1 S + \theta_2 \ln(S),$$

---

**Step 0.** Initialization:

**Step 0a.** Initialize $V_t^{\pi,0}, \ t \in \mathcal{T}$.

**Step 0b.** Set $n = 1$.

**Step 0c.** Initialize $S_0^1$.

**Step 1.** Do for $n = 1, 2, \ldots, N$:

**Step 2.** Do for $m = 1, 2, \ldots, M$:

**Step 3.** Choose a sample path $\omega^m$.

**Step 4:** Initialize $\hat{v}^m = 0$

**Step 5:** Do for $t = 0, 1, \ldots, T$:

**Step 5a.** Solve:

$$x_t^{n,m} = \arg\max_{x_t \in \mathcal{X}_t^{n,m}} \left(C_t(S_t^{n,m}, x_t) + \gamma V_t^{\pi,n-1}(S^{M,x}(S_t^{n,m}, x_t))\right) \quad (8.17)$$

**Step 5b.** Compute:

$$\begin{aligned} S_t^{x,n,m} &= S^{M,x}(S_t^{n,m}, x_t^{n,m}) \\ S_{t+1}^{n,m} &= S^{M,W}(S_t^{x,n,m}, W_{t+1}(\omega^m)). \end{aligned}$$

**Step 6.** Do for $t = T-1, \ldots, 0$:

**Step 6a.** Accumulate the path cost (with $\hat{v}_T^m = 0$)

$$\hat{v}_t^m = C_t(S_t^{n,m}, x_t^m) + \gamma \hat{v}_{t+1}^m$$

**Step 6b.** Update approximate value of the policy starting at time $t$:

$$\bar{V}_{t-1}^{n,m} \leftarrow U^V(\bar{V}_{t-1}^{n,m-1}, S_{t-1}^{x,n,m}, \hat{v}_t^m) \quad (8.18)$$

where we typically use $\alpha_{m-1} = 1/m$.

**Step 7.** Update the policy value function

$$V_t^{\pi,n}(S_t^x) = \bar{V}_t^{n,M}(S_t^x) \ \ \forall t = 0, 1, \ldots, T$$

**Step 8.** Return the value functions $(V_t^{\pi,N})_{t=1}^T$.

---

**Figure 8.5** Approximate policy iteration using value function-based policies.

where $S$ is some scalar (how much oil we have in inventory, how many hospital beds are occupied) and $X^\pi(S)$ is a scalar action (how much oil to purchase or what price to charge, how many patients to admit).

(d) Neural networks - An even more general form of regression model is a neural network which can be trained to learn which action we should take given a state.

(e) A myopic policy with a parameter vector $\pi$ - There is an entire community which optimizes stochastic systems which are operated using a myopic policy that is influenced by one or more parameters. In section 3.5, we introduced a replenishment problem where we ordered $Q - R_t$ if $R_t < q$ and 0 otherwise (the policy was the parameters $(q, Q)$). We may wish to determine where to locate emergency response facilities, where the "policy" is the location of the facilities (once we locate the facilities, we assign events to the closest facility).

The first two policies (value function and lookup-table) can be viewed as two ways of representing a lookup-table policy (one explicit, the other implicit). For ADP, it does not make sense to store an explicit action for every state (there are too many states).

The regression function and neural network are both statistical methods for computing an action from a state (typically both are continuous). While the functions are estimated using a value function, the regression function (or neural network) is not explicitly a function of the value function approximation (as it was with option (a)). When we use a regression function, there are two small changes. First, step 5a would become

**Step 5a.** Solve

$$x_t^{n,m} = X_t^{\pi,n-1}(S^{n,m}).$$

$X_t^{\pi,n-1}(S^{n,m})$ refers to the regression function. We fit the regression by generating state-action pairs and then using regression techniques to fit the parameters. For this, we would insert a new step [7a] after step 7 which might look like the following:

**Step 7a.** Generate one or more state-action pairs. For example, we can compute a new set of actions

$$x_t^{n,m} = \arg\max_{x_t \in \mathcal{X}_t^{n,m}} \left( C_t(S_t^{n,m}, x_t) + \gamma V_t^{\pi,n-1}(S^{M,x}(S_t^{n,m}, x_t)) \right)$$

for all $m$. Now use the pairs $(S^{n,m}, x^{n,m})$ to update the regression function that describes the policy $X^{\pi,n-1}$, producing $X^{\pi,n}$.

## 8.5 MONTE CARLO VALUE AND POLICY ITERATION

In the discussion above, we were a little vague as to how decisions were being made as we moved forward in time. We were effectively assuming that decisions were determined by some "policy" $\pi$ that we normally represent as a decision function $X^\pi$. When we compute $\sum_{\tau=t}^{T} \lambda^{\tau-t} D_{t+\tau}$, we effectively update our estimate of the value function *when we follow policy* $\pi$. But, we make decisions by solving approximations of the form given in equation (8.1). As a result, we mix the estimation of the value of being in a state and the process of choosing decisions based on the value.

It might be a bit cleaner if we wrote our decision problem using

$$X_t^\pi(S_t^n) = \arg\max_{x_t \in \mathcal{X}_t} \left( C(S_t^n, x_t) + \gamma V^\pi(S^{M,x_t}(S_t^n, x_t)) \right)$$

for a given function $V^\pi(S)$. Fixing $V^\pi$ determines the decision we are making given our state and information, so fixing $V^\pi$ is equivalent to fixing the decision function (or equivalently, the policy we are following). We can then use any of the family of temporal differencing methods to estimate the value of being in a state while following policy $\pi$. After a while, our estimates of the value of being in a state will start to converge, although the quality of our decisions is not changing (because we are not changing $V^\pi$). We can reasonably assume that we will make better decisions if the functions $V^\pi$ are close to our estimates of the value of being in a state, given by our best estimates $\bar{V}$. So, we could periodically set $V^\pi = \bar{V}$ and then start the process all over again.

This strategy represents a process of estimating value functions (value iteration) and then updating the rules that we use to make decisions (policy iteration). The result is the Monte Carlo version of the hybrid value-policy iteration, which we presented in chapter 3. The overall algorithm is summarized in figure 8.6. Here, we use $n$ to index our policy-update iterations, and we use $m$ to index our value-update iterations. The optimal value of the

---

**Step 0.** Initialization:

**Step 0a.** Initialize $V_t^{\pi,0}, \quad t \in \mathcal{T}$.

**Step 0b.** Set $n = 1$.

**Step 0c.** Initialize $S_0^1$.

**Step 1.** Do for $n = 1, 2, \ldots, N$:

**Step 2.** Do for $m = 1, 2, \ldots, M$:

**Step 3.** Choose a sample path $\omega^m$.

**Step 4.** Do for $t = 1, 2, \ldots, T$:

**Step 4a.** Solve:

$$\hat{v}_t^m = \max_{x_t \in \mathcal{X}_t^m} \left( C_t(S_t^m, x_t) + \gamma V_t^{\pi,n-1}(S^{M,x}(S_t^m, x_t)) \right) \quad (8.19)$$

and let $x_t^m$ be the value of $x_t$ that solves (8.19).

**Step 4b.** Compute:

$$S_{t+1}^m = S^M(S_t^m, x_t^m, W_{t+1}(\omega^m)).$$

**Step 4c.** Update the value function:

$$\bar{V}_{t-1}^n \leftarrow U^V(\bar{V}_{t-1}^{n-1}, S_{t-1}^m, \hat{v}_t^m)$$

**Step 5.** Update the policy:

$$V_t^{\pi,n}(S_t) = \bar{V}_t^M(S_t) \quad \forall S_t.$$

**Step 6:** Return the value functions $(V_t^{\pi,N})_{t=1}^T$.

---

**Figure 8.6** Hybrid value/policy iteration in a Monte Carlo setting.

inner iteration limit $M$ is an open question. The figure assumes that we are using a TD(0) algorithm for finding updates (step $4a$), but this can easily be replaced with a general TD($\lambda$) procedure.

## 8.6 THE ACTOR-CRITIC PARADIGM

It is very popular in some communities to view approximate dynamic programming in terms of an "actor" and a "critic." Simply put, the actor is a policy and the critic is the mechanism for updating the policy, typically through updates to the value function approximation.

In this setting, a decision function that chooses a decision given the state is known as an *actor*. The process that determines the contribution (cost or reward) from a decision is known as the *critic*, from which we can compute a value function. The interaction of making decisions and updating the value function is referred to as an actor-critic framework. The slight change in vocabulary brings out the observation that the techniques of approximate dynamic programming closely mimic human behavior. This is especially true when we drop any notion of costs or contributions and simply work in terms of succeeding (or winning) and failing (or losing).

The policy iteration algorithm in figure 8.5 is a nice illustration of the actor-critic paradigm. The decision function is equation (8.17), where $V^{\pi,n-1}$ determines the policy (in this case). This is the actor. Equation (8.18), where we update our estimate of the value of the policy, is the critic. We fix the actor for a period of time and perform repeated iterations where we try to estimate value functions given a particular actor (policy). From time to time, we stop and use our value function to modify our behavior (something critics like to do). In this case, we update the behavior by replacing $V^\pi$ with our current $\bar{V}$.

In other settings, the policy is a rule or function that does not directly use a value function (such as $V^\pi$ or $\bar{V}$). For example, if we are driving through a transportation network (or traversing a graph) the policy might be of the form "when at node $i$, go next to node $j$." As we update the value function, we may decide the right policy at node $i$ is to traverse to node $k$. Once we have updated our policy, the policy itself does not directly depend on a value function.

Another example might arise when determining how much of a resource we should have on hand. We might solve the problem by maximizing a function of the form $f(x) = \beta_0 - \beta_1(x - \beta_2)^2$. Of course, $\beta_0$ does not affect the optimal quantity. We might use the value function to update $\beta_0$ and $\beta_1$. Once these are determined, we have a function that does not itself directly depend on a value function.

## 8.7 BIAS IN VALUE FUNCTION ESTIMATION

There are two sources of bias that need to be recognized when testing an approximate dynamic programming algorithm. One typically tends to underestimate the value function while the other overestimates, but they hardly cancel.

### 8.7.1 Value iteration bias

The most widely used updating mechanism for a value function (since it is the simplest) is to estimate the value of being in a state using

$$\hat{v}_t^n = \max_{x \in \mathcal{X}} \left( C(S_t, x_t) + \gamma \bar{V}_t^{n-1}(S_t^x(S_t, x_t)) \right), \tag{8.20}$$

after which we would use $\hat{v}_t^n$ to update the value function approximation

$$\bar{V}_{t-1}^n(S_{t-1}^x) = (1 - \alpha_{n-1})\bar{V}_{t-1}^{n-1}(S_{t-1}^x) + \alpha_{n-1}\hat{v}_t^n.$$

This is the single-pass procedure described in figure 8.1. Since $\bar{V}_t^{n-1}(S_t^x)$ is only an estimate of the value of being in state $S_t^x$, $\hat{v}_t^n$ is biased, which means we are trying to use biased observations to estimate $\bar{V}_{t-1}^n(S_{t-1}^x)$.

The bias is well known, particularly in the artificial intelligence community which works on problems where it takes a number of steps to earn a reward. The effect is illustrated in figure 8.3, where there are five steps before earning a reward of 1 (which we always earn). In this illustration, there are no decisions and the contribution is zero for every other time period. A stepsize of $1/n$ was used throughout.

The table illustrates that the rate of convergence for $\bar{V}_0$ is dramatically slower than for $\bar{V}_4$. The reason is that as we smooth $\hat{v}_t$ into $\bar{V}_{t-1}$, the stepsize has a discounting effect. The problem is most pronounced when the value of being in a state at time $t$ depends on contributions that are a number of steps into the future (imagine the challenge of training a value function to play the game of chess). For problems with long horizons, and in particular those where it takes many steps before receiving a reward, this bias can be so serious that it can appear that ADP simply does not work. To counter this bias requires using an inflated stepsize such as the search-then-converge formula (see section 6.2).

| Iteration | $\bar{V}_0$ | $\hat{v}_1$ | $\bar{V}_1$ | $\hat{v}_2$ | $\bar{V}_2$ | $\hat{v}_3$ | $\bar{V}_3$ | $\hat{v}_4$ | $\bar{V}_4$ | $\hat{v}_5$ |
|---|---|---|---|---|---|---|---|---|---|---|
| 0 | 0.000 | | 0.000 | | 0.000 | | 0.000 | | 0.000 | 1 |
| 1 | 0.000 | 0.000 | 0.000 | 0.000 | 0.000 | 0.000 | 0.000 | 0.000 | 1.000 | 1 |
| 2 | 0.000 | 0.000 | 0.000 | 0.000 | 0.000 | 0.000 | 0.500 | 1.000 | 1.000 | 1 |
| 3 | 0.000 | 0.000 | 0.000 | 0.000 | 0.167 | 0.500 | 0.667 | 1.000 | 1.000 | 1 |
| 4 | 0.000 | 0.000 | 0.042 | 0.167 | 0.292 | 0.667 | 0.750 | 1.000 | 1.000 | 1 |
| 5 | 0.008 | 0.042 | 0.092 | 0.292 | 0.383 | 0.750 | 0.800 | 1.000 | 1.000 | 1 |
| 6 | 0.022 | 0.092 | 0.140 | 0.383 | 0.453 | 0.800 | 0.833 | 1.000 | 1.000 | 1 |
| 7 | 0.039 | 0.140 | 0.185 | 0.453 | 0.507 | 0.833 | 0.857 | 1.000 | 1.000 | 1 |
| 8 | 0.057 | 0.185 | 0.225 | 0.507 | 0.551 | 0.857 | 0.875 | 1.000 | 1.000 | 1 |
| 9 | 0.076 | 0.225 | 0.261 | 0.551 | 0.587 | 0.875 | 0.889 | 1.000 | 1.000 | 1 |
| 10 | 0.095 | 0.261 | 0.294 | 0.587 | 0.617 | 0.889 | 0.900 | 1.000 | 1.000 | 1 |

**Table 8.3** Effect of stepsize on backward learning

The bias in table 8.3 arises purely because it takes a number of iterations before $\bar{V}_t$ accurately estimates the future (discounted) rewards. Since we always earn a reward of 1, we quickly see the bias produced by the interaction of backward learning and the stepsize. A different bias arises when we use a double-pass procedure (figure 8.2), or any variation of policy iteration where we are computing the updates $\hat{v}_t^n$ by simulating a suboptimal policy. The bias is more subtle and arises purely because we are simulating a suboptimal policy (in table 8.3, we do not have to make any decisions, which means we are technically following an optimal policy).

### 8.7.2 Statistical bias in the max operator

A much more subtle bias arises because we are taking the maximum over a set of random variables. Again consider our basic updating equation (8.20). We are computing $\hat{v}_t^n$ by choosing the best of a set of decisions which depend on $\bar{V}_t^{n-1}(S_t^x)$. The problem is that the estimates $\bar{V}_t^{n-1}(S_t^x)$ are random variables. In the best of circumstances, assume that $\bar{V}_t^{n-1}(S_t^x)$ is an unbiased estimate of the true value $V_t(S_t^x)$ of being in (post-decision) state $S_t^x$. Because it is still a statistical estimate with some degree of variation, some of the estimates will be too high while others will be too low. If a particular action takes us to a state where the estimate just happens to be too high (due to statistical variation), then we are more likely to choose this as the best action and use it to compute $\hat{v}_t^n$.

To illustrate, assume we have to choose a decision $d \in \mathcal{D}$, where $C(S, d)$ is the contribution earned by using decision $d$ (given that we are in state $S$) which then takes us to state $S^x(S, d)$ where we receive an estimated value $\bar{V}(S^x(S, d))$. Normally, we would update the value of being in state $S$ by computing

$$\hat{v}^n = \max_{d \in \mathcal{D}} \left( C(S, d) + \bar{V}^{n-1}(S^x(S, d)) \right).$$

We would then update the value of being in state $S$ using our standard update formula

$$\bar{V}^n(S) = (1 - \alpha_{n-1})\bar{V}^{n-1}(S) + \alpha_{n-1}\hat{v}^n.$$

Since $\bar{V}^{n-1}(S^x(S, d))$ is a random variable, sometimes it will overestimate the true value of being in state $S^x(S, d)$ while other times it will underestimate the true value. Of course, we are more likely to choose an action that takes us to a state where we have overestimated the value.

We can quantify the error due to statistical bias as follows. Fix the iteration counter $n$ (so that we can ignore it), and let

$$U_d = C(S,d) + \bar{V}(S^x(S,d))$$

be the estimated value of using decision $d$. The statistical error, which we represent as $\beta$, is given by

$$\beta = \mathbb{E}\{\max_{d\in\mathcal{D}} U_d\} - \max_{d\in\mathcal{D}} \mathbb{E}U_d. \tag{8.21}$$

The first term on the right-hand side of (8.21) is the expected value of $\bar{V}(S)$, which is computed based on the best observed value. The second term is the correct answer (which we can only find if we know the true mean). We can get an estimate of the difference by using a strategy known as the "plug-in principle." We assume that $\mathbb{E}U_d = \bar{V}(S^x(S,d))$, which means that we assume that the estimates $\bar{V}(S^x(S,d))$ are correct, and then try to estimate $\mathbb{E}\{\max_{d\in\mathcal{D}} U_d\}$. Thus, computing the second term in (8.21) is easy.

The challenge is computing $\mathbb{E}\{\max_{d\in\mathcal{D}} U_d\}$. We assume that while we have been computing $\bar{V}(S^x(S,d))$, we have also been computing $\bar{\sigma}^2(d) = Var(U_d) = Var\big(\bar{V}(S^x(S,d))\big)$. Using the plug-in principle, we are going to assume that the estimates $\bar{\sigma}^2(d)$ represent the true variances of the value function approximations. Computing $\mathbb{E}\{\max_{d\in\mathcal{D}} U_d\}$ for more than a few decisions is computationally intractable, but we can use a technique called the Clark approximation to provide an estimate. This strategy finds the exact mean and variance of the maximum of two normally distributed random variables, and then assumes that this maximum is also normally distributed. Assume the decisions can be ordered so that $\mathcal{D} = \{1, 2, \ldots, |\mathcal{D}|\}$. Now let

$$\bar{U}_2 = \max\{U_1, U_2\}.$$

We can compute the mean and variance of $\bar{U}_2$ as follows. First compute

$$a^2 = \sigma_1^2 + \sigma_2^2 - 2\sigma_1\sigma_2\rho_{12}$$

where $\sigma_1^2 = Var(U_1)$, $\sigma_2^2 = Var(U_2)$, and $\rho_{12}$ is the correlation coefficient between $U_1$ and $U_2$ (we allow the random variables to be correlated, but shortly we are going to approximate them as being independent). Next find

$$z = (\mu_1 - \mu_2)/a.$$

where $\mu_1 = \mathbb{E}U_1$ and $\mu_2 = \mathbb{E}U_2$. Now let $\Phi(z)$ be the cumulative standard normal distribution (that is, $\Phi(z) = \mathbb{P}[Z \le z]$ where $Z$ is normally distributed with mean 0 and variance 1), and let $\phi(z)$ be the standard normal density function. If we assume that $U_1$ and $U_2$ are normally distributed (a reasonable assumption when they represent sample estimates of the value of being in a state), then it is a straightforward exercise to show that

$$\mathbb{E}\bar{U}_2 = \mu_1\Phi(z) + \mu_2\Phi(-z) + a\phi(z) \tag{8.22}$$

$$Var(\bar{U}_2) = \left[(\mu_1^2+\sigma_1^2)\Phi(z) + (\mu_1^2+\sigma_2^2)\Phi(-z) + (\mu_1+\mu_2)a\phi(z)\right] - (\mathbb{E}\bar{U}_2)^2. \tag{8.23}$$

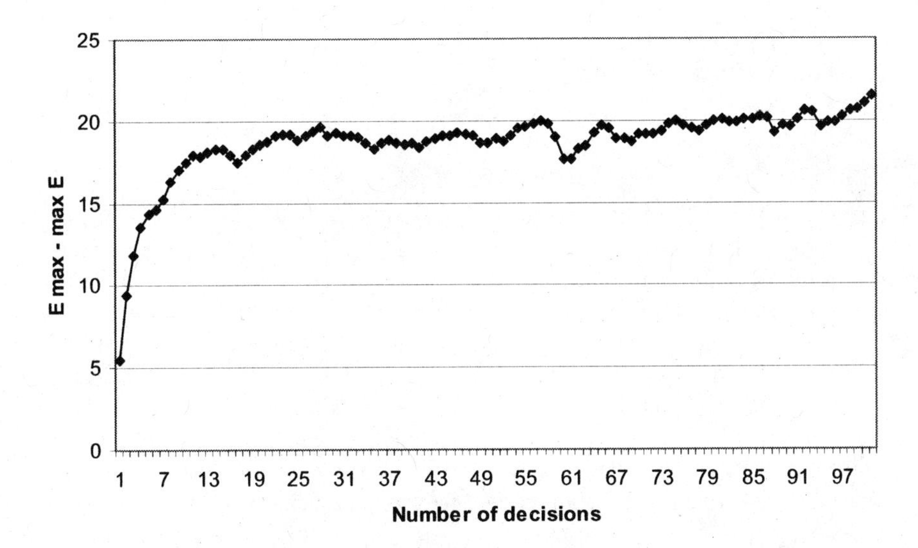


**Figure 8.7** $\mathbb{E}\max_d U_d - \max_d \mathbb{E}U_d$ for 100 decisions, averaged over 30 sample realizations. The standard deviation of all sample realizations was 20.

Now assume that we have a third random variable, $U_3$, where we wish to find $\mathbb{E}\max\{U_1, U_2, U_3\}$. The Clark approximation solves this by using

$$\begin{aligned} \bar{U}_3 &= \mathbb{E}\max\{U_1, U_2, U_3\} \\ &\approx \mathbb{E}\max\{U_3, \bar{U}_2\}, \end{aligned}$$

where we assume that $\bar{U}_2$ is normally distributed with mean given by (8.22) and variance given by (8.23). For our setting, it is unlikely that we would be able to estimate the correlation coefficient $\rho_{12}$ (or $\rho_{23}$), so we are going to assume that the random estimates are independent. This idea can be repeated for large numbers of decisions by using

$$\begin{aligned} \bar{U}_d &= \mathbb{E}\max\{U_1, U_2, \ldots, U_d\} \\ &\approx \mathbb{E}\max\{U_d, \bar{U}_{d-1}\}. \end{aligned}$$

We can apply this repeatedly until we find the mean of $\bar{U}_{|\mathcal{D}|}$, which is an approximation of $\mathbb{E}\{\max_{d\in\mathcal{D}} U_d\}$. This, in turn, allows us to compute an estimate of the statistical bias $\beta$ given by equation (8.21).

Figure 8.7 plots $\beta = \mathbb{E}\max_d U_d - \max_d \mathbb{E}U_d$ as it is being computed for 100 decisions, averaged over 30 sample realizations. The standard deviation of each $U_d$ was fixed at $\sigma = 20$. The plot shows that the error increases steadily until the set $\mathcal{D}$ reaches about 20 or 25 decisions, after which it grows much more slowly. Of course, in an approximate dynamic programming application, each $U_d$ would have its own standard deviation which would tend to decrease as we sample a decision repeatedly (a behavior that the approximation above captures nicely).

This brief analysis suggests that the statistical bias in the max operator can be significant. However, it is highly data dependent. If there is a single dominant decision, then the error will be negligible. The problem only arises when there are many (as in 10 or more) decisions that are competitive, and where the standard deviation of the estimates is not small relative to the differences between the means. Unfortunately, this is likely to be the case in most large-scale applications (if a single decision is dominant, then it suggests that the solution is probably obvious).

### 8.7.3 Remarks

The relative magnitudes of value iteration bias over statistical bias will depend on the nature of the problem. If we are using a pure forward pass (TD(0)), and if the value of being in a state at time $t$ reflects rewards earned over many periods into the future, then the value iteration bias can be substantial (especially if the stepsize is too small). Value iteration bias has long been recognized in the dynamic programming community. By contrast, statistical bias appears to have received almost no attention, and as a result we are not aware of any research addressing this problem. We suspect that statistical bias is likely to inflate value function approximations fairly uniformly, which means that the impact on the policy may be quite small. However, if the goal is to obtain the value function itself (for example, to estimate the value of an asset or a contract), then the bias can distort the results.

## 8.8 STATE SAMPLING STRATEGIES

Dynamic programming is at its best when decisions can be made at one point in time based on a value function approximation of the future. In the early iterations, we generally have to depend on fairly poor approximations of the value function. The problem we typically encounter is that the states we visit depend on our value function approximations, which in turn depend on the states we visit. If we depend purely on our value function approximations to determine which state we visit next, it is very easy to get caught in a local optimum, where poor approximations of the value of being in some states prevent us from revisiting those states.

In this section, we review some strategies for overcoming this classic chicken-and-egg problem. In later chapters, we partially overcome these problems by using different methods for approximating the value function and by exploiting problem structure.

### 8.8.1 Sampling all states

If we are managing a single resource, our state space is given by the size of the attribute space $|\mathcal{A}|$. While this can be extremely large for more complex resources, for many problems, it may be small enough that we can enumerate and sample at every iteration. If this is the case, it may be possible for us to use the classical backward dynamic programming techniques described in chapter 3. However, we may still encounter our "second curse of dimensionality": computing the expectation (when managing a single resource, the action space is typically not too large).

For example, consider our nomadic trucker example (section 5.3.3). Here, the attribute vector might consist of just the location (a typical discretization of the continental United States is 100 regions), but it may include other attributes such as a fleet type (for this discussion, assume there are five fleet types), producing an attribute space of 500, which is not large at all. Each time the trucker enters a region, he faces the random realization of market demands out of that region to all 100 regions (a demand can originate and terminate in the same region). Taking expectations over this much larger vector of market demands is where the problem can become quite hard.

Since the attribute space is not that large, it is not unreasonable to think of looking over all the states of our resource and using Monte Carlo methods to sample the value of being in each state at every iteration. An illustration of such an algorithm for approximating the value functions is given in figure 8.8. On the 3-GHz chips available at the time of this writing, this technique can be used effectively when the attribute space ranges up to 50,000 or more, but it would generally be impractical if the attribute space were in the millions.

---

**Step 0.** Initialize $\bar{V}_t^0, \; t \in \mathcal{T}$.

**Step 1.** Do while $n \leq N$:

  **Step 2.** Do for $t = 1, \ldots, T$:

    **Step 3.** Choose $\omega_t$.

    **Step 4.** Do for all $a \in \mathcal{A}$:

      **Step 4a.** Set $R_{ta} = 1$.

      **Step 4b.** Solve:

$$\hat{v}_t^n = \max_{x_t \in \mathcal{X}_t^n} C_t(R_{t-1}^n, x_t) + \gamma \bar{V}_t^{n-1}(R^M(R_{t-1}^n, x_t))$$

      **Step 4c.** Update the value function:

$$\bar{V}_{t-1}^n \leftarrow U^V(\bar{V}_{t-1}^{n-1}, R_{t-1}^n, \hat{v}_t^n)$$

---

**Figure 8.8** A single-resource learning algorithm sampling the entire attribute space.

Algorithms which loop over all states are referred to as *synchronous* because we are synchronizing the updates across all the states.

If we are dealing with more than one resource at a time, we have to work in terms of the resource state space rather than the attribute space. Multiple resource problems typically exhibit extremely large state spaces (even "small" problems can have state spaces that are in the $10^{10} - 10^{100}$ range). In chapter 12, we describe techniques for solving multiple resource problems that involve successively approximating single resource problems.

### 8.8.2 Tree search

It may not be practical to search the entire attribute space simply because it is far too large or because enumerating all combinations of the attribute vector produces attributes that would simply not happen in practice. If we are not able to sample the entire attribute space, we face the problem that we may have to make a decision about what to do with a resource based on a very poor estimate of the value of being in a particular state in the future.

A strategy for overcoming this is to use tree search, which is illustrated using our dynamic traveling salesman in figure 8.9. Assume we have a resource with initial attribute $a_0$ at location A, and we are contemplating a decision $d_0$ that would produce a resource with attribute $a_1 = a^M(a_0, d_0)$ at location B, whose value is approximated as $\bar{V}(a_1)$. Assume now that we have no observations of the attribute $a_1$, or we have decided that our estimate is not good enough (we revisit this topic in much greater depth in chapter 10). We can quickly get an estimate of a resource being in a particular state by searching outward from $a_1$, and evaluating each decision $d_1 \in \mathcal{D}_{a^M(a_1,d_1)}$ (location C in our figure). To evaluate decision $d_1$, we will need an estimate of the value of the resource with attribute $a_2(a_1, d_1)$ produced by each decision $d_1$.

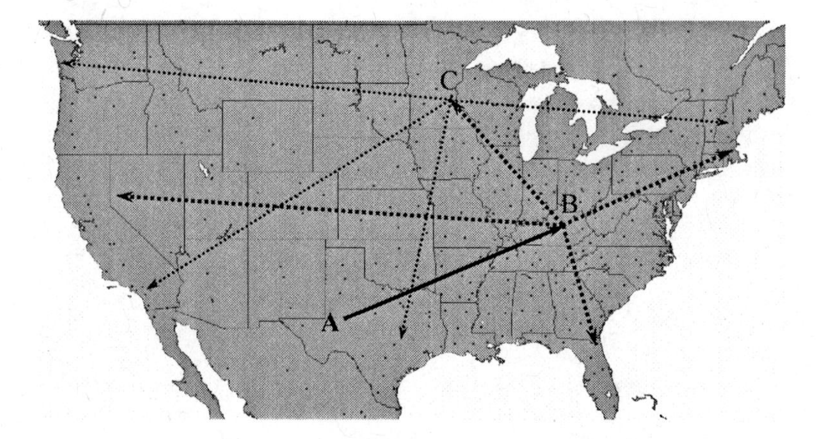


**Figure 8.9** Illustration of a tree search, branching forward in time along multiple trajectories.

---

**Procedure EvaluateV(a)**

**Step 0.** Initialize: Given initialize attribute $a$, increment tree depth $m = m + 1$.

**Step 1.** For all $d \in \mathcal{D}$:

**If** $m < M$ do:

**Step 2a.** Set $a' = a^M(a, d)$.

**Step 2b.** If $\bar{V}^n(a') = $ null: % Use tree search to evaluate the value of the entity at the destination.

$$\bar{V}^n(a') = \text{EvaluateV(a')}.$$

**Step 2c.** Find the value of the decision:

$$\hat{v}_d = C(a, d) + \gamma \bar{V}^n(a').$$

**Else:** % If we have reached the end of our tree search, just use the immediate contribution.

$$\hat{v}_d = C(a, d)$$

**Endif**

**Step 3.** Find the value of the best decision using

$$\text{EvaluateV} = \max_{d \in \mathcal{D}} \hat{v}_d$$

---

**Figure 8.10** A tree search algorithm.

Tree search is used only as a method to obtain an estimate of the value of being in a state. For each decision $d$ which takes us to a state $a' = a^M(a, d)$, we may have to use tree search to obtain an estimate of the value of being in state $a'$. We do not use tree search to actually solve the dynamic program.

An outline of a tree search algorithm is shown in figure 8.10. These algorithms are most naturally implemented as recursive procedures. Thus, as we enter the procedure $EvaluateV(a)$ with an attribute vector $a$, we enumerate the set of potential decisions $\mathcal{D}$. If the depth of our tree search is less than $M$, we use our terminal attribute function $a^M(a, d)$ to determine the attribute of the resource after being acted on by decision $d$ and call the function over again.

---

**Procedure MyopicRollOut(a)**

**Step 0.** Initialize: Given initial attribute $a$, increment tree depth $m = m + 1$.

**Step 1.** Find the best myopic decision using

$$\bar{d} = \max_{d \in \mathcal{D}} C(a, d).$$

**Step 2.** If $m < M$, do:

**If** $m < M$**:** Do:

**Step 3a.** Set $a' = a^M(a, \bar{d})$.

**Step 3b.** If $\bar{V}^n(a') =$ null, we have

$$\bar{V}^n(a') = \text{MyopicRollOut(a')}.$$

**Step 3c.** Find the value of the decision:

$$\hat{v} = C(a, \bar{d}) + \gamma \bar{V}^n(a').$$

**Else:**

$$\hat{v} = C(a, \bar{d}).$$

**Endif**

**Step 4.** Approximate the value of the state using:

$$\text{MyopicRollOut} = \hat{v}.$$

---

**Figure 8.11** Approximating the value of being in a state using a myopic policy.

Tree search is, of course, a brute force strategy. It will work well for problems with a small number of decisions for each state, but where the state space may still be quite large. It can also work well for problems where reasonable approximations can be obtained with very shallow trees (our nomadic trucker problem would be a good example of this problem class). Deeper trees are needed if we need a number of steps before learning whether a decision is good or bad. For most problems, tree search has to be significantly truncated to keep run times reasonable, which means that it is an approximation of last resort.

### 8.8.3 Rollout heuristics

For problems where the decision sets are of moderate to large sizes and where a very shallow tree does not work well, another strategy is to use an approximate policy to guide the choice of decisions. The easiest way to illustrate the idea is by using a myopic policy, which is depicted in figure 8.11.

The advantage of this strategy is that it is simple and fast. It provides a better approximation than using a pure myopic policy since in this case, we are using a myopic policy to provide a rough estimate of the value of being in a state. This can work quite well, but it can work very poorly. The performance is very situation-dependent.

It is also possible to choose a decision using the current value function approximation, as in

$$\bar{d} = \max_{d \in \mathcal{D}} \left( C(a, d) + \gamma \bar{V}^{n-1}(a^M(a, d)) \right).$$

We note that we are most dependent on this logic in the early iterations when the estimates $\bar{V}^n$ may be quite poor, but some estimate may be better than nothing.

## 8.9 STARTING AND STOPPING

Two challenges we face in approximate dynamic programming is getting started, which means dealing with very poor approximations of the value function, and stopping (when have we "converged" ?). Solving these problems tends to be unique to each problem, but it is important to recognize that they need to be addressed.

### 8.9.1 Getting through the early iterations

One issue that always arises in approximate dynamic programming is that you have to get through the early iterations when the value function approximation is very inaccurate. While it may be possible to start with an initial value function approximation that is fairly good, it is often the case that we have no idea what the value function should be and we simply use zero (or some comparable default approximation). After a few iterations, we may have updated the approximation with a few observations (allowing us to update the parameter vector $\theta$), but the approximation may be quite poor.

As a general rule, it is better to use a relatively simple approximation in the early iterations. The problem is that as the algorithm progresses, we may stall out at a poor solution. We would like to design a value function approximation that allows us to produce an accurate approximation, but this may require estimating a large number of parameters.

In section 7.1, we introduced the idea of estimating the value of being in a state at different levels of aggregation. Rather than use any single level of aggregation, we showed that we could estimate the value of being in a state by using a weighted sum of estimates at different levels of aggregation. Figure 8.12 shows what happens when we use a purely aggregate estimate, a purely disaggregate estimate, and a weighted combination. The purely aggregate estimate produces much faster initial convergence, reflecting the fact that an aggregate estimate of the value function is much easier to estimate with a few observations. The problem is that this estimate does not work as well in the long run.

Using a purely disaggregate estimate produces slow initial convergence, but ultimately provides a much better objective function in the later iterations. However, using a weighted combination creates faster initial convergence and the best results over all. The lesson of this demonstration is that it can be better to use a simpler approximation in the early iteration, and transition to a more refined approximation as the algorithm progresses.

This idea can be applied in a variety of settings. Consider the approximation we used in section 7.3.6 to model the value of a portfolio allocation given by $R_t$ which also depends on two financial statistics, $f_{t1}$ and $f_{t2}$. We can immediately produce value function approximations at three levels of aggregation: a) $\bar{V}(R_t|\theta)$, b) $\bar{V}(R_t|\theta(f_{t1}))$, and c) $\bar{V}(R_t|\theta(f_{t1}, f_{t2}))$. The first value function approximation completely ignores the two financial statistics. The second computes a parameter vector $\theta$ for each value of $f_{t1}$. The third computes a parameter vector $\theta$ for each combination of $f_{t1}$ and $f_{t2}$. Depending on how $f_{t1}$ and $f_{t2}$ are discretized, we could easily be computing thousands of estimates of $\theta$. As a result, the quality of the estimate of $\theta(f_{t1}, f_{t2})$ for a particular combination of $f_{t1}$ and $f_{t2}$ could be very inaccurate due to statistical error. We handle this by using a weighted sum of approximations, such as

$$\bar{V}(S_t|\theta) = w_1\bar{V}(R_t|\theta) + w_2\bar{V}(R_t|\theta(f_{t1})) + w_3\bar{V}(R_t|\theta(f_{t1}, f_{t2})).$$

The weights can be computed using the techniques described in section 7.1.3. Computing the weights in this way produces a natural transition from simpler to more complex models with nominal computational effort.

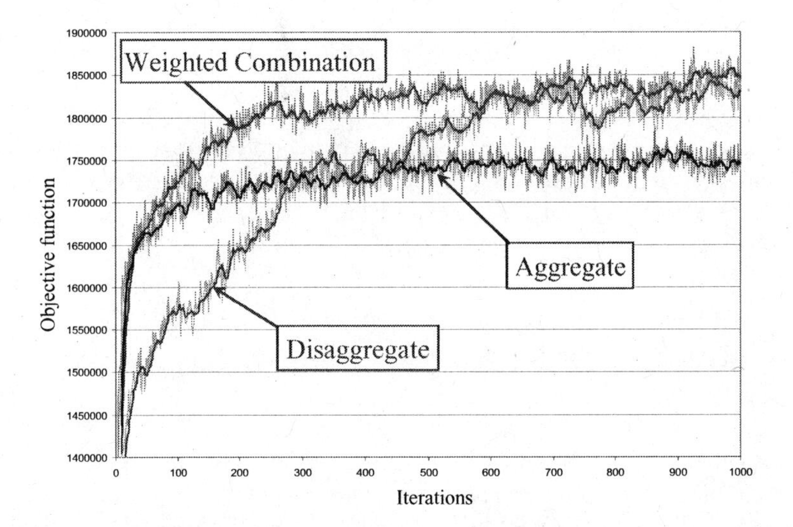


**Figure 8.12** Objective function produced with (a) a purely aggregate value function approximation, (b) a purely disaggregate one, and (c) one using a weighted combination of aggregate and disaggregate approximations.

It is easy to overlook the importance of using simpler functional approximations in the early iterations. A more complex function can be hard to estimate, producing large statistical errors in the beginning. It can be much more effective to do a better job (statistically speaking) of estimating a simpler function than to do a poor job of trying to estimate a more sophisticated function.

## 8.9.2 Convergence issues

Approximate value functions can work extremely well. But if the approximations do not reasonably capture the true behavior of the value function, they can also work terribly, even diverging (that is, the parameter vector $\bar{\theta}^n$ is constantly moving, possibly swinging wildly or continually growing). The problem is that our regression model might be a very poor approximation, possibly because it just does not capture the behavior of the problem or because early samples produced extremely poor estimates of the parameters. The value of regression models is that we can approximate the value of being in every state with a relatively small number of parameters. The problem with regression models is that an errant update to one parameter can produce dramatic swings in the value of many states. One update can do a lot more damage with a regression model than with a lookup-table representation.

What does work is if we hold the policy constant while we estimate the parameter vector. We do this by making a decision using

$$\hat{v}_t^n = \max_{x_t \in \mathcal{X}_t^n} \left( C_t(S_t^n, x_t) + \gamma \sum_{f \in \mathcal{F}} \theta_f^\pi \phi_f(S_t^n) \right),$$

where $\theta^\pi$ represents a policy which we do not change after each iteration. Instead, by locking in the parameter vector (which has the effect of fixing the policy), we can focus on the statistical problem of finding a good estimate for $\bar{\theta}^n$ (given $\theta^\pi$). If we do this, then we do have a guarantee that $\bar{\theta}^n$ will converge to the best possible set of parameters (again, given $\theta^\pi$).

Of course, we are not that interested in just approximating the value of a single policy. But this suggests an algorithm that is similar to our hybrid value/policy iteration algorithm (see the algorithm in figure 3.6). Using this policy, we hold $\theta^\pi$ constant for $M$ iterations, and then update the policy by setting $\theta^\pi = \bar{\theta}^n$. Again, if the regression model is a particularly poor approximation, we still have no guarantee of convergence, but this is likely to stabilize the algorithm.

## 8.10 A TAXONOMY OF APPROXIMATE DYNAMIC PROGRAMMING STRATEGIES

There are a wide variety of approximate dynamic programming algorithms. The algorithms reviewed in this chapter provide a sample of some of the variations. It is easier to get a sense of the diversity of strategies by creating a taxonomy based on the different choices that must be made in the design of an algorithm. Below is a summary of the dimensions of an algorithmic strategy. Note that some of the dimensions involve issues that are covered in later chapters (these are indicated as appropriate).

(1) Choice of state variable for developing a value function approximation (see chapter 5):

  (a) Use traditional pre-decision state variable.
  (b) Use state-action combination (estimate the value of a state-action pair using $Q$-learning).
  (c) Use the post-decision state variable.

(2) State sampling (see chapter 10):

  (a) Asynchronous (choose specific states). Asynchronous state sampling has a number of variants:
    i) Pure exploitation - Choose the states based on a greedy policy.
    ii) Decision-based exploration - From a given state, choose a decision at random which produces a downstream state.
    iii) Uniform state-based sampling - Randomly choose a state from a uniform distribution.
    iv) Nonuniform state-based sampling - Choose a state based on an exogenous probability distribution.
  (b) Synchronous (loop over all states).

(3) Average vs. marginal values - Depending on the application we may (see chapter 11):

  (a) Estimate the value of being in a state (may be used for single or multiple resource management problems).
  (b) Estimate the marginal value of an additional resource with a particular attribute (multiple resource management problems).

(4) Representation of the value function (see chapter 7):

(a) Use a discrete lookup-table representation. Variations include:

i) The value function is represented using the original state variable.

ii) The value function is represented using an aggregated state variable.

iii) The value function is represented using a weighted combination of a family of aggregations.

(b) Use a continuous approximation parameterized by a lower-dimensional parameter vector $\theta$. For resource allocation problems, these approximations may be (see chapter 11) as follows:

i) Linear in $R_{ta}$.

ii) Nonlinear (or piecewise linear) and separable in $R_{ta}$.

iii) More general functions.

(c) For more general problems, we may represent the state variable using a smaller set of basis functions $(\phi_f(S_t))_{f\in\mathcal{F}}$. These can be divided into two broad classes:

i) The approximation is linear in the parameters, which is to say that $\bar{V}(S_t) = \sum_{f\in\mathcal{F}} \theta_f \phi_f(S_t)$.

ii) The approximation is nonlinear in the parameters.

(5) Obtaining a Monte Carlo update of a value function from a state $s$ ($\hat{v}(s)$):

(a) Perform a single-step update. If we use a post-decision state variable, this would be computed using $\hat{v}_{t-1}(s_{t-1}) = \max_x \big(C(s_{t-1}, \omega, x) + \gamma \bar{V}_t(S_t(s_{t-1}, \omega, x))\big)$. This is TD(0), which we have represented as our single-pass algorithm.

(b) Simulate a policy over the remainder of the horizon (starting at time $t$), and use the value of this policy (or the derivative) to update the value function.

(c) Simulate a policy over the remainder of the horizon (starting at time $t$), but discount contributions $C_{t'}(S_{t'}, x_{t'})$ received at time $t'$ by $\lambda^{t'-t}$ (this is TD($\lambda$)).

(d) Simulate a policy for up to $T^{ph}$ time periods into the future, computing temporal differences factored by $\lambda$. The last temporal difference includes a value function approximation. Variations:

i) Use the trajectory to update only the starting state.

ii) Use all the intermediate partial trajectories to update all the states visited along the trajectory.

(6) Number of samples - We have to decide how well to estimate the value of the policy produced by the current value function approximation (see chapters 6 and 7):

(a) Update the value function after a single realization (single step, $T$ steps, or until the end of the horizon).

(b) Average (or smooth) $N$ forward trajectories before updating $\bar{v}$.

(c) Average (or smooth) an infinite number of forward trajectories (value iteration with exact function updates).

(7) Updating the value function - we compute a sample observation $\hat{v}$ which is then used to update the current estimate $\bar{v}$. We may update $\bar{v}$ using two strategies (see chapter 7):

(a) Stochastic gradient updates using a single-sample realization. There are two variations here:

  i) If we are using a lookup-table representation, this involves smoothing the new estimate $\bar{v}(s)$ with the current approximation $\bar{v}(s)$.

  ii) If we are using a parameterized approximation $\bar{V}(s|\theta)$, our stochastic gradient algorithm would update $\theta$.

(b) Parameter estimation using (recursive) regression methods.

(c) Use multiple realizations of $\hat{v}$ to find an updated parameter vector (typically $\theta$) by solving an optimization problem that produces the best parameter vector over the set of observations $(\hat{v})_i$.

## 8.11 WHY DOES IT WORK**

Although most of the early literature focuses on infinite horizon problems, it was not uncommon for authors to effectively assume that the problem was finite by introducing the idea of a zero-cost absorbing state, and then assuming that all policies were guaranteed to enter this state. These were termed "proper policies" (Bertsekas & Tsitsiklis (1996)). The bibliographic notes point to some of the key convergence proofs that have been obtained in what remains a very young field. The biggest weakness of many of these proofs is that they depend on assumptions that states are being visited infinitely often. Such assumptions (which might require introducing periodic exploration steps) do nothing to improve the performance of an algorithm since if we are using ADP, then typically our state spaces (and possibly action spaces) are extremely large.

## 8.12 BIBLIOGRAPHIC NOTES

Section 8.1 - This section builds on the foundational material from the field of approximate dynamic programming. Bertsekas & Tsitsiklis (1996) and Sutton & Barto (1998) are important early introductions to this field. The idea of backpropagation through time was introduced in Werbos (1974) (see also Werbos (1994)).

Section 8.2 - $Q$-learning was first introduced in Watkins' Ph.D. dissertation (Watkins (1989)), and later introduced in a journal publication in Watkins & Dayan (1992). Tsitsiklis (1994) and Jaakkola et al. (1994) provide a more formal proof of convergence using the theory of stochastic approximation methods. Readers interested in additional discussion of the topic should see the discussions in Bertsekas & Tsitsiklis (1996) and Sutton & Barto (1998). An application of $Q$-learning to elevator group control is given in Crites & Barto (1994). Even-Dar & Mansour (2004) provides a discussion of learning rates.

Section 8.3 - Temporal difference learning was first proposed in Sutton (1988) (based on the author's dissertation Sutton (1984)), with a proof of convergence for TD(0). This proof was extended informally for general TD($\lambda$) in Dayan (1992). These proofs were done without establishing the relationship to the elegant theory of stochastic approximation methods. This relationship was exploited in Jaakkola et al. (1994) and Tsitsiklis (1994) (the latter treatment is also presented in Bertsekas & Tsitsiklis (1996) (chapter 4)). These proofs applied to lookup-table approximations of the value function. de Farias & Van Roy (2000) prove the existence of fixed-points for TD learning.

Section 8.4 - Different forms of policy iteration are popular in approximate dynamic programming where we simulate a policy for a number of iterations in order to obtain a good value function approximation. Whereas classical value iteration will use a single update, approximate dynamic programming will often simulate a policy for a number of iterations to counteract the effect of randomness. In a recent monograph, Chang et al. (2007) describe several algorithms for performing policy iteration using Monte Carlo methods. The hybrid use of value and policy iteration is basically a Monte Carlo adaptation of the classical hybrid algorithm given in section 3.6. This strategy has been investigated in the ADP community under the name of "optimistic policy iteration," described in Tsitsiklis (2002).

Section 8.6 - The term "actor-critic" was first used in Widrow et al. (1973), with significant early developments in Werbos (1987), Werbos (1990) and Werbos (1992*a*). Overviews of adaptive-critic designs are given by Prokhorov & Wunsch (1997) and Konda & Tsitsiklis (2003). Lendaris & Neidhoefer (2004) provides an excellent tutorial on the use of adaptive critics for control.

Section 8.7 - Bias due to maximizing a stochastic function is discussed in Ormoneit & Sen (2002) where a different formula for the bias is presented. The Clark approximation (section 8.7.2 is given in Clark (1961), but more modern treatments of this issue (but not in the context of approximate dynamic programming) are found in Coles (2001) and Ross (2005). This issue was brought to our attention by Drake Jr. (2005).

Section 8.8 - Most of the strategies in this section are widely used in various forms in the artificial intelligence literature. Tree search is widely used in the artificial intelligence community. Roll-out heuristics, in the context of approximate dynamic programming, were first introduced by Bertsekas & Castanon (1999).

## PROBLEMS

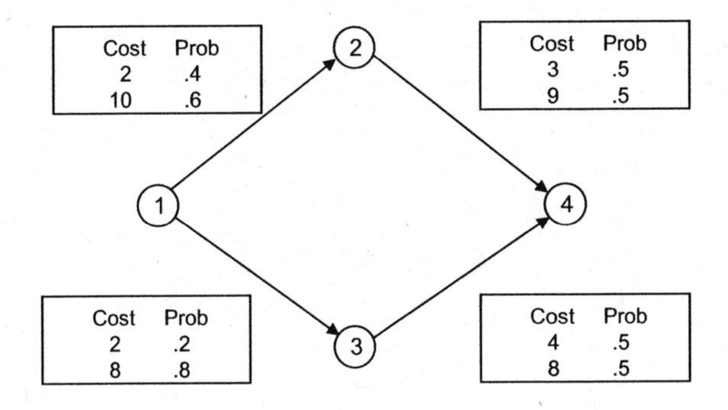


**8.1** The network above has an origin at node 1 and a destination at node 4. Each link has two possible costs with a probability of each outcome.

(a) Write out the dynamic programming recursion to solve the shortest path problem from 1 to 4. Assuming that the driver does not see the cost on a link until he arrives at the link (that is, he will not see the cost on link (2,4) until he arrives to node 2). Solve the dynamic program and give the expected cost of getting from 1 to 4.

(b) Set up and solve the dynamic program to find the expected shortest path from 1 to 4 assuming the driver sees all the link costs before he starts the trip.

(c) Set up and solve the dynamic program to find the expected shortest path assuming the driver does not see the cost on a link until after he traverses the link.

(d) Give a set of inequalities relating the results from parts (a), (b), and (c) and provide a coherent argument to support your relationships.

**8.2** Here we are going to again solve a variant of the asset selling problem using a post-decision state variable, but this time we are going to use asynchronous state sampling (in chapter 4 we used synchronous approximate dynamic programming). We assume we are holding a real asset and we are responding to a series of offers. Let $\hat{p}_t$ be the $t^{th}$ offer, which is uniformly distributed between 500 and 600 (all prices are in thousands of dollars). We also assume that each offer is independent of all prior offers. You are willing to consider up to 10 offers, and your goal is to get the highest possible price. If you have not accepted the first nine offers, you must accept the $10^{th}$ offer.

(a) Implement an approximate dynamic programming algorithm using *asynchronous* state sampling, initializing all value functions to zero. Using 100 iterations, write out your estimates of the value of being in each state immediately after each offer. Use a stepsize rule of $5/(5+n-1)$. Summarize your estimate of the value of each state after each offer.

(b) Compare your results against the estimates you obtain using synchronous sampling. Which produces better results?

**8.3** The taxicab problem is a famous learning problem in the artificial intelligence literature. The cab enters a grid (below), and at each cell, it can go up/down or left/right. The challenge is to teach it to find the exit as quickly as possible. Write an approximate

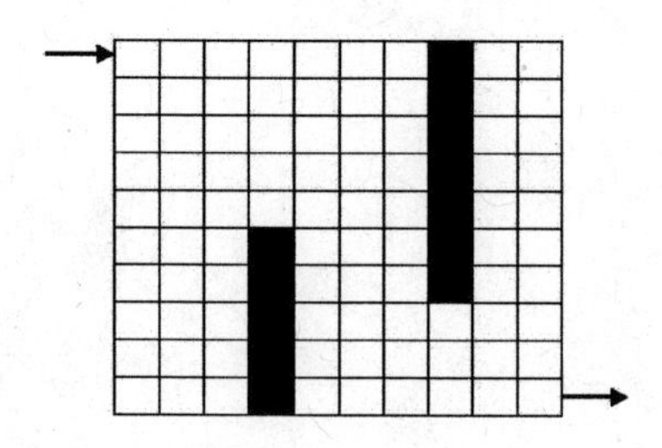

dynamic programming algorithm to learn the best path to the exit where at every iteration you are allowed to loop over every cell and update the value of being in that cell.

**8.4** Repeat the taxicab exercise, but now assume you can only update the value of the cell that you are in, using the following initialization strategies:

(a) Initialize your value function with 0 everywhere.

(b) Initialize your value function where the value of being in cell $(i, j)$ is $10 - i$, where $i$ is the row, and $i = 1$ is the bottom row.

(c) Initialize your value function using the formula $(10 - i) + j$. This estimate pushes the system to look down and to the right.

**8.5** Repeat the derivation of (8.14) but this time introduce a discount factor $\gamma$. Show that you obtain the updating equation given in (8.15).

**8.6** **Nomadic trucker project** - We are going to create a version of the nomadic trucker problem using randomly generated data. Assume our trucker may visit any of 20 cities in a set $\mathcal{I}$. The trucker may move loaded from $i$ to $j$ earning a positive reward $r_{ij}$ as long as there is a demand $\hat{D}_{tij}$ at time $t$ to move a load from $i$ to $j$. Alternatively, the trucker may move empty from $i$ to $j$ at a cost $c_{ij}$.

Assume that the demands $\hat{D}_{tij}$ follow a Poisson distribution with mean $\lambda_{ij}$. Randomly generate these means by first generating a parameter $\rho_i$ for each $i$, where $\rho_i$ is uniformly distributed between 0 and 1. Then set $\lambda_{ij} = \theta\rho_i(1 - \rho_j)$, where $\theta$ is a scaling parameter (for this exercise, use $\theta = 2$). Let $d_{ij}$ be the distance between $i$ and $j$. Randomly generate distances from a uniform distribution between 100 and 1500 miles. Now let $r_{ij} = 4\rho_i d_{ij}$ and $c_{ij} = 1.2d_{ij}$. Assume that if our trucker is in location $i$, he can only serve demands out of location $i$, and that any demands not served at one point in time are lost. Further assume that it takes one day (with one time period per day) to get from location $i$ to location $j$ (regardless of the distance). We wish to solve our problem over a horizon $T = 21$ days.

At location $i$, our trucker may choose to move loaded to $j$ if $\hat{D}_{ij} > 0$, or move empty to $j$. Let $x^L_{tij} = 1$ if he moves loaded from $i$ to $j$ on day $t$, and 0 otherwise. Similar, let $x^E_{tij} = 1$ if he moves empty from $i$ to $j$ on day $t$, and 0 otherwise. Of course, $\sum_j (x^L_{tij} + x^E_{tij}) = 1$. We make the decision by solving

$$\max_x \sum_j \left((r_{ij} + \bar{V}^{n-1}(j))x^D_{tij} + (-c_{ij} + \bar{V}^{n-1}(j))x^E_{tij}\right)$$

subject to the constraint that $x^L_{ij} = 0$ if $\hat{D}_{tij} = 0$.

(a) Assume our trucker starts in city 1. Use a single-pass algorithm using a stepsize of $\alpha_{n-1} = a/(a + n - 1)$ with $a = 10$. Train the value functions for 1000 iterations. Then, holding the value functions constant, perform an additional 1000 simulations, and report the mean and standard deviation of the profits, as well as the number of times the trucker visits each city.

(b) Repeat part (a), but this time insert a loop over all cities, so that for each iteration $n$ and time $t$, we pretend that we are visiting every city to update the value of being in the city. Again perform 1000 iterations to estimate the value function (this will now take 10 times longer), and then perform 1000 testing iterations.

(c) Repeat part (a), but this time, after solving the decision problem for location $i$ and updating the value function, randomly choose the next city to visit.

(d) Compare your results in terms of solution quality and computational requirements (measured by how many times you solve a decision problem).

# CHAPTER 9

# INFINITE HORIZON PROBLEMS

Infinite horizon problems arise in any setting where the parameters of the underlying processes do not vary over time. These models are particularly useful in applications where the primary interest is in understanding the properties of the problem, rather than in operational solutions. They tend to be computationally much easier because we do not have to estimate value functions at each point in time, reducing the number of parameters we have to estimate. For the problems that cannot be solved exactly, we can apply the techniques of approximate dynamic programming that we have initially presented in the context of finite horizon problems. There are, however, some computational issues that arise in the context of infinite horizon problems.

Typically, adapting the algorithms that we presented in chapter 8 to infinite horizon problems is fairly straightforward. Often, all that is required is dropping the time index. This is one reason our presentation started with the time-dependent (finite horizon) version of the problem. It is much easier to convert a time-indexed version of the algorithm to steady state (where there is no time indexing) than the other way around. For this reason, we begin our presentation by briefly summarizing some of the same algorithms that use discrete lookup-table forms of the value function. We then show how to make the transition to the infinite horizon setting.

## 9.1 FROM FINITE TO INFINITE HORIZON

Our algorithmic strategy is going to closely mimic the algorithms used for finite horizon problems. When we solved a finite horizon problem, we solved recursions of the form

$$V_t(S_t) = \max_{x \in \mathcal{X}_t} \left( C_t(S_t, x_t) + \gamma \mathbb{E}\left\{ V_{t+1}(S_{t+1}) | S_t \right\} \right),$$

where $S_{t+1} = S^M(S_t, x_t, W_{t+1})$ is the state at time $t+1$ given the state at time $t$, the decision $x_t$ made at time $t$, and the information that arrives during time interval $t+1$ given by $W_{t+1}$. Using the post-decision state variable gives us

$$V^x_{t-1}(S^x_{t-1}) = \mathbb{E}\left\{ \max_{x \in \mathcal{X}_t} \left( C_t(S_t, x_t) + \gamma V^x_t(S^x_t) \right) | S^x_{t-1} \right\},$$

where $S_t = S^{M,W}(S^x_{t-1}, W_t)$ and $S^x_t = S^{M,x}(S_t, x_t)$. Using the post-decision form, we replace the value function with our approximation, giving us a decision function that looks like

$$X^\pi_t(S_t) = \arg\max_{x \in \mathcal{X}^n_t} \left( C_t(S^n_t, x_t) + \gamma \bar{V}^{n-1}_t(S^x_t) \right),$$

where $S^n_t = S^{M,W}(S^x_{t-1}, W_t(\omega^n))$ is computed from $S^x_{t-1}$ using a sample realization $W_t(\omega^n)$. Recall also that $\mathcal{X}^n_t$ depends on $S^n_t$.

When we solve a steady-state problem, we want to solve recursions of the form

$$V(S) = \max_{x \in \mathcal{X}} \left( C(S, x) + \gamma \mathbb{E}\left\{ V(S^M(S, x, W)) | S \right\} \right),$$

where $W$, of course, is the random variable within the expectation. Using the post-decision state variable, our decision function would look like

$$X^\pi(S^n) = \arg\max_{x \in \mathcal{X}} \left( C(S^n, x) + \gamma V(S^x) \right),$$

where $S = S^{M,W}(S^x, W(\omega^n))$ is the pre-decision state computed from a post-decision state $S^x$ and a sample realization $W(\omega^n)$. Finally, in approximate dynamic programming we substitute in an approximate value function producing

$$X^\pi(S^n) = \arg\max_{x \in \mathcal{X}} \left( C(S^n, x) + \gamma \bar{V}^{n-1}(S^x) \right).$$

We would use information from this problem to update our approximate value function at the previous pre-decision state $S^x$.

A sketch of a basic approximate dynamic programming algorithm for infinite horizon problems is given in figure 9.1. Note that in step 3 we choose a random outcome $\omega^n$. This means a sample realization of the information that will arrive during a single period. Contrast this with our finite horizon algorithms where we would choose a sample path, which is still denoted $\omega^n$. However, the sample path refers to a realization of all the information over all the time periods within the planning horizon.

## 9.2 ALGORITHMIC STRATEGIES

The solution of infinite horizon problems is primarily an adaptation of the techniques for finite horizon problems (chapter 8). Often, steady-state models look the same as finite horizon models, but without the time index. In some cases, however, the transition to the infinite horizon setting introduces some subtleties.

---

**Step 0.** Initialization:

**Step 0a.** Initialize $\bar{V}^0(s)$ for all states $s \in \mathcal{S}$.

**Step 0b.** Initialize $S^0$.

**Step 0c.** Let $n = 1$.

**Step 1.** Solve

$$x^n = \arg\max_{x \in \mathcal{X}^n} \left( C(S^n, x) + \gamma \bar{V}^{n-1}(S^{M,x}(S^n, x)) \right). \tag{9.1}$$

**Step 2.** Use the results to update the approximation $\bar{V}^{n-1}(S)$ around the point $S = S^n$ (using any of a variety of algorithms).

**Step 3.** Choose $\omega^{n+1}$ and compute $S^{n+1} = S^M(S^n, x^n, W(\omega^{n+1}))$.

**Step 4.** Let $n = n + 1$. If $n < N$, go to step 1.

---

**Figure 9.1** A basic approximate dynamic programming algorithm for infinite horizon problems.

### 9.2.1 Value iteration

Figure 9.2 describes an infinite-horizon adaptation of value iteration using approximate dynamic programming. When we applied this idea to finite horizon problems, we would start in state $S^x_{t-1}$, sample $W_t(\omega^n)$, and then find the pre-decision state $S^n_t$. After finding a estimate of the value of being in state $S^n_t$, given by $\hat{v}^n_t$, we would then update the post-decision value function at state $S^x_{t-1}$.

In our infinite horizon model, we lose the time index so we have to be careful to realize that we are starting at a post-decision state $S^{x,n-1}$, we are then sampling $W(\omega^n)$ which takes us to the pre-decision state $S^n$. We then use the sample estimate of the value of being in state $S^n$ to update the value function around $S^{x,n-1}$. Note that the iteration counter $n$ plays the same role that the time index $t$ served for finite horizon problems.

Proofs of convergence of approximate value iteration for discrete state and discrete action problems generally require assuming that every state will be visited infinitely often. We can accomplish this by looping over all the states in every iteration and sampling their value. Needless to say, this only works for problems with small state spaces, and typically, we would try to use exact methods for these problems. It is possible that we cannot use exact techniques if the "second curse of dimensionality," the exogenous information process, makes the expectation computationally intractable, or if we are missing a probability model of the information process. Instead, authors will introduce strategies that insure that every state is visited with a strictly positive probability (which may be small). While these ideas will provide a proof of convergence, they may not be computationally effective since the rate of convergence can be extremely slow.

### 9.2.2 Policy iteration

In section 8.4, we presented a form of a policy iteration algorithm for finite horizon problems. This algorithm featured the ability to sample a trajectory until the end of the horizon multiple times, building up an estimate of the value of a policy before updating the policy. The problem with the algorithm we presented (in figure 8.5) was that we did not estimate the value of being in every state. While this is more practical for large problems, it left us without the ability to say anything about whether the algorithm actually worked.

---

**Step 0.** Initialization:

**Step 0a.** Initialize $\bar{V}^0$.

**Step 0b.** Initialize $S^1$.

**Step 0c.** Set $n = 1$.

**Step 1.** Choose a sample $\omega^n$.

**Step 2a.** Solve

$$\hat{v}^n = \max_{x \in \mathcal{X}^n} \left( C(S^n, x) + \gamma \bar{V}^{n-1}(S^{M,x}(S^n, x)) \right) \tag{9.2}$$

and let $x^n$ be the value of $x$ that solves (9.2).

**Step 2b.** Update the value function

$$\bar{V}^n \leftarrow U^V(\bar{V}^{n-1}, S^{x,n-1}, \hat{v}^n) \tag{9.3}$$

**Step 2c.** Update the states

$$\begin{aligned} S^{x,n} &= S^{M,x}(S^n, x^n), \\ S^n &= S^{M,W}(S^{x,n-1}, W(\omega^n)). \end{aligned}$$

**Step 3.** Increment $n$. If $n \leq N$ go to Step 1.

**Step 4.** Return the value functions $\bar{V}^N$.

---

**Figure 9.2** Generic approximate dynamic programming algorithm for infinite horizon problems

The limitation (from a theoretical perspective) of our algorithm for finite horizon problems is that we would have to loop over all possible states for all possible time periods. In figure 9.3 we describe a version of an approximate dynamic programming algorithm for policy iteration for an infinite horizon problem. Readers should note that we have tried to index variables in a way that shows how they are changing (do they change with outer iteration $n$? inner iteration $m$? the forward look-ahead counter $t$?). This does not mean that it is necessary to store, for example, each state or decision for every $n$, $m$, and $t$.

It is unlikely that anyone would actually implement this algorithm, but it helps to illustrate the choices that can be made when designing a policy iteration algorithm in an approximate setting. The algorithm features four nested loops. The innermost loop steps forward and backward in time from an initial state $S^{n,0}$. The purpose of this loop is to obtain an estimate of the value of a path. Normally, we would choose $T$ large enough so that $\gamma^T$ is quite small (thereby approximating an infinite path). The next outer loop repeats this process $M$ times to obtain a statistically reliable estimate of the value of a policy (determined by $\bar{V}^{\pi,n}$). The third loop performs policy updates (in the form of updating the value function). The fourth and most outer loop considers all possible states. In a more practical implementation, we might choose states at random rather than looping over all states.

The algorithm helps us illustrate different variables. First, if we let $T \to \infty$, we are evaluating a true infinite horizon policy. If we simultaneously let $M \to \infty$, then $\bar{v}^n$ approaches the exact, infinite horizon value of the policy $\pi$ determined by $\bar{V}^{\pi,n}$. Thus, for $M = T = \infty$, we have a Monte Carlo-based version of policy iteration.

Of course, this is impractical. We can choose a finite value of $T$ that produces values $\hat{v}^{n,m}$ that are close to the infinite horizon results. We can also choose finite values of $M$, including $M = 1$. When we use finite values of $M$, this means that we are updating the

---

**Step 0.** Initialization:

**Step 0a.** Initialize $\bar{V}^{\pi,0}$.

**Step 0b.** Set a look-ahead parameter $T$ and inner iteration counter $M$.

**Step 0c.** Set $n = 1$.

**Step 1.** Loop over all states $S_0^{x,n}$:

**Step 2.** Do for $m = 1, 2, \ldots, M$:

**Step 3.** Choose a sample path $\omega^m$ (a sample realization over the lookahead horizon $T$).

**Step 4.** Set the starting state $S_0^{n,m} = S^{M,W}(S^{x,n}, W_0(\omega^m))$.

**Step 5.** Do for $t = 0, 1, \ldots, T$:

**Step 5a.** Compute

$$x_t^{n,m} = \arg\max_{x_t \in \mathcal{X}_t^{n,m}} \left(C(S_t^{n,m}, x_t) + \gamma \bar{V}^{\pi,n-1}(S^{M,x}(S_t^{n,m}, x_t))\right).$$

**Step 5b.** Compute

$$S_{t+1}^{n,m} = S^M(S_t^{n,m}, x_t^{n,m}, W_{t+1}(\omega^m)).$$

**Step 6.** Initialize $\hat{v}_{T+1}^{n,m} = 0$.

**Step 7:** Do for $t = T, T-1, \ldots, 0$:

$$\hat{v}_t^{n,m} = C(S_t^{n,m}, x_t^{n,m}) + \gamma \hat{v}_{t+1}^{n,m}.$$

**Step 8.** Update the approximate value of the policy:

$$\bar{v}^{n,m} = (1 - \alpha_{m-1})\bar{v}^{n,m-1} + \alpha_{m-1}\hat{v}_0^{n,m},$$

where typically $\alpha_{m-1} = 1/m$ (the observations $\hat{v}^{n,m}$ are drawn from the same distribution).

**Step 9.** Update the value function at $S^{x,n}$:

$$\bar{V}^n \leftarrow U^V(\bar{V}^{n-1}, S^{x,n}, \hat{v}^{n,M}).$$

**Step 10.** Update the policy:

$$\bar{V}^{\pi,n} = \bar{V}^n.$$

**Step 11.** Set $n = n + 1$. If $n < N$, go to Step 1.

**Step 12.** Return the value functions $(\bar{V}^{\pi,N})$.

---

**Figure 9.3** A policy iteration algorithm for steady-state problems

policy before we have fully evaluated the policy. This variant is known in the literature as *optimistic policy iteration* because rather than wait until we have a true estimate of the value of the policy, we update the policy after each sample (presumably, although not necessarily, producing a better policy). Students may also think of this as a form of partial policy evaluation, not unlike the hybrid value/policy iteration described in section 3.6. Optimistic policy iteration ($M = 1$) is one of the few variations of approximate dynamic programming which produces a provably convergent algorithm. But it does require synchronous state sampling (where we loop over all possible states). Nice, but not very practical.

### 9.2.3 Temporal-difference learning

As we learned with finite horizon problems, there is a close relationship between value iteration and a particular case of temporal difference learning. To see this, we start by rewriting equation (9.3) as

$$\begin{aligned}\bar{V}^n(s^n) &= \bar{V}^{n-1}(s^n) - \alpha_{n-1}\left(\bar{V}^{n-1}(s^n) - \hat{v}^n\right)\\ &= \bar{V}^{n-1}(s^n) - \alpha_{n-1}\left[\bar{V}^{n-1}(s^n) - \left(C(s^n,x^n) + \gamma\bar{V}^{n-1}(S^{M,x}(s^n,x^n))\right)\right]\\ &= \bar{V}^{n-1}(s^n) + \alpha_{n-1}D^n,\end{aligned}$$

where

$$D^n = C(s^n,x^n) + \gamma\bar{V}^{n-1}(S^{M,x}(s^n,x^n)) - \bar{V}^{n-1}(s^n)$$

is our temporal difference (see equations (8.11) and (8.14)). Thus, our basic value iteration algorithm is the same as TD(0).

We can perform updates using a general TD($\lambda$) strategy as we did for finite horizon problems. However, there are some subtle differences. With finite horizon problems, it is common to assume that we are estimating a different function $\bar{V}_t$ for each time period $t$. As we step through time, we obtain information that can be used for a value function at a *specific* point in time. With stationary problems, each transition produces information that can be used to update the value function, which is then used in all future updates. By contrast, if we update $\bar{V}_t$ for a finite horizon problem, then this update is not used until the next forward pass through the states.

In chapter 8, we found that we could write our updating logic as

$$\bar{V}_t^n(s_t) = \bar{V}_t^{n-1}(s_t) + \alpha_{n-1}\left(\sum_{\tau=t}^{T}\gamma^{\tau-t}\left(C_\tau(s_\tau,x_\tau) + \gamma\bar{V}_{\tau+1}^{n-1}(s_{\tau+1}) - \bar{V}_\tau^{n-1}(s_\tau)\right) - \bar{V}_T^{n-1}(s_T)\right). \tag{9.4}$$

We then defined the temporal differences

$$D_\tau = C_\tau(s_\tau,x_\tau) + \gamma\bar{V}_{\tau+1}^{n-1}(s_{\tau+1}) - \bar{V}_\tau^{n-1}(s_\tau),$$

which allowed us to rewrite the updating equation (9.4) using

$$\bar{V}_t^n(s_t) = \bar{V}_t^{n-1}(s_t) + \alpha_{n-1}\sum_{\tau=t}^{T}\gamma^{\tau-t}D_\tau.$$

Finally, we introduced an artificial discount factor $\lambda$, giving us

$$\bar{V}_t^n(s_t) = \bar{V}_t^{n-1}(s_t) + \alpha_{n-1}\sum_{\tau=t}^{T}(\gamma\lambda)^{\tau-t}D_\tau.$$

When we move to infinite horizon problems, we drop the indexing by $t$. Instead of stepping forward in time, we step through iterations, where at each iteration we generate a temporal difference

$$D^n = C(s^n,x^n) + \gamma\bar{V}^{n-1}(S^{M,x}(s^n,x^n)) - \bar{V}^{n-1}(s^n).$$

To do a proper update of the value function at each state, we would have to use an infinite series of the form

$$\bar{V}^n(s) = \bar{V}^{n-1}(s) + \alpha_n\sum_{m=0}^{\infty}(\gamma\lambda)^m D^{n+m}, \tag{9.5}$$

---

**Step 0.** Initialization:

**Step 0a.** Initialize $\bar{V}^0(S)$ for all $S$.

**Step 0b.** Initialize the post-decision state $S^{x,0}$.

**Step 0c.** Set $n = 1$.

**Step 1.** Choose $\omega^n$ and compute the pre-decision state:

$$S^n = S^{M,w}(S^{x,n}, W(\omega^n)).$$

**Step 2.** Solve

$$x^n = \arg\max_{x \in \mathcal{X}^n} \left( C(S^n, x) + \gamma \bar{V}^{n-1}(S^{M,x}(S^n, x)) \right). \tag{9.8}$$

**Step 3.** Compute the temporal difference for this step:

$$D^n = C(S^n, x^n) + \gamma \left( \bar{V}^{n-1}(S^{M,x}(S^n, x^n)) - \bar{V}^{n-1}(S^n) \right).$$

**Step 4.** Update $\bar{V}$ for $m = n, n-1, \ldots, 1$:

$$\bar{V}^n(S^m) = \bar{V}^{n-1}(S^m) + (\gamma\lambda)^{n-m} D^n. \tag{9.9}$$

**Step 5.** Compute $S^{x,n+1} = S^{M,x}(S^n, x^n)$.

**Step 6.** Let $n = n + 1$. If $n < N$, go to step 1.

---

**Figure 9.4** A TD($\lambda$) algorithm for infinite horizon problems.

where we can use any initial starting state $s^0 = s$. Of course, we would use the same update for each state $s^m$ that we visit, so we might write

$$\bar{V}^n(s^m) = \bar{V}^{n-1}(s^m) + \alpha_n \sum_{n=m}^{\infty} (\gamma\lambda)^{(n-m)} D^n. \tag{9.6}$$

Equations (9.5) and (9.6) both imply stepping forward in time (presumably a "large" number of iterations) and computing temporal differences before performing an update. A more natural way to run the algorithm is to do the updates incrementally. After we compute $D^n$, we can update the value function at each of the previous states we visited. So, at iteration $n$, we would execute

$$\bar{V}^n(s^m) := \bar{V}^n(s^m) + \alpha_n (\gamma\lambda)^{n-m} D^m, \quad m = n, n-1, \ldots, 1. \tag{9.7}$$

We can now use the temporal difference $D^n$ to update the estimate of the value function for every state we have visited up to iteration $n$.

Figure 9.4 outlines the basic structure of a TD($\lambda$) algorithm for an infinite horizon problem. Step 1 begins by computing the first post-decision state, after which step 2 makes a single step forward. After computing the temporal-difference in step 3, we traverse previous states we have visited in Step 4 to update their value functions.

In step 3, we update all the states $(S^m)_{m=1}^n$ that we have visited up to then. Thus, at iteration $n$, we would have simulated the partial update

$$\bar{V}^n(S^0) = \bar{V}^{n-1}(S^0) + \alpha_{n-1} \sum_{m=0}^{n} (\gamma\lambda)^m D^m. \tag{9.10}$$

This means that at any iteration $n$, we have updated our values using biased sample observations (as is generally the case in value iteration). We avoided this problem for finite horizon problems by extending out to the end of the horizon. We can obtain unbiased updates for infinite horizon problems by assuming that all policies eventually put the system into an "absorbing state." For example, if we are modeling the process of holding or selling an asset, we might be able to guarantee that we eventually sell the asset.

One subtle difference between temporal difference learning for finite horizon and infinite horizon problems is that in the infinite horizon case, we may be visiting the same state two or more times on the same sample path. For the finite horizon case, the states and value functions are all indexed by the time that we visit them. Since we step forward through time, we can never visit the same state at the same point in time twice in the same sample path. By contrast, it is quite easy in a steady-state problem to revisit the same state over and over again. For example, we could trace the path of our nomadic trucker, who might go back and forth between the same pair of locations in the same sample path. As a result, we are using the value function to determine what state to visit, but at the same time we are updating the value of being in these states.

### 9.2.4 Q-learning

Q-learning in steady state is exactly analogous to Q-learning for time-dependent problems (section 8.2). A generic Q-learning algorithm is given in figure 9.5. We continue to represent our system as being in state $s^n$, but we revert to using a discrete decision $d$ (rather than $x$) since $Q$-learning is completely impractical if the decision is a vector (as $x$ often is).

---

**Step 0.** Initialization:

**Step 0a.** Initialize an approximation for the value function $\bar{Q}^0(s, d)$ for all states $s$ and decisions $d$.

**Step 0b.** Initialize $\bar{V}_0 = \max_{d \in \mathcal{D}} \bar{Q}^0(S, d)$.

**Step 0c.** Fix the starting state $S^1$.

**Step 0c.** Let $n = 1$.

**Step 1.** Determine

$$d^n = \arg\max_{d \in \mathcal{D}} \bar{Q}^{n-1}(S^n, d).$$

**Step 2.** Choose $\omega^n$ and find

$$\hat{Q}^n(S^n, d^n) = C(S^n, d^n) + \gamma \bar{V}^{n-1}(S^M(S^n, d^n, \omega^n))$$

**Step 3.** Update $\bar{Q}^{n-1}$ and $\bar{V}^{n-1}$ using

$$\bar{Q}^n(S^n, d^n) = (1 - \alpha_{n-1})\bar{Q}^{n-1}(S^n, d^n) + \alpha_{n-1}\hat{Q}^n(S^n, d^n),$$
$$\bar{V}^n(S^n) = (1 - \alpha_{n-1})\bar{V}^{n-1}(S^n) + \alpha_{n-1}\hat{Q}^n(S^n, d^n).$$

**Step 4.** Compute a new sample $\hat{\omega}^{n+1}$.

**Step 5.** Compute $S^{n+1} = S^M(S^n, d^n, W(\hat{\omega}^{n+1}))$.

**Step 6.** Let $n = n + 1$. If $n < N$, go to step 1.

---

**Figure 9.5** A $Q$-learning algorithm for a steady-state problem.

### 9.2.5 The linear programming method with approximate value functions

In section 3.7, we showed that the determination of the value of being in each state can be found by solving the following linear program

$$\min_{v} \sum_{s \in \mathcal{S}} \beta_s v(s) \tag{9.11}$$

subject to

$$v(s) \geq C(s,x) + \gamma \sum_{s' \in \mathcal{S}} p(s'|s,x) v(s') \quad \text{for all } s \text{ and } x. \tag{9.12}$$

The problem with this formulation arises because it requires that we enumerate the state space to create the value function vector $(v(s))_{s \in \mathcal{S}}$. Furthermore, we have a constraint for each state-action pair, a set that will be huge even for relatively small problems.

We can partially solve this problem by replacing the discrete value function with a regression function such as

$$\bar{V}(s|\theta) = \sum_{f \in \mathcal{F}} \theta_f \phi_f(s).$$

where $(\phi_f)_{f \in \mathcal{F}}$ is an appropriately designed set of basis functions. This produces a revised linear programming formulation

$$\min_{\theta} \sum_{s \in \mathcal{S}} \beta_s \sum_{f \in \mathcal{F}} \theta_f \phi_f(s)$$

subject to:

$$v(s) \geq C(s,x) + \gamma \sum_{s' \in \mathcal{S}} p(s'|s,x) \sum_{f \in \mathcal{F}} \theta_f \phi_f(s') \quad \text{for all } s \text{ and } x.$$

This is still a linear program, but now the decision variables are $(\theta_f)_{f \in \mathcal{F}}$ instead of $(v(s))_{s \in \mathcal{S}}$. Note that rather than use a stochastic iterative algorithm, we obtain $\theta$ directly by solving the linear program.

We still have a problem with a huge number of constraints. Since we no longer have to determine $|\mathcal{S}|$ decision variables (in (9.11)-(9.12) the parameter vector $(v(s))_{s \in \mathcal{S}}$ represents our decision variables), it is not surprising that we do not actually need all the constraints. One strategy that has been proposed is to simply choose a random sample of states and actions. Given a state space $\mathcal{S}$ and set of actions (decisions) $\mathcal{X}$, we can randomly choose states and actions to create a smaller set of constraints.

Some care needs to be exercised when generating this sample. In particular, it is important to generate states roughly in proportion to the probability that they will actually be visited. Then, for each state that is generated, we need to randomly sample one or more actions. The best strategy for doing this is going to be problem-dependent.

This technique has been applied to the problem of managing a network of queues. Figure 9.6 shows a queueing network with three servers and eight queues. A server can serve only one queue at a time. For example, server A might be a machine that paints components one of three colors (say, red, green, and blue). It is best to paint a series of parts red before switching over to blue. There are customers arriving exogenously (denoted by the arrival

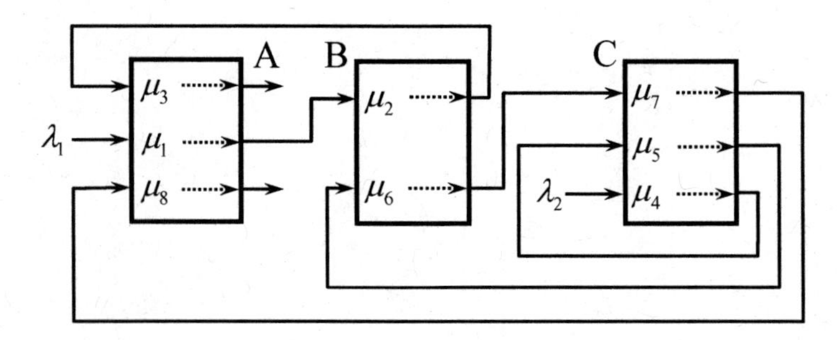


**Figure 9.6** Queueing network with three servers serving a total of eight queues, two with exogenous arrivals ($\lambda$) and six with arrivals from other queues (from de Farias & Van Roy (2003)).

rates $\lambda_1$ and $\lambda_2$). Other customers arrive from other queues (for example, departures from queue 1 become arrivals to queue 2). The problem is to determine which queue a server should handle after each service completion.

If we assume that customers arrive according to a Poisson process and that all servers have negative exponential service times (which means that all processes are memoryless), then the state of the system is given by

$$S_t = R_t = (R_{ti})_{i=1}^8,$$

where $R_{ti}$ is the number of customers in queue $i$. Let $\mathcal{K} = \{1, 2, 3\}$ be our set of servers, and let $a_t$ be the attribute vector of a server given by $a_t = (k, q_t)$, where $k$ is the identity of the server and $q_t$ is the queue being served at time $t$. Each server can only serve a subset of queues (as shown in figure 9.6). Let $\mathcal{D} = \{1, 2, \ldots, 8\}$ represent a decision to serve a particular queue, and let $\mathcal{D}_a$ be the decisions that can be used for a server with attribute $a$. Finally, let $x_{tad} = 1$ if we decide to assign a server with attribute $a$ to serve queue $d \in \mathcal{D}_a$.

The state space is effectively infinite (that is, too large to enumerate). But we can still sample states at random. Research has shown that it is important to sample states roughly in proportion to the probability they are visited. We do not know the probability a state will be visited, but it is known that the probability of having a queue with $r$ customers (when there are Poisson arrivals and negative exponential servers) follows a geometric distribution. For this reason, it was chosen to sample a state with $r = \sum_i R_{ti}$ customers with probability $(1 - \gamma)\gamma^r$, where $\gamma$ is a discount factor (a value of 0.95 was used).

Further complicating this problem class is that we also have to sample actions. Let $\mathcal{X}$ be the set of all feasible values of the decision vector $x$. The number of possible decisions for each server is equal to the number of queues it serves, so the total number of values for the vector $x$ is $3 \times 2 \times 3 = 18$. In the experiments for this illustration, only 5,000 states were sampled (in portion to $(1 - \gamma)\gamma^r$), but all the actions were sampled for each state, producing 90,000 constraints.

Once the value function is approximated, it is possible to simulate the policy produced by this value function approximation. The results were compared against two myopic policies: serving the longest queue, and first-in, first-out (that is, serve the customer who

| Policy | Cost |
|---|---|
| ADP | 33.37 |
| Longest | 45.04 |
| FIFO | 45.71 |

**Table 9.1** Average cost estimated using simulation (from de Farias & Van Roy (2003)).

had arrived first). The costs produced by each policy are given in table 9.1, showing that the ADP-based strategy significantly outperforms these other policies.

Considerably more numerical work is needed to test this strategy on more realistic systems. For example, for systems that do not exhibit Poisson arrivals or negative exponential service times, it is still possible that sampling states based on geometric distributions may work quite well. More problematic is the rapid growth in the feasible region $\mathcal{X}$ as the number of servers, and queues per server, increases.

An alternative to using constraint sampling is an advanced technique known as column generation. Instead of generating a full linear program which enumerates all decisions (that is, $v(s)$ for each state), and all constraints (equation (9.12)), it is possible to generate sequences of larger and larger linear programs, adding rows (constraints) and columns (decisions) as needed. These techniques are beyond the scope of our presentation, but readers need to be aware of the range of techniques available for this problem class.

## 9.3 STEPSIZES FOR INFINITE HORIZON PROBLEMS

Care has to be used in the choice of stepsize for infinite horizon problems. In chapter 8 (section 8.7.1), we demonstrated the degree to which the standard $1/n$ stepsize rule slowed the rate of convergence of the value function for the earliest time periods. The problem is that the stepsize not only smooths over randomness, it also produces a discounting effect as values are passed backward through time. This effect becomes much more pronounced with infinite horizon problems.

To illustrate, assume that we have a simple problem with no decisions (so we do not need to maximize anything), and for simplicity, no randomness. Classical value iteration would update the value function using

$$\bar{v}^n = C(S) + \gamma \bar{v}^{n-1},$$

where $C(S)$ is our one-period contribution function. We can see that we are simply computing $C(S) \sum_{n=0}^{\infty} \gamma^n$, or

$$v^* = \frac{C(S)}{1-\gamma}.$$

If $\gamma = 0.90$, $v^* = 10C(S)$. Classical value iteration would give us a very accurate estimate of this quantity quite quickly.

Now assume that we can only measure $C(S)$ with noise, which means that at iteration $n$, we observe a random contribution $\hat{c}^n$ where $\mathbb{E}\hat{c}^n = C(S)$. The correct value of $v^*$ is unchanged, but as a result of the noise in our measurement of the contribution, we can no longer use value iteration. Instead, we would compute a sample observation of the value

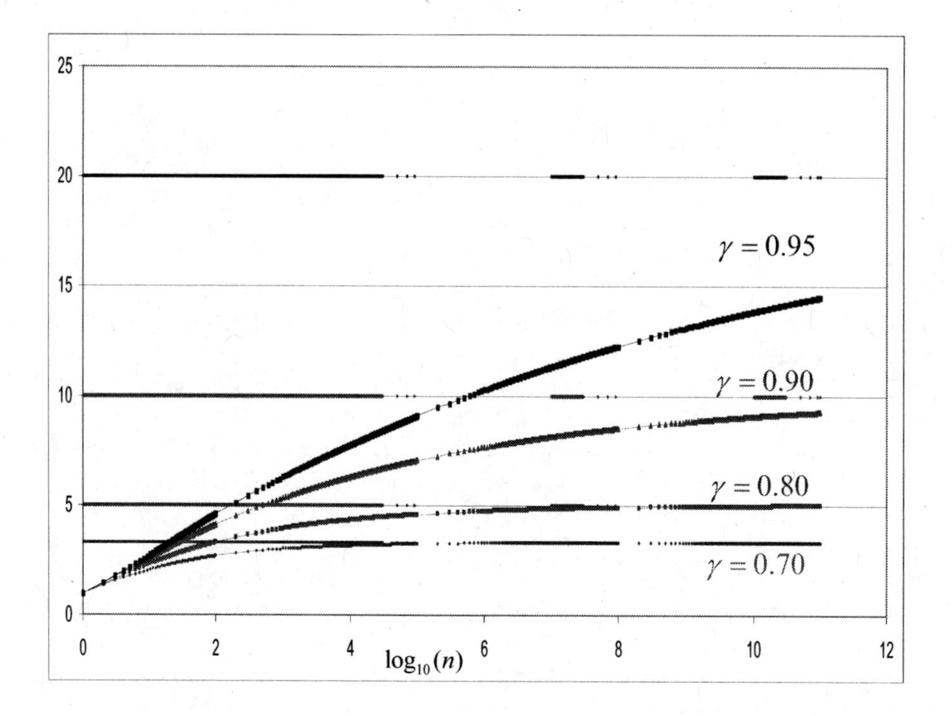


**Figure 9.7** $\bar{v}^n$ plotted against $\log_{10}(n)$ when we use a $1/n$ stepsize rule for updating.

function

$$\hat{v}^n = \hat{c}^n + \gamma \bar{v}^{n-1}.$$

We would then perform smoothing to obtain an updated estimate of the value, giving us

$$\bar{v}^n = (1 - \alpha_{n-1})\bar{v}^{n-1} + \alpha_{n-1}\hat{v}^n. \tag{9.13}$$

A simple way to demonstrate the rate of convergence is to disregard the randomness and simply let $\hat{c}^n = c$, which means that in the limit as $n$ goes to infinity, we would get $\bar{v}^n \rightarrow v^* = c/(1-\gamma)$. Although equation (9.13) will eventually produce this result, the rate at which it happens is astonishingly slow. Figure 9.7 is a plot of $\bar{v}^n$ as a function of $\log_{10}(n)$ for $\gamma = 0.7, 0.8, 0.9$, and $0.95$, where we have set $c = 1$. For $\gamma = 0.90$, we need $10^{10}$ iterations to get $\bar{v}^n = 9$, which means we are still 10 percent from the optimal. For $\gamma = 0.95$, we are not even close to converging after 100 billion iterations.

It is possible to derive compact bounds, $\nu^L(n)$ and $\nu^U(n)$ for $\bar{v}^n$ where

$$\nu^L(n) < \bar{v}^n < \nu^U(n).$$

These are given by

$$\nu^L(n) = \frac{c}{1-\gamma}\left(1 - \left(\frac{1}{1+n}\right)^{1-\gamma}\right), \tag{9.14}$$

$$\nu^U(n) = \frac{c}{1-\gamma}\left(1 - \frac{1-\gamma}{\gamma n} - \frac{1}{\gamma n^{1-\gamma}}\left(\gamma^2 + \gamma - 1\right)\right). \tag{9.15}$$

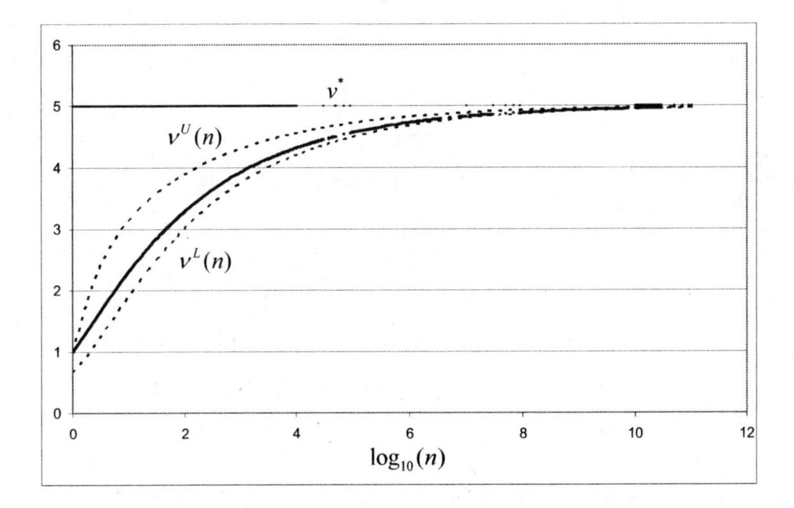


**Figure 9.8** $v^* = 5, \nu^U(n), \bar{v}^n$ and $\nu^L(n)$ for $\gamma = 0.8$.

It is also possible to show that $\nu^U(n) \leq v^*$ as long as $(\gamma^2 + \gamma - 1) \geq 0$, or $\gamma \geq 0.618$. Figure 9.8 shows the bounds $\nu^U(n)$ and $\nu^L(n)$ along with $\bar{v}^n$ and the limiting point $v^*$ for $\gamma = 0.8$.

Using the formula for the lower bound (which is fairly tight when $n$ is large enough that $\bar{v}^n$ is close to $v^*$), we can derive the number of iterations to achieve a particular degree of accuracy. Let $c = 1$, which means that $v^* = 1/(1-\gamma)$. For a value of $v < 1/(1-\gamma)$, we would need at least $n(v)$ to achieve $\bar{v}^* = v$, where $n(v)$ is found (from (9.14)) to be

$$n(v) \geq [1 - (1-\gamma)v]^{-1/(1-\gamma)}. \tag{9.16}$$

If $\gamma = 0.9$, we would need $n(v) = 10^{20}$ iterations to reach a value of $v = 9.9$, which gives us a one percent error. On a 3-GHz chip, assuming we can perform one iteration per clock cycle (that is, $3 \times 10^9$ iterations per second), it would take 1,000 years to achieve this result. A discount factor of 0.8 requires $10^{10}$ iterations. There are many applications which require several seconds or even several minutes to perform a single iteration.

These results apply to virtually any problem. As we learned in chapter 3, after enough iterations, a Markov decision process will evolve to a single policy and a steady-state distribution across the states. As a result, the expected reward per iteration becomes a constant. The message is clear: if you are solving an infinite horizon problem using ADP, do not use $1/n$ for a stepsize. We recommend using the bias-adjusted Kalman filter stepsize rule (equation (6.62)).

## 9.4 ERROR MEASURES

If you are implementing an ADP algorithm, you might start asking for some way to evaluate the accuracy of your procedure. This question can be posed in two broad ways: (1) How good is your policy? (2) How good is your value function approximation? Section 4.7.4 addressed the issue of evaluating a policy. In this section, we turn to the question of evaluating the quality of the value function approximation. There are applications such as pricing an asset or estimating the probability of winning a game or reaching a goal where the primary interest is in the value function itself.

Let $\bar{V}(S)$ be an approximation of the value of being in state $S$. The approximation could have a lookup-table format, or might be a regression of the form

$$\bar{V}(S) = \sum_{f \in \mathcal{F}} \bar{\theta}_f \phi_f(S).$$

Now imagine that we have access to the true value of being in state $S$, which we denote by $V(s)$. Typically this would happen when we have a problem that we can solve exactly using the techniques of chapter 3, and we are now testing an ADP algorithm to evaluate the effectiveness of a particular approximation strategy. If we want an aggregate measure, we could use

$$\nu^{\text{avg}} = \frac{1}{|\mathcal{S}|} \sum_{s \in \mathcal{S}} |V(s) - \bar{V}(s)|.$$

$\nu^{\text{avg}}$ expresses the error as an average over all the states. The problem is that it puts equal weight on all the states, regardless of how often they are visited. Such an error is unlikely to be very interesting.

This measure highlights the problem of determining how to weight the states. Assume that while determining $V(s)$ we also find $p^*(s)$, which is the steady-state value of being in state $s$ (under an optimal policy). If we have access to this measure, we might use

$$\nu^{\pi^*} = \sum_{s \in \mathcal{S}} p^*(s) |V(s) - \bar{V}(s)|,$$

where $\pi^*$ represents the fact that we are following an optimal policy. The advantage of this error measure is that $p^*(s)$ is independent of the approximation strategy used to determine $\bar{V}(s)$. If we want to compare different approximation strategies, where each produces different estimates for $\bar{V}(s)$, then we get a more stable estimate of the quality of the value function.

Such an error measure is not as perfect as it seems. Assume we have two approximation strategies which produce the approximations $\bar{V}^{(1)}(s)$ and $\bar{V}^{(2)}(s)$. Also let $\bar{p}^{(1)}(s)$ and $\bar{p}^{(2)}(s)$ be, respectively, estimates of the steady-state probabilities of being in state $s$ produced by each policy. If a particular approximation has us visiting certain states more than others, then it is easy to argue that the value functions produced by this strategy should be evaluated primarily based on the frequency with which we visit states under that policy. For example, if we wish to obtain an estimate of the value of the dynamic program (using the first approximation), we would use

$$\bar{V}^{(1)} = \sum_{s \in \mathcal{S}} p^{(1)}(s) \bar{V}^{(1)}(s).$$

For this reason, it is fair to measure the error based on the frequency with which we actually visit states under the approximation, leading to

$$\nu^{(1)} = \sum_{s \in \mathcal{S}} p^{(1)} |V(s) - \bar{V}^{(1)}(s)|.$$

If we compute $\bar{V}^{(2)}$ and $\nu^{(2)}$, then if $\nu^{(1)} < \nu^{(2)}$, we could conclude that $\bar{V}^{(1)}$ is a better approximation than $\bar{V}^{(2)}$, but it would have to be understood that this means that $\bar{V}^{(1)}$ does a better job of estimating the value of states that are visited under policy 1 than the job that $\bar{V}^{(2)}$ does in estimating the value of states that are visited under policy 2.

## 9.5 DIRECT ADP FOR ON-LINE APPLICATIONS

There are many applications where we have a process that is actually happening in the physical world. It could be real-time changes in prices, daily changes in inventories, or the process of monitoring a physical plant (e.g., electric power plant) on a minute-by-minute basis. These are referred to as *on-line* applications. Assume that we have some decision to make (do we sell the stock? Should we order more inventory? Should we change the controls to run the power plant?). Normally, we use various transition functions to capture the effect of the decision, but in an on-line application, we have the opportunity to simply observe the effect of our decision. We do not have to predict the next state - we can simply observe it. In fact, it might be the case that we do not even know the transition function. For complex problems such as an electric power plant, it might be so complicated that we do not even know the transition function.

Situations where we do not know the transition function give rise to what is often referred to as *model-free* dynamic programming, where "model" refers to the transition function (the engineering community often refers to the transition function as the system model or plant model). However, we are not just observing the state transition. We are also observing (possibly indirectly) the exogenous information process. The exogenous information process may be buried in the transition function (we can observe the state we transition to, which reflects both our decision and the exogenous information) or we can observe the exogenous information directly (e.g., the market price of electricity).

There are many applications where we know the transition function, but we do not know the probability distribution of the exogenous information. For example, if the problem involves spot prices for electricity, the state is the price, $p_t$. The transition function is simply $p_{t+1} = p_t + \hat{p}_{t+1}$, where $\hat{p}_{t+1}$ is the exogenous information. We may have no idea of the probability distribution of $\hat{p}_{t+1}$. We can observe it, but we do not have a mathematical model describing it.

With on-line applications, it is generally the case that we do not need either the transition function or a model of the exogenous information process. As a result, the term "model free" is often confused with these two issues.

## 9.6 FINITE HORIZON MODELS FOR STEADY-STATE APPLICATIONS

It is easy to assume that if we have a problem with stationary data (that is, all random information is coming from a distribution that is not changing over time), then we can solve the problem as an infinite horizon problem, and use the resulting value function to produce a policy that tells us what to do in any state. If we can, in fact, find the optimal value function for every state, this is true.

There are many applications of infinite horizon models to answer policy questions. Do we have enough doctors? What if we increase the buffer space for holding customers in a queue? What is the impact of lowering transaction costs on the amount of money a mutual fund holds in cash? What happens if a car rental company changes the rules allowing rental offices to give customers a better car if they run out of the type of car that a customer reserved? These are all dynamic programs controlled by a constraint (the size of a buffer or the number of doctors), a parameter (the transaction cost), or the rules governing the physics of the problem (the ability to substitute cars). We may be interested in understanding the behavior of such a system as these variables are adjusted. For infinite horizon problems that are too complex to solve exactly, ADP offers a way to approximate these solutions.

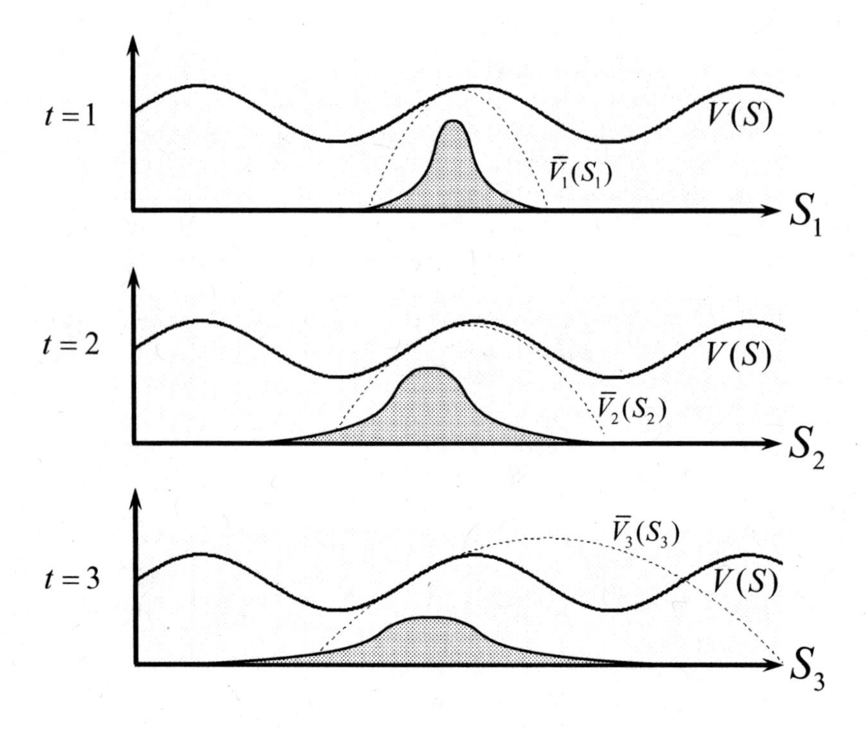


**Figure 9.9** Exact value function (sine curve) and value function approximations for $t = 1, 2, 3$, which change with the probability distribution of the states that we can reach from $S_0$.

Infinite horizon models also have applications in operational settings. Assume that we have a problem governed by stationary processes. We could solve the steady-state version of the problem, and use the resulting value function to define a policy that would work from any starting state. This works if we have, in fact, found at least a close approximation of the optimal value function for any starting state. However, if you have made it this far in this book, then that means you are interested in working on problems where the optimal value function cannot be found for all states. Typically, we are forced to approximate the value function, and it is always the case that we do the best job of fitting the value function around states that we visit most of the time.

When we are working in an operational setting, then we start with some known initial state $S_0$. From this state, there are a range of "good" decisions, followed by random information, that will take us to a set of states $S_1$ that is typically heavily influenced by our starting state. Figure 9.9 illustrates the phenomenon. Assume that our true, steady-state value function approximation looks like the sine function. At time $t = 1$, the probability distribution of the state $S_t$ that we can reach is shown as the shaded area. Assume that we have chosen to fit a quadratic function of the value function, using observations of $S_t$ that we generate through Monte Carlo sampling. We might obtain the dotted curve labeled as $\bar{V}_1(S_1)$, which closely fits the true value function around the states $S_1$ that we have observed.

For times $t = 2$ and $t = 3$, the distribution of states $S_2$ and $S_3$ that we actually observe grows wider and wider. As a result, the best fit of a quadratic function spreads as well. So, even though we have a steady-state problem, the best value function approximation depends on the initial state $S_0$ and how many time periods into the future that we are projecting. Such problems are best modeled as finite horizon problems, but only because we are forced to approximate the problem.

## 9.7 WHY DOES IT WORK?**

Much of the theory of approximate dynamic programming has been developed in the context of infinite horizon problems. Important references are provided in the bibliographic notes. Infinite horizon problems have long been popular in the research community, partly because they are more compact and partly because of the elegant theory that evolved in the original study of Markov decision processes (3). While the theory of infinite horizon problems is more interesting when we use expectations, proving convergence results is harder since we do not have the benefit of the end of the horizon.

## 9.8 BIBLIOGRAPHIC NOTES

Section 9.2 - Approximate dynamic programming is most often described in the context of infinite horizon with discrete states and actions. Sutton & Barto (1998) provides a clear algorithmic descriptions for this class of algorithms. Bertsekas & Tsitsiklis (1996) also provide descriptions of the principles behind these algorithms.

Section 9.2.3 - Boyan (2002) and Bertsekas et al. (2004) describe temporal-difference learning using continuous value functions.

Section 9.2.5 - Schweitzer & Seidmann (1985) was the first to propose approximation methods in the context of infinite horizon problems using the linear programming method. de Farias & Van Roy (2003) describe how basis functions can be used in the context of the linear programming method and describe an application to the control of a small queueing network. de Farias & Van Roy (2004) provide convergence results when constraint sampling is used. Adelman (2003) and Adelman (2004) applies the linear programming method to the stochastic inventory routing problem, where vehicles are routed to replenish known inventories that are used to satisfy random demands. Adelman (to appear) uses column generation to reduce the size of the linear program in a revenue management application. The use of constraint sampling and column generation remains an open question for other problem classes.

Section 9.3 - This section is based on Frazier & Powell (2007). A thorough analysis of the rate of convergence of $Q$-learning is given in Even-Dar & Mansour (2007) and Szepesvari (1998), although these results are more complex.

Section 9.5 - See Si et al. (2004*b*) for a discussion of direct approximate dynamic programming.

## PROBLEMS

**9.1** You have two two-state Markov chains, with transition matrices

| Action 1 | | |
|---|---|---|
| From-to | 1 | 2 |
| 1 | 0.2 | 0.8 |
| 2 | 0.7 | 0.3 |

| Action 2 | | |
|---|---|---|
| From-to | 1 | 2 |
| 1 | 0.5 | 0.5 |
| 2 | 0.6 | 0.4 |

The reward for each state-action pair is given by

| | Action | |
|---|---|---|
| State | 1 | 2 |
| 1 | 5 | 10 |
| 2 | 10 | 15 |

(a) Solve the problem exactly using value iteration. Plot the value of being in each state as a function of the number of iterations, and give the optimal policy (the best action for each state) at optimality.

(b) Use the approximate dynamic programming algorithm in figure 9.1 to estimate the value of being in each state and estimate the optimal policy. Use 1000 Monte Carlo samples, and plot the estimate of the value function after each iteration.

**9.2** You are holding an asset which has an initial market price $p_0 = 100$. The price evolves each time period according to $p_t = p_{t-1} + \epsilon_t$ where $\epsilon_t$ is the discrete uniform distribution taking values from -5 to 5 (each with probability $1/11$). There is no limit to when you have to sell the asset. Let $x_t$ be the decision variable, where $x_t = 1$ means sell and $x_t = 0$ means hold, and let $\gamma$ be the discount factor. We wish to maximize $\mathbb{E}\sum_{t=0}^{\infty} \gamma^t x_t p_t$.

(a) Set up the optimality equations and show that there exists a price $\bar{p}$ where we will sell if $p_t > \bar{p}$.

(b) Design and implement an approximate dynamic programming algorithm to find $\bar{p}$ using the stepsize rule $\alpha_{n-1} = 8/(7+n)$. Provide a plot of your estimate of $\bar{p}$ (derived from your value function approximation) at each iteration.

**9.3** Consider the infinite horizon problem that we introduced in section 9.3 where we want to use value iteration to compute

$$v^* = \mathbb{E}\sum_{t=0}^{\infty} \gamma^t \hat{C}_t,$$

where $\hat{C}_t = \hat{C}$ is a random variable with mean $c$. We want to approximate $v^*$ using value iteration with the iteration

$$\begin{aligned} \hat{v}^n &= \hat{C}^n + \gamma \bar{v}^{n-1}, \\ \bar{v}^n &= (1 - \alpha_{n-1} \bar{v}^{n-1} + \alpha_{n-1} \hat{v}^n. \end{aligned}$$

(a) Using a spreadsheet, compute $\bar{v}^n$ assuming that $\hat{C}^n = 1$, $\gamma = .9$, for $n = 1, \ldots, 50,000$ using a stepsize $\alpha_{n-1} = 1/n$. Plot your results and compare to the correct solution, $v^* = 1/(1-\gamma) = 10$.

(b) Repeat (a) using a stepsize $\alpha_{n-1} = a/(a+n)$ for $a = 5, 10$ and 20. Compare your results to what you obtained with $a = 1$.

(c) Repeat the exercise assuming that $\hat{C}$ is described by a continuous uniform distribution between 0 and 100.

**9.4 Infinite horizon nomadic trucker project** - Do the nomadic trucker problem described in exercise 8.6 as an infinite horizon problem, using a discount factor of 0.80.

# CHAPTER 10

# EXPLORATION VS. EXPLOITATION

A fundamental challenge with approximate dynamic programming is that our ability to estimate a value function may require that we visit states just to estimate the value of being in the state. Should we make a decision because we think it is the best decision (based on our current estimate of the values of states the decision may take us to), or do we make a decision just to try something new? This is a decision we face in day-to-day life, so it is not surprising that we face this problem in our algorithms.

This choice is known in the approximate dynamic programming literature as the "exploration vs. exploitation" problem. Do we make a decision to explore a state? Or do we "exploit" our current estimates of downstream values to make what we think is the best possible decision? It can cost time and money to visit a state, so we have to consider the future value of action in terms of improving future decisions.

When we are using approximate dynamic programming, we always have to think about how well we know the value of being in a state, and whether a decision will help us learn this value better. The science of learning in approximate dynamic programming is quite young, with many unanswered questions. This chapter provides a peek into the issues and some of the simpler results that can be put into practice.

## 10.1 A LEARNING EXERCISE: THE NOMADIC TRUCKER

A nice illustration of the exploration versus exploitation problem is provided by our nomadic trucker example. Assume that the only attribute of our nomadic trucker is his location. Thus, $a = \{i\}$, where $i \in \mathcal{I}$ is a location. We let $b$ be the attributes of a demand. For

*Approximate Dynamic Programming.* By Warren B. Powell

this problem, $b = (i, j)$ where $i$ is the origin of the order and $j$ is the destination. At any location, we have two types of choices:

$\mathcal{D}^M$ = The set of locations to which a driver can move empty. Every element of $\mathcal{D}^M$ corresponds to a location in $\mathcal{I}$.

$\mathcal{D}^D$ = The set of demand types a customer may cover, where every element of $\mathcal{D}^D$ corresponds to a demand type $b$.

$\mathcal{D}$ = $\mathcal{D}^M \cup \mathcal{D}^D$.

The demands are defined by

$\hat{D}_{tb}$ = The number of customer demands with attribute $b$ that become known during time interval $t$.

We represent decisions using the indicator variable

$$x_d = \begin{cases} 1 & \text{if we choose decision } d, \\ 0 & \text{otherwise.} \end{cases}$$

As the driver arrives to location $i$, he sees the demands in $\hat{D}_t$. The driver may choose to serve one of these orders, thereby earning a positive revenue, or he may choose to move empty to another location (that is, he may choose a decision $d \in \mathcal{D}^M$). Included in the set $\mathcal{D}^M$ is location $i$, representing a decision to stay in the same location for another time period. Each decision earns a contribution $C(a, d)$ which is positive if $d \in \mathcal{D}^D$ and negative or zero if $d \in \mathcal{D}^M$.

The driver makes his decision by solving

$$x^n = \arg\max_x \sum_{d \in \mathcal{D}} \left(C(a^n, d) + \gamma \bar{V}^{n-1}(a^M(a^n, d))\right) x_d.$$

subject to

$$\begin{aligned} \sum_{d \in \mathcal{D}} x_d &= 1, \\ x_d &\leq \hat{D}_{tb}, \\ x_d &\geq 0. \end{aligned}$$

Let $d^n$ be the decision corresponding to $x^n_d = 1$. Here, $a^M(a, d)$ tells us the destination that results from making a decision. For this illustration, we assume the attribute transition function is deterministic. $\bar{V}^{n-1}$ is our estimate of the value of being in this state. After making a decision, we compute

$$\hat{v}^n_a = C(a^n, d^n) + \gamma \bar{V}^{n-1}(a^M(a^n, d^n)),$$

and then update our value function using

$$\bar{V}^n(a) = \begin{cases} (1 - \alpha_{n-1}) \bar{V}^{n-1}(a) + \alpha_{n-1} \hat{v}^n_a & \text{if } a = a^M(a^n, d^n), \\ \bar{V}^{n-1}(a) & \text{otherwise.} \end{cases}$$

We start by initializing the value of being in each location to zero, and use a pure exploitation strategy. If we simulate 500 iterations of this process, we produce the pattern

10.1a: Low initial estimate of the value function.

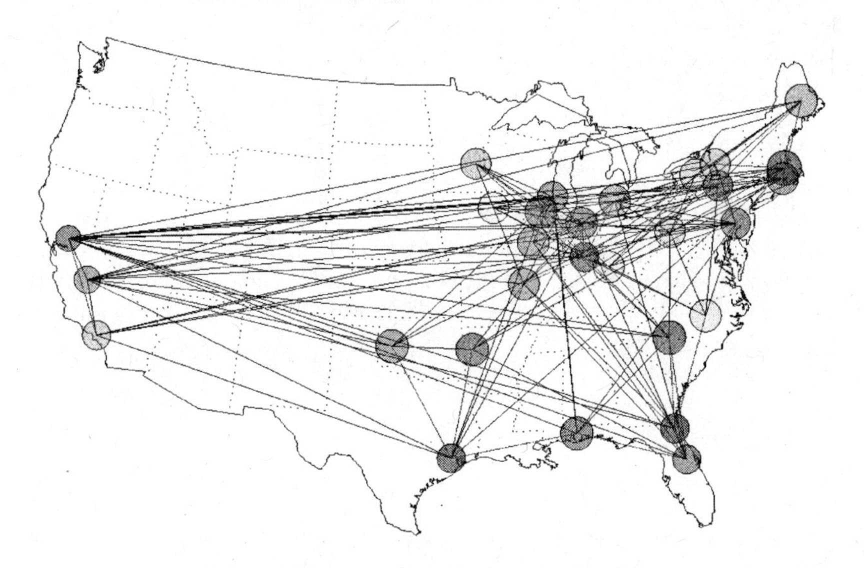

10.1b: High initial estimate of the value function.

**Figure 10.1** The effect of value function initialization on search process. Case (a) uses a low initial estimate and produces limited exploration; case (b) uses a high initial estimate, which forces exploration of the entire state space.

shown in figure 10.1a. Here, the circles at each location are proportional to the value $\bar{V}^{500}(a)$ of being in that location. The small circles indicate places where the trucker never visited. Out of 30 cities, our trucker has ended up visiting seven.

An alternative strategy is to initialize $\bar{V}^0(a)$ to a large number. For our illustration, where rewards tends to average several hundred dollars per iteration (we are using a discount factor of 0.80), we might initialize the value function to \$2000 which is higher than we would

| Logic | Profits |
|---|---|
| Low initial estimate | 120 |
| High initial estimate | 208 |
| Explore all | 248 |

**Table 10.1** Comparison of pure exploitation with low and high initial estimates to one which uses exploration.

expect the optimal solution to be. Using the same strategy, visiting a state generally produces a reduction in the estimate of the value of being in the state. Not surprisingly, the logic tends to favor visiting locations we have never visited before (or have visited the least). The resulting behavior is shown in figure 10.1b. Here, the pattern of lines shows that after 500 iterations, we have visited every city.

How do these strategies compare? We also ran an experiment where we estimated the value functions by using a pure exploration strategy, where we ran five iterations of sampling every single location. Then, for all three methods of estimating the value function, we simulated the policy produced by these value functions for 200 iterations. The results are shown in table 10.1. The results show that for this example, pure exploitation with a high initial estimate for the value function works better than when we use a low initial estimate, but estimating the value functions using a pure exploration strategy works best of all.

The difficulty with using a high initial estimate for the value function is that it leads us to visit every state (at least, every state that we can reach). If the problem has a large state space, this will produce very slow convergence since we have to visit many states just to determine that the states should not have been visited. In effect, exploitation with a high initial value will behave (at least initially) like pure exploration. For applications such as the nomadic trucker (where the state space is relatively small), this works quite well. But the logic will not generalize to more complex problems.

## 10.2 LEARNING STRATEGIES

Much of the challenge of estimating a value function is identical to that facing any statistician trying to fit a model to data. The biggest difference is that in dynamic programming, we may choose what data to collect by controlling what states to visit. Further complicating the problem is that it takes time (and may cost money) to visit these states to collect the information. Do we take the time to visit the state and better learn the value of being in the state? Or do we live with what we know?

Below we review several simple strategies, any of which can be effective for specific problem classes.

### 10.2.1 Pure exploration

Here, we use an exogenous process (such as random selection) to choose either a state to visit, or a state-action pair (which leads to a state). Once in a state, we sample information and obtain an estimate of the value of being in the state which is then used to update our estimate.

In a pure exploration strategy, we can guarantee that we visit every state, or at least have a chance of visiting every state. This property is often used to produce convergence proofs. We need to remember that some problems have $10^{100}$ states or more, so even if we run a million iterations, we may sample only a fraction of the complete state space. For problems with large state spaces, random exploration (that is, choosing states or actions purely at random) is unlikely to work well. This is a reminder that provably convergent algorithms can work terribly in practice.

The amount of exploration we undertake depends in large part on the cost of collecting the information (how much time does it take to run each iteration) and the value of the information we collect. It is important to explore the most important states (of course, we may not know which states are important).

### 10.2.2 Pure exploitation

A pure exploitation strategy assumes that we have to make decisions by solving

$$x_t^n = \arg\max_{x_t \in \mathcal{X}_t} \left( C(S_t^n, x_t) + \bar{V}_t^{n-1}(S^{M,x}(S_t^n, x_t)) \right).$$

This decision then determines the next state that we visit. Some authors refer to this as a *greedy* strategy, since we are doing the best that we think we can given what we know.

A pure exploitation strategy may be needed for practical reasons. For example, consider a large resource allocation problem where we have a resource vector $R_t = (R_{ta})_{a \in \mathcal{A}}$ which we act on with a decision vector $x_t = (x_{tad})_{a \in \mathcal{A}, d \in \mathcal{D}}$. For some applications, the dimensionality of $R_t$ may be in the thousands, while $x_t$ may be in the tens of thousands. For problems of this size, exploration may be of little or no value (what is the point in randomly sampling even a million states out of a population of $10^{100}$?). Exploitation strategies focus our attention on states that offer some likelihood of being states we are interested in.

The problem with pure exploitation is that it is quite easy to become stuck in a local solution simply because we have poor estimates of the value of being in some states. While it is not hard to construct small problems where this problem is serious, the errors can be substantial on virtually any problem that lacks specific structure that can be exploited to ensure convergence. As a rule, optimal solutions are not available for large problems, so we have to be satisfied with doing the best we can. But just because your algorithm appears to have converged, do not fool yourself into believing that you have reached an optimal, or even near-optimal, solution.

### 10.2.3 Mixed exploration and exploitation

A common strategy is to mix exploration and exploitation. We might specify an exploration rate $\rho$ where $\rho$ is the fraction of iterations where decisions should be chosen at random (exploration). The intuitive appeal of this approach is that we maintain a certain degree of forced exploration, while the exploitation steps focuses attention on the states that appear to be the most valuable.

This strategy is particularly popular for proofs of convergence because it guarantees that in the limit, all (reachable) states will be visited infinitely often. This property is then used to prove that estimates will reach their true value. Unfortunately, "provably convergent" algorithms can produce extremely poor solutions. A proof of convergence may bring some level of comfort, but in practice it may provide almost no guarantee at all of a high quality solution.

In practice, using a mix of exploration steps only adds value for problems with relatively small state or action spaces. The only exception arises when the problem lends itself to an approximation which is characterized by a relatively small number of parameters. Otherwise, performing, say, 1000 exploration steps for a problem with $10^{100}$ states may provide little or no practical value.

### 10.2.4 Boltzmann exploration

The problem with exploration steps is that you are choosing a decision $d \in \mathcal{D}$ at random (we use $d$ instead of $x$ since this idea really only works for small action spaces). Sometimes this means that you are choosing really poor decisions where you are learning nothing of value. An alternative is *Boltzmann exploration* where from state $s$, a decision $d$ is chosen with a probability proportional to the estimated value of a decision. For example, let $Q(s,d) = C(s,d) + \bar{V}^n(s,d)$ be the value of choosing decision $d$ when we are in state $s$. Using Boltzmann exploration, if we are in state $s$ at iteration $n$ we would choose decision $d$ with probability

$$P^n(s,d) = \frac{e^{Q(s,d)/T}}{\sum_{d' \in \mathcal{D}} e^{Q(s,d')/T}}.$$

$T$ is known as the temperature, since in physics (where this idea has its origins), electrons at high temperatures are more likely to bounce from one state to another (we generally replace $1/T$ with a scaling factor $\beta$). As the parameter $T$ increases, the probability of choosing different decisions becomes more uniform. As $T \rightarrow 0$, the probability of choosing the best decision approaches 1.0. It makes sense to start with $T$ relatively large and steadily decrease it as the algorithm progresses.

It is common to write $P^n(s,d)$ in the form

$$P^n(s,d) = \frac{e^{\beta^n(s)Q(S,d)}}{\sum_{d' \in \mathcal{D}} e^{\beta^n(s)Q(S,d')}}. \tag{10.1}$$

where $\beta^n(s)$ is a state-specific scaling coefficient. Let $N^n(s)$ be the number of visits to state $s$ after $n$ iterations. We can ensure convergence if we choose $\beta^n(s)$ so that $P^n(s,d) \geq c/n$ for some $c \leq 1$ (which we choose below). This means that we want

$$\begin{aligned}
\frac{e^{\beta^n(s)Q(S,d)}}{\sum_{d' \in \mathcal{D}} e^{\beta^n(s)Q(S,d')}} &\geq \frac{c}{N^n(s)}, \\
N^n(s)e^{\beta^n(s)Q(S,d)} &\geq c \sum_{d' \in \mathcal{D}} e^{\beta^n(s)Q(S,d')}.
\end{aligned}$$

Let $|\mathcal{D}|$ be the number of decisions $d$, and let

$$d^{max}(S) = \arg\max_{d \in \mathcal{D}} Q(S,d)$$

be the decision that returns the highest value of $Q(S,d)$. Then, we can require that $\beta^n(s)$ satisfy

$$\begin{aligned}
N^n(s)e^{\beta^n(s)Q(S,d)} &\geq c|\mathcal{D}|e^{\beta^n(s)Q(S,d^{max}(S))}, \\
\frac{N^N(S)}{c|\mathcal{D}|} &\geq e^{\beta^n(s)(Q(S,d^{max}(S))-Q(S,d))}, \\
\ln N^n(s) - \ln c|\mathcal{D}| &\geq \beta^n(s)(Q(S,d^{max}(S)) - Q(S,d)).
\end{aligned}$$

Now choose $c = 1/|\mathcal{D}|$ so that $\ln c|\mathcal{D}| = 0$. Solving for $\beta^n(s)$ gives us

$$\beta^n(s) = N^n(s)/\delta Q(s),$$

where $\delta Q = \max_{d\in\mathcal{D}} |Q(S, d^{max}(S)) - Q(S,d)|$. Since $\delta Q$ is bounded, this choice guarantees that $\beta^n(s) \rightarrow \infty$, which means that in the limit, we will be only visiting the best decision (a pure greedy policy).

Boltzmann exploration provides for a more rational choice of decision which focuses our attention on better decisions, but provides for some degree of exploration. The probability of exploring decisions that look really bad is small, limiting the amount of time that we spend exploring poor decisions (but this assumes we have reasonable estimates of the value of these state/action pairs). Those which appear to be of lower value are selected with a lower probability. We focus our energy on the decisions that appear to be the most beneficial, but provide for intelligent exploration. However, this idea only works for problems where the number of decisions $\mathcal{D}$ is relatively small, and it requires that we have some sort of estimate of the values $Q(S,d)$.

### 10.2.5 Epsilon-greedy exploration

We can apply a principle we used in Boltzmann exploration to produce a somewhat more sophisticated form of exploration. Instead of executing an exploration step with a fixed probability $\rho$, we can explore with a probability $\rho^n(s)$ when we are in state $s$ after $n$ iterations using

$$\rho^n(s) \quad = \quad \epsilon^n(s).$$

Let $\epsilon^n(s) = c/N^n(s)$ where $0 < c < 1$. As before, $N^n(s)$ is the number of times we have visited state $s$ by iteration $n$. When we explore, we will choose an action $d$ with probability $1/|\mathcal{D}|$. We choose to explore with probability $\epsilon^n(s)$. This means that the probability we choose decision $d$ when we are in state $s$, given by $P^n(s,d)$, is at least $\epsilon^n(s)/|\mathcal{D}|$. This guarantees that we will visit every state infinitely often since

$$\sum_{n=1}^{\infty} P^n(s,d) = \sum_{n=1}^{\infty} \epsilon^n(s)/|\mathcal{D}| = \infty.$$

As with Boltzmann exploration, this logic depends on fairly small action spaces.

### 10.2.6 Remarks

The tradeoff between exploration and exploitation is illustrated in figure 10.2 where we are estimating the value of being in each state for a small problem with a few dozen states. For this problem, we are able to compute the exact value function, which allows us to compute the value of a policy using the approximate value function as a percentage of the optimal. This graph nicely shows that pure exploration has a much faster initial rate of convergence, whereas the pure exploitation policy works better as the function becomes more accurate.

This behavior, however, is very problem-dependent. The value of any exploration strategy drops as the number of parameters increases. If a mixed strategy is used, the best fraction of exploration iterations depends on the characteristics of each problem and may be difficult to ascertain without access to an optimal solution. Tests on smaller, computationally tractable problems (where exploration is more useful) will not tell us the right balance for larger problems.

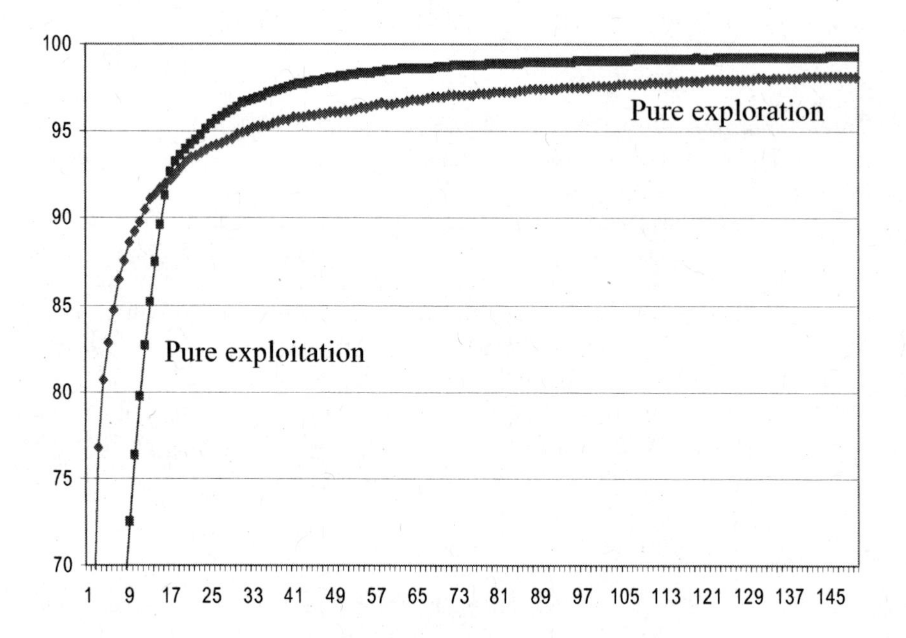


**Figure 10.2** Pure exploration outperforms pure exploitation initially, but slows as the iterations progress.

## 10.3 A SIMPLE INFORMATION ACQUISITION PROBLEM

Consider the situation of a company selling a product at a price $p$ during time period $t$. Assume that production costs are negligible, and that the company wants to sell the product at a price that maximizes revenue. Let $p*$ be this price, which is unknown. Further assume that the lost revenue (per unit sold) is approximated by $\beta(p_t - p^*)^2$ which, of course, can only be computed if we actually knew the optimal price.

In any given time period (e.g., a month) the company may conduct market research at a cost of \$$c$ per unit sold (assume the company continues to sell the product during this time). When the company conducts a market research study, it obtains an imperfect estimate of the optimal price which we denote $\hat{p}_t = p^* + \epsilon_t$ where the errors $\epsilon_t$ are independent and normally distributed with $\mathbb{E}\epsilon = 0$ and $Var(\epsilon) = \sigma^2$. Let $x_t = 1$ if the company conducts a market research study during time period $t$, and 0 otherwise. We assume that our ability to estimate the correct price is independent of our pricing policy. For this reason, the market research strategy, captured by $x = (x_t)_t$, is independent of the actual observations (and is therefore deterministic). Our goal is to minimize expected costs (lost revenue plus marketing costs) per unit over a finite horizon $t = 1, 2, \ldots, T$.

Since each market research study gives us an unbiased estimate of the true optimal price, it makes sense for us to set our price to be the average over all the market research studies. Let

$$n_t = \sum_{t'=1}^{t} x_{t'}$$

be the number of market research studies we have performed up to (and including) time $t$. Thus

$$\bar{p}_t = \begin{cases} \frac{n_t - 1}{n_t}\bar{p}_{t-1} + \frac{1}{n_t}\hat{p}_t & \text{if } x_t = 1, \\ \bar{p}_{t-1} & \text{otherwise.} \end{cases}$$

$\bar{p}_t$ is an unbiased estimate of $p^*$ with variance

$$\bar{\sigma}^2 = \frac{\sigma^2}{n_t},$$

where we assume for simplicity that $\sigma^2$ is known. We note that our lost revenue function was conveniently chosen so that

$$\mathbb{E}\beta(\bar{p}_t - p^*)^2 = \beta\bar{\sigma}^2.$$

Since our decisions $x_t$ are independent of the state of our system, we can formulate the optimization problem for choosing $x$ as follows:

$$\begin{aligned} \min_x F(x) &= \mathbb{E}\sum_{t=1}^{T}\left(\beta(\bar{p}_t - p^*)^2 + cx_t\right) \\ &= \sum_{t=1}^{T}\left(\beta\bar{\sigma}^2 + cx_t\right). \end{aligned}$$

We use the intuitive result (which the reader is expected to prove in exercise 10.4) that we should perform market research for $\tau$ time periods and then stop. This means that $x_t = 1,\ t = 1, 2, \ldots, \tau$ with $x_t = 0,\ t > \tau$, which also implies that $n_t = t$ for $t \leq \tau$. Using this behavior, we may simplify $F(x)$ to be

$$F(\tau) = \sum_{t=1}^{\tau}\left(\beta\frac{\sigma^2}{t} + c\right) + \sum_{t=\tau+1}^{T}\beta\frac{\sigma^2}{\tau}.$$

We can solve this easily if we treat time as continuous, which allows us to write $F(\tau)$ as

$$\begin{aligned} F(\tau) &= \int_{t=1}^{\tau}\left(\beta\frac{\sigma^2}{t} + c\right)dt + \int_{t=\tau}^{T}\beta\frac{\sigma^2}{\tau}\,dt \\ &= \left(\beta\sigma^2\ln t + ct\right)\Big|_1^{\tau} + \beta\frac{\sigma^2}{\tau}(t)\Big|_{\tau}^{T} \\ &= \beta\sigma^2\ln\tau + c(\tau - 1) + \beta\frac{\sigma^2}{\tau}(T - \tau). \end{aligned}$$

Differentiating with respect to $\tau$ and setting the result equal to zero gives

$$\frac{\partial F(\tau)}{\partial\tau} = \beta\sigma^2\frac{1}{\tau} + c - \beta\sigma^2\frac{T}{\tau^2} = 0.$$

Finding the optimal point $\tau^*$ to stop collecting information requires solving

$$c\tau^2 + \beta\sigma^2\tau - \beta\sigma^2 T = 0.$$

Applying the familiar solution to quadratic equations and recognizing that we are interested in a positive solution, gives

$$\tau = \frac{-\beta\sigma^2 + \sqrt{(\beta\sigma^2)^2 + 4c\beta\sigma^2 T}}{2c}.$$

We see from this expression that the amount of time we should be collecting information increases with $\sigma^2$, $\beta$, and $T$ and decreases with $c$, as we would expect. If there is no noise ($\sigma^2 = 0$), then we should not collect any information. Most importantly, it highlights the concept that there is an optimal strategy for collecting information, and that we should collect more information when our level of uncertainty is higher. The next section extends this basic idea to a more general (but still restrictive) class of problems.

## 10.4 GITTINS INDICES AND THE INFORMATION ACQUISITION PROBLEM

For the most part, the best balance of exploration and exploitation is ad hoc, problem-dependent and highly experimental. There is, however, one body of theory that offers some very important insights into how to best make the tradeoff between exploring and exploiting. This theory is often referred to as *multiarmed bandits*, which is the name given to the underlying mathematical model, or *Gittins indices*, which refers to an elegant method for solving the problem.

There are two perspectives of the information acquisition problem. First, this is a problem in its own right that can be modeled and solved as a dynamic program. There are many situations where the problem is to collect information, and we have to balance the cost of collecting information against the benefits of using it. The market research problem in section 10.3 is one such example. The second perspective is that this is a problem that arises in virtually every approximate dynamic programming problem. As we illustrated with our nomadic trucking application (which is, after all, a generic discrete dynamic program), we have to visit states to estimate the value of being in a state. We depend on our value function approximation $\bar{V}_t(S_t^x)$ to approximate the value of a decision that takes us to state $S_t^x$. Our estimate may be low, but the only way we are going to find out is to actually visit the state.

In this section, we begin with a pure application of information acquisition and then make the transition to its application in a broad range of dynamic programming applications.

### 10.4.1 Foundations

Consider the problem faced by a gambler playing a set of slot machines (often referred to as "one-armed bandits") in a casino. Now pretend that the probability of winning is different for each slot machine, but we do not know what these probabilities are. We can, however, obtain information about the probabilities by playing a machine and watching the outcomes. Because our observations are random, the best we can do is obtain statistical estimates of the probabilities, but as we play a machine more, the quality of our estimates improves.

Since we are looking at a set of slot machines, the problem is referred to as the multiarmed bandit problem. This is a pure exercise in information acquisition, since after every round, our player is faced with the same set of choices. Contrast this situation with most dynamic programs which involve allocating an asset where making a decision changes the attribute (state) of the asset. In the multiarmed bandit problem, after every round the player faces the same decisions with the same rewards. All that has changed is what she *knows* about the system.

This problem, which is extremely important in approximate dynamic programming, provides a nice illustration of what might be called the *knowledge state* (some refer to it as the *information state*). The difference between the state of the resource (in this case, the player) and the state of what we know has confused authors since Bellman first encountered the issue. The vast majority of papers in dynamic programming implicitly assume that the state variable is the state of the resource or the physical state of the system. This is precisely the reason that our presentation in chapter 5 adopted the term "resource state" to be clear about what we were referring to. In other areas of engineering, we might use the term "physical state."

In our multiarmed bandit problem, let $\mathcal{I}$ be the set of slot machines, and let $W_i$ be the random variable that gives the amount that we win if we play the $i^{th}$ bandit. Most of our presentation assumes that $W_i$ is normally distributed. Let $\theta_i$ be the true mean of $W_i$ (which is unknown) and let $\sigma_i^2$ be the variance (which we may assume is known or unknown). Let $(\bar{\theta}_i^n, \hat{\sigma}_i^{2,n})$ be our estimate of the mean and variance of $W_i$ after $n$ iterations. Under our assumption of normality, the mean and variance completely determine the distribution.

We next need to specify our transition equations. When we were managing physical assets, we used equations such as $R_{t+1} = [R_t + x_t - D_{t+1}]^+$ to capture the quantity of assets available. In our bandit problem, we have to show how the estimates of the parameters of the distribution evolve over time. Now let $x_i^n = 1$ if we play the $i^{th}$ slot machine during the $n^{th}$ round, and let $W_i^n$ be the amount that we win in this round. Note that this represents a departure from our traditional indexing style, since we choose $x^n$ and then learn $W^n$. Also let

$$N_i^n = \sum_{i=1}^{n} x_i^n$$

be the total number of times we sampled the $i^{th}$ machine. Since the observations $(W_i^{n'})_{n'=1}^n$ come from the same distribution, the best estimate of the mean is a simple average, which can be computed recursively using

$$\bar{\theta}_i^n = \begin{cases} \frac{N_i^n - 1}{N_i^n} \bar{\theta}_i^{n-1} + \frac{1}{N_i^n} W_i^n & \text{If } x_i^n = 1 \\ \bar{\theta}_i^{n-1} & \text{Otherwise.} \end{cases} \tag{10.2}$$

Similarly, we would estimate the variance of $W_i$ using

$$\hat{\sigma}_i^{2,n} = \begin{cases} \frac{N_i^n - 2}{N_i^n - 1} \hat{\sigma}_i^{2,n-1} + \frac{1}{N_i^n} (W_i^n - \bar{\theta}_i^{n-1})^2 & \text{if } x_i^n = 1 \text{ and } N_i^n \geq 2, \\ \hat{\sigma}_i^{2,n-1} & \text{if } x_i^n = 0. \end{cases} \tag{10.3}$$

Of course, $N_i^n$ can be updated using

$$N_i^n = N_i^{n-1} + x_i^n. \tag{10.4}$$

We are more interested in the variance of $\bar{\theta}_i^n$, which is given by

$$\bar{\sigma}_i^{2,n} = \frac{1}{N_i^n} \hat{\sigma}_i^{2,n}. \tag{10.5}$$

The state of our system (that is, our "state of knowledge") is $S^n = (\bar{\theta}^n, \hat{\sigma}^{2,n}, N^n)$ where each quantity is a vector over all slot machines. The transition function consists of equations

(10.2), (10.3), and (10.4). We would write the transition function, using our indexing notation, as $S^n = S^M(S^{n-1}, x^n, W^n)$.

We note that $\hat{\sigma}_i^{2,n} = \infty$ if $N_i^n$ is 0 or 1. This assumes that we do not know anything about the distribution before we start collecting data. It is often the case when we are working on information acquisition problems that we are willing to use some prior distribution.

One challenge in using (10.3) to estimate the variance, especially for larger problems, is that the number of observations $N_i^n$ may be quite small, and often zero or 1. A reasonable approximation may be to assume (at least initially) that the variance is the same across the slot machines. In this case, we could estimate a single population variance using

$$\hat{\sigma}^{2,n} = \frac{n-2}{n-1}\hat{\sigma}^{2,n-1} + \frac{1}{n}\sum_{i\in\mathcal{I}} x_i^n (W_i^n - \bar{\theta}_i^{n-1})^2,$$

which is updated after every play. The variance of $\bar{\theta}_i^n$ would then be given by

$$\bar{\sigma}_i^{2,n} = \frac{1}{N_i^n}\hat{\sigma}^{2,n}.$$

Even if significant differences are suspected between different choices, it is probably a good idea to use a single population variance unless $N_i^n$ is at least 10.

Under the assumption of normality, $S^n = (\bar{\theta}^n, \hat{\sigma}^{2,n}, N^n)$ is our state variable, where equations (10.2) and (10.3) represent our transition function. We do not have a resource (physical) state variable because our "resource" (the player) is always able to play the same machines after every round, without affecting the reward structure. Some authors (including Bellman) refer to $S^n$ as the *hyperstate*, but given our definitions (see section 5.4), this is a classic state variable since it captures everything we need to know to model the future evolution of our system.

Given this model, it would appear that we have a classic dynamic program. We have a $3|\mathcal{I}|$-dimensional state variable, consisting of the vectors $\bar{\theta}^n$, $\hat{\sigma}^{2,n}$ and $N^n$. In addition, two of these vectors are continuous. Even if we could model $\bar{\theta}^n$ and $\hat{\sigma}^{2,n}$ as discrete, we have a multidimensional state variable with all the computational challenges this entails.

In a landmark paper (Gittins & Jones (1974)), it was shown that this problem could be solved as a series of one-dimensional problems using an index policy. That is, it is possible to compute a number $\nu_i$ for each bandit $i$, using information about only this bandit. It is then optimal to choose which bandit to play next by simply finding the largest $\nu_i$ for all $i \in \mathcal{I}$. This is known as an index policy, and the values $\nu_i$ are widely known as *Gittins indices*.

### 10.4.2 Basic theory of Gittins indices

Assume we face the choice of playing a single slot machine, or stopping and converting to a process that pays a fixed reward $\rho$ in each time period until infinity. If we choose to stop sampling and accept the fixed reward, the total future reward is $\rho/(1-\gamma)$. Alternatively, if we play the slot machine, we not only win a random amount $W$, we also learn something about the parameter $\theta$ that characterizes the distribution of $W$ (for our presentation, $\mathbb{E}W = \theta$, but $\theta$ could be a vector of parameters that characterizes the distribution of $W$). $\bar{\theta}^n$ represents our state variable, and the optimality equations are

$$V(\bar{\theta}^n, \rho) = \max\left[\rho + \gamma V(\bar{\theta}^n, \rho), C(\bar{\theta}^n) + \gamma \mathbb{E}\left\{V(\bar{\theta}^{n+1}, \rho) \middle| \bar{\theta}^n\right\}\right], \quad (10.6)$$

where we have written the value function to express the dependence on $\rho$. $C(\bar{\theta}^n) = \mathbb{E}W = \bar{\theta}^n$ is our expected reward given our estimate $\bar{\theta}^n$. We use the format $C(\bar{\theta}^n)$ for consistency with our earlier models.

Since we have an infinite horizon problem, the value function must satisfy the optimality equations

$$V(\bar{\theta}, \rho) = \max\left[\rho + \gamma V(\bar{\theta}, \rho), C(\bar{\theta}) + \gamma \mathbb{E}\left\{V(\bar{\theta}', \rho) \middle| \bar{\theta}\right\}\right],$$

where $\bar{\theta}'$ is defined by equation (10.2). It can be shown that if we choose to stop sampling in iteration $n$ and accept the fixed payment $\rho$, then that is the optimal strategy for all future rounds. This means that starting at iteration $n$, our optimal future payoff (once we have decided to accept the fixed payment) is

$$\begin{aligned} V(\bar{\theta}, \rho) &= \rho + \gamma\rho + \gamma^2\rho + \cdots \\ &= \frac{\rho}{1-\gamma}, \end{aligned}$$

which means that we can write our optimality recursion in the form

$$V(\bar{\theta}^n, \rho) = \max\left[\frac{\rho}{1-\gamma}, C(\bar{\theta}^n) + \gamma \mathbb{E}\left\{V(\bar{\theta}^{n+1}, \rho) \middle| \bar{\theta}^n\right\}\right]. \tag{10.7}$$

Now for the magic of Gittins indices. Let $\nu$ be the value of $\rho$ which makes the two terms in the brackets in (10.7) equal. That is,

$$\frac{\nu}{1-\gamma} = C(\bar{\theta}) + \gamma \mathbb{E}\left\{V(\bar{\theta}', \nu) \middle| \bar{\theta}\right\}. \tag{10.8}$$

$\nu$ depends on our current estimate of the mean, $\bar{\theta}$, the estimate of the variance $\bar{\sigma}^2$, and the number of observations $n$ we have made of the process. We express this dependence by writing the index as $\nu(\bar{\theta}, \bar{\sigma}^2, n)$. Next assume that we have a family of slot machines $\mathcal{I}$, and let $\nu_i(\bar{\theta}_i, \bar{\sigma}_i^2, N_i^n)$ be the value of $\nu$ that we compute for each slot machine $i \in \mathcal{I}$, where $N_i^n$ is the number of times we have played slot machine $i$ by iteration $n$. An optimal policy for selecting slot machines is to choose the slot machine with the highest value for $\nu_i(\bar{\theta}_i, \bar{\sigma}_i^2, N_i^n)$. Such policies are known as *index policies*, and for this problem, the parameters $\nu_i(\bar{\theta}_i, \bar{\sigma}_i^2, N_i^n)$ are called Gittins indices.

To put this solution in more familiar notation, imagine that we are facing the problem of choosing a decision $d \in \mathcal{D}$, and let $x_d = 1$ if we choose decision $d$, and 0 otherwise. Further assume that at each iteration, we face the same decisions, and we are learning the value of making a decision just as we were learning the value of a reward from using a slot machine. Let $\nu_d(\bar{\theta}_d, \bar{\sigma}_d^2, N_d^n)$ be the "value" of making decision $d$, given our current belief (captured by $(\bar{\theta}_d, \bar{\sigma}_d^2)$) about the potential reward we would receive from this decision. When we ignore the value of acquiring information (as we have done in our presentation of approximate dynamic programming algorithms up to now), we would make a decision by solving

$$\max_{\{x | \sum_d x_d = 1\}} \sum_{d \in \mathcal{D}} \bar{\theta}_d x_d.$$

This solution might easily lead us to avoid a decision that might be quite good, but which we currently think is poor (and we are unwilling to learn anything more about the decision). Gittins theory tells us to solve

$$\max_{\{x | \sum_d x_d = 1\}} \sum_{d \in \mathcal{D}} \nu_d(\bar{\theta}_d, \bar{\sigma}_d^2, N_d^n) x_d,$$

which is the same as choosing the largest value of $\nu_d(\bar{\theta}_d, \bar{\sigma}^2_d, N^n_d)$.

The computation of Gittins indices highlights a subtle issue when computing expectations for information-collection problems. The proper computation of the expectation needed to solve the optimality equations requires, in theory, knowledge of exactly the distribution that we are trying to compute. To illustrate, the expected winnings are given by $C(\bar{\theta}^n) = \mathbb{E}W = \theta$, but $\theta$ is unknown. Instead, we adopt a Bayesian approach that our expectation is computed with respect to the distribution *we believe* to be true. Thus, at iteration $n$ we believe that our winnings are normally distributed with mean $\bar{\theta}^n$, so we would use $C(\bar{\theta}^n) = \bar{\theta}^n$. The term $\mathbb{E}\left\{ V(\bar{\theta}^{n+1}, \rho) \middle| \bar{\theta}^n \right\}$ captures what we believe the effect of observing $W^{n+1}$ will have on our estimate $\bar{\theta}^{n+1}$, but this belief is based on what we think the distribution of $W^{n+1}$ is, rather than the true distribution.

The beauty of Gittins indices (or any index policy) is that it reduces $N$-dimensional problems into a series of one-dimensional problems. The problem is that solving equation (10.7) (or equivalently, (10.8)) offers its own challenges. Finding $\nu(\bar{\theta}, \bar{\sigma}^2, n)$ requires solving the optimality equation in (10.7) for different values of $\rho$ until (10.8) is satisfied. In addition, this has to be done for different values of $\bar{\theta}$ and $n$.. Although algorithmic procedures have been designed for this, they are not simple.

### 10.4.3 Gittins indices for normally distributed rewards

The calculation of Gittins indices is simplified for special classes of distributions. In this section, we consider the case where the observations of rewards $W$ are normally distributed. This is the case we are most interested in, since in the next section we are going to apply this theory to the problem where the unknown reward is, in fact, the value of being in a future state $\bar{V}(s')$. Since $\bar{V}(s')$ is computed using averages of random observations, the distribution of $\bar{V}(s')$ will be closely approximated by a Normal distribution.

Students learn in their first statistics course that normally distributed random variables satisfy a nice property. If $Z$ is normally distributed with mean 0 and variance 1 and if

$$X = \mu + \sigma Z$$

then $X$ is normally distributed with mean $\mu$ and variance $\sigma^2$. This property simplifies what are otherwise difficult calculations about probabilities of events. For example, computing $\mathbb{P}[X \geq x]$ is difficult because the normal density function cannot be integrated analytically. Instead, we have to resort to numerical procedures. But because of the above translationary and scaling properties of normally distributed random variables, we can perform the difficult computations for the random variable $Z$ (the "standard normal deviate"), and use this to answer questions about any random variable $X$. For example, we can write

$$\begin{aligned} \mathbb{P}[X \geq x] &= \mathbb{P}\left[\frac{X-\mu}{\sigma} \geq \frac{x-\mu}{\sigma}\right] \\ &= \mathbb{P}\left[Z \geq \frac{x-\mu}{\sigma}\right]. \end{aligned}$$

Thus, the ability to answer probability questions about $Z$ allows us to answer the same questions about any normally distributed random variable.

The same property applies to Gittins indices. Although the proof requires some development, it is possible to show that

$$\nu(\bar{\theta}, \bar{\sigma}^2, n) = \bar{\theta} + \bar{\sigma}\nu(0, 1, n).$$

Thus, we have only to compute a "standard normal Gittins index" for problems with mean 0 and variance 1, and $n$ observations. Since these "standard normal" indices are special, we denote them by

$$\Gamma(n) = \nu(0, 1, n).$$

Unfortunately, as of this writing, there do not exist easy-to-use software utilities for computing standard Gittins indices. The situation is similar to doing statistics before computers when students had to look up the cumulative distribution for the standard normal deviate in the back of a statistics book. Table 10.2 is exactly such a table for Gittins indices. The table gives indices for both the variance-known and variance-unknown cases. In the variance-known case, we assume that $\sigma^2$ is given, which allows us to calculate the variance of the estimate for a particular slot machine just by dividing by the number of observations.

Given access to a table of values, applying Gittins indices becomes quite simple. Instead of choosing the option with the highest $\bar{\theta}_i^n$ (which we would do if we were ignoring the value of collecting information), we choose the option with the highest value of

$$\bar{\theta}_i^n + \bar{\sigma}_i^n \Gamma(N_i^n),$$

where $\Gamma(N_i^n)$ is the Gittins index for a mean 0, variance 1 problem (we anticipate that in time, software libraries will compute this for us, just as utilities now exist for computing the cumulative standard normal distribution).

Instead of using a table, a common strategy is to set $\Gamma(n)$ heuristically. The simplest strategy is to let $\Gamma(n) = \Gamma$, where $\Gamma$ is a parameter that has to be tuned. Even with a fixed value for $\Gamma$, we still obtain the behavior that our estimate of the standard deviation $\bar{\sigma}^n(a^M(a, d))$ decreases as we sample it more often. Alternatively, we could approximate $\Gamma(n)$ using an expression such as

$$\Gamma(n) = \frac{\rho^G}{\sqrt{n}} \tag{10.9}$$

which captures the behavior that $\Gamma(n)$ approximately decreases according to $\sqrt{n}$. Here, $\rho^G$ is a parameter that has to be tuned.

## 10.5 VARIATIONS

The basic intuition behind Gittins indices has spawned several variations that mimic the basic structure of Gittins indices without the sophisticated theory. To simplify our presentation, we return to our bandit problem where $\bar{\theta}_d^n$ is our estimate of the random variable $W_d$ which gives our winnings if we choose decision $d$, and $\bar{\sigma}_d^n$ is the standard deviation of $\bar{\theta}_d^n$. These are updated each iteration using equations (10.2) to (10.5).

We briefly review two closely related ideas that have been proposed to solve the exploration problem.

### 10.5.1 Interval estimation

Interval estimation sets the value of a decision to the $90^{th}$ or $95^{th}$ percentile of the estimate of the value of a decision. Thus, our decision problem would be

$$\max_{\{x \mid \sum_d x_d = 1\}} \sum_{d \in \mathcal{D}} \left( \bar{\theta}_d^n + \bar{\sigma}_d^n z_{\alpha/2} \right) x_d.$$

| | Discount factor | | | |
|---|---|---|---|---|
| | Known variance | | Unknown variance | |
| Observations | 0.95 | 0.99 | 0.95 | 0.99 |
| 1 | 0.9956 | 1.5758 | - | - |
| 2 | 0.6343 | 1.0415 | 10.1410 | 39.3343 |
| 3 | 0.4781 | 0.8061 | 1.1656 | 3.1020 |
| 4 | 0.3878 | 0.6677 | 0.6193 | 1.3428 |
| 5 | 0.3281 | 0.5747 | 0.4478 | 0.9052 |
| 6 | 0.2853 | 0.5072 | 0.3590 | 0.7054 |
| 7 | 0.2528 | 0.4554 | 0.3035 | 0.5901 |
| 8 | 0.2274 | 0.4144 | 0.2645 | 0.5123 |
| 9 | 0.2069 | 0.3808 | 0.2353 | 0.4556 |
| 10 | 0.1899 | 0.3528 | 0.2123 | 0.4119 |
| 20 | 0.1058 | 0.2094 | 0.1109 | 0.2230 |
| 30 | 0.0739 | 0.1520 | 0.0761 | 0.1579 |
| 40 | 0.0570 | 0.1202 | 0.0582 | 0.1235 |
| 50 | 0.0464 | 0.0998 | 0.0472 | 0.1019 |
| 60 | 0.0392 | 0.0855 | 0.0397 | 0.0870 |
| 70 | 0.0339 | 0.0749 | 0.0343 | 0.0760 |
| 80 | 0.0299 | 0.0667 | 0.0302 | 0.0675 |
| 90 | 0.0267 | 0.0602 | 0.0269 | 0.0608 |
| 100 | 0.0242 | 0.0549 | 0.0244 | 0.0554 |

**Table 10.2** Gittins indices $\Gamma(n)$ for the case of observations that are normally distributed with mean 0, variance 1, from Gittins (1989)

This is identical to the adjustment made in Gittins exploration, but instead of the Gittins index $\Gamma(n)$, we use $z_{\alpha/2}$ where $\alpha = .05$ if we want to use the $90^{th}$ percentile or $\alpha = .025$ if we want to use the $95^{th}$ percentile. Unlike the Gittins index, this number does not decline with the number of iterations (something that we achieved with the approximation for $\Gamma(n)$ in equation (10.9)).

### 10.5.2 Upper confidence bound sampling algorithm

The upper confidence bound (UCB) uses the same idea as Gittins exploration or interval estimation, but with a somewhat different scaling factor for the standard deviation. If $n$ is our iteration counter (which means the total number of samples across all decisions), and $N_d^n$ is the number of times we have sampled decision $d$ after $n$ iterations, then the UCB algorithm makes a decision by solving

$$\max_{\{x|\sum_d x_d=1\}} \sum_{d\in\mathcal{D}} \left( \bar{\theta}_d^n + W^{max}\sqrt{\frac{2\ln n}{N_d}} \right) x_d.$$

where $W^{max}$ is the maximum possible winnings. Instead of $W^{max}$, we might use an estimate of the $95^{th}$ percentile for $W$ (not the $95^{th}$ percentile of $\bar{\theta}^n$).

## 10.6 THE KNOWLEDGE GRADIENT ALGORITHM

Assume that we know that we are going to have a total of $N$ trials to sample a discrete set of $\mathcal{D}$ alternatives. Let $\bar{\theta}_d^N$ be our estimate of the value of decision $d$ after all the trials have been completed. Let $\pi$ be a policy that describes how we are going to go about sampling each decision, and let $\bar{\theta}_d^{\pi,N}$ be the estimate of the value of decision $d$ after $N$ iterations when we used policy $\pi$. Our goal is to find a policy $\pi$ that solves

$$V = \max_{\pi} \mathbb{E}\{\max_d \bar{\theta}_d^{\pi,N}\}.$$

Stating the problem in this way closely matches how most dynamic programming algorithms are run. That is, we will run $N$ training iterations, after which we have to live with our estimates.

At iteration $n$, we again define our state variable as

$$S^n = (\bar{\theta}_d^n, \hat{\sigma}_d^{2,n}, N_d^n)_{d\in\mathcal{D}}.$$

The optimality equation for $V^n(S^n)$ is given by

$$V^n(S^n) = \max_{\{x|\sum_d x_d=1\}} \mathbb{E}\{V^{n+1}(S^M(S^n, x, W^{n+1}))|S^n\}, \tag{10.10}$$

where $W^{n+1}$ is the information that we learn by making decision $x$, and $S^M(S^n, x, W^{n+1})$ represents our updating equations (10.2) - (10.5).

Let $\pi_0$ be a policy where we do not collect any additional information. Now let $V^{\pi_0}(S^n)$ be our estimate of $V^N$ given what we know at iteration $n$ (captured by $S^n$) and assuming that we follow policy $\pi_0$ (which means we do not learn anything more). This means that

$$V^{\pi_0}(S^n) = \mathbb{E}\left\{ \max_{\{x|\sum_d x_d=1\}} \sum_d \bar{\theta}_d^n x_d \right\}.$$

Now consider a decision that maximizes the value of our decision measured by its impact on $V$. That is, we want to make a decision that has the highest marginal impact on the value of our knowledge (hence “knowledge gradient”). Thus, the knowledge gradient policy is defined by

$$X^{KG}(S^n) = \arg\max_{\{x|\sum_d x_d=1\}} \mathbb{E}\left[V^{\pi_0}(S^M(S^n, x, W^{n+1})) - V^{\pi_0}(S^n)|S^n\right].$$

Of course, $V^{\pi_0}(S^n)$ is a constant given $S^n$, but expressing it in this way captures the marginal value of the decision relative to what we know now.

Computing a knowledge gradient policy is extremely easy. We assume that all rewards are normally distributed, and that we start with an initial estimate of the mean and variance of the value of decision $d$, given by

$$\bar{\theta}_d^0 = \text{The initial estimate of the expected reward from making decision } d,$$
$$\bar{\sigma}_d^0 = \text{The initial estimate of the standard deviation of } \bar{\theta}_d^0.$$

Assume that each time we make a decision we receive a reward given by

$$W_d = \mu_d + \varepsilon,$$

where $\mu_d$ is the true expected reward from decision $d$ (which is unknown) and $\varepsilon$ is the measurement error with standard deviation $\sigma^\varepsilon$ (which we assume is known).

Assume that $(\bar{\theta}_d^n, \bar{\sigma}_d^{2,n})$ is the mean and variance after $n$ observations. If we make decision $d$ and observe a reward $W_d(\omega^n)$, we can use Bayesian updating (which produces estimates that are most likely given what we knew before and what we just learned) to obtain new estimates of the mean and variance for decision $d$. To update the mean, we use a weighted sum of the previous estimate, $\bar{\theta}_d^{n-1}$, and the latest observation, $W_d(\omega^n)$, where the weights are inversely proportional to the variance of each estimate. The updated variance is given by

$$\begin{aligned}\bar{\sigma}_d^{2,n} &= \left((\bar{\sigma}_d^{2,n-1})^{-1} + (\sigma^\varepsilon)^{-2}\right)^{-1} \\ &= \frac{(\bar{\sigma}_d^{2,n-1})}{1 + \bar{\sigma}_d^{2,n-1}/(\sigma^\varepsilon)^2}.\end{aligned} \tag{10.11}$$

Thus, if $\bar{\sigma}_d^{2,n-1} = 20^2$ and $(\sigma^\varepsilon)^2 = 40^2$, then

$$\begin{aligned}\bar{\sigma}_d^{2,n} &= \frac{20^2}{1 + \frac{20^2}{40^2}} \\ &= 320 = (17.89)^2.\end{aligned}$$

The updated mean can be written

$$\begin{aligned}\bar{\theta}_d^n &= \frac{(\bar{\sigma}_d^{2,n-1})^{-1}\bar{\theta}_d^{n-1} + (\sigma^\varepsilon)^{-2}W_d(\omega^n)}{(\bar{\sigma}_d^{2,n-1})^{-1} + (\sigma^\varepsilon)^{-2}} \\ &= \bar{\sigma}_d^{2,n}\left((\bar{\sigma}_d^{2,n-1})^{-1}\bar{\theta}_d^{n-1} + (\sigma^\varepsilon)^{-2}W_d(\omega^n)\right).\end{aligned}$$

The expression for updating $\bar{\theta}_d$ simply uses a weighted average of $\bar{\theta}_d^{n-1}$ and $W_d(\omega^n)$, where the weights are proportional to the inverse variances. If we designate the weights $w_1$ and $w_2$, we get

$$\begin{aligned}w_1 &= \frac{(\bar{\sigma}_d^{2,n-1})^{-1}}{(\bar{\sigma}_d^{2,n-1})^{-1} + (\sigma^\varepsilon)^{-2}} \\ &= 0.80, \\ w_2 &= \frac{(\sigma^\varepsilon)^{-2}}{(\bar{\sigma}_d^{2,n-1})^{-1} + (\sigma^\varepsilon)^{-2}} \\ &= 0.20.\end{aligned}$$

Continuing our example, if $\bar{\theta}_d^{n-1} = 200$ and $W_d(\omega^n) = 250$, then

$$\begin{aligned}\bar{\theta}_d^n &= w_1\bar{\theta}_d^{n-1} + w_2 W_d(\omega^n) \\ &= 0.80(200) + 0.20(250) \\ &= 210.\end{aligned}$$

Of course, these updates occur only if we select decision $d$ (that is, if $x_d^n = 1$). Otherwise, the estimates are unchanged.

We next find the variance of the *change* in our estimate of $\theta_d$ assuming we choose to sample decision $d$ in iteration $n$. For this we define

$$\tilde{\sigma}_d^{2,n} = Var[\bar{\theta}_d^{n+1} - \bar{\theta}_d^n | S^n]. \tag{10.12}$$

With a little work, we can write $\tilde{\sigma}_d^{2,n}$ in different ways, including

$$\tilde{\sigma}_d^{2,n} = \bar{\sigma}_d^{2,n-1} - \bar{\sigma}_d^{2,n}, \tag{10.13}$$

$$= \frac{(\bar{\sigma}_d^{2,n-1})}{1 + (\sigma^\varepsilon)^2/\bar{\sigma}_d^{2,n-1}}. \tag{10.14}$$

Equation (10.13) expresses the (perhaps unexpected) result that $\tilde{\sigma}_d^{2,n}$ measures the change in the estimate of the standard deviation of the reward from decision $d$ from iteration $n-1$ to $n$. Equation (10.14) closely parallels equation (10.11). Using our numerical example, equations (10.13) and (10.14) both produce the result

$$\begin{aligned}\tilde{\sigma}_d^{2,n} &= 400 - 320 = 80 \\ &= \frac{40^2}{1 + \frac{10^2}{40^2}} = 80.\end{aligned}$$

We next compute

$$\zeta_d^n = -\left|\frac{\bar{\theta}_d^n - \max_{d' \neq d} \bar{\theta}_{d'}^n}{\tilde{\sigma}_d^n}\right|.$$

$\zeta_d^n$ is called the *normalized influence* of decision $d$. It measures the number of standard deviations from the current estimate of the value of decision $d$, given by $\bar{\theta}_d^n$, and the best alternative other than decision $d$. We then find

$$f(\zeta) = \zeta\Phi(\zeta) + \phi(\zeta),$$

where $\Phi(\zeta)$ and $\phi(\zeta)$ are, respectively, the cumulative standard normal distribution and the standard normal density. Thus, if $Z$ is normally distributed with mean 0, variance 1, $\Phi(\zeta) = \mathbb{P}[Z \leq \zeta]$ while

$$\phi(\zeta) = \frac{1}{\sqrt{2\pi}} \exp\left(-\frac{\zeta^2}{2}\right).$$

The knowledge gradient algorithm chooses the decision $d$ with the largest value of $\nu_d^{KG,n}$ given by

$$\nu_d^{KG,n} = \tilde{\sigma}_d^n f(\zeta_d^n).$$

The knowledge gradient algorithm is quite simple to implement. Table 10.3 illustrates a set of calculations for a problem with five options. $\bar{\theta}$ represents the current estimate of the value of each decision, while $\bar{\sigma}$ is the current standard deviation of each $\bar{\theta}$. Options 1, 2 and 3 have the same value for $\bar{\sigma}$, but with increasing values of $\bar{\theta}$. The table illustrates that when the variance is the same, the knowledge gradient prefers the decisions that appear to be the best. Decisions 3 and 4 have the same value of $\bar{\theta}$, but decreasing values of $\bar{\sigma}$, illustrating that the knowledge gradient prefers decisions with the highest variance. Finally, decision 5 appears to be the best of all the decisions, but has the lowest variance (meaning that we have the highest confidence in this decision). The knowledge gradient is the smallest for this decision out of all of them.

| Decision | $\bar{\theta}$ | $\bar{\sigma}$ | $\tilde{\sigma}$ | $\zeta$ | $f(\zeta)$ | KG index |
|---|---|---|---|---|---|---|
| 1 | 1.0 | 2.5 | 1.336 | -1.497 | 0.030 | 0.789 |
| 2 | 1.5 | 2.5 | 1.336 | -1.122 | 0.066 | 1.754 |
| 3 | 2.0 | 2.5 | 1.336 | -0.748 | 0.132 | 3.516 |
| 4 | 2.0 | 2.0 | 1.155 | -0.866 | 0.107 | 2.467 |
| 5 | 3.0 | 1.0 | 0.707 | -1.414 | 0.036 | 0.503 |

**Table 10.3** The calculations behind the knowledge gradient algorithm

Figure 10.3 compares the performance of the knowledge gradient algorithm against interval estimation (using $\alpha/2 = .025$), Boltzmann exploration, Gittins exploration, uniform exploration and pure exploitation. The results show steady improvement as we collect more observations. Knowledge gradient and interval estimation work the best for this problem. However, it is possible to fool interval estimation by providing an option where the interval is very tight (for an option that is good but not the best). Both of these methods outperform Gittins exploration.

We note that Gittins exploration was designed for "on-line" learning problems, where we earn a reward equal to our measurement $W_d(\omega^n)$. By contrast, the knowledge gradient is designed for "off-line" learning, where we are spending $N$ iterations learning the best value, after which we earn a reward equal to the best value of $\bar{\theta}_d^N$.

## 10.7 INFORMATION ACQUISITION IN DYNAMIC PROGRAMMING

Up to now, we have focused on the problem of determining how to collect information to make a choice which returns an uncertain reward. This is an important problem in its own right, but it turns out that this is something that we are doing all the time in approximate dynamic programming. The difference is that instead of receiving an uncertain contribution, we are making a decision that takes us to a future state that has an uncertain value (captured by our value function approximation).

We can illustrate this using our nomadic trucker example. Assume we are in state $a$, and we need to choose a decision $d \in \mathcal{D}$. If we ignored the value of collecting information, we would solve

$$x^n = \arg\max_{\{x:\sum_d x_d = 1\}} \sum_{d\in\mathcal{D}} x_d\big(C(a,d) + \gamma\bar{V}^{n-1}(a^M(a,d))\big).$$

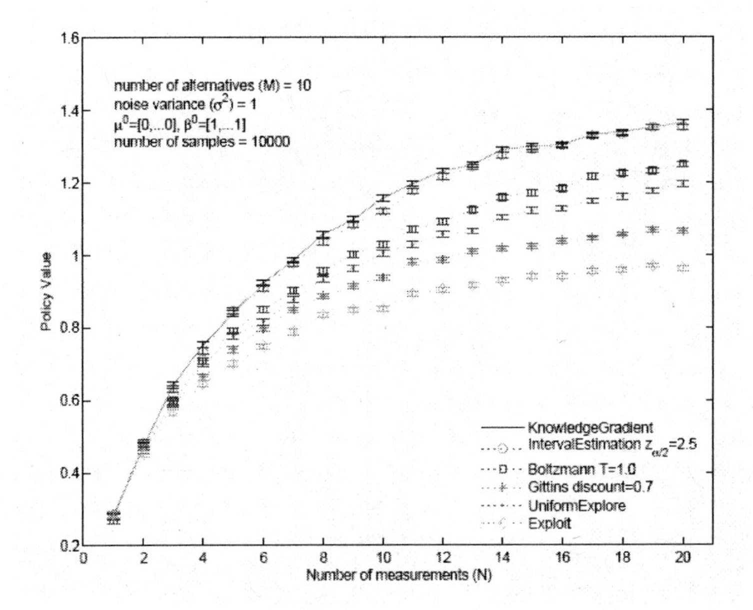


**Figure 10.3** Comparison of knowledge gradient against other policies as a function of the number of observations.

To put this in the context of the bandit problem, we would let

$$\bar{\theta}_d^{n-1} = C(a,d) + \gamma \bar{V}^{n-1}(a^M(a,d)).$$

In this setting, our only source of uncertainty is the estimate of the value function $\bar{V}^{n-1}(a^M(a,d))$ (we can think of other applications where $C(a,d)$ is also uncertain and we have to estimate it). If the variance of $\bar{V}^{n-1}(a^M(a,d))$ is high enough, we might want to visit the state just to learn more about the value of being in the state.

More generally, if we are in state $S$ and take action $x$, the value of this action is given by

$$\bar{\theta}^{n-1}(S,x) = C(S,x) + \gamma \bar{V}^{n-1}(S^{M,x}(S,x)).$$

We encounter three problems when we try to design exploration strategies in the context of general dynamic programs:

1) The value of improving the estimate of the value of a downstream state. Classic exploration strategies (Boltzmann exploration, mixed exploration and epsilon-greedy exploration) do not consider the variance in the estimate of the value of the state $S^{M,x}(S,x)$ that we next visit as a result of our action.

2) The problem of vectors. Imagine that $x$ is a vector, which makes the action space (feasible region) $\mathcal{X}$ virtually infinite in size. It is usually the case that the state is also a vector (with a similarly large state space) for these problems. Choosing an action at random teaches us virtually nothing.

3) Bias due to learning. It is often the case that the more we visit a state, the higher the value. This arises partly because the value is trying to represent a sum of future

rewards (this is particularly pronounced if we are using a single-pass, or TD(0), algorithm), and partly because the value of contributions in the future depends on the quality of the policy (which presumably improves with each iteration).

The first two issues effectively eliminate classical exploration strategies as an option, despite their popularity in the machine learning literature. Boltzmann exploration and exploration strategies where we choose an action at random do not consider the variance in our estimate of the downstream state. Visiting a state to reduce this variance is the central reason for exploring. In addition, if our action is a vector, choosing one decision out of an effectively infinite set of actions is simply not going to teach us anything.

Given these issues, we feel that the most promising methods for exploration are those that assign a value to a decision that captures the benefit from reducing the uncertainty in the estimate of the value of being in the downstream state. Gittins indices, interval estimation, and upper confidence bounding all accomplish this in the same way - adding a bonus equal to a certain number of standard deviations above the point estimate. Some refer to this as the "uncertainty bonus." These methods differ only in how they compute how many standard deviations to use in the bonus. Given the maturity of Gittins indices, we refer to this class of methods as *Gittins exploration*.

Knowledge gradients offer an alternative method for putting a value on a decision. The knowledge gradient is a value, but it measures the value of learning, rather than using an estimate of the value of a decision (plus an "uncertainty bonus"). As of this writing, we do not have any experience using knowledge gradients in a dynamic programming setting.

### 10.7.1 Gittins exploration

If we apply the basic idea behind the Gittins index theory, we would make decisions by solving

$$x^n = \underset{\{x:\sum_d x_d = 1\}}{\arg\max} \sum_{d \in \mathcal{D}} x_d\big(C(s,x) + \gamma \bar{V}^n(S^M(s,x)) + \bar{\sigma}^n(S^M(s,x))\Gamma(n)\big), \quad (10.15)$$

where $\bar{\sigma}^n(S^M(s,x))$ is our estimate of the standard deviation of $\bar{V}(S^M(s,x))$, and $\Gamma(n)$ is the standard normal Gittins index which tells us how many standard deviations away from the mean we should consider (which is a function of the number of observations $n$). This is a purely heuristic use of Gittins indices, and as a result it is common to let $\Gamma(n) = \Gamma$, a constant that does not depend on $n$ (and which has to be tuned for a particular application).

The beauty of equation (10.15) is that it can be used as if we are pursuing an exploitation strategy with a modified contribution function that captures the value of collecting information for future decisions. Unlike combined exploration/exploitation policies, we never choose an action purely at random (regardless of how bad we think it might be). However, it produces a behavior similar to Boltzmann exploration in that we are more likely to make a decision when the reward is highly uncertain (Boltzmann exploration does not even consider the degree of uncertainty).

We refer to this heuristic as *Gittins exploration*. It captures the property that it is a form of exploitation policy, in that at no time are we ever choosing states or actions at random. We are solving a modified optimization problem which includes an additional term that captures the value of gathering information. This term rewards visiting states where there is a high level of uncertainty in the estimate, but will not visit states unless they are competitive with other states (we will not visit a terrible state just to get a better estimate of how terrible it is).

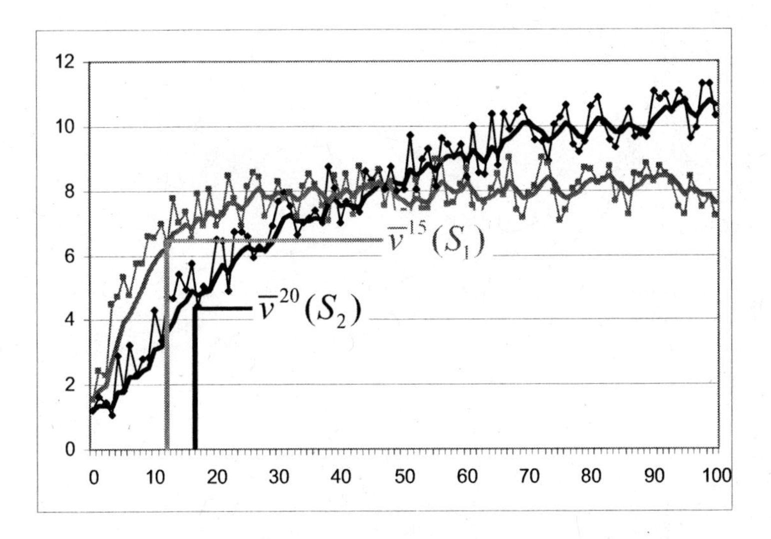


**Figure 10.4** The value of being in state $S_1$ after 15 iterations, versus the value of being in state $S_2$ after 20 iterations.

### 10.7.2 The problem of bias and learning

Perhaps the biggest challenge with all the methods we have introduced for exploration is that they depend on the assumption that the observations of the value of a decision are unbiased. When learning a value function, it is not uncommon for the value to grow steadily; the more you visit a state, the higher the estimate of the value.

The situation is illustrated in figure 10.4, which shows the value of being in state $S_1$ after 15 iterations (given by $\bar{v}^{15}(S_1)$), and state $S_2$ after 20 iterations (given by $\bar{v}^{20}(S_2)$). Not only is $\bar{v}^{15}(S_1)$ higher, it presumably also has a higher variance (since we have visited it fewer times than state $S_2$). All of the exploration methods described earlier would favor decisions that take us to state $S_1$ either because it looks better, or because it has a higher variance. But as the graph shows, if we were to visit state $S_2$ enough, it would eventually have a higher value.

One way to overcome this bias is to use large values of $\Gamma$ in Gittins exploration. Whereas classic Gittins theory would suggest values of $\Gamma$ less than 1 (see table 10.2), much larger values will help overcome the bias due to learning. Experiments with Gittins exploration using the nomadic trucker problem, shown in figure 10.5, demonstrated that the best results were obtained for $\Gamma = 10$, but values as high as 20 and 50 outperformed pure exploitation (equivalent to $\Gamma = 0$). However, it is important to emphasize that this behavior is very problem-dependent, and it would be necessary to tune $\Gamma$ for each application.

One way to overcome the bias in the estimate of the value function is to simulate an information-collecting policy using any of the previously described algorithms to make decisions. We can simulate this policy for several steps (or until the end of the horizon) to obtain a better sample estimate of the value of being in a state (much as we did with tree search in section 8.8.2).

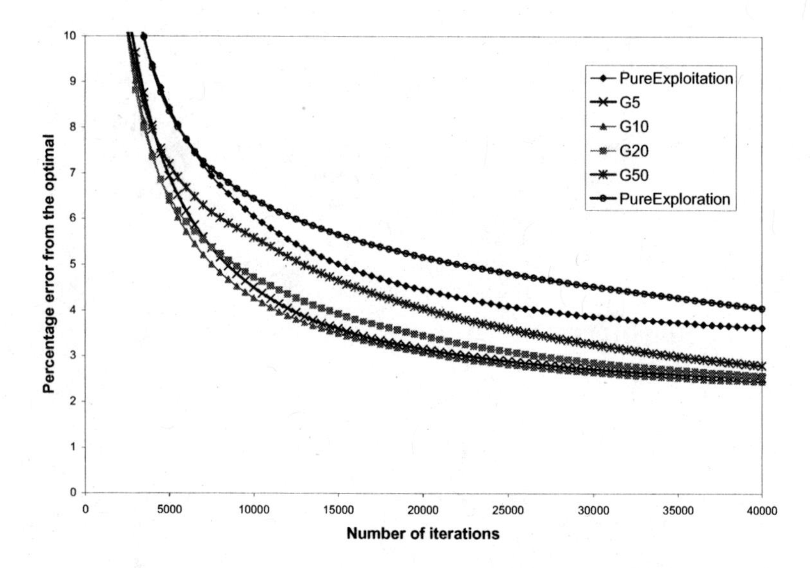


**Figure 10.5** The heuristic application of Gittins to a multiattribute resource management problem produces more accurate value functions than either pure exploitation or pure exploration policies (from George (2005)).

## 10.8 BIBLIOGRAPHIC NOTES

Section 10.2 - A nice introduction to various learning strategies is contained in Kaelbling (1993) and Sutton & Barto (1998). Thrun (1992) contains a good discussion of exploration in the learning process. The discussion of Boltzmann exploration and epsilon-greedy exploration is based on Singh et al. (2000).

Section 10.4 - What came to be known as "Gittins indices" was first introduced in Gittins & Jones (1974) to solve bandit problems (see DeGroot (1970) for a discussion of bandit problems before the development of Gittins indices). This was more thoroughly developed in Gittins (1979), Gittins (1981), and Gittins (1989). Whittle (1982*b*) and Ross (1983) provide very clear tutorials on Gittins indices, helping to launch an extensive literature on the topic (see, for example, Lai & Robbins (1985), Berry & Fristedt (1985), and Weber (1992)).

Section 10.5 - Interval estimation is due to Kaelbling (1993). The upper confidence bound is due to Chang et al. (2007).

Section 10.6 - The knowledge gradient algorithm was introduced by Frazier et al. (2007).

Section 10.7.2 - Chang et al. (2007) present the upper confidence bounding method in the context of a dynamic program which requires searching forward in a tree to help develop the bound. This is the only formal exploration method that we have seen proposed to account for learning bias in dynamic programs.

## PROBLEMS

**10.1** Joe Torre, manager of the great Yankees, has to struggle with the constant game of guessing who his best hitters are. The problem is that he can only observe a hitter if he puts him in the order. He has four batters that he is looking at. The table below shows their actual batting averages (that is to say, batter 1 will produce hits 30 percent of the time, batter 2 will get hits 32 percent of the time, and so on). Unfortunately, Joe does not know these numbers. As far as he is concerned, these are all .300 hitters.

For each at bat, Joe has to pick one of these hitters to hit. Table 10.4 below shows what would have happened if each batter were given a chance to hit (1 = hit, 0 = out). Again, Joe does not get to see all these numbers. He only gets to observe the outcome of the hitter who gets to hit.

Assume that Joe always lets the batter hit with the best batting average. Assume that he uses an initial batting average of .300 for each hitter (in case of a tie, use batter 1 over batter 2 over batter 3 over batter 4). Whenever a batter gets to hit, calculate a new batting average by putting an 80 percent weight on your previous estimate of his average plus a 20 percent weight on how he did for his at bat. So, according to this logic, you would choose batter 1 first. Since he does not get a hit, his updated average would be $0.80(.200) + .20(0) = .240$. For the next at bat, you would choose batter 2 because your estimate of his average is still .300, while your estimate for batter 1 is now .240.

After 10 at bats, who would you conclude is your best batter? Comment on the limitations of this way of choosing the best batter. Do you have a better idea? (It would be nice if it were practical.)

| | Actual batting average | | | |
|---|---|---|---|---|
| | 0.300 | 0.320 | 0.280 | 0.260 |
| Day | | Batter | | |
| | A | B | C | D |
| 1 | 0 | 1 | 1 | 1 |
| 2 | 1 | 0 | 0 | 0 |
| 3 | 0 | 0 | 0 | 0 |
| 4 | 1 | 1 | 1 | 1 |
| 5 | 1 | 1 | 0 | 0 |
| 6 | 0 | 0 | 0 | 0 |
| 7 | 0 | 0 | 1 | 0 |
| 8 | 1 | 0 | 0 | 0 |
| 9 | 0 | 1 | 0 | 0 |
| 10 | 0 | 1 | 0 | 1 |

**Table 10.4** Data for problem 10.1

**10.2** There are four paths you can take to get to your new job. On the map, they all seem reasonable, and as far as you can tell, they all take 20 minutes, but the actual times vary quite a bit. The value of taking a path is your current estimate of the travel time on that path. In the table below, we show the travel time on each path if you had travelled that path. Start with an initial estimate of each value function of 20 minutes with your tie-breaking rule to use the lowest numbered path. At each iteration, take the path with the best estimated value, and update your estimate of the value of the path based on your experience. After 10 iterations, compare your estimates of each path to the estimate you obtain by averaging the "observations" for each path over all 10 days. Use a constant stepsize of 0.20. How well did you do?

| | Paths | | | |
|---|---|---|---|---|
| Day | 1 | 2 | 3 | 4 |
| 1 | 37 | 29 | 17 | 23 |
| 2 | 32 | 32 | 23 | 17 |
| 3 | 35 | 26 | 28 | 17 |
| 4 | 30 | 35 | 19 | 32 |
| 5 | 28 | 25 | 21 | 26 |
| 6 | 24 | 19 | 25 | 31 |
| 7 | 26 | 37 | 33 | 30 |
| 8 | 28 | 22 | 28 | 27 |
| 9 | 24 | 28 | 31 | 30 |
| 10 | 33 | 29 | 17 | 29 |

**10.3** We are going to try again to solve our asset selling problem, We assume we are holding a real asset and we are responding to a series of offers. Let $\hat{p}_t$ be the $t^{th}$ offer, which is uniformly distributed between 500 and 600 (all prices are in thousands of dollars). We also assume that each offer is independent of all prior offers. You are willing to consider up to 10 offers, and your goal is to get the highest possible price. If you have not accepted the first nine offers, you must accept the $10^{th}$ offer.

(a) Write out the decision function you would use in an approximate dynamic programming algorithm in terms of a Monte Carlo sample of the latest price and a current estimate of the value function approximation.

(b) Write out the updating equations (for the value function) you would use after solving the decision problem for the $t^{th}$ offer.

(c) Implement an approximate dynamic programming algorithm using *synchronous* state sampling. Using 1000 iterations, write out your estimates of the value of being in each state immediately after each offer. For this exercise, you will need to discretize prices for the purpose of approximating the value function. Discretize the value function in units of 5 dollars.

(d) From your value functions, infer a decision rule of the form "sell if the price is greater than $\bar{p}_t$."

**10.4** Prove our intuitive claim in section 10.3 that it is optimal to first collect information and then, at some time $\tau$, stop collecting information and use it. This means that if $x_t = 0$ for some time $t$, then $x_{t+1} = 0$ (and so on). [Hint: This can be proved with an interchange argument. Assume that $x^1$ is one policy with, for some $t$, $x_t = 0$ and $x_{t+1} = 1$, and let $x^2$ be the same policy except that $x_t = 1$ and $x_{t+1} = 0$ (we have simply exchanged the decision to collect information at time $t+1$ to time $t$). Show that $x^2$ is better than $x^1$. Use this to prove your result.]

**10.5** Assume you are considering five options. The actual value $\theta_d$, the initial estimate $\bar{\theta}_d^0$ and the initial standard deviation $\bar{\sigma}_d^0$ of each $\bar{\theta}_d^0$ are given in table 10.5. Perform 20 iterations of each of the following algorithms:

(a) Gittins exploration using $\Gamma(n) = 2$.

(b) Interval exploration using $\alpha/2 = .025$.

(c) The upper confidence bound algorithm using $W^{max} = 6$.

(d) The knowledge gradient algorithm.

(e) Pure exploitation.

(f) Pure exploration.

Each time you sample a decision, randomly generate an observation $W_d = \theta_d + \sigma^\varepsilon Z$ where $\sigma^\varepsilon = 1$ and $Z$ is normally distributed with mean 0 and variance 1. [Hint: You can generate random observations of $Z$ in Excel by using *=NORMSINV(RAND())*.]

| Decision | $\theta$ | $\bar{\theta}^0$ | $\bar{\sigma}^0$ |
|---|---|---|---|
| 1 | 1.4 | 1.0 | 2.5 |
| 2 | 1.2 | 1.2 | 2.5 |
| 3 | 1.0 | 1.4 | 2.5 |
| 4 | 1.5 | 1.0 | 1.5 |
| 5 | 1.5 | 1.0 | 1.0 |

**Table 10.5** Data for exercise 10.5

| Decision | $\theta$ | $\bar{\theta}^0$ | $\bar{\sigma}^0$ |
|---|---|---|---|
| 1 | 100 | 100 | 20 |
| 2 | 80 | 100 | 20 |
| 3 | 120 | 100 | 20 |
| 4 | 110 | 100 | 10 |
| 5 | 60 | 100 | 30 |

**Table 10.6** Data for exercise 10.6

**10.6** Repeat exercise 10.5 using the data in table 10.6, with $\sigma^\varepsilon = 10$.

**10.7** Repeat exercise 10.5 using the data in table 10.7, with $\sigma^\varepsilon = 20$.

| Decision | $\theta$ | $\bar{\theta}^0$ | $\bar{\sigma}^0$ |
|---|---|---|---|
| 1 | 120 | 100 | 30 |
| 2 | 110 | 105 | 30 |
| 3 | 100 | 110 | 30 |
| 4 | 90 | 115 | 30 |
| 5 | 80 | 120 | 30 |

**Table 10.7** Data for exercise 10.7

**10.8 Nomadic trucker project revisited** - We are going to again solve the nomadic trucker problem first posed in exercise 8.6, but this time we are going to experiment with our more advanced information acquisition policies. Assume that we are solving the finite horizon problem with a discount factor of $\gamma = .1$. Compare the following policies for determining the next state to visit:

(a) Pure exploitation - This is the policy that you would have implemented in exercise 8.6.

(b) Pure exploration - After solving a decision problem (and updating the value of being in a city), choose the next city at random.

(c) Boltzmann exploration - Use equation 10.1 to determine the next city to visit.

(d) Gittins exploration - Use equation (10.15) to decide which city to visit next, and use equation (10.9) to compute $\Gamma(n)$. You will have to experiment to find the best value of $\rho^G$.

(e) Interval estimation - Use $\alpha/2 = .025$.

(f) The knowledge gradient policy.

**10.9** Repeat exercise 10.8 using a discount factor $\gamma = 0.8$. How does this affect the behavior of the search process? How does it change the performance of the different methods for collecting information?

# CHAPTER 11

# VALUE FUNCTION APPROXIMATIONS FOR SPECIAL FUNCTIONS

In chapter 7, we introduced general purpose approximation tools for approximating value functions without assuming any special structural properties. In this chapter, we focus on approximating value functions that arise in settings where we have to determine the right quantity. For example, if $R$ is the amount of resource available (water, oil, money, vaccines) and $V(R)$ is the value of having $R$ units of our resource, we often find that $V(R)$ might be linear (or approximately linear), nonlinear (concave), piecewise linear, or in some cases, simply something continuous. Value functions with this structure yield to special approximation strategies.

We consider a series of approximation strategies of increasing sophistication:

Linear approximations - These are typically the simplest nontrivial approximations that work well when the functions are approximately linear over the range of interest. It is important to realize that we mean "linear in the state" as opposed to the more classical "linear in the parameters" model that we considered earlier.

Separable, piecewise linear, concave (convex if minimizing) - These functions are especially useful when we are interested in integer solutions. Separable functions are relatively easy to estimate and offer special structural properties when solving the optimality equations.

Auxiliary functions - This is a special class of algorithms that fixes an initial approximation and uses stochastic gradients to adjust the function.

General nonlinear regression equations - Here, we bring the full range of tools available from the field of regression.

Cutting planes - This is a technique for approximating multidimensional, piecewise linear functions that has proven to be particularly powerful for multistage linear programs such those that arise in dynamic resource allocation problems.

An important dimension of this chapter will be our use of derivatives to estimate value functions, rather than just the value of being in a state. When we want to determine how much oil should be sent to a storage facility, what matters most is the marginal value of additional oil. For some problem classes, this is a particularly powerful device that dramatically improves convergence.

## 11.1 VALUE FUNCTIONS VERSUS GRADIENTS

It is common in dynamic programming to talk about the problem of estimating the value of being in a state. There are many applications where it is more useful to work with the derivative or gradient of the value function. In one community, where "heuristic dynamic programming" represents approximate dynamic programming based on estimating the value of being in a state, "dual heuristic programming" refers to approximating the gradient.

We are going to use the context of resource allocation problems to illustrate the power of using the gradient. In principal, the challenge of estimating the slope of a function is the same as that of estimating the function itself (the slope is simply a different function). However, there can be important, practical advantages to estimating slopes. First, if the function is approximately linear, it may be possible to replace estimates of the value of being in each state (or set of states) with a single parameter which is the estimate of the slope of the function. Estimating constant terms is typically unnecessary.

A second and equally important difference is that if we estimate the value of being in a state, we get one estimate of the value of being in a state when we visit that state. When we estimate a gradient, we get an estimate of a derivative for each type of resource. For example, if $R_t = (R_{ta})_{a\in\mathcal{A}}$ is our resource vector and $V_t(R_t)$ is our value function, then the gradient of the value function with respect to $R_t$ would look like

$$\nabla_{R_t} V_t(R_t) = \begin{pmatrix} \hat{v}_{ta_1} \\ \hat{v}_{ta_2} \\ \vdots \\ \hat{v}_{ta_{|\mathcal{A}|}} \end{pmatrix},$$

where

$$\hat{v}_{ta_i} = \frac{\partial V_t(R_t)}{\partial R_{ta_i}}.$$

There may be additional work required to obtain each element of the gradient, but the incremental work can be far less than the work required to get the value function itself. This is particularly true when the optimization problem naturally returns these gradients (for example, dual variables from a linear program), but this can even be true when we have to resort to numerical derivatives. Once we have all the calculations to solve a problem once, solving small perturbations can be relatively inexpensive.

There is one important problem class where finding the value of being in a state is equivalent to finding the derivative. That is the case of managing a single resource (see section 5.3.1). In this case, the state of our system (the resource) is the attribute vector $a$, and we are interested in estimating the value $V(a)$ of our resource being in state $a$. Alternatively, we can represent the state of our system using the vector $R_t$, where $R_{ta} = 1$ indicates that our resource has attribute $a$ (we assume that $\sum_{a \in \mathcal{A}} R_{ta} = 1$). In this case, the value function can be written

$$V_t(R_t) = \sum_{a \in \mathcal{A}} v_{ta} R_{ta}.$$

Here, the coefficient $v_{ta}$ is the derivative of $V_t(R_t)$ with respect to $R_{ta}$.

In a typical implementation of an approximate dynamic programming algorithm, we would only estimate the value of a resource when it is in a particular state (given by the vector $a$). This is equivalent to finding the derivative $\hat{v}_a$ only for the value of $a$ where $R_{ta} = 1$. By contrast, computing the gradient $\nabla_{R_t} V_t(R_t)$ implicitly assumes that we are computing $\hat{v}_a$ for each $a \in \mathcal{A}$. There are some algorithmic strategies (we will describe an example of this in section 11.6) where this assumption is implicit in the algorithm. Computing $\hat{v}_a$ for all $a \in \mathcal{A}$ is reasonable if the attribute state space is not too large (for example, if $a$ is a physical location among a set of several hundred locations). If $a$ is a vector, then enumerating the attribute space can be prohibitive (it is, in effect, the "curse of dimensionality" revisited).

Given these issues, it is critical to first determine whether it is necessary to estimate the slope of the value function, or the value function itself. The result can have a significant impact on the algorithmic strategy.

## 11.2 LINEAR APPROXIMATIONS

There are a number of problems where we are allocating resources of different types. As in the past, we let $a$ be the attributes of a resource and $R_{ta}$ be the quantity of resources with attribute $a$ in our system at time $t$ with $R_t = (R_{ta})_{a \in \mathcal{A}}$. $R_t$ may describe our investments in different resource classes (growth stocks, value stocks, index funds, international mutual funds, domestic stock funds, bond funds). Or $R_t$ might be the amount of oil we have in different reserves or the number of people in a management consulting firm with particular skill sets. We want to make decisions to acquire or sell resources of each type, and we want to capture the impact of decisions now on the future through a value function $V_t(R_t)$.

Rather than attempt to estimate $V_t(R_t)$ for each value of $R_t$, it may make more sense to estimate a linear approximation of the value function with respect to the resource vector. Linear approximations can work well when the single-period contribution function is continuous and increases or decreases monotonically over the range we are interested in (the function may or may not be differentiable). They can also work well in settings where the value function increases or decreases monotonically, even if the value function is neither convex nor concave, nor even continuous.

To illustrate, consider the problem of purchasing a commodity. Let

$\hat{D}_t$ = The random demand during time interval $t$,

$R_t$ = The resources on hand at time $t$ just before we make an ordering decision,

$x_t$ = The quantity ordered at time $t$ to be used during time interval $t + 1$,

$R_t^x$ = The resources available just after we make a decision,

$\hat{p}_t$ = The market price for selling commodities during time interval $t$,

$c_t$ = The purchase cost for commodities purchased at time $t$.

At time $t$, we know the price $\hat{p}_t$ and demand $\hat{D}_t$ for time interval $t$, but we have to choose how much to order for the next time interval. The transition equations are given by

$$\begin{aligned} R_t^x &= R_t + x_t, \\ R_{t+1} &= [R_t^x - \hat{D}_{t+1}]^+. \end{aligned}$$

The value of being in state $R_t$ is given by

$$V_t(R_t) = \max_{x_t} \mathbb{E}\big(\hat{p}_{t+1} \min\{R_t + x_t, \hat{D}_{t+1}\} - c_t x_t + V_t(R_t + x_t)\big). \tag{11.1}$$

Now assume that we introduce a linear value function approximation

$$\bar{V}_t(R_t^x) \approx \bar{v}_t R_t^x.$$

The resulting approximation can be written

$$\begin{aligned} \widetilde{V}_t(R_t) &= \max_{x_t} \mathbb{E}\big(\hat{p}_{t+1} \min\{R_t + x_t, \hat{D}_{t+1}\} - c_t x_t + \bar{v}_t R_t^x\big) \\ &= \max_{x_t} \mathbb{E}\big(\hat{p}_{t+1} \min\{R_t + x_t, \hat{D}_{t+1}\} - c_t x_t + \bar{v}_t(R_t + x_t)\big). \end{aligned} \tag{11.2}$$

We assume that we can compute, or at least approximate, the expectation in equation (11.2). If this is the case, we may approximate the gradient at iteration $n$ using a numerical derivative, as in

$$\hat{v}_t = \widetilde{V}_t(R_t + 1) - \widetilde{V}_t(R_t).$$

We now use $\hat{v}_t$ to update the value function $\bar{V}_{t-1}$ using

$$\bar{v}_{t-1} \leftarrow (1 - \alpha)\bar{v}_{t-1} + \alpha \hat{v}_t.$$

Normally, we would use $\hat{v}_t$ to update $\bar{V}_{t-1}(R_{t-1}^x)$ around the previous post-decision state variable $R_{t-1}^x$. Linear approximations, however, are a special case, since the slope is the same for all $R_{t-1}^x$, which means it is also the same for $R_{t-1} = R_{t-1}^x - x_{t-1}$.

Linear approximations are useful in two settings. First, the value function may be approximately linear over the range that we are interested in. Imagine, for example, that you are trying to decide how many shares of stock you want to sell, where the range is between 0 and 1,000. As an individual investor, it is unlikely that selling all 1,000 shares will change the market price. However, if you are a large mutual fund and you are trying to decide how many of your 50 million shares you want to sell, it is quite likely that such a high volume would, in fact, move the market price. When this happens, we need a nonlinear function.

A second use of linear approximations arises when managing resources such as people and complex equipment such as locomotives or aircraft. Let $a$ be the attributes of a resource and $R_{ta}$ be the number of resources with attribute $a$ at time $t$. Then it is likely that $R_{ta}$ will be 0 or 1, implying that a linear function is all we need. For these problems, a linear value function is particularly convenient because it means we need one parameter, $\bar{v}_{ta}$, for each attribute $a$.

## 11.3 PIECEWISE LINEAR APPROXIMATIONS

There is a vast range of problems where we have to estimate the value of having a quantity $R$ of some resource (where $R$ is a scalar). We might want to know the value of having $R$ dollars in a budget, $R$ pieces of equipment, or $R$ units of some inventory. $R$ may be discrete or continuous, but we are going to focus on problems where $R$ is either discrete or is easily discretized.

Assume now that we want to estimate a function $V(R)$ that gives the value of having $R$ resources. There are applications where $V(R)$ increases or decreases monotonically in $R$. There are other applications where $V(R)$ is piecewise linear, concave (or convex) in $R$, which means the slopes of the function are monotonically decreasing (if the function is concave) or increasing (if it is convex). When the function (or the slopes of the function) is steadily increasing or decreasing, we would say that the function is *monotone*. If the function is increasing in the state variable, we might say that it is "monotonically increasing," or that it is *isotone* (although the latter term is not widely used). To say that a function is "monotone" can mean that it is monotonically increasing or decreasing.

Assume we have a function that is monotonically decreasing, which means that while we do not know the value function exactly, we know that $V(R+1) \leq V(R)$ (for scalar $R$). If our function is piecewise linear concave, then we will assume that $V(R)$ refers to the slope at $R$ (more precisely, to the right of $R$). Assume our current approximation $\bar{V}^{n-1}(R)$ satisfies this property, and that at iteration $n$, we have a sample observation of $V(R)$ for $R = R^n$. If our function is piecewise linear concave, then $\hat{v}^n$ would be a sample realization of a derivative of the function. If we use our standard updating algorithm, we would write

$$\bar{V}^n(R^n) = (1 - \alpha_{n-1})\bar{V}^{n-1}(R^n) + \alpha_{n-1}\hat{v}^n.$$

After the update, it is quite possible that our updated approximation no longer satisfies our monotonicity property. One way to maintain monotonicity is through the use of a *leveling algorithm*, which works as follows

$$\bar{V}^n(r) = \begin{cases} (1 - \alpha_{n-1})\bar{V}^{n-1}(R^n) + \alpha_{n-1}\hat{v}^n & \text{if } r = R^n, \\ \bar{V}^n(r) \vee \{(1 - \alpha_{n-1})\bar{V}^{n-1}(R^n) + \alpha_{n-1}\hat{v}^n\} & \text{if } r > R^n, \\ \bar{V}^n(r) \wedge \{(1 - \alpha_{n-1})\bar{V}^{n-1}(R^n) + \alpha_{n-1}\hat{v}^n\} & \text{if } r < R^n, \end{cases} \tag{11.3}$$

where $x \wedge y = \max\{x, y\}$, and $x \vee y = \min\{x, y\}$. Equation (11.3) starts by updating the slope $\bar{V}^n(r)$ for $r = R^n$. We then want to make sure that the slopes are declining. So, if we find a slope to the right that is larger, we simply bring it down to our estimated slope for $r = R^n$. Similarly, if there is a slope to the left that is smaller, we simply raise it to the slope for $r = R^n$. The steps are illustrated in figure 11.1.

The leveling algorithm is easy to visualize, but it is unlikely to be the best way to maintain monotonicity. For example, we may update a value at $r = R^n$ for which there are very few observations. But because it produces an unusually high or low estimate, we find ourselves simply forcing other slopes higher or lower just to maintain monotonicity.

A more elegant strategy is the SPAR algorithm which works as follows. Assume that we start with our original set of values $(\bar{V}^{n-1}(r))_{r \geq 0}$, and that we sample $r = R^n$ and obtain an estimate of the slope $\hat{v}^n$. After the update, we obtain the set of values (which we

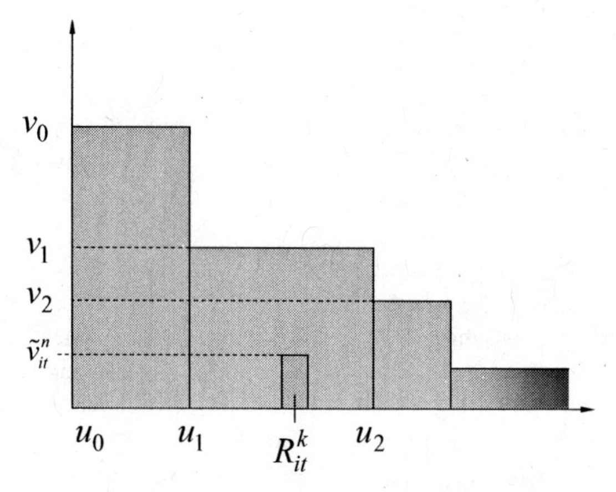


11.1a: Initial monotone function.

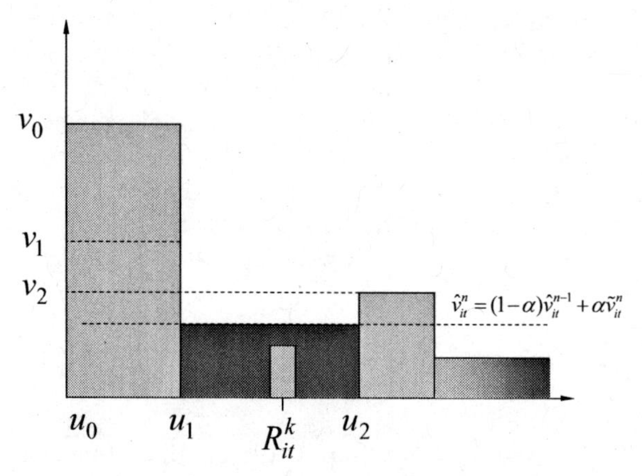


11.1b: After update of a single segment.

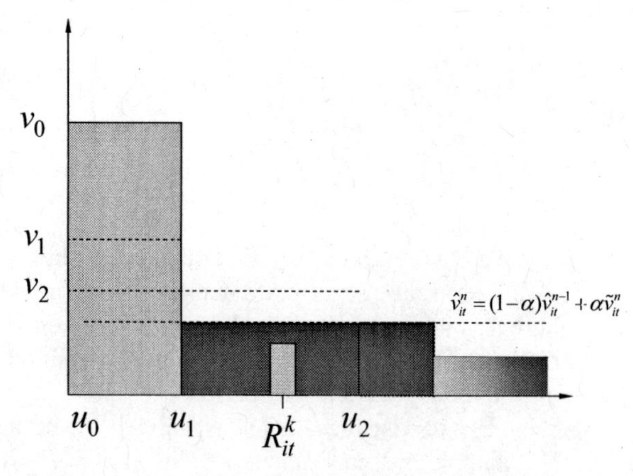


11.1c: After leveling operation.

**Figure 11.1** Steps of the leveling algorithm. Figure 11.1a shows the initial monotone function, with the observed $R$ and observed value of the function $\hat{v}$. Figure 11.1b shows the function after updating the single segment, producing a non-monotone function. Figure 11.1c shows the function after monotonicity restored by leveling the function.

store temporarily in the function $\bar{y}^n(r)$)

$$\bar{y}^n(r) = \begin{cases} (1-\alpha_{n-1})\bar{V}^{n-1}(r) + \alpha_{n-1}\hat{v}^n, & r = R^n \\ \bar{V}^{n-1}(r) & \text{otherwise.} \end{cases} \tag{11.4}$$

If $\bar{y}^n(r) \geq \bar{y}^n(r+1)$ for all $r$, then we are in good shape. If not, then either $\bar{y}^n(R^n) < \bar{y}^n(R^n+1)$ or $\bar{y}^n(R^n-1) < \bar{y}^n(R^n)$. We can fix the problem by solving the projection problem

$$\min_v \|v - \bar{y}^n\|^2. \tag{11.5}$$

subject to

$$v(r+1) - v(r) \quad \leq \quad 0. \tag{11.6}$$

Solving this projection is especially easy. Imagine that after our update, we have a violation to the left. The projection is achieved by averaging the updated cell with all the cells to the left that create a monotonicity violation. This means that we want to find the largest $i \leq R^n$ such that

$$\bar{y}^n(i-1) \geq \frac{1}{R^n - i + 1}\sum_{r=i}^{R^n} \bar{y}^n(r).$$

In other words, we can start by averaging the values for $R^n$ and $R^n - 1$ and checking to see if we now have a concave function. If not, we keep lowering the left end of the range until we either restore monotonicity or reach $r = 0$. If our monotonicity violation is to the right, then we repeat the process to the right.

The steps of the algorithm are given in figure 11.2, with an illustration given in figure 11.3. We start with a monotone set of values (a), then update one of the values to produce a monotonicity violation (b), and finally average the violating values together to restore monotonicity (c).

There are a number of variations of these algorithms that help with convergence. For example, in the SPAR algorithm, we can solve a weighted projection that gives more weight

---

**Step 0** Initialize $\bar{V}^0$ and set $n = 1$.

**Step 1** Sample $R^n$.

**Step 2** Observe a sample of the value function $\hat{v}^n$.

**Step 3** Calculate the vector $y^n$ as follows

$$y^n(r) = \begin{cases} (1-\alpha_{n-1})V_{R^n}^{n-1} + \alpha_{n-1}\hat{v}^n & \text{if } r = R^n, \\ v^{n-1}(r) & \text{otherwise} \end{cases}$$

**Step 4** Project the updated estimate onto the space of monotone functions:

$$v^n = \Pi(y^n),$$

by solving (11.5)-(11.6). Increase $n$ by one and go to Step 1.

---

**Figure 11.2** The learning form of the separable, projective approximation routine (SPAR).

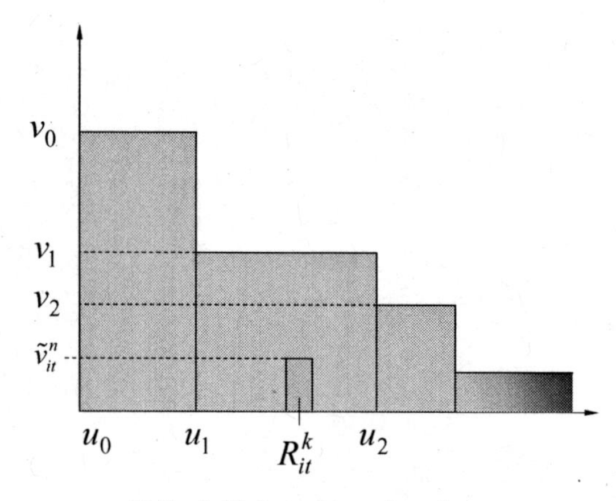


11.3a: Initial monotone function.

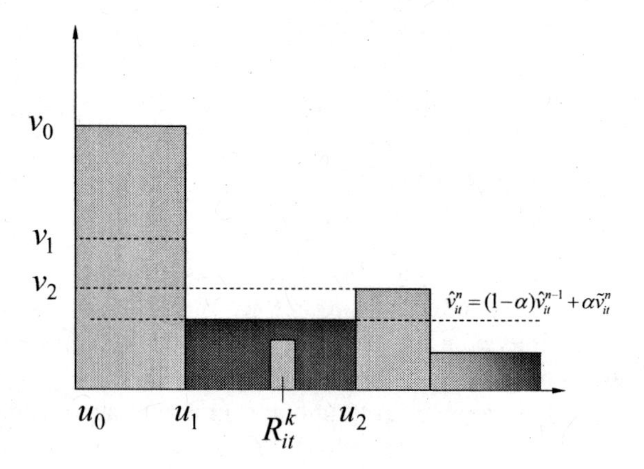


11.3b: After update of a single segment.

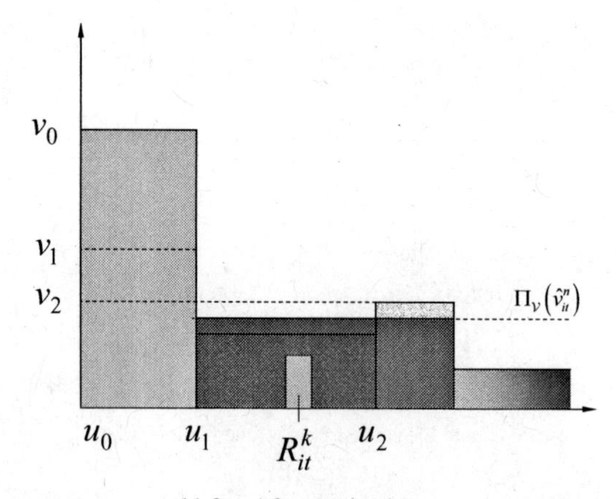


11.3c: After projection.

**Figure 11.3** Steps of the SPAR algorithm. Figure 11.3a shows the initial monotone function, with the observed $R$ and observed value of the function $\hat{v}$. Figure 11.3b shows the function after updating the single segment, producing a non-monotone function. Figure 11.3c shows the function after the projection operation.

to slopes that have received more observations. To do this, we weight each value of $\bar{y}(r)$ by the number of observations that segment $r$ has received when computing $\bar{y}^n(i-1)$.

A particularly useful variation is to perform an initial update (when we compute $\bar{y}$) over a wider interval than just $r = R^n$. Assume we are given a parameter $\delta^0$ which has been chosen so that it is approximately 20 to 50 percent of the maximum value that $R^n$ might take. Now compute $\bar{y}(r)$ using

$$\bar{y}^n(r) = \begin{cases} (1-\alpha_{n-1})\bar{V}^{n-1}(r) + \alpha_{n-1}\hat{v}^n, & R^n - \delta^n \le r \le R^n + \delta^n, \\ \bar{V}^{n-1}(r) & \text{otherwise.} \end{cases}$$

Here, we are using $\hat{v}^n$ to update a wider range of the interval. We then apply the same logic for maintaining monotonicity (concavity if these are slopes). We start with the interval $R^n \pm \delta^0$, but we have to periodically reduce $\delta^0$. We might, for example, track the objective function (call it $F^n$), and update the range using

$$\delta^n = \begin{cases} \delta^{n-1} & \text{If } F^n \ge F^{n-1} - \epsilon, \\ \max\{1, .5\delta^{n-1}\} & \text{otherwise.} \end{cases}$$

While the rules for reducing $\delta^n$ are generally ad hoc, we have found that this is critical for fast convergence. The key is that we have to pick $\delta^0$ so that it plays a critical scaling role, since it has to be set so that it is roughly on the order of the maximum value that $R^n$ can take.

## 11.4 THE SHAPE ALGORITHM

A particularly simple algorithm for approximating continuous value functions starts with an initial approximation and then "tilts" this function to improve the approximation. The concept is most effective if it is possible to build an initial approximation, perhaps using some simplifications, that produces a "pretty good" solution.

### 11.4.1 The basic idea

The idea works as follows. Assume that we are trying to solve the stochastic optimization problem we first saw in section 6.1 which is given by

$$\min_{x \in \mathcal{X}} \mathbb{E}F(x, W), \tag{11.7}$$

where $x$ is our decision variable and $W$ is a random variable. We already know that we can solve this problem using a stochastic gradient algorithm which takes the form

$$x^n = x^{n-1} - \alpha_{n-1}\nabla_x F(x^{n-1}, W(\omega^n)). \tag{11.8}$$

The challenge we encounter with this algorithm is that the units of the decision variable and the gradient may be different, which means that our stepsize has to be properly scaled. Also, while these algorithms may work in the limit, convergence can be slow and choppy.

Now assume that we have a rough approximation of $\mathbb{E}F(x, W)$ as a function of $x$, which we call $\bar{F}^0(x)$. For example, if $x$ is a scalar and we think that the optimal solution is approximately equal to $a$, we might let $\bar{F}^0(x) = (x-a)^2$. The SHAPE algorithm iteratively adjusts this initial approximation by first finding a stochastic gradient $\nabla_x F(x^{n-1}, W(\omega^n))$, and then computing an updated approximation using

$$\bar{F}^n(x) = \bar{F}^{n-1}(x) + \alpha_{n-1} \underbrace{\left(\nabla_x F(x^{n-1}, W(\omega^n)) - \nabla_x \bar{F}^{n-1}(x^{n-1})\right)}_{I} x.$$

Our updated approximation is the original approximation plus a linear correction term. Note that $x$ is a variable while the term $I$ is a constant. Since the correction term has the same units as the approximation, we can assume that $0 < \alpha_n \leq 1$. The linear correction term is simply the difference between the slope of the original function at a sample realization $W(\omega^n)$ and the exact slope of the approximation at $x^{n-1}$. Typically, we would pick an approximation that is easy to differentiate.

The SHAPE algorithm is depicted graphically in figure 11.4. Figure 11.4a shows the initial true function and the approximation $\bar{F}^0(x)$. Figure 11.4b shows the gradient of each function at the point $x^{n-1}$, with the difference between the two slopes. If there were no noise, and if our approximation were exact, these two gradients would match (at least at this point). Finally, figure 11.4c shows the updated approximation after the correction has been added in. At this point, the two slopes will match only if the stepsize $\alpha_{n-1} = 1$.

We can illustrate the SHAPE algorithm using a simple numerical example. Assume that our problem is to solve

$$\max_{x \geq 0} \mathbb{E}\bar{F}(x, W) = \mathbb{E}\left\{\frac{1}{2}\ln(x + W) - 2(x + W)\right\},$$

where $W$ represents random measurement error, which is normally distributed with mean 0 and variance 4. Now, assume that we start with a convex approximation such as

$$\bar{F}^0(x) = 6\sqrt{x} - 2x.$$

We begin by obtaining the initial solution $x^0$

$$x^0 = \arg\max_{x \geq 0} \left(6\sqrt{x} - 2x\right).$$

Note that our solution to the approximate problem may be unbounded, requiring us to impose artificial limits. Since our approximation is concave, we can set the derivative equal to zero to find

$$\begin{aligned} \nabla \bar{F}^0(x) &= 3/\sqrt{x} - 2 \\ &= 0, \end{aligned}$$

which gives us $x^0 = 2.25$. Since $x^0 \geq 0$, it is optimal. To find the stochastic gradient, we have to sample the random variable $W$. Assume that $W(\omega^1) = 1.75$. Our stochastic gradient is then

$$\begin{aligned} \nabla \bar{F}(x, W(\omega^1)) &= \frac{1}{2(x^0 + W(\omega^1))} - 2 \\ &= \frac{1}{2(2.25 + 1.75))} \\ &= 0.1250. \end{aligned}$$

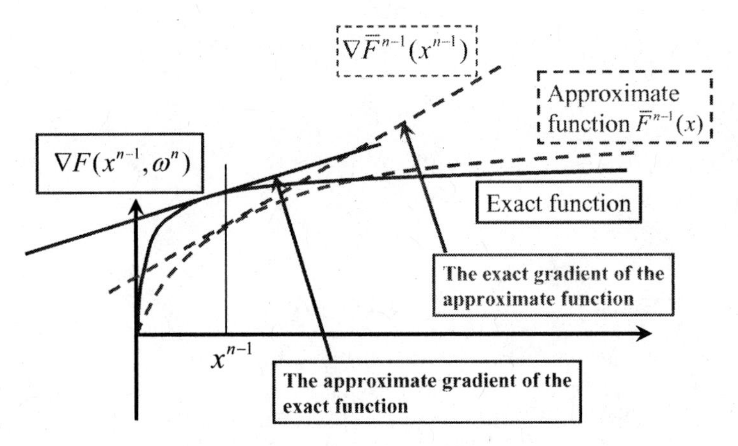


11.4a: True function and initial approximation.

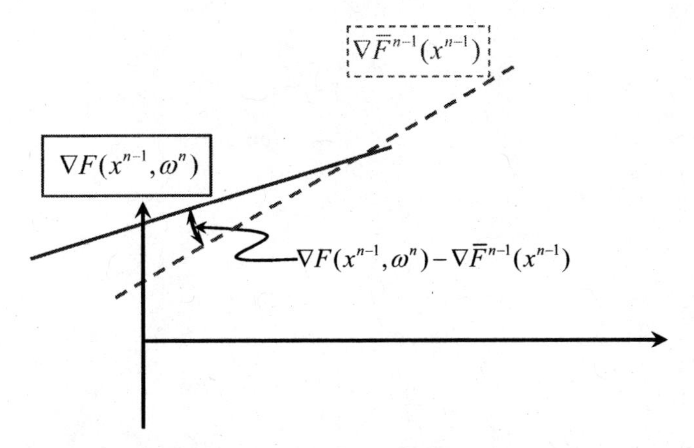


11.4b: Difference between stochastic gradient of true function and actual gradient of approximation.

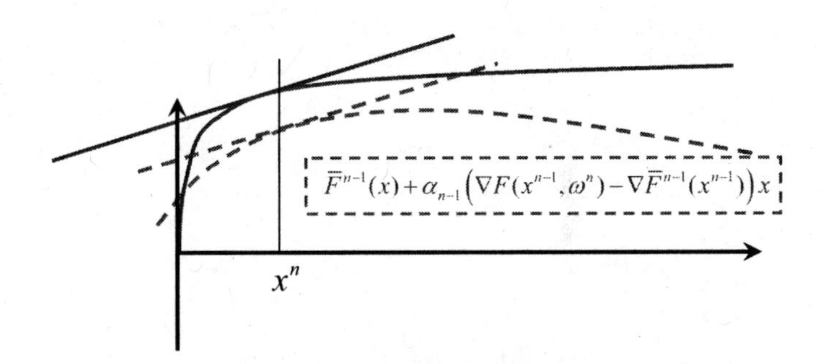


11.4c: Updated approximation.

**Figure 11.4** Illustration of the steps of the SHAPE algorithm.

Thus, while we have found the optimal solution to the approximate problem (which produces a zero slope), our estimate of the slope of the true function is positive, so we update $\bar{F}^1(s)$ with the adjustment

$$\begin{aligned}\bar{F}^1(x) &= 6\sqrt{x} - 2x - \alpha_0(0.1250 - 0)x \\ &= 6\sqrt{x} - 3.125x,\end{aligned}$$

where we have used $\alpha_0 = 1$.

### 11.4.2 A dynamic programming illustration

Now consider what happens when we have to solve a dynamic program (approximately). For this illustration, assume that $R_t$ and $x_t$ are scalars, although everything we are going to do works fine with vectors. Still using the value function around the post-decision state, we would start with the previous post-decision state $R^x_{t-1}$. We then use our sample path $\omega^n$ to find the pre-decision state $R^n_t = R^{M,W}(R^x_{t-1}, W_t(\omega^n))$ and solve

$$\widetilde{V}^n_t(R^n_t) = \max_{x_t \in \mathcal{X}^n_t} \left(C(R^n_t, x_t) + \bar{V}^{n-1}_t(R^{M,x}(R^n_t, x_t))\right).$$

We need the derivative of $\widetilde{V}^n_t(R^n_t)$. There are different ways of getting this depending on the structure of the optimization problem, but for the moment we will simply let

$$\hat{v}^n_t = \widetilde{V}^n_t(R^n_t + 1) - \widetilde{V}^n_t(R^n_t).$$

It is important to remember that imbedded in the calculation of $\widetilde{V}^n_t(R^n_t + 1)$ is the need to reoptimize $x_t$. Recall that we use $\hat{v}^n_t$, which is computed at $R_t = R^n_t$, to update $\bar{V}_{t-1}(R^x_{t-1})$ as a function of the post-decision state $R^x_{t-1}$. Using the SHAPE algorithm, we update our approximation using

$$\bar{V}^n_{t-1}(R^x_{t-1}) = \bar{V}^{n-1}_{t-1}(R^x_{t-1}) + \alpha_{n-1}\left(\hat{v}^n_t - \nabla_R \bar{V}^{n-1}_{t-1}(R^x_{t-1})\right)R^x_{t-1}.$$

The steps of the SHAPE algorithm are given in figure 11.5.

It is useful to consider how decisions are made using the SHAPE algorithm versus a stochastic gradient procedure. Using SHAPE, we would choose $x_t$ using

$$x^n_t = \arg\max_{x_t \in \mathcal{X}^n_t} \left(C(R^n_t, x_t) + \bar{V}^{n-1}_t(R^{M,x}(R^n_t, x_t))\right). \tag{11.9}$$

Contrast determining the decision $x^n_t$ from (11.9) to equation (11.8), which depends heavily on the choice of stepsize. This property is true of most algorithms that involve finding nonlinear approximations of the value function.

## 11.5 REGRESSION METHODS

As in chapter 7 we can create regression models where the basis functions are manipulations of the number of resources of each type. For example, we might use

$$\bar{V}(R) = \theta_0 + \sum_{a \in \mathcal{A}} \theta_{1a} R_a + \sum_{a \in \mathcal{A}} \theta_{2a} R_a^2, \tag{11.10}$$

**Step 0.** Initialization:

**Step 0a.** Initialize $\bar{V}_t^0, \; t \in \mathcal{T}$.

**Step 0b.** Set $n = 1$.

**Step 0c.** Initialize $R_0^1$.

**Step 1.** Choose a sample path $\omega^n$.

**Step 2.** Do for $t = 0, 1, 2, \ldots, T$:

**Step 2a.** Solve

$$\widetilde{V}_t^n(R_t^n) = \max_{x_t \in \mathcal{X}_t^n} \left( C(R_t^n, x_t) + \bar{V}_t^{n-1}(R^{M,x}(R_t^n, x_t)) \right)$$

and let $x_t^n$ be the value of $x_t$ that solves the maximization problem.

**Step 2b.** Compute the derivative

$$\hat{v}_t^n = \widetilde{V}_t^n(R_t^n + 1) - \widetilde{V}_t^n(R_t^n)$$

**Step 2c.** If $t > 0$ update the value function

$$\bar{V}_{t-1}^n(R_{t-1}^x) = \bar{V}_{t-1}^{n-1}(R_{t-1}^x) + \alpha_{n-1}\left(\hat{v}_t^n - \nabla_R \bar{V}_{t-1}^{n-1}(R_{t-1}^x)\right) R_{t-1}^x.$$

**Step 2d.** Update the states:

$$\begin{aligned} R_t^{x,n} &= R^{M,x}(R_t^n, x_t^n), \\ R_{t+1}^n &= R^{M,W}(R_t^{x,n}, W_{t+1}(\omega^n)). \end{aligned}$$

**Step 3.** Increment $n$. If $n \leq N$ go to Step 1.

**Step 4.** Return the value functions $(\bar{V}_t^N)_{t=1}^T$.

**Figure 11.5** The SHAPE algorithm for approximate dynamic programming.

where $\theta = (\theta_0, (\theta_{1a})_{a \in \mathcal{A}}, (\theta_{2a})_{a \in \mathcal{A}})$ is a vector of parameters that are to be determined. The choice of explanatory terms in our approximation will generally reflect an understanding of the properties of our problem. For example, equation (11.10) assumes that we can use a mixture of linear and separable quadratic terms. A more general representation is to assume that we have developed a family $\mathcal{F}$ of basis functions $(\phi_f(R))_{f \in \mathcal{F}}$. Examples of a basis function are

$$\begin{aligned} \phi_f(R) &= R_{a_f}^2, \\ \phi_f(R) &= \left( \sum_{a \in \mathcal{A}_f} R_a \right)^2 \quad \text{for some subset } \mathcal{A}_f, \\ \phi_f(R) &= (R_{a_1} - R_{a_2})^2, \\ \phi_f(R) &= |R_{a_1} - R_{a_2}|. \end{aligned}$$

A common strategy is to capture the number of resources at some level of aggregation. For example, if we are purchasing emergency equipment, we may care about how many pieces we have in each region of the country, and we may also care about how many pieces of a type of equipment we have (regardless of location). These issues can be captured using a family of aggregation functions $G_f$, $f \in \mathcal{F}$, where $G_f(a)$ aggregates an attribute vector $a$ into a space $\mathcal{A}^{(f)}$ where for every basis function $f$ there is an element $a_f \in \mathcal{A}^{(f)}$. Our basis function might then be expressed using

$$\phi_f(R) = \sum_{a \in \mathcal{A}} 1_{\{G_f(a)=a_f\}} R_a.$$

As we originally introduced in section 7.2.2, the explanatory variables used in the examples above, which are generally referred to as independent variables in the regression literature, are typically referred to as basis functions by the approximate dynamic programming community. A basis function can be linear, nonlinear separable, nonlinear nonseparable, and even nondifferentiable, although the nondifferentiable case will introduce additional technical issues. The challenge, of course, is that it is the responsibility of the modeler to devise these functions for each application. We have written our basis functions purely in terms of the resource vector, but it is possible for them to be written in terms of other parameters in a more complex state vector, such as asset prices.

Given a set of basis functions, we can write our value function approximation as

$$\bar{V}(R|\theta) = \sum_{f \in \mathcal{F}} \theta_f \phi_f(R). \tag{11.11}$$

It is important to keep in mind that $\bar{V}(R|\theta)$ (or more generally, $\bar{V}(S|\theta)$), is any functional form that approximates the value function as a function of the state vector parameterized by $\theta$. Equation (11.11) is a classic linear-in-the-parameters function. We are not constrained to this form, but it is the simplest and offers some algorithmic shortcuts.

The issues that we encounter in formulating and estimating $\bar{V}(R|\theta)$ are the same that any student of statistical regression would face when modeling a complex problem. The major difference is that our data arrives over time (iterations), and we have to update our formulas recursively. Also, it is typically the case that our observations are nonstationary. This is particularly true when an update of a value function depends on an approximation of the value function in the future (as occurs with value iteration or any of the TD($\lambda$) classes of algorithms). When we are estimating parameters from nonstationary data, we do not want to equally weight all observations.

The problem of finding $\theta$ can be posed in terms of solving the following stochastic optimization problem

$$\min_\theta \mathbb{E} \frac{1}{2}(\bar{V}(R|\theta) - \hat{V})^2.$$

We can solve this using a stochastic gradient algorithm, which produces updates of the form

$$\begin{aligned} \bar{\theta}^n &= \bar{\theta}^{n-1} - \alpha_{n-1}(\bar{V}(R^n|\bar{\theta}^{n-1}) - \hat{V}(\omega^n))\nabla_\theta \bar{V}(R^n|\theta^n) \\ &= \bar{\theta}^{n-1} - \alpha_{n-1}(\bar{V}(R^n|\bar{\theta}^{n-1}) - \hat{V}(\omega^n)) \begin{pmatrix} \phi_1(R^n) \\ \phi_2(R^n) \\ \vdots \\ \phi_F(R^n) \end{pmatrix}. \end{aligned}$$

If our value function is linear in $R_t$, we would write

$$\bar{V}(R|\theta) = \sum_{a \in \mathcal{A}} \theta_a R_a.$$

In this case, our number of parameters has shrunk from the number of possible realizations of the entire vector $R_t$ to the size of the attribute space (which, for some problems, can still be large, but nowhere near as large as the original state space). For this problem, $\phi(R^n) = R^n$.

It is not necessarily the case that we will always want to use a linear-in-the-parameters model. We may consider a model where the value increases with the number of resources, but at a declining rate that we do not know. Such a model could be captured with the representation

$$\bar{V}(R|\theta) = \sum_{a \in \mathcal{A}} \theta_{1a} R_a^{\theta_{2a}},$$

where we expect $\theta_2 < 1$ to produce a concave function. Now, our updating formula will look like

$$\begin{aligned}
\theta_1^n &= \theta_1^{n-1} - \alpha_{n-1}(\bar{V}(R^n|\bar{\theta}^{n-1}) - \hat{V}(\omega^n))R^{\theta_2^n}, \\
\theta_2^n &= \theta_2^{n-1} - \alpha_{n-1}(\bar{V}(R^n|\bar{\theta}^{n-1}) - \hat{V}(\omega^n))R^{\theta_2^n}, \ln R^n
\end{aligned}$$

where we assume the exponentiation operator in $R^{\theta_2^n}$ is performed componentwise.

We can put this updating strategy in terms of temporal differencing. As before, the temporal difference is given by

$$D_\tau = C_\tau(R_\tau, x_{\tau+1}) + \bar{V}_{\tau+1}^{n-1}(R_{\tau+1}) - \bar{V}_\tau^{n-1}(R_\tau).$$

The original parameter updating formula (equation (8.14)) when we had one parameter per state now becomes

$$\bar{\theta}^n = \bar{\theta}_t^{n-1} + \alpha_{n-1} \sum_{\tau=t}^{T} \lambda^{\tau-t} D_\tau \nabla_\theta \bar{V}(R^n|\bar{\theta}^n).$$

It is important to note that in contrast with most of our other applications of stochastic gradients, updating the parameter vector using gradients of the objective function requires mixing the units of $\theta$ with the units of the value function. In these applications, the stepsize $\alpha_{n-1}$ has to also perform a scaling role.

## 11.6 CUTTING PLANES*

Cutting planes represent a powerful strategy for representing concave (or convex if we are minimizing), piecewise-linear functions for multidimensional problems. This method evolved originally not as a method for approximating dynamic programs, but instead as a technique for solving linear programs in the presence of uncertainty. In the 1950's, the research community recognized that many optimization problems involve different forms of uncertainty, with the most common being the challenge of allocating resources now to serve demands in the future that have not yet been realized. For this reason, a subcommunity within math programming, known as the stochastic programming community, has developed a rich theory and some powerful algorithmic strategies for handling uncertainty within linear programs and, more recently, integer programs.

Historically, dynamic programming has been viewed as a technique for small, discrete optimization problems, while stochastic programming has been the field that handles uncertainty within math programs (which are typically characterized by high-dimensional decision vectors and large numbers of constraints). The connections between stochastic programming and dynamic programming, historically viewed as diametrically competing frameworks, have been largely overlooked. This section is designed to bridge the gap between stochastic programming and approximate dynamic programming. Our presentation is facilitated by notational decisions (in particular the use of $x$ as our decision vector) that we made in the beginning of the book.

The material in this section is somewhat more advanced and requires a fairly strong working knowledge of linear programming and duality theory. Our presentation is designed primarily to establish the relationship between the fields of approximate dynamic programming and stochastic programming.

We begin our presentation by introducing a classical model in stochastic programming known as the two-stage resource allocation problem.

### 11.6.1 A two-stage resource allocation problem

We use the same notation we have been using for resource allocation problems, where

$$\begin{aligned} R_{ta} &= \text{The number of resources with attribute } a \in \mathcal{A} \text{ in the system at time } t, \\ R_t &= (R_{ta})_{a \in \mathcal{A}}, \\ D_{tb} &= \text{The number of demands of type } b \in \mathcal{B} \text{ in the system at time } t, \\ D_t &= (D_{tb})_{b \in \mathcal{B}}. \end{aligned}$$

$R_0$ and $D_0$ capture the initial state of our system. Decisions are represented using

$\mathcal{D}^D$ = Decision to satisfy a demand with attribute $b$ (each decision $d \in \mathcal{D}^D$ corresponds to a demand attribute $b_d \in \mathcal{B}$),

$\mathcal{D}^M$ = Decision to modify a resource (each decision $d \in \mathcal{D}^M$ has the effect of modifying the attributes of the resource). $\mathcal{D}^M$ includes the decision to "do nothing,"

$\mathcal{D}$ = $\mathcal{D}^D \cup \mathcal{D}^M$.

$x_{tad}$ = The number of resources that initially have attribute $a$ that we act on with decision $d$,

$x_t$ = $(x_{tad})_{a \in \mathcal{A}, d \in \mathcal{D}}$.

$x_0$ is the vector of decisions we have to make now given $S_0 = (R_0, D_0)$. The post-decision state is given by

$R^x_{0a'}$ = The number of resources with attribute $a'$ available in the second stage after we have made our original decision $x_0$

$$= \sum_{a \in \mathcal{A}} \sum_{d \in \mathcal{D}} \delta_{a'}(a,d) x_{0ad},$$

$$R^x_0 = \Delta x_0,$$

where $\Delta$ is a matrix with $\delta_{a'}(a,d)$ in row $a'$, column $(a,d)$. Before we solve the second-stage problem (at time $t = 1$), we observe random changes to the resource vector as well as random new demands. For our illustration, we assume that unsatisfied demands from the first stage are lost. This means that

$$\begin{aligned} R_1 &= R^x_0 + \hat{R}_1, \\ D_1 &= \hat{D}_1. \end{aligned}$$

The resource vector $R_1$ can be used to satisfy demands that first become known during time interval 1. In the second stage, we need to choose $x_1$ to minimize the cost function $C_1(x_1)$. At this point, we assume that our problem is finished. Thus, the problem over both stages would be written

$$\max_{x_0, x_1} \left(C_0(x_0) + \mathbb{E}C_1(x_1)\right). \tag{11.12}$$

This problem has to be solved subject to the first-stage constraints:

$$\sum_{d \in \mathcal{D}} x_{0ad} = R_{0a}, \tag{11.13}$$

$$\sum_{a \in \mathcal{A}} x_{0ad} \leq \hat{D}_{0b_d} \quad d \in \mathcal{D}^D, \tag{11.14}$$

$$x_{0ad} \geq 0, \tag{11.15}$$

$$\sum_{a \in \mathcal{A}} \sum_{d \in \mathcal{D}} \delta_{a'}(a,d) x_{0ad} - R^x_{0a'} = 0. \tag{11.16}$$

The second-stage decisions have to be made subject to constraints for each outcome $\omega \in \Omega$, given by

$$\sum_{d \in \mathcal{D}} x_{1ad}(\omega) = R_{1a}(\omega) = R^x_{0a} + \hat{R}_{1a}(\omega), \tag{11.17}$$

$$\sum_{a \in \mathcal{A}} x_{1ad}(\omega) \leq \hat{D}_{1b_d}(\omega) \quad d \in \mathcal{D}^D, \tag{11.18}$$

$$x_{1ad}(\omega) \geq 0. \tag{11.19}$$

Constraint (11.13) limits our decisions by the resources that are initially available. We may use these resources to satisfy initial demands, contained in $\hat{D}_0$, where flows are limited by constraint (11.14). Constraint (11.15) enforces nonnegativity, while constraint (11.16) defines the post-decision resource vector $R^x_0$.

The constraints for the second stage are similar to those for the first stage, except that there is a set of constraints for every outcome $\omega$. Note that equation (11.12) is written as if we are using $x_1$ as the state variable for the second stage. Again, this is the standard notational style of stochastic programming.

Equations (11.12)-(11.19) describe a fairly general model. It is useful to see the model formulated at a detailed level, but for what we are about to do, it is convenient to express it in matrix form.

**First-stage constraints:**

$$A_0 x_0 = R_0, \tag{11.20}$$

$$B_0 x_0 \leq \hat{D}_0, \tag{11.21}$$

$$x_0 \geq 0, \tag{11.22}$$

$$\Delta x_0 - R^x_0 = 0. \tag{11.23}$$

**Second-stage constraints:**

$$A_1 x_1(\omega) = R_1(\omega) \quad \forall \omega \in \Omega, \tag{11.24}$$

$$B_1 x_1(\omega) \leq \hat{D}_1(\omega) \quad \forall \omega \in \Omega, \tag{11.25}$$

$$x_1(\omega) \geq 0 \quad \forall \omega \in \Omega. \tag{11.26}$$

We note that in the classical language of math programming, $R_0^x$ is a decision variable defined by equation (11.23). In dynamic programming, we view $R_0^x = R^{M,x}(R_0, x_0)$ as a function that depends on $R_0$ and $x_0$.

The second-stage contribution function depends on the first-stage allocation, $R_0^x(x_0)$, so we can write it as

$$V_0(R_0^x) = \mathbb{E}C_1(x_1),$$

which allows us to rewrite (11.12) as

$$\max_{x_0} \big(C_0(x_0) + V_0(R_0^x)\big). \tag{11.27}$$

This shows that our two-stage problem consists of a one-period contribution function (using the information we know now) and a value function that captures the expected contribution, which depends on the decisions that we made in the first stage.

If we were to solve this two-stage problem using approximate dynamic programming, we would progress as depicted in figure 11.6. Figure 11.6a shows a classic two-stage problem with a first stage (deterministic) decision with random outcomes in the second stage. Figure 11.6b shows that we will make the first-stage decision using an approximation of the second stage. Finally, 11.6c illustrates solving the second stage using a sample realization, producing the dual variables which can be used to update the value function approximation used in the first stage.

We pause to note a major stylistic departure between stochastic and dynamic programming. In writing equation (11.27), it is clear that we have written our expected value function in terms of the resource state variable $R_0^x$. Of course, $R_0^x$ is a function of $x_0$, which means we could write (11.27) as

$$\max_{x_0} \big(C_0(x_0) + V_0(x_0)\big). \tag{11.28}$$

This is the style favored by the stochastic programming community, where $V_0(x_0)$ is referred to as the *recourse function* which is typically denoted $Q$. Mathematically, $V_0(x_0)$ and $V_0(R_0^x)$ are equivalent, but computationally, they can be quite different. If $|\mathcal{A}|$ is the size of our resource attribute space and $|\mathcal{D}|$ is the number of types of decisions, the dimensionality of $x_0$ is typically on the order of $|\mathcal{A}| \times |\mathcal{D}|$ whereas the dimensionality of $R_0^x$ will be on the order of $|\mathcal{A}|$. As a rule, it is computationally much easier to approximate $V_0(R_0^x)$ than $V_0(x_0)$ simply because functions of lower-dimensional variables are easier to approximate than functions of higher-dimensional ones. In some of our discussions in this chapter, we adopt the convention of writing $V_0$ as a function of $x_0$ in order to clarify the relationship with stochastic programming.

As we have seen in this volume, approximating the value function involves exploiting structure in the state variable. The stochastic programming community does not use the term "state variable," but some authors will use the term *tenders*, which is to say that the first stage "tenders" $R_0^x$ to the second stage. However, whereas the ADP community puts tremendous energy into exploiting the structure of the state variable when developing an approximation, state variables play either a minor or nonexistent role in stochastic programming.

Figure 11.6 illustrates solving the first-stage problem using a particular approximation for the second stage. The stochastic programming community has developed algorithmic strategies which solve these problems optimally (although the term "optimal" carries different meanings). The next two sections describe algorithmic strategies that have been developed within the stochastic programming community.

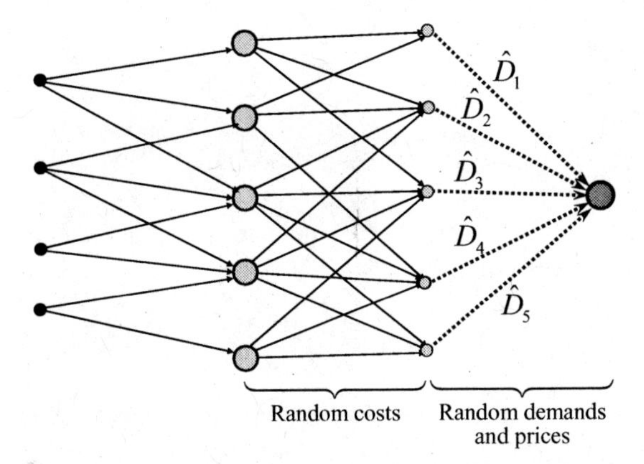


11.6a: The two-stage problem with stochastic second-stage data.

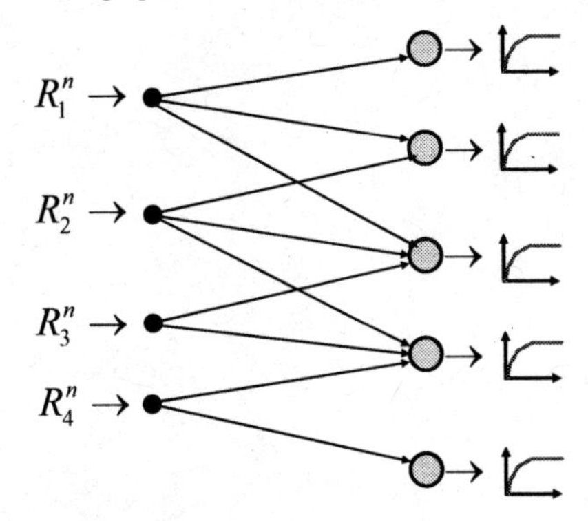


11.6b: Solving the first stage using a separable, piecewise linear approximation of the second stage.

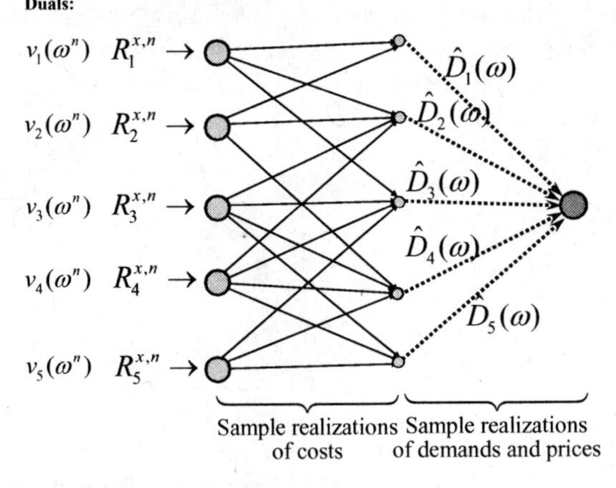


11.6c: Solving a Monte Carlo realization of the second stage and obtaining dual variables.

**Figure 11.6** Steps in estimating separable, piecewise-linear approximations for two-stage stochastic programs.

### 11.6.2 Benders' decomposition

Benders' decomposition is most easily described in the context of our two-stage resource allocation problem. Initially, we have to solve

$$\max_{x_0} \left(c_0 x_0 + V_0(R_0^x)\right) \tag{11.29}$$

subject to constraints (11.20) - (11.23). We note in passing that the stochastic programming community would normally write (11.29) in the form

$$\max_{x_0} \left(c_0 x_0 + V_0(x_0)\right),$$

which is mathematically correct (after all, $R_0^x$ is a function of $x_0$), but we prefer the form in (11.29) to capture the second-stage dependence on $R_0^x$.

As before, the value function is given by

$$V_0(R_0^x) = \sum_{\omega \in \Omega} p(\omega) V_0(R_0^x, \omega),$$

where $V_0(R_0^x, \omega)$ is given by

$$V_0(R_0^x, \omega) = \max_{x_1(\omega)} c_1 x_1(\omega) \tag{11.30}$$

subject to constraints (11.24) - (11.26). For ease of reference, we write the second-stage constraints as

$$\begin{array}{rcll} & & & \text{Dual} \\ A_1 x_1(\omega) & = & \Delta x_0 + \hat{R}_1(\omega), & \hat{v}_1^R(\omega), \\ B_1 x_1(\omega) & \leq & \hat{D}_1(\omega), & \hat{v}_1^D(\omega), \\ x_1(\omega) & \geq & 0, & \end{array}$$

where $\hat{v}^R(\omega)$ and $\hat{v}^D(\omega)$ are the dual variables for the resource constraints and demand constraints under outcome (scenario) $\omega$. For our discussion, we assume that we have a discrete set of sample outcomes $\Omega$, where the probability of outcome $\omega$ is given by $p(\omega)$. The linear programming dual of the second-stage problem takes the form

$$z^*(x_0, \omega) = \min_{\hat{v}_1^R(\omega), \hat{v}_1^D(\omega)} (\Delta x_0 + \hat{R}_1(\omega))^T \hat{v}_1^R(\omega) + (\hat{D}_1(\omega))^T \hat{v}_1^D(\omega) \tag{11.31}$$

subject to

$$A_1^T \hat{v}_1^R(\omega) + B_1^T \hat{v}_1^D(\omega) \geq c_1, \tag{11.32}$$

$$\hat{v}_1^D(\omega) \geq 0. \tag{11.33}$$

The dual is also a linear program, and the optimal $\hat{v}_1^*(\omega) = (\hat{v}_1^{R*}(\omega), \hat{v}_1^{D*}(\omega))$ must occur at one of a set of vertices, which we denote by $\mathcal{V}_1$. Note that the set of feasible dual solutions $\mathcal{V}_1$ is independent of $\omega$, a property that arises because we have assumed that our only source of randomness is in the right-hand side constraints (we would lose this if our costs were random).

Since $z^*(x_0, \omega)$ is the optimal solution of the dual, we have

$$z^*(x_0, \omega) = (\Delta x_0 + \hat{R}_1(\omega))^T \hat{v}_1^{R*}(\omega) + (\hat{D}_1(\omega))^T \hat{v}_1^{D*}(\omega).$$

Next define

$$z(x_0, \hat{v}_1, \omega) = (\Delta x_0 + \hat{R}_1(\omega))^T \hat{v}_1^R + (\hat{D}_1(\omega))^T \hat{v}_1^D.$$

Note that while $z(x_0, \hat{v}_1, \omega)$ depends on $\omega$, it is written for any set of duals $\hat{v}_1 \in \mathcal{V}_1$ (we do not need to index them by $\omega$). We observe that

$$\begin{aligned} z^*(x_0, \omega) &= z(x_0, \hat{v}_1^*(\omega), \omega) \\ &\leq z(x_0, \hat{v}_1, \omega) \quad \forall \hat{v}_1 \in \mathcal{V}_1, \end{aligned} \tag{11.34}$$

since $z^*(x_0, \omega)$ is the best we can do. Furthermore, equation (11.34) is true for all $\hat{v}_1 \in \mathcal{V}_1$, and all outcomes $\omega$. We know from the theory of linear programming that our primal must always be less than or equal to our dual, which means that

$$\begin{aligned} V_0(x_0, \omega) &\leq z(x_0, \hat{v}_1, \omega) \;\; \forall \hat{v}_1 \in \mathcal{V}_1, \;\; \omega \in \Omega, \\ &= z^*(x_0, \omega), \;\; \forall \omega \in \Omega. \end{aligned}$$

This means that for all $\hat{v}_1 \in \mathcal{V}_1$,

$$V_0(x_0) \leq \sum_{\omega \in \Omega} p(\omega) \left( (\Delta x_0 + \hat{R}_1(\omega))^T \hat{v}_1^R + (\hat{D}_1(\omega))^T \hat{v}_1^D \right), \tag{11.35}$$

where the inequality (11.35) is tight for $\hat{v}_1 = \hat{v}_1^*(\omega)$. Now let

$$\alpha_1(\hat{v}_1) = \sum_{\omega \in \Omega} p(\omega) (\hat{R}_1(\omega)^T \hat{v}_1^R + \hat{D}_1(\omega)^T \hat{v}_1^D), \tag{11.36}$$

$$\beta_1(\hat{v}_1) = \sum_{\omega \in \Omega} p(\omega) \Delta^T \hat{v}_1^R. \tag{11.37}$$

For historical reasons, we use $\alpha$ as a coefficient rather than a stepsize in our discussion of Benders' decomposition. This allows us to write (11.35) in the more compact form

$$V_0(x_0) \leq \alpha_1(\hat{v}_1) + (\beta_1(\hat{v}_1))^T x_0. \tag{11.38}$$

The right-hand side of (11.38) is called a *cut* since it is a plane (actually, an $n$-dimensional hyperplane) that represents an upper bound on the value function. Using these cuts, we can replace (11.29) with

$$\max_{x_0} \left( c_0 x_0 + z \right) \tag{11.39}$$

subject to (11.20)-(11.23) plus

$$z \leq \alpha_1(\hat{v}_1) + (\beta_1(\hat{v}_1))^T x_0 \;\; \forall \hat{v}_1 \in \mathcal{V}_1. \tag{11.40}$$

Unfortunately, equation (11.40) can be computationally problematic. The set $\mathcal{V}_1$ may be extremely large, so enforcing this constraint for each vertex $\hat{v}_1 \in \mathcal{V}_1$ is prohibitive. Furthermore, even if $\Omega$ is finite, it may be quite large, making summations over the elements in $\Omega$ expensive for some applications.

A simple strategy overcomes the problem of enumerating the set of dual vertices. Assume that after iteration $n-1$ we have a set of cuts $(\alpha_1^m, \beta_1^m)_{m=1}^n$ that we have generated from previous iterations. Using these cuts, we solve equation (11.39):

$$z \leq \alpha_1^m + (\beta_1^m)^T x_0, \quad m = 1, 2, \ldots, n-1. \tag{11.41}$$

to obtain a first-stage solution $x_0^n$. Given $x_0^n$, we then solve the dual (11.31) to obtain $\hat{v}_1^n(\omega)$ (the optimal duals) for each $\omega \in \Omega$. Using this information, we obtain $\alpha_1^n$ and $\beta_1^n$. Thus, instead of having to solve (11.39) subject to the entire set of constraints (11.40), we use only the cuts in equation (11.41).

This algorithm is known as the "L-shaped" algorithm (more precisely, "L-shaped decomposition"). For a finite $\Omega$, it has been proven to converge to the optimal solution. The problem is the requirement that we calculate $\hat{v}_1^*(\omega)$ for all $\omega \in \Omega$, which means we must solve a linear program for each outcome at each iteration. For most problems, this is computationally pretty demanding.

### 11.6.3 Variations

Two variations of this algorithm that avoid the computational burden of computing $\hat{v}_1^*(\omega)$ for each $\omega$ have been proposed. These algorithms vary only in how the cuts are computed and updated. The first is known as *stochastic decomposition*. At iteration $n$, after solving the first-stage problem we would solve (for sample realization $\omega^n$)

$$\left(\hat{v}_1^{R,n}, \hat{v}_1^{D,n}\right) = \underset{\hat{v}_1^R(\omega^n), \hat{v}_1^D(\omega^n)}{\arg\min} \quad (\Delta x_0^n + \hat{R}_1(\omega^n))^T \hat{v}_1^R(\omega^n) + \hat{D}_1(\omega^n)^T \hat{v}_1^D(\omega^n) \tag{11.42}$$

for a single outcome $\omega^n$. We then update all previous cuts by first computing for $m = 1, 2, \ldots, n$

$$(\nu_m^{R,n}, \nu_m^{D,n}) = \underset{\hat{v}_1^R, \hat{v}_1^D}{\arg\min} \left( (\Delta x_0^n + \hat{R}_1(\omega^m))^T \hat{v}_1^R + \hat{D}_1^T(\omega^m) \hat{v}_1^D \,\middle|\, (\hat{v}_1^R, \hat{v}_1^D) \in \mathcal{V}^n \right).$$

We then compute the next cut using

$$\alpha_n^n = \frac{1}{n} \sum_{m=1}^{n} \left( \hat{R}_1(\omega^m)^T \nu_m^{R,n} + \hat{D}_1(\omega^m)^T \nu_m^{D,n} \right), \tag{11.43}$$

$$\beta_n^n = \frac{1}{n} \sum_{m=1}^{n} \Delta^T \nu_m^{R,n}. \tag{11.44}$$

Then, all the previous cuts are updated using

$$\alpha_m^n = \frac{n-1}{n} \alpha_m^{n-1}, \quad \beta_m^n = \frac{n-1}{n} \beta_m^{n-1}, \quad m = 1, \ldots, n-1. \tag{11.45}$$

The complete algorithm is given in figure 11.7. The beauty of stochastic decomposition is that we never have to loop over all outcomes, and we solve a single linear program for the second stage at each iteration. There is a requirement that we loop over all the cuts that we have previously generated (in equation (11.45)), but the calculation here is fairly trivial (contrast the need to solve a complete linear program for every outcome in the L-shaped algorithm).

A second variation, called the CUPPS algorithm (for "cutting plane and partial sampling" algorithm), finds new dual extreme points using

$$\alpha_n^n = \frac{1}{|\Omega|} \sum_{\omega \in \Omega} \left( \hat{R}_1(\omega)^T \nu^{R,n}(\omega) + \hat{D}_1(\omega^m)^T \nu^{D,n}(\omega) \right), \tag{11.46}$$

$$\beta_n^n = \frac{1}{|\Omega|} \sum_{\omega \in \Omega} \Delta^T \nu^{R,n}(\omega), \tag{11.47}$$

---

**Step 0.** Set $\mathcal{V}^0 = \phi, n = 1$

**Step 1.** Solve the following master problem:

$$x_0^n = \arg\max_{x_0 \in \mathcal{X}_0} \left(c_0 x_0 + z\right)$$

where $\mathcal{X}_0 = \{A_0 x_0 = R_0, B_0 x_0 \leq \hat{D}_0, \; z \leq \alpha_m^n + \beta_m^n x_0, \; m = 1, \ldots, n-1, \; x_0 \geq 0\}$.

**Step 2.** Sample $\omega^n \in \Omega$ and solve the second-stage problem:

$$\max\{c_1 x_1 : A_1 x_1 = \Delta x_0^n + \hat{R}_1(\omega^n), B_1 x_1 \leq \hat{D}_1(\omega^n), \; x_1 \geq 0\}$$

to obtain the optimal dual solution from equation (11.42) and store the dual vertex

$$\mathcal{V}^n \leftarrow \mathcal{V}^{n-1} \cup (\hat{v}_1^{R,n}, \hat{v}_1^{D,n}).$$

**Step 3.** Update the cuts:

**Step 3a.** Compute for $m = 1, \ldots, n$:

$$(\nu_m^{R,n}, \nu_m^{D,n}) = \arg\min_{\hat{v}_1^R, \hat{v}_1^D} \left( (\Delta x_0^n + \hat{R}_1(\omega^m))^T \hat{v}_1^R + \hat{D}_1^T(\omega^m)\hat{v}_1^D \;\middle|\; (\hat{v}_1^R, \hat{v}_1^D) \in \mathcal{V}^n \right)$$

**Step 3b.** Construct the coefficients $\alpha_n^n$ and $\beta_n^n$ of the $n^{th}$ cut to be added to the master problem using (11.43) - (11.44).

**Step 3c.** Update the previously generated cuts by

$$\alpha_m^n = \frac{n-1}{n}\alpha_m^{n-1}, \quad \beta_m^n = \frac{n-1}{n}\beta_m^{n-1}, \qquad m = 1, \ldots, n-1.$$

**Step 4.** Check for convergence; if not, set $n = n + 1$ and return to Step 1.

---

**Figure 11.7** Sketch of the stochastic decomposition algorithm.

where for each $\omega \in \Omega$, we compute the duals

$$(\nu^{R,n}, \nu^{D,n})(\omega) = \arg\min_{\hat{v}_1^R, \hat{v}_1^D} \left( (\Delta x_0^n + \hat{R}_1(\omega^m))^T \hat{v}_1^R + \hat{D}_1^T(\omega^m)\hat{v}_1^D \;\middle|\; (\hat{v}_1^R, \hat{v}_1^D) \in \mathcal{V}^n \right). \tag{11.48}$$

Equations (11.46) - (11.48) represent the primary computational burden of CUPPS (outside of solving the linear program for each time period in each iteration). These have to be computed for each sample realization in $\Omega$. Stochastic decomposition, by contrast, requires that we update all previously generated cuts, a step that CUPPS does not share. The steps of this algorithm are given in figure 11.8.

L-shaped decomposition, stochastic decomposition, and the CUPPS algorithm offer three contrasting strategies for generating cuts, which are illustrated in figure 11.9. L-shaped decomposition is computationally the most demanding, but it produces tight cuts and will, in general, produce the fastest convergence (measured in terms of the number of iterations). The cuts produced by stochastic decomposition are neither tight nor valid, but steadily converge to the correct function. The cuts generated by CUPPS are valid but not tight.

Stochastic decomposition and CUPPS require roughly the same amount of work as any of the other approximation strategies we have presented for approximate dynamic programming, but it offers a proof of convergence. Notice also that it does not require exploration, but it does require that we generate a dual variable for every second-stage constraint (which requires that our attribute space $\mathcal{A}$ be small enough to enumerate). The

**Step 0.** Set $\mathcal{V}^0 = \phi, n = 1$

**Step 1.** Solve the following master problem:

$$x_0^n = \arg\max_{x_0 \in \mathcal{X}_0} (c_0 x_0 + z)$$

where $\mathcal{X}_0 = \{A_0 x_0 = R_0, B_0 x_0 \leq \hat{D}_0,\ z \leq \alpha^m + \beta^m x_0,\ m = 1, \ldots, n-1,\ x_0 \geq 0\}$.

**Step 2.** Sample $\omega^n \in \Omega$ and solve the second-stage problem:

$$\max\{c_1 x_1 : A_1 x_1 = \Delta x_0^n + \hat{R}_1(\omega^n), B_1 x_1 \leq \hat{D}_1(\omega^n),\ x_1 \geq 0\}$$

to obtain the optimal dual solution from equation (11.42) and store the dual vertex

$$\mathcal{V}^n \leftarrow \mathcal{V}^{n-1} \cup (\hat{v}_1^{R,n}, \hat{v}_1^{D,n}).$$

**Step 3.** Construct the coefficients of the $n^{th}$ cut to be added to the master problem using (11.46) and (11.47).

**Step 4.** Check for convergence; if not, set $n = n + 1$ and return to Step 1.

**Figure 11.8** Sketch of the CUPPS algorithm.

real question is rate of convergence. The next section provides some experimental work that hints at answers to this question, although the real answer depends on the characteristics of individual problems.

### 11.6.4 Experimental comparisons

For resource allocation problems, it is important to evaluate approximations two ways. The first is to consider "two-stage" problems (a terminology from stochastic programming) where we make a decision, see a random outcome, make one more decision, and then stop. For resource allocation problems, the first decision consists of meeting known demands and then acting on resources so that they are in position to satisfy demands that will become known in the second stage (at time $t = 1$). Two-stage problems are important because we solve multiperiod (multistage) problems as sequences of two-stage problems.

Topaloglu (2001) reports on an extensive set of experimental comparisons of separable, piecewise linear value function approximations (denoted the "SPAR" algorithm since it uses the projection operator for maintaining concavity described in section 11.3) and variations of Benders' decomposition (which can be viewed as nonseparable value function approximations). These variations include L-shaped (which requires solving a linear program for every sample outcome), CUPPS (which requires looping over all sample outcomes and performing a trivial calculation), and stochastic decomposition (which only requires looping over the sample outcomes that we have actually sampled, performing modest calculations for each outcome). Stochastic decomposition can be viewed as a form of approximate dynamic programming (without a state variable). All of our experiments were done (by necessity) with a finite set of outcomes. L-shaped is computationally expensive, but for smaller problems could be generally counted on to produce the optimal solution which was then used to evaluate the quality of the solutions produced by the other algorithms.

The experiments were conducted for a distribution problem where resources were distributed to a set of locations, after which they were used to satisfy a demand. The number of locations determines the dimensionality of $R_0^x$ and $R_1$. The problems tested used 10,

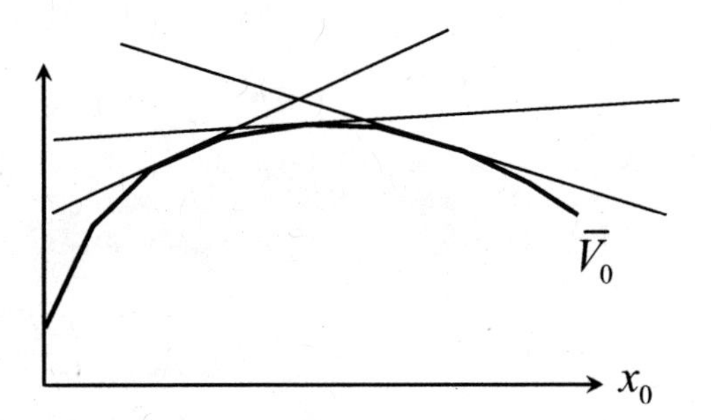


11.9a: Cuts generated by L-shaped decomposition.

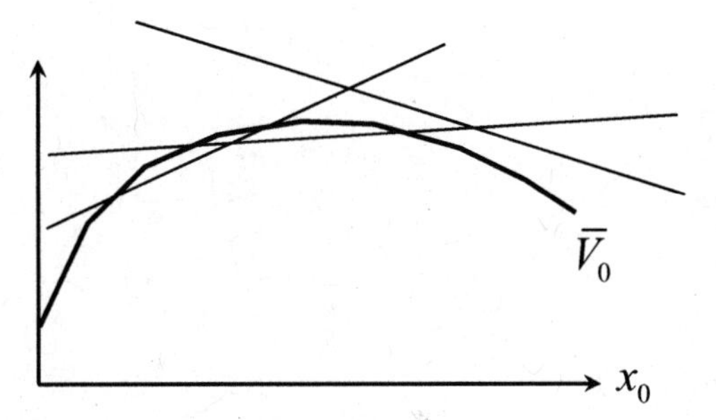


11.9b: Cuts generated by the stochastic decomposition algorithm.

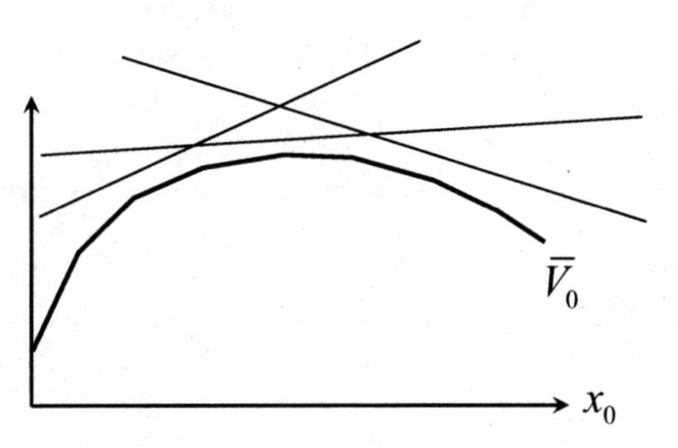


11.9c: Cuts generated by the CUPPS algorithm.

**Figure 11.9** Illustration of the cuts generated by (a) L-shaped decomposition, (b) stochastic decomposition, and (c) CUPPS.

25, 50, and 100 locations. L-shaped decomposition found the optimal solution for all the problems except the one with 100 locations.

The results are shown in table 11.1, which shows the percentage distance from the optimal solution for all three algorithms as a function of the number of iterations. The results show that L-shaped decomposition generally has the fastest rate of convergence (in terms of iterations), but the computational demand of solving linear programs for

| Locations | Method | Number of independent demand observations | | | | | | | |
|---|---|---|---|---|---|---|---|---|---|
| | | 25 | 50 | 100 | 250 | 500 | 1000 | 2500 | 5000 |
| 10 | SPAR | 18.65 | 12.21 | 7.07 | 2.25 | 0.48 | 0.28 | 0.04 | 0.15 |
| | L-Shaped | 3.30 | 0.19 | 0.00 | | | | | |
| | CUPPS | 5.84 | 4.19 | 0.50 | 0.29 | 0.00 | | | |
| | SD | 45.45 | 9.18 | 11.35 | 2.35 | 2.30 | 1.12 | 0.30 | 0.26 |
| 25 | SPAR | 11.73 | 4.04 | 2.92 | 0.89 | 0.34 | 0.13 | 0.19 | 0.06 |
| | L-Shaped | 19.88 | 15.98 | 2.14 | 0.11 | 0.00 | | | |
| | CUPPS | 8.27 | 13.40 | 4.33 | 4.02 | 1.47 | 0.16 | 0.00 | |
| | SD | 40.55 | 29.79 | 22.22 | 12.80 | 4.24 | 4.80 | 1.06 | 0.95 |
| 50 | SPAR | 9.99 | 2.60 | 1.18 | 0.48 | 0.26 | 0.30 | 0.12 | 0.05 |
| | L-Shaped | 42.56 | 20.30 | 6.07 | 1.49 | 0.52 | 0.04 | 0.00 | |
| | CUPPS | 34.93 | 9.91 | 19.30 | 11.71 | 5.09 | 1.38 | 0.32 | 0.00 |
| | SD | 43.18 | 29.81 | 17.94 | 8.09 | 5.91 | 6.25 | 2.73 | 1.02 |
| 100* | SPAR | 8.74 | 4.61 | 1.20 | 0.45 | 0.16 | 0.05 | 0.01 | 0.00 |
| | L-Shaped | 74.52 | 29.79 | 26.21 | 7.30 | 2.32 | 0.85 | 0.11 | 0.02 |
| | CUPPS | 54.59 | 35.54 | 23.99 | 17.58 | 14.68 | 14.13 | 5.36 | 0.91 |
| | SD | 62.63 | 34.82 | 40.73 | 12.14 | 15.22 | 17.43 | 17.49 | 9.42 |

* Optimal solution not found, figures represent the deviation from the best objective value known.

**Table 11.1** Numerical results for two-stage distribution problems while varying the number of locations. The table gives the percent over optimal (where available) for SPAR (separable, piecewise linear value functions), along with three flavors of Benders' decomposition: L-shaped, CUPPS, and stochastic decomposition (SD) (from Topaloglu (2001)).

every outcome usually makes this computationally intractable for most applications. All the methods based on Benders' decomposition show progressively slower convergence as the number of locations increase, suggesting that this method will struggle with higher dimensional state variables. As the problems became larger, the separable approximation showed much faster convergence with near optimal performance in the limit.

It is impossible to make general conclusions about the performance of one method over another, since this is an experimental question which depends on the characteristics of the datasets used for testing. However, it appears from these experiments that all flavors of Benders' decomposition become much slower as the dimensionality of the state variable grows. Stochastic decomposition and approximate dynamic programming (SPAR) require approximately the same computational effort per iteration. SD is dramatically faster than L-shaped decomposition in terms of the work required per iteration, but the price is that it will have a slower rate of convergence (in terms of iterations). However, even L-shaped decomposition exhibits a very slow rate of convergence, especially as the problem size grows (where problem size is measured in terms of the dimensionality of the state variable).

Benders' decomposition (in one of its various forms) offers significant value since it provides a provably convergent algorithm. SPAR (which uses piecewise linear, separable value function approximations) has been found to work on large-scale resource allocation problems, but the error introduced by the use of a separable approximation will be problem-dependent.

## 11.7 WHY DOES IT WORK?**

### 11.7.1 Proof of the SHAPE algorithm

The convergence proof of the SHAPE algorithm is a nice illustration of a martingale proof. We start with the Martingale convergence theorem (Doob (1953), Neveu (1975), and Taylor (1990)), which has been the basis of convergence proofs for stochastic subgradient methods (as illustrated in Gladyshev (1965)). We then list some properties of the value function approximation $\hat{V}(\cdot)$ and produce a bound on the difference between $x^n$ and $x^{n+1}$. Finally, we prove the theorem. This section is based on Cheung & Powell (2000).

Let $\omega^n$ be the information that we sample in iteration $n$, and let $\omega = (\omega^1, \omega^2, \ldots)$ be an infinite sequence of observations of sample information where $\omega \in \Omega$. Let $h^n = \omega^1, \omega^2, \ldots, \omega^n$ be the history up to (and including) iteration $n$. Let $\mathfrak{F}$ be the $\sigma$-algebra on $\Omega$, and let $\mathfrak{F}^n \subseteq \mathfrak{F}^{n+1}$ be the sequence of increasing sub-sigma-algebras on $\Omega$ representing the information we know up through iteration $n$.

We assume the following:

**(A.1)** $\mathcal{X}$ is convex and compact.

**(A.2)** $\mathbb{E}V(x, W)$ is convex, finite and continuous on $\mathcal{X}$.

**(A.3)** $g^n$ is bounded such that $\|g^n\| \leq c_1$.

**(A.4)** $\hat{V}^n(x)$ is strongly convex, meaning that

$$\hat{V}^n(y) - \hat{V}^n(x) \geq \hat{v}^n(x)(y - x) + b\|x - y\|^2, \tag{11.49}$$

where $b$ is a positive number that is a constant throughout the optimization process. The term $b\|x - y\|^2$ is used to ensure that the slope $\hat{v}^n(x)$ is a monotone function of $x$. When we interchange $y$ and $x$ in (11.49) and add the resulting inequality to (11.49), we obtain

$$(\hat{v}^n(y) - \hat{v}^n(x))(y - x) \geq 2b\|x - y\|^2. \tag{11.50}$$

**(A.5)** The stepsizes $\alpha_n$ are $\mathfrak{F}^n$ measurable and satisfy

$$0 < \alpha_n < 1, \quad \sum_{n=0}^{\infty} \mathbb{E}\{(\alpha_n)^2\} < \infty$$

and

$$\sum_{n=0}^{\infty} \alpha_n = \infty \quad a.s.$$

**(A.6)** $\hat{V}^0(x)$ is bounded and continuous, and $\hat{v}^0(x)$ (the derivative of $\hat{V}^0(x)$) is bounded for $x \in \mathcal{X}$.

Note that we require that the expected sum of squares be bounded whereas we must impose the almost sure condition that the sum of stepsizes be infinite. We now state our primary theorem.

**Theorem 11.7.1** *If (A.1) - (A.6) are satisfied, then the sequence $x^n$, produced by algorithm SHAPE, converges almost surely to an optimal solution $x^* \in \mathcal{X}^*$ of problem (11.7).*

The proof illustrates several proof techniques that are commonly used for these problems. Students interested in doing more fundamental research in this area may be able to use some of the devices in their own work.

To prove the theorem, we need to use the Martingale convergence theorem and two lemmas.

### *Martingale convergence theorem*

A sequence of random variables $\{W^n\}$ that are $\mathfrak{F}^n$-measurable is said to be a *supermartingale* if the sequence of conditional expectations $\mathbb{E}\{W^{n+1} \mid \mathfrak{F}^n\}$ exists and satisfies

$$\mathbb{E}\{W^{n+1} \mid \mathfrak{F}^n\} \leq W^n.$$

**Theorem 11.7.2** *(p. 26, Neveu (1975)) Let $W^n$ be a positive supermartingale. Then, $W^n$ converges to a finite random variable a.s.*

From the definition, $W^n$ is essentially the stochastic analog of a decreasing sequence.

### *Property of Approximations*

In addition to equations (11.49) – (11.50) of assumption (A.4), the optimal solution for problem (11.7) at iteration $n$ can be characterized by the variational inequality

$$\hat{v}^n(x^n)(x - x^n) \geq 0 \quad \forall x \in \mathcal{X}. \tag{11.51}$$

Furthermore, at iteration $k + 1$,

$$\left(\hat{v}^n(x^{n+1}) + \alpha_n(g^n - \hat{v}^n(x^n))\right)(x - x^{n+1}) \geq 0 \quad \forall x \in \mathcal{X}. \tag{11.52}$$

The first lemma below provides a bound on the difference between two consecutive solutions. The second lemma establishes that $V^n(x)$ is bounded.

**Lemma 11.7.1** *The solutions $x^n$ produced by algorithm SHAPE satisfy*

$$\|x^n - x^{n+1}\| \leq \frac{\alpha_n}{2b}\|g^n\|.$$

*where $b$ satisfies equation (11.49).*

**Proof:** Substituting $x$ by $x^n$ in (11.52), we have

$$\alpha_n \left(g^n - \hat{v}^n(x^n)\right)(x^n - x^{n+1}) \geq \hat{v}^n(x^{n+1})(x^{n+1} - x^n).$$

Rearranging the terms, we obtain

$$\begin{aligned}
\alpha_n g^n(x^n - x^{n+1}) &\geq \hat{v}^n(x^{n+1})(x^{n+1} - x^n) - \alpha_n \hat{v}^n(x^{n+1} - x^n) \\
&= (\hat{v}^n(x^{n+1}) - \hat{v}^n(x^n))(x^{n+1} - x^n) \\
&\quad +(1 - \alpha_n)\hat{v}^n(x^n)(x^{n+1} - x^n).
\end{aligned}$$

Combining (11.50), (11.51), and $0 < \alpha_n < 1$ gives us

$$\begin{aligned} \alpha_n g^n(x^n - x^{n+1}) &\geq 2b\|x^n - x^{n+1}\|^2 + (1-\alpha_n)\hat{v}^n(x^n)(x^{n+1} - x^n) \\ &\geq 2b\|x^n - x^{n+1}\|^2. \end{aligned}$$

Applying Schwarz's inequality, we have that

$$\alpha_n \|g^n\| \cdot \|x^n - x^{n+1}\| \geq \alpha_n g^n(x^n - x^{n+1}) \geq 2b\|x^n - x^{n+1}\|^2.$$

Dividing both sides by $\|x^n - x^{n+1}\|$, it follows that $\|x^n - x^{n+1}\| \leq \frac{\alpha_n}{2b}\|g^n\|$.

**Lemma 11.7.2** *The approximation function $\hat{V}^n(x)$ in iteration $n$ can be written as*

$$\hat{V}^n(x) = \hat{V}^0(x) + r^n x,$$

*where $r^n$ is a finite vector.*

**Proof:** The algorithm proceeds by adding linear terms to the original approximation. Thus, at iteration $n$, the approximation is the original approximations plus the linear term

$$\hat{V}^n(x) = \hat{V}^0(x) + r^n x \tag{11.53}$$

where $r^n$ is the sum of the linear correction terms. We just have to show that $r^n$ is finite.

When taking the first derivative of $\hat{V}^n(x)$ in equation (11.53), we have

$$\hat{v}^n(x) = \hat{v}^0(x) + r^n.$$

With that, we can write $\hat{V}^{n+1}(x)$ in terms of $\hat{V}^0(x)$

$$\begin{aligned} \hat{V}^{n+1}(x) &= \hat{V}^n(x) + \alpha_n(g^n - \hat{v}^n(x))x \\ &= \hat{V}^0(x) + r^n x + \alpha_n(g^n - \hat{v}^n(x))x \\ &= \hat{V}^0(x) + r^n x + \alpha_n(g^n - \hat{v}^0(x_k) - r^n)x. \end{aligned}$$

Therefore, $r^{n+1}$ and $r^n$ are related as follows:

$$r^{n+1} = \alpha_n(g^n - \hat{v}^0(x_k)) + (1-\alpha_n)r^n. \tag{11.54}$$

So, the total change in our initial approximation is a weighted sum of $g^n - \hat{v}^0(x^n)$ and the current cumulative change. Since both $g^n$ and $\hat{v}^0(x^n)$ are finite, there exists a finite, positive vector such that

$$\hat{d} \geq \max_k |g^n - \hat{v}^0(x^n)|. \tag{11.55}$$

We can now use induction to show that $r^n \leq \hat{d}$ for all $n$. For $n = 1$, we have $r^1 = a_0(g^0 - \hat{v}^0_0) \leq a_0\hat{d}$. Since $a_0 < 1$ and is positive, we have $r^1 \leq \hat{d}$. Assuming that $r^n \leq \hat{d}$, we want to show $r^{n+1} \leq \hat{d}$. By using this assumption and the definition of $\hat{d}$, equation (11.54) implies that

$$r^{n+1} \leq \alpha_n|g^n - \hat{v}^0(x_k)| + (1-\alpha_n)r^n \leq \alpha_n\hat{d} + (1-\alpha_n)\hat{d} = \hat{d}.$$

We now return to our main result.

## Proof of theorem 11.7.1

Our algorithm proceeds in three steps. First we establish a supermartingale that provides a basic convergence result. Then, we show that there is a convergent subsequence. Finally, we show that the entire sequence is convergent. For simplicity, we write $\hat{v}^n = \hat{v}^n(x^n)$.
*Step 1: Establish a supermartingale for theorem 11.7.2.*
Let $T^n = \hat{V}^n(x^*) - \hat{V}^n(x^n)$, and consider the difference of $T^{n+1}$ and $T^n$:

$$\begin{aligned} T^{n+1} &- T^n \\ &= \hat{V}^{n+1}(x^*) - \hat{V}^{n+1}(x^{n+1}) - \hat{V}^n(x^*) + \hat{V}^n(x^n) \\ &= \hat{V}^n(x^*) + \alpha_n(g^n - \hat{v}^n)x^* - \hat{V}^n(x^{n+1}) - \alpha_n(g^n - \hat{v}^n)x^{n+1} - \hat{V}^n(x^*) + \hat{V}^n(x^n). \end{aligned}$$

If we write $x^* - x^{n+1}$ as $x^* - x^n + x^n - x^{n+1}$, we get

$$\begin{aligned} T^{n+1} &- T^n \\ &= \underbrace{\hat{V}^n(x^n) - \hat{V}^n(x^{n+1}) - \alpha_n\hat{v}^n\,(x^n - x^{n+1})}_{\text{(I)}} - \underbrace{\alpha_n\hat{v}^n\,(x^* - x^n)}_{\text{(II)}} \\ &\quad + \underbrace{\alpha_n g^n(x^* - x^n)}_{\text{(III)}} + \underbrace{\alpha_n g^n(x^n - x^{n+1})}_{\text{(IV)}}. \end{aligned}$$

Consider each part individually. First, by the convexity of $\hat{V}^n(x)$, it follows that

$$\begin{aligned} \hat{V}^n(x^n) - \hat{V}^n(x^{n+1}) &\le \hat{v}^n\,(x^n - x^{n+1}) \\ &= (1-\alpha_n)\hat{v}^n\,(x^n - x^{n+1}) + \alpha_n\hat{v}^n\,(x^n - x^{n+1}). \end{aligned}$$

From equation (11.51) and $0 < \alpha_n < 1$, we know that (I) $\le 0$. Again, from equation (11.51) and $0 < \alpha_n < 1$, we show that (II) $\ge 0$.

For (III), by the definition that $g^n \in \partial V(x^n, \omega^{n+1})$,

$$g^n(x^* - x^n) \le V(x^*, \omega^{n+1}) - V(x^n, \omega^{n+1}),$$

where $V(x, \omega^{n+1})$ is the recourse function given outcome $\omega^{n+1}$.

For (IV), lemma 11.7.1 implies that

$$\alpha_n g^n(x^n - x^{n+1}) \le \alpha_n\|g^n\| \cdot \|x^n - x^{n+1}\| \le \frac{(\alpha_n)^2\|g^n\|^2}{2b} \le \frac{(\alpha_n)^2 c_1^2}{2b}.$$

Therefore, the difference $T^{n+1} - T^n$ becomes

$$T^{n+1} - T^n \le -\alpha_n\left(V(x^n, \omega^{n+1}) - V(x^*, \omega^{n+1})\right) + \frac{(\alpha_n)^2 c_1^2}{2b}. \tag{11.56}$$

Taking the conditional expectation with respect to $\mathfrak{F}^n$ on both sides, it follows that

$$\mathbb{E}\{T^{n+1}|\mathfrak{F}^n\} \le T^n - \alpha_n(\bar{V}(x^n) - \bar{V}(x^*)) + \frac{(\alpha_n)^2 c_1^2}{2b}, \tag{11.57}$$

where $T^n, \alpha_n$ and $x_k$ on the right-hand side are deterministic given the conditioning on $\mathfrak{F}^n$. We replace $V(x, \omega^{n+1})$ (for $x = x^n$ and $x = x^*$) with its expectation $\bar{V}(x)$ since conditioning on $\mathfrak{F}^n$ tells us nothing about $\omega^{n+1}$. Since $\bar{V}(x^n) - \bar{V}(x^*) \ge 0$, the sequence

$$W^n = T^n + \frac{c_1^2}{2b}\sum_{i=k}^{\infty} a_i^2$$

is a positive supermartingale. Theorem 11.7.2 implies the almost sure convergence of $W^n$. Thus,

$$T^n \rightarrow T^* \text{ a.s.}$$

*Step 2: Show that there exists a subsequence $n_j$ of $n$ such that $x^{n_j} \rightarrow x^* \in \mathcal{X}^*$ a.s.*

Summing equation (11.56) over $n$ up to $N$ and canceling the alternating terms of $T^n$ gives

$$T^{N+1} - T^0 = -\sum_{n=0}^{N} \alpha_n (V(x^n, \omega^{n+1}) - V(x^*, \omega^{n+1})) + \sum_{n=0}^{N} \frac{(\alpha_n)^2 c_1^2}{2b}.$$

Take expectations of both sides. For the first term on the right-hand side, we take the conditional expectation first conditioned on $\mathfrak{F}^n$ and then over all $\mathfrak{F}^n$, giving us

$$\begin{aligned}
\mathbb{E}\{T^{N+1} - T^0\} &= -\sum_{n=0}^{N} \mathbb{E}\left\{\mathbb{E}\{\alpha_n (V(x^n, \omega^{n+1}) - V(x^*, \omega^{n+1}))|\mathfrak{F}^n\}\right\} \\
&\quad + \mathbb{E}\left\{\sum_{n=0}^{N} \frac{(\alpha_n)^2 c_1^2}{2b}\right\} \\
&= -\sum_{n=0}^{N} \mathbb{E}\left\{\alpha_n (\bar{V}(x^n) - \bar{V}(x^*))\right\} + \frac{c_1^2}{2b} \sum_{n=0}^{N} \mathbb{E}\left\{(\alpha_n)^2\right\}.
\end{aligned}$$

Taking the limit as $N \rightarrow \infty$ and using the finiteness of $T^n$ and $\sum_{n=0}^{\infty} \mathbb{E}\{\alpha_n^2\}$, we have

$$\sum_{n=0}^{\infty} \mathbb{E}\left\{\alpha_n (\bar{V}(x^n) - \bar{V}(x^*))\right\} < \infty.$$

Since $\bar{V}(x^n) - \bar{V}(x^*) \geq 0$ and $\sum_{n=0}^{\infty} \alpha_n = \infty$ (a.s.), there exists a subsequence $n_j$ of $n$ such that

$$\bar{V}(x^{n_j}) \rightarrow \bar{V}(x^*) \text{ a.s.}$$

By continuity of $\bar{V}$, this sequence converges. That is,

$$x^{n_j} \rightarrow x^* \in \mathcal{X}^* \text{ a.s.}$$

*Step 3: Show that $x^n \rightarrow x^* \in \mathcal{X}^*$ a.s.*

Consider the convergent subsequence $x^{n_j}$ in Step 2. By using the expression of $\hat{V}^n$ in lemma 11.7.2, we can write $T^{n_j}$ as

$$\begin{aligned}
T^{n_j} &= \hat{V}^{n_j}(x^{n_j}) - \hat{V}^{n_j}(x^*) \\
&= \hat{V}^0(x^{n_j}) - \hat{V}^0(x^*) + r^{n_j}(x^{n_j} - x^*) \\
&\leq \hat{V}^0(x^{n_j}) - \hat{V}^0(x^*) + r^{n_j}|x^{n_j} - x^*| \\
&\leq \hat{V}^0(x^{n_j}) - \hat{V}^0(x^*) + \hat{d}|x^{n_j} - x^*|,
\end{aligned}$$

where $\hat{d}$ is the positive, finite vector defined in (11.55). When $x^{n_j} \rightarrow x^*$, both the terms $\hat{d}|x^{n_j} - x^*|$ and $\hat{V}^0(x^{n_j}) - \hat{V}^0(x^*)$ (by continuity of $\hat{V}^0$) go to 0. Since $T^{n_j}$ is positive, we obtain that $T^{n_j} \rightarrow 0$ a.s. Combining this result and the result in Step 1 (that $T^n$ converges to a unique nonnegative $T^*$ a.s.), we have $T^n \rightarrow T^* = 0$ a.s. Finally, we know from the strong convexity of $\hat{V}^n(\cdot)$ that

$$T^n = \hat{V}^n(x^n) - \hat{V}^n(x^*) \geq b\|x^n - x^*\|^2 \geq 0.$$

Therefore, $x^n \rightarrow x^*$ a.s.

### 11.7.2 The projection operation

Let $v_s^{n-1}$ be the value (or the marginal value) of being in state $s$ at iteration $n-1$ and assume that we have a function where we know that we should have $v_{s+1}^{n-1} \geq v_s^{n-1}$ (we refer to this function as *monotone*). For example, if this is marginal values, we would expect this if we were describing a concave function. Now assume that we have a sample realization $\hat{v}_s^n$ which is the value (or marginal value) of being in state $s$. We would then smooth this new observation with the previous estimates using

$$z_s^n = \begin{cases} (1-\alpha_{n-1})v_s^{n-1} + \alpha_{n-1}\hat{v}^n & \text{if } s = s^n, \\ v_s^n & \text{otherwise.} \end{cases} \tag{11.58}$$

Since $\hat{v}_s^n$ is random, we cannot expect $z_s^n$ to also be monotone. In this section, we want to restore monotonicity by defining an operator $v^n = \Pi_V(z)$ where $v_{s+1}^n \geq v_s^n$. There are several ways to do this. In this section, we define the operator $v = \Pi_V(z)$, which takes a vector $z$ (which is not necessarily monotone) and produces a monotone vector $v$. If we wish to find $v$ that is as close as possible to $z$, we would solve

$$\min \frac{1}{2}\|v - z\|^2$$
$$\text{subject to: } v_{s+1} - v_s \leq 0, \quad s = 0, \ldots, M. \tag{11.59}$$

Assume that $v_0$ is bounded above by $B$, and $v_{M+1}$, for $s < M$, is bounded from below by $-B$. Let $\lambda_s \geq 0$, $s = 0, 1, \ldots, M$ be the Lagrange multipliers associated with equation (11.59). It is easy to see that the optimality equations are

$$v_s = z_s + \lambda_s - \lambda_{s-1}, \quad s = 1, 2, \ldots, M, \tag{11.60}$$
$$\lambda_s(v_{s+1} - v_s) = 0, \quad s = 0, 1, \ldots, M. \tag{11.61}$$

Let $i_1, \ldots, i_2$ be a sequence of states where

$$v_{i_1-1} > v_{i_1} = v_{i_1+1} = \cdots = c = \cdots = v_{i_2-1} = v_{i_2} > v_{i_2+1}.$$

We can then add equations (11.60) from $i_1$ to $i_2$ to yield

$$c = \frac{1}{i_2 - i_1 + 1} \sum_{s=i_1}^{i_2} z_s.$$

If $i_1 = 1$, then $c$ is the smaller of the above and $B$. Similarly, if $i_2 = M$, then $c$ is the larger of the above and $-B$.

We also note that $v^{n-1} \in \mathcal{V}$ and $z^n$ computed by (11.58) differs from $v^{n-1}$ in just one coordinate. If $z^n \notin \mathcal{V}$ then either $z_{s^n-1}^n < z_{s^n}^n$, or $z_{s^n+1}^n > z_{s^n}^n$.

If $z^n_{s^n-1} < z^n_{s^n}$, then we we need to find the largest $1 < i \leq s^n$ where

$$z^n_{i-1} \geq \frac{1}{s^n - i + 1} \sum_{s=i}^{s^n} z^n_s.$$

If $i$ cannot be found, then we use $i = 1$. We then compute

$$c = \frac{1}{s^n - i + 1} \sum_{s=i}^{s^n} z^n_s$$

and let

$$v^{n+1}_j = \min(B, c), \quad j = i, \dots, s^n.$$

We have $\lambda_0 = \max(0, c - B)$, and

$$\lambda_s = \begin{cases} 0 & s = 1, \dots, i-1, \\ \lambda_{s-1} + z_s - v_s & s = i, \dots, s^n - 1, \\ 0 & s = s^n, \dots, M. \end{cases}$$

It is easy to show that the solution found and the Lagrange multipliers satisfy equations (11.60) -(11.61).

If $z^n_{s^n} < z^n_{s^n+1}$, then the entire procedure is basically the same with appropriate inequalities reversed.

## 11.8 BIBLIOGRAPHIC NOTES

Section 11.1 - The decision of whether to estimate the value function or its derivative is often overlooked in the dynamic programming literature, especially within the operations research community. In the controls community, use of gradients is sometimes referred to as *dual heuristic dynamic programming* (see, for example, Venayagamoorthy et al. (2002)).

Section 11.3 - The theory behind the projective SPAR algorithm is given in Powell et al. (2004). A proof of convergence of the leveling algorithm is given in Topaloglu & Powell (2003).

Section 11.4 - The SHAPE algorithm was introduced by Cheung & Powell (2000).

Section 11.6 - The first paper to formulate a math program with uncertainty appears to be Dantzig & Ferguson (1956). For a broad introduction to the field of stochastic optimization, see Ermoliev (1988) and Pflug (1996). For complete treatments of the field of stochastic programming, see Infanger (1994), Kall & Wallace (1994), Birge & Louveaux (1997), and Kall & Mayer (2005). For an easy tutorial on the subject, see Sen & Higle (1999). A very thorough introduction to stochastic programming is given in Ruszczyński & Shapiro (2003). Mayer (1998) provides a detailed presentation of computational work for stochastic programming. There has been special interest in the types of network problems we have considered (see Wallace (1987), Wallace (1986*a*) and Wallace (1986*b*)). Rockafellar & Wets (1991) presents specialized algorithms for stochastic programs formulated using scenarios. This modeling framework has been of particular interest in the are of financial portfolios (Mulvey & Ruszczyński (1995)). Benders' decomposition for two-stage stochastic programs was first proposed by Van Slyke & Wets (1969) as the "L-shaped" method. Higle & Sen (1991) introduce stochastic decomposition, which is a Monte-Carlo based algorithm that is most similar in spirit to approximate dynamic programming. Chen & Powell (1999) present a variation of Benders that falls between stochastic decomposition and the L-shaped method. The relationship between Benders' decomposition and dynamic programming is often overlooked. A notable exception is Pereira & Pinto (1991), which uses Benders to solve a resource allocation problem arising in the management of reservoirs. This paper presents Benders as a method for avoiding the curse of dimensionality of dynamic programming. For an excellent review of Benders' decomposition for multistage problems, see Ruszczyński (2003). Benders has been extended to multistage problems in Birge (1985), Ruszczyński (1993), and Chen & Powell (1999), which can be viewed as a form of approximate dynamic programming using cuts for value function approximations.

Section 11.7.1 - The proof of convergence of the SHAPE algorithm is based on Cheung & Powell (2000).

Section 11.7.2 - The proof of the projection operation is based on Powell et al. (2004).

## PROBLEMS

**11.1** Consider a newsvendor problem where we solve

$$\max_x \mathbb{E}F(x, \hat{D}),$$

where

$$F(x, \hat{D}) = p\min(x, \hat{D}) - cx.$$

We have to choose a quantity $x$ before observing a random demand $\hat{D}$. For our problem, assume that $c = 1$, $p = 2$, and that $\hat{D}$ follows a discrete uniform distribution between 1 and 10 (that is, $\hat{D} = d, d = 1, 2, \ldots, 10$ with probability 0.10). Approximate $\mathbb{E}F(x, \hat{D})$ as a piecewise linear function using the methods described in section 11.3, using a stepsize $\alpha_{n-1} = 1/n$. Note that you are using derivatives of $F(x, \hat{D})$ to estimate the slopes of the function. At each iteration, randomly choose $x$ between 1 and 10. Use sample realizations of the gradient to estimate your function. Compute the exact function and compare your approximation to the exact function.

**11.2** Repeat exercise 11.1, but this time approximate $\mathbb{E}F(x, \hat{D})$ using a linear approximation:

$$\bar{F}(x) = \theta x.$$

Compare the solution you obtain with a linear approximation to what you obtained using a piecewise-linear approximation. Now repeat the exercise using demands that are uniformly distributed between 500 and 1000. Compare the behavior of a linear approximation for the two different problems.

**11.3** Repeat exercise 11.1, but this time approximate $\mathbb{E}F(x, \hat{D})$ using the SHAPE algorithm. Start with an initial approximation given by

$$\bar{F}^0(x) = \theta_0(x - \theta_1)^2.$$

Use the recursive regression methods of sections 11.5 and 7.3 to fit the parameters. Justify your choice of stepsize rule. Compute the exact function and compare your approximation to the exact function.

**11.4** Repeat exercise 11.1, but this time approximate $\mathbb{E}F(x,\hat{D})$ using the regression function given by

$$\bar{F}(x) = \theta_0 + \theta_1 x + \theta_2 x^2.$$

Use the recursive regression methods of sections 11.5 and 7.3 to fit the parameters. Justify your choice of stepsize rule. Compute the exact function and compare your approximation to the exact function. Estimate your value function approximation using two methods:

(a) Use observations of $F(x,\hat{D})$ to update your regression function.

(b) Use observations of the derivative of $F(x,\hat{D})$, so that $\bar{F}(x)$ becomes an approximation of the derivative of $\mathbb{E}F(x,\hat{D})$.

**11.5** Approximate the function $\mathbb{E}F(x,\hat{D})$ in exercise 11.1, but now assume that the random variable $\hat{D} = 1$ (that is, it is deterministic). Using the following approximation strategies:

(a) Use a piecewise linear value function approximation. Try using both left and right derivatives to update your function.

(b) Use the regression $\bar{F}(x) = \theta_0 + \theta_1 x + \theta_2 x^2$.

**11.6** We are going to solve the basic asset acquisition problem (section 2.2.5) where we purchase assets (at a price $p^p$) at time $t$ to be used in time interval $t+1$. We sell assets at a price $p^s$ to satisfy the demand $\hat{D}_t$ that arises during time interval $t$. The problem is to be solved over a finite time horizon $T$. Assume that the initial inventory is 0 and that demands follow a discrete uniform distribution over the range $[0, D^{max}]$. The problem parameters are given by

$$\begin{aligned} \gamma &= 0.8, \\ D^{max} &= 10, \\ T &= 20, \\ p^p &= 5, \\ p^s &= 8. \end{aligned}$$

Solve this problem by estimating a piecewise linear value function approximation (section 11.3). Choose $\alpha_{n+1} = a/(a+n)$ as your stepsize rule, and experiment with different values of $a$ (such as 1, 5, 10, and 20). Use a single-pass algorithm, and report your profits (summed over all time periods) after each iteration. Compare your performance for different stepsize rules. Run 1000 iterations and try to determine how many iterations are needed to produce a good solution (the answer may be substantially less than 1000).

**11.7** Repeat exercise 11.6, but this time use the SHAPE algorithm to approximate the value function. Use as your initial value function approximation the function

$$\bar{V}_t^0(R_t) = \theta_0(R_t - \theta_2)^2.$$

For each of the exercises below, you may have to tweak your stepsize rule. Try to find a rule that works well for you (we suggest stick with a basic $a/(a+n)$ strategy). Determine an appropriate number of training iterations, and then evaluate your performance by averaging results over 100 iterations (testing iterations) where the value function is not changed.

(a) Solve the the problem using $\theta_0 = 1, \theta_1 = 5$.

(b) Solve the the problem using $\theta_0 = 1, \theta_1 = 50$.

(c) Solve the the problem using $\theta_0 = 0.1, \theta_1 = 5$.

(d) Solve the the problem using $\theta_0 = 10, \theta_1 = 5$.

(e) Summarize the behavior of the algorithm with these different parameters.

**11.8** Repeat exercise 11.6, but this time assume that your value function approximation is given by

$$\bar{V}_t^0(R_t) = \theta_0 + \theta_1 R_t + \theta_2 R_t^2.$$

Use the recursive regression techniques of sections 11.5 and 7.3 to determine the values for the parameter vector $\theta$.

**11.9** Repeat exercise 11.6, but this time assume you are solving an infinite horizon problem (which means you only have one value function approximation).

**11.10** Repeat exercise 11.8, but this time assume an infinite horizon.

**11.11** Repeat exercise 11.6, but now assume the following problem parameters:

$$\begin{aligned} \gamma &= 0.99, \\ T &= 200, \\ p^p &= 5, \\ p^s &= 20. \end{aligned}$$

For the demand distribution, assume that $\hat{D}_t = 0$ with probability 0.95, and that $\hat{D}_t = 1$ with probability 0.05. This is an example of a problem with low demands, where we have to hold inventory for a fairly long time.

# CHAPTER 12

# DYNAMIC RESOURCE ALLOCATION PROBLEMS

There is a vast array of problems that fall under the umbrella of "resource allocation." We might be managing a team of medical specialists who have to respond to emergencies, or technicians who have to provide local support for the installation of sophisticated medical equipment. Alternatively, a transportation company might be managing a fleet of vehicles, or an investment manager might be trying to determine how to allocate funds between different asset classes. We can even think of playing a game of backgammon as a problem of managing a single resource (the board), although in this chapter we are only interested in problems that involve multiple resources, where the quantity of flow is an important element of the decision variable.

Aside from the practical importance of these problems, this problem class provides a special challenge. In addition to the usual challenge of making decisions over time under uncertainty (the theme of this entire book) we now have to deal with the fact that our decision $x_t$ is suddenly of very high dimensionality, requiring that we use the tools of math programming (in particular, linear programming). For practical applications, it is not hard to find problems in this class where $x_t$ has thousands, or even tens of thousands, of dimensions. This problem class clearly suffers from all three "curses of dimensionality."

We illustrate the ideas by starting with a scalar problem, after which we move to a sequence of multidimensional (and in several cases, very high-dimensional) resource allocation problems. Each problem offers different features, but they can all be solved using the ideas we have been developing throughout this volume.

## 12.1 AN ASSET ACQUISITION PROBLEM

Perhaps one of the simplest resource allocation problems is the basic asset acquisition where we acquire assets (money, oil, water, aircraft) to satisfy a future, random demand. In our simple illustration, we purchase assets at time $t$ at a price which can be used to satisfy demands in the next time period (which means we do not know what they are). If we buy too much, the left-over inventory is held to future time periods. If we cannot satisfy all the demand, we lose revenue, but we do not allow ourselves to hold these demands to be satisfied at a later time period.

We use the simple model presented in section 2.2.5 so that we can focus our attention on using derivatives to estimate a piecewise linear function in the context of a dynamic program.

### 12.1.1 The model

Our model is formulated as follows:

$R_t$ = The assets on hand at time $t$ before we make a new ordering decision, and before we have satisfied any demands arising in time interval $t$,

$x_t$ = The amount of product purchased at time $t$ to be used during time interval $t+1$,

$\hat{D}_t$ = The random demands that arise between $t-1$ and $t$,

$\hat{R}_t$ = Random, exogenous changes to our asset levels (donations of blood, theft of product, leakage).

New assets are purchased at a fixed price $p^p$ and are sold at a fixed price $p^s > p^p$. The contribution earned during time interval $t$ is given by

$$C_t(R_t, x_t) = p^s \min\{R_t, \hat{D}_t\} - p^p x_t.$$

The transition function for $R_t$ is given by

$$\begin{aligned} R_t^x &= R_t - \min\{R_t, \hat{D}_t\} + x_t, \\ R_{t+1} &= R_t^x + \hat{R}_{t+1}, \end{aligned}$$

where we assume that any unsatisfied demands are lost to the system.

This problem can be solved using Bellman's equation. For this problem, $R_t$ is our state variable. Let $V_t(R_t^x)$ be the value of being in post-decision state $R_t^x$. Following our usual approximation strategy, the decision problem at time $t$, given that we are in state $R_t^n$, is given by

$$x_t^n = \arg\max_{x_t \in \mathcal{X}_t^n} \left( C_t(R_t^n, x_t) + \gamma \bar{V}_t^{n-1}(R_t^x) \right). \tag{12.1}$$

Recall that solving (12.1) is a form of decision function that we represent by $X_t^\pi(R_t)$, where the policy is to solve (12.1) using the value function approximation $\bar{V}_t^{n-1}(R_t^x)$. Assume, for example, that our approximation is of the form

$$\bar{V}_t^{n-1}(R_t^x) = \theta_0 + \theta_1 R_t^x + \theta_2 (R_t^x)^2,$$

where $R_t^x = R_t - \min\{R_t, \hat{D}_t\} + x_t$. In this case, (12.1) looks like

$$x_t^n = \arg\max_{x_t \in \mathcal{X}_t^n} \left( p^s \min\{R_t, \hat{D}_t\} - p^p x_t + \gamma(\theta_0 + \theta_1 R_t^x + \theta_2 (R_t^x)^2) \right). \tag{12.2}$$

We find the optimal value of $x_t^n$ by taking the derivative of the objective function and setting it equal to zero, giving us

$$\begin{aligned} 0 &= \frac{dC_t(R_t^n, x_t)}{dx_t} + \gamma \frac{d\bar{V}_t^{n-1}(R_t^x)}{dx_t} \\ &= -p^p + \gamma \frac{d\bar{V}_t^{n-1}(R_t^x)}{dR_t^x} \frac{dR_t^x}{dx_t} \\ &= -p^p + \gamma(\theta_1 + 2\theta_2 R_t^x) \\ &= -p^p + \gamma(\theta_1 + 2\theta_2(R_t - \min\{R_t, \hat{D}_t\} + x_t)), \end{aligned}$$

where we used $dR_t^x/dx_t = 1$. Solving for $x_t$ gives

$$x_t^n = \frac{1}{2\theta_2}\left(\frac{p^p}{\gamma} - \theta_1\right) - (R_t^n - \min\{R_t^n, \hat{D}_t^n\}).$$

Of course, we assume that $x_t^n \geq 0$.

### 12.1.2 An ADP algorithm

The simplest way to compute a gradient is using a pure forward-pass implementation. At time $t$, we have to solve

$$\tilde{V}_t^n(R_t^n) = \max_{x_t}\left(p^s \min\{R_t, \hat{D}_t\} - p^p x_t + \gamma \bar{V}_t^{n-1}(R_t^x)\right).$$

Here, $\tilde{V}_t^n(R_t^n)$ is just a placeholder. We are going to compute a derivative using a finite difference. We begin by recalling that the pre-decision resource state is given by

$$\begin{aligned} R_t^n &= R^{M,W}(R_{t-1}^{x,n}, W_t(\omega^n)) \\ &= R_{t-1}^{x,n} + \hat{R}_t^n. \end{aligned}$$

Now let

$$R_t^{n+} = R^{M,W}(R_{t-1}^{x,n} + 1, W_t(\omega^n))$$

be the pre-decision resource vector when $R_{t-1}^{x,n}$ is incremented by one. We next compute a derivative using the finite difference

$$\hat{v}_t^n = \tilde{V}_t^n(R_t^{n+}) - \tilde{V}_t^n(R_t^n).$$

Note that when we solve the perturbed value $\tilde{V}_t^n(R_t^{n+})$ we have to re-optimize $x_t$. For example, it is entirely possible that if we increase $R_{t-1}^{x,n}$ by one then $x_t^n$ will decrease by one. As is always the case with a single-pass algorithm, $\hat{v}_t^n$ depends on $\bar{V}_t^n(R_t^x)$ and, as a result will typically be biased.

Once we have a sample estimate of the gradient $\hat{v}_t^n$, we next have to update the value function. We can represent this updating process generically using

$$\bar{V}_{t-1}^n = U^V(\bar{V}_{t-1}^{n-1}, R_{t-1}^{x,n}, \hat{v}_t^n).$$

The actual updating process depends on whether we are using a piecewise linear approximation, the SHAPE algorithm, or a general regression equation. Remember that we are using $\hat{v}_t^n$ to update the post-decision value function at time $t-1$.

The complete algorithm is outlined in figure 12.1, which is an adaptation of our original single-pass algorithm. A simple but critical conceptual difference is that we are now explicitly assuming that we are using a continuous functional approximation.

**Step 0.** Initialize:

**Step 0a.** Initialize $\bar{V}_t^0(S_t^x)$ for all time periods $t$.

**Step 0b.** Initialize $S_0^1$.

**Step 0c.** Let $n = 1$.

**Step 1.** Choose $\omega^n$.

**Step 2.** Do for $t = 0, 1, 2, \ldots, T$:

**Step 2a.** Let the state variable be

$$S_t^n = (R_t^n, \hat{D}_t(\omega^n))$$

and let

$$S_t^{n+} = (R_t^n + 1, \hat{D}_t(\omega^n))$$

**Step 2b.** Solve:

$$\widetilde{V}_t^n(S_t^n) = \max_{x_t} \left(p^s \min\{R_t, \hat{D}_t(\omega^n)\} - p^p x_t + \gamma \bar{V}_t^{n-1}(R_t^x)\right)$$

where $R_t^x = R_t - \min\{R_t, \hat{D}_t\} + x_t$. Let $x_t^n$ be the value of $x_t$ that solves the maximization problem. Also find $\widetilde{V}_t^n(S_t^{n+})$.

**Step 2c.** Compute

$$\hat{v}_t^n = \widetilde{V}_t^n(S_t^{n+}) - \widetilde{V}_t^n(S_t^n).$$

**Step 2d.** If $t > 0$ update the value function:

$$\bar{V}_{t-1}^n = U^V(\bar{V}_{t-1}^{n-1}, S_{t-1}^{x,n}, \hat{v}_t^n).$$

**Step 2e.** Update the states:

$$\begin{aligned} S_t^{x,n} &= S^{M,x}(S_t^n, x_t), \\ S_{t+1}^n &= S^{M,W}(S_t^{x,n}, W_{t+1}(\omega^n)). \end{aligned}$$

**Step 3.** Let $n = n + 1$. If $n < N$, go to step 1.

**Figure 12.1** An approximate dynamic programming algorithm for the asset acquisition problem

### 12.1.3 How well does it work?

This approximation strategy is provably convergent. It also works in practice (!!). Unfortunately, while a proof of convergence can be reassuring, it is not hard to find provably convergent algorithms for problems that would never actually work (typically, because the rate of convergence is far too slow).

The asset acquisition problem is simple enough that we can find the optimal solution using the exact methods provided in chapter 3. For our approximation, we used a piecewise linear value function approximation. We used a pure exploitation strategy since the concave structure of the value function will eventually push us toward the correct solution.

Figure 12.2 shows the solution quality for a finite horizon problem (undiscounted). We are within five percent of optimal after about 50 iterations, and within one percent after 100 iterations. For many problems in practice (where we do not even know the optimal solution) this is extremely good.

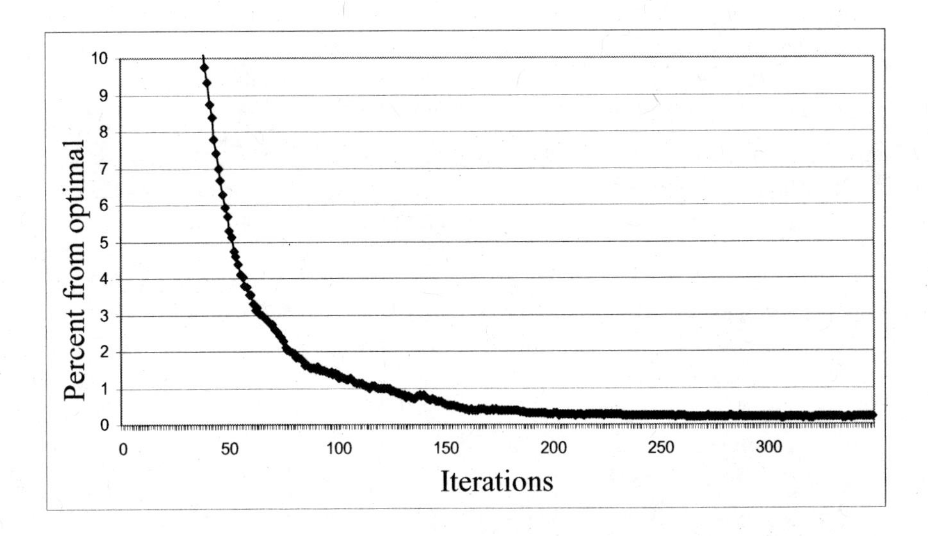


**Figure 12.2** The percent from optimal for the simple asset acquisition problem using a piecewise linear value function approximation.

### 12.1.4 Variations

This problem class is special because we are controlling the quantity of a single type of asset to serve random demands. Using approximate dynamic programming to fit piecewise linear, value function approximations produces an algorithm that is provably convergent (at least for the finite horizon case) without the need for exploration. We can use our current value function approximation to determine our decision at time $t$. Even if the value function is off, it will eventually converge to the correct value function (for the states that an optimal policy visits infinitely often).

This problem class seems quite simple, but it actually includes a number of variations that arise in industry.

- Cash balance for mutual funds - Imagine that we want to determine how much cash to have on hand in a mutual fund in order to handle requests for redemptions. If we have too much cash, we can invest it in equities do obtain a higher rate of return, but this return fluctuates and there are transaction costs for moving money between cash and equities. The state variable has to include not just how much cash we have on hand, but also information on the state of the stock market and interest rates. This information is part of the state variable, but evolves exogenously.

- Gas storage - Companies will store natural gas in coal mines, buying gas when the price is low and selling it when it is high. Gas can be purchased from multiple suppliers at prices that fluctuate over time. Gas may be sold at time $t$ to be delivered at a time $t' > t$ (when demand is higher).

- Managing vaccine inventories - Imagine that we have a limited supply of vaccines which have to be made available for the highest risk segments of the population. The strategy has to balance the degree of risk now, against potential future needs for the vaccine. Instead of satisfying a single demand, we face different types of demands with different levels of importance.

- Maintaining spares - There are many problems where we have to maintain spares (spare engines for business aircraft, spare transformers for electric power grids). We have to decide how many new spares to purchase, and whether to satisfy a demand against holding inventory for future (and potentially more important) demands.

Some of these problems can be solved using a minor variation of the algorithmic strategy that we described in this section. But this is not always the case. Consider a problem where we are managing the inventory of a single product (water, oil, money, spares) given by a scalar $R_t$, but where our decisions have to consider other variables (stock market, weather, technology parameters) that evolve exogenously. Let $\tilde{S}_t$ be the vector of variables that influence the behavior of our system (which means they belong to the state variable) but which evolve exogenously. Our state variable is given by

$$S_t = (R_t, \tilde{S}_t).$$

For example, $\tilde{S}_t$ might be temperature (if we have an energy problem), rainfall (if we are planning the storage of water in reservoirs), or perhaps the S & P 500 stock index and interest rates. $\tilde{S}_t$ might be a scalar, but it might consist of several variables.

We can still approximate the problem using a piecewise linear value function, but instead of one function, we have to estimate a family of functions which we would represent using $\bar{V}_t(R_t|\tilde{S}_t)$. It is just that instead of using $\bar{V}_t(R_t^x)$, we have to use $\bar{V}_t(R_t^x|\tilde{S}_t)$. The algorithm is still provably convergent, but computationally can be much more difficult. When $\tilde{S}_t$ is a vector, we encounter a curse-of-dimensionality problem if we estimate one function for each possible value of $\tilde{S}_t$ (a classic lookup-table representation). Instead of estimating one piecewise linear value function $\bar{V}_t(R_t^x)$, we might have to estimate hundreds or thousands (or more). If $\tilde{S}_t$ is a vector, we might need to look into other types of functional representations.

## 12.2 THE BLOOD MANAGEMENT PROBLEM

The problem of managing blood inventories serves as a particularly elegant illustration of a resource allocation problem. We are going to start by assuming that we are managing inventories at a single hospital, where each week we have to decide which of our blood inventories should be used for the demands that need to be served in the upcoming week.

We have to start with a bit of background about blood. For the purposes of managing blood inventories, we care primarily about blood type and age. Although there is a vast range of differences in the blood of two individuals, for most purposes doctors focus on the eight major blood types: $A+$ (" A positive"), $A-$ ("A negative"), $B+$, $B-$, $AB+$, $AB-$, $O+$, and $O-$. While the ability to substitute different blood types can depend on the nature of the operation, for most purposes blood can be substituted according to table 12.1.

A second important characteristic of blood is its age. The storage of blood is limited to six weeks, after which it has to be discarded. Hospitals need to anticipate if they think they can use blood before it hits this limit, as it can be transferred to blood centers which monitor inventories at different hospitals within a region. It helps if a hospital can identify blood it will not need as soon as possible so that the blood can be transferred to locations that are running short.

One mechanism for extending the shelf-life of blood is to freeze it. Frozen blood can be stored up to 10 years, but it takes at least an hour to thaw, limiting its use in emergency situations or operations where the amount of blood needed is highly uncertain. In addition, once frozen blood is thawed it must be used within 24 hours.

| Donor | Recipient | | | | | | | |
|---|---|---|---|---|---|---|---|---|
| | $AB+$ | $AB-$ | $A+$ | $A-$ | $B+$ | $B-$ | $O+$ | $O-$ |
| $AB+$ | X | | | | | | | |
| $AB-$ | X | X | | | | | | |
| $A+$ | X | | X | | | | | |
| $A-$ | X | X | X | X | | | | |
| $B+$ | X | | | | X | | | |
| $B-$ | X | X | | | X | X | | |
| $O+$ | X | | X | | X | | X | |
| $O-$ | X | X | X | X | X | X | X | X |

**Table 12.1** Allowable blood substitutions for most operations, 'X' means a substitution is allowed (from Cant (2006)).

### 12.2.1 A basic model

We can model the blood problem as a heterogeneous resource allocation problem. We are going to start with a fairly basic model which can be easily extended with almost no notational changes. We begin by describing the attributes of a unit of stored blood using

$$a = \begin{pmatrix} a_1 \\ a_2 \end{pmatrix} = \begin{pmatrix} \text{Blood type } (A+, A-, \ldots) \\ \text{Age (in weeks)} \end{pmatrix},$$

$$\mathcal{A} = \text{Set of all attribute types.}$$

We will limit the age to the range $0 \leq a_2 \leq 6$. Blood with $a_2 = 6$ (which means blood that is already six weeks old) is no longer usable. We assume that decision epochs are made in one-week increments.

Blood inventories, and blood donations, are represented using

$$\begin{aligned} R_{ta} &= \text{Units of blood available to be assigned or held at time } t, \\ R_t &= (R_{ta})_{a \in \mathcal{A}}, \\ \hat{R}_{ta} &= \text{Number of new units of blood donated between } t-1 \text{ and } t, \\ \hat{R}_t &= (\hat{R}_{ta})_{a \in \mathcal{A}}. \end{aligned}$$

The attributes of demand for blood are given by

$$b = \begin{pmatrix} b_1 \\ b_2 \\ b_3 \end{pmatrix} = \begin{pmatrix} \text{Blood type of patient} \\ \text{Surgery type: urgent or elective} \\ \text{Is substitution allowed?} \end{pmatrix},$$

$$\mathcal{B} = \text{Set of all demand types } b.$$

The attribute $b_3$ captures the fact that there are some operations where a doctor will not allow any substitution. One example is childbirth, since infants may not be able to handle a different blood type, even if it is an allowable substitute. For our basic model, we do not allow unserved demand in one week to be held to a later week. As a result, we need only model new demands, which we accomplish with

$$\begin{aligned} \hat{D}_{tb} &= \text{Units of demand with attribute } b \text{ that arose between } t-1 \text{ and } t, \\ \hat{D}_t &= (\hat{D}_{tb})_{b \in \mathcal{B}}. \end{aligned}$$

We act on blood resources with decisions given by:

$$\begin{aligned}
\mathcal{D}^D &= \text{The set of decisions to satisfy demands, where each element } d \in \mathcal{D}^D \text{ corresponds to a demand with attribute } b_d \in \mathcal{B}, \\
d^\phi &= \text{Decision to hold blood in inventory ("do nothing")}, \\
\mathcal{D} &= \mathcal{D}^D \cup d^\phi, \\
x_{tad} &= \text{Number of units of blood with attribute } a \text{ that we act on with a decision of type } d, \\
x_t &= (x_{tad})_{a \in \mathcal{A}, d \in \mathcal{D}}.
\end{aligned}$$

The feasible region $\mathcal{X}_t$ is defined by the following constraints:

$$\sum_{d \in \mathcal{D}} x_{tad} = R_{ta}, \tag{12.3}$$

$$\sum_{a \in \mathcal{A}} x_{tad} \leq \hat{D}_{tb_d}, \quad d \in \mathcal{D}^D, \tag{12.4}$$

$$x_{tad} \geq 0. \tag{12.5}$$

Blood that is held simply ages one week, but we limit the age to six weeks. Blood that is assigned to satisfy a demand can be modeled as being moved to a blood-type sink, denoted, perhaps, using $a_{t,1} = \phi$ (the null blood type). The attribute transition function $a^M(a_t, d_t)$ is given by

$$a_{t+1} = \begin{pmatrix} a_{t+1,1} \\ a_{t+1,2} \end{pmatrix} = \begin{cases} \begin{pmatrix} a_{t,1} \\ \min\{6, a_{t,2}+1\} \end{pmatrix}, & d_t = d^\phi, \\ \begin{pmatrix} \phi \\ - \end{pmatrix}, & d_t \in \mathcal{D}^D. \end{cases}$$

To represent the transition function, it is useful to define

$$\begin{aligned}
\delta_{a'}(a,d) &= \begin{cases} 1 & a_t^x = a' = a^M(a_t, d_t), \\ 0 & \text{otherwise}, \end{cases} \\
\Delta &= \text{Matrix with } \delta_{a'}(a,d) \text{ in row } a' \text{ and column } (a,d).
\end{aligned}$$

We note that the attribute transition function is deterministic. A random element would arise, for example, if inspections of the blood resulted in blood that was less than six weeks old being judged to have expired. The resource transition function can now be written

$$\begin{aligned}
R^x_{ta'} &= \sum_{a \in \mathcal{A}} \sum_{d \in \mathcal{D}} \delta_{a'}(a,d) x_{tad}, \\
R_{t+1,a'} &= R^x_{ta'} + \hat{R}_{t+1,a'}.
\end{aligned}$$

In matrix form, these would be written

$$R^x_t = \Delta x_t, \tag{12.6}$$

$$R_{t+1} = R^x_t + \hat{R}_{t+1}. \tag{12.7}$$

Figure 12.3 illustrates the transitions that are occurring in week $t$. We either have to decide which type of blood to use to satisfy a demand (figure 12.3a), or to hold the blood

| Condition | Description | Value |
|---|---|---|
| if $d = d^\phi$ | Holding | 0 |
| if $a_1 = b_1$ when $d \in \mathcal{D}^D$ | No substitution | 0 |
| if $a_1 \neq b_1$ when $d \in \mathcal{D}^D$ | Substitution | -10 |
| if $a_1 =$ O- when $d \in \mathcal{D}^D$ | O- substitution | 5 |
| if $b_2 =$ Urgent | Filling urgent demand | 40 |
| if $b_2 =$ Elective | Filling elective demand | 20 |

**Table 12.2** Contributions for different types of blood and decisions

until the following week. If we use blood to satisfy a demand, it is assumed lost from the system. If we hold the blood until the following week, it is transformed into blood that is one week older. Blood that is six weeks old may not be used to satisfy any demands, so we can view the bucket of blood that is six weeks old as a sink for unusable blood (the value of this blood would be zero). Note that blood donations are assumed to arrive with an age of 0. The pre- and post-decision state variables are given by

$$\begin{aligned} S_t &= (R_t, \hat{D}_t), \\ S_t^x &= (R_t^x). \end{aligned}$$

There is no real "cost" to assigning blood of one type to demand of another type (we are not considering steps such as spending money to encourage additional donations, or transporting inventories from one hospital to another). Instead, we use the contribution function to capture the preferences of the doctor. We would like to capture the natural preference that it is generally better not to substitute, and that satisfying an urgent demand is more important than an elective demand. For example, we might use the contributions described in table 12.2. Thus, if we use $O-$ blood to satisfy the needs for an elective patient with $A+$ blood, we would pick up a -\$10 contribution (penalty since it is negative) for substituting blood, a +\$5 for using $O-$ blood (something the hospitals like to encourage), and a +\$20 contribution for serving an elective demand, for a total contribution of +\$15.

The total contribution (at time $t$) is finally given by

$$C_t(S_t, x_t) = \sum_{a \in \mathcal{A}} \sum_{d \in \mathcal{D}} c_{tad} x_{tad}.$$

As before, let $X_t^\pi(S_t)$ be a policy (some sort of decision rule) that determines $x_t \in \mathcal{X}_t$ given $S_t$. We wish to find the best policy by solving

$$\max_{\pi \in \Pi} \mathbb{E} \sum_{t=0}^{T} \gamma^t C_t(S_t, x_t). \tag{12.8}$$

The most obvious way to solve this problem is with a simple myopic policy, where we maximize the contribution at each point in time without regard to the effect of our decisions on the future. We can obtain a family of myopic policies by adjusting the one-period contributions. For example, our bonus of \$5 for using $O-$ blood (in table 12.2), is actually a type of myopic policy. We encourage using $O-$ blood since it is generally more available than other blood types. By changing this bonus, we obtain different types of myopic

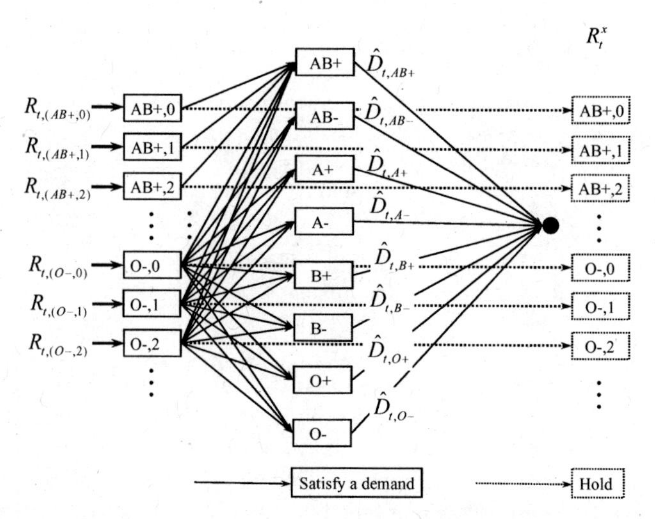


12.3a - Assigning blood supplies to demands in week $t$. Solid lines represent assigning blood to a demand, dotted lines represent holding blood.

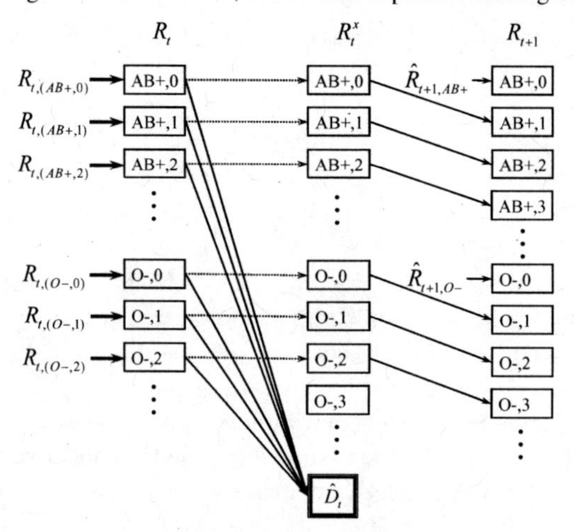


12.3b - Holding blood supplies until week $t + 1$.

**Figure 12.3** The assignment of different blood types (and ages) to known demands in week $t$ (12.3a), and holding blood until the following week (12.3b).

policies that we can represent by the set $\Pi^M$, where for $\pi \in \Pi^M$ our decision function would be given by

$$X_t^\pi(S_t) = \arg\max_{x_t \in \mathcal{X}_t} \sum_{a \in \mathcal{A}} \sum_{d \in \mathcal{D}} c_{tad} x_{tad}. \tag{12.9}$$

The optimization problem in (12.9) is a simple linear program (known as a "transportation problem"). Solving the optimization problem given by equation (12.8) for the set $\pi \in \Pi^M$ means searching over different values of the bonus for using $O-$ blood.

### 12.2.2 An ADP algorithm

As a traditional dynamic program, the optimization problem posed in equation (12.8) is quite daunting. The state variable $S_t$ has $|\mathcal{A}| + |\mathcal{B}| = 8 \times 6 + 8 \times 2 \times 2 = 80$ dimensions. The random variables $\hat{R}$ and $\hat{D}$ also have a combined 80 dimensions. The decision vector $x_t$ has $27 + 8 = 35$ dimensions. Using the classical techniques of chapter 3, this problem looks hopeless. Using the methods that we have presented up to now, obtaining effective solutions to this problem is fairly straightforward.

We follow our standard solution strategy (such as the single-pass algorithm described in figure 8.1) and determine the allocation vector $x_t$ using

$$x_t^n = \arg\max_{x_t \in \mathcal{X}_t^n} \left( C_t(S_t^n, x_t) + \gamma \bar{V}_t^{n-1}(R_t^x) \right), \tag{12.10}$$

where $R_t^x = R^M(R_t, x_t)$ is given by equation (12.6). The first (and most important) challenge we face is identifying an appropriate approximation strategy for $\bar{V}_t^{n-1}(R_t^x)$. A simple and effective approximation is to use separable, piecewise linear approximations, which is to say

$$\bar{V}_t(R_t^x) = \sum_{a \in \mathcal{A}} \bar{V}_{ta}(R_{ta}^x),$$

where $\bar{V}_{ta}(R_{ta}^x)$ is a scalar, piecewise, linear function. It is easy to show that the value function is concave (as well as piecewise linear), so each $\bar{V}_{ta}(R_{ta}^x)$ should also be concave. Without loss of generality, we can assume that $\bar{V}_{ta}(R_{ta}^x) = 0$ for $R_{ta}^x = 0$, which means the function is completely characterized by its set of slopes. We can write the function using

$$\bar{V}_{ta}^{n-1}(R_{ta}^x) = \left( \sum_{r=1}^{\lfloor R_{ta}^x \rfloor} \bar{v}_{ta}^{n-1}(r-1) + (R_{ta}^x - \lfloor R_{ta}^x \rfloor) \bar{v}_{ta}^{n-1}(\lfloor R_{ta}^x \rfloor) \right), \tag{12.11}$$

where $\lfloor R \rfloor$ is the largest integer less than or equal to $R$. As we can see, this function is determined by the set of slopes $(\bar{v}_{ta}^{n-1}(r))$ for $r = 0, 1, \ldots, R^{max}$, where $R^{max}$ is an upper bound on the number of resources of a particular type.

Assuming we can estimate this function, the optimization problem that we have to solve (equation (12.10)) is the fairly modest linear program shown in figure 12.4. As with figure 12.3, we have to consider both the assignment of different types of blood to different types of demand, and the decision to hold blood. To simplify the figure, we have collapsed the network of different demand types into a single aggregate box with demand $\hat{D}_t$. This network would actually look just like the network in figure 12.3a. The decision to hold blood has to consider the value of a type of blood (including its age) in the future, which we are approximating using separable, piecewise linear value functions. Here, we use a standard modeling trick that converts the separable, piecewise linear value function approximations into a series of parallel links from each node representing an element of $R_t^x$ into a supersink. Piecewise linear functions are not only easy to solve (we just need access to a linear programming solver), they are easy to estimate. In addition, for many problem classes (but not all) they have been found to produce very fast convergence with high quality solutions.

With this decision function, we would follow the basic algorithmic strategy described in figure 8.1. Aside from the customization of the value function approximation, the biggest difference is in how the value functions are updated, a problem that is handled in the next

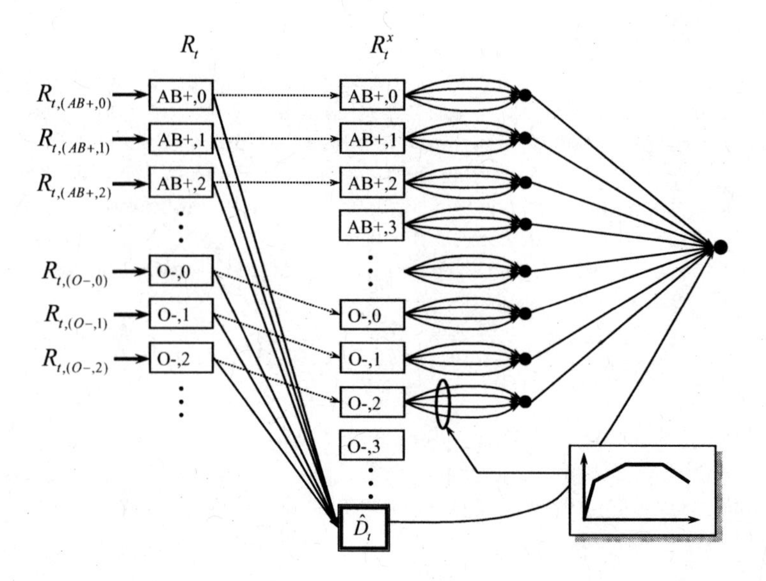


**Figure 12.4** Network model for time $t$ with separable, piecewise linear value function approximations.

section. Of course we could use other nonlinear approximation strategies, but this would mean that we would have to solve a nonlinear programming problem instead of a linear program. While this can, of course, be done, it complicates the strategy for updating the value function approximation.

For most operational applications, this problem would be solved over a finite horizon (say, 10 weeks) giving us a recommendation of what to do right now. Alternatively, we can solve an infinite horizon version of the model by simply dropping the $t$ index on the value function approximation.

### 12.2.3 Updating the value function approximation

Since our value function depends on slopes, we want to use derivatives to update the value function approximation. Let

$$\widetilde{V}_t(S_t) = \max_{x_t \in \mathcal{X}_t^n} \left( C_t(S_t^n, x_t) + \gamma \bar{V}_t^{n-1}(R_t^x) \right)$$

be the value of our decision function at time $t$. But what we want is the derivative with respect to the previous post-decision state $R_{t-1}^x$. Recall that $S_t = (R_t, D_t)$ depends on $R_t = R_{t-1}^x + \hat{R}_t$. We want to find the gradient $\nabla \widetilde{V}_t(S_t)$ with respect to $R_{t-1}^x$. We apply the chain rule to find

$$\frac{\partial \widetilde{V}_t(S_t)}{\partial R_{t-1,a}^x} = \sum_{a' \in \mathcal{A}} \frac{\partial \widetilde{V}_t(S_t)}{\partial R_{ta'}} \frac{\partial R_{ta'}}{\partial R_{t-1,a}^x}.$$

If we decide to hold an additional unit of blood of type $a_{t-1}$, then this produces an additional unit of blood of type $a_t = a^M(a_{t-1}, d)$ where $a^M(a, d)$ is our attribute transition function and $d$ is our decision to hold the blood until the next week. For this application, the only difference between $a_{t-1}$ and $a_t$ is that the blood has aged by one week. We also have to model the fact that there may be exogenous changes to our blood supply (donations, deliveries from other hospitals, as well as blood that has gone bad). This means that

$$R_{ta_t} = R^x_{t-1,a_{t-1}} + \hat{R}_{ta_t}.$$

Note that we start with $R^x_{t-1,a_{t-1}}$ units of blood with attribute $a$, and then add the exogenous changes $\hat{R}_{ta_t}$ to blood of type $a_t$. This allows us to compute the derivative of $R_t$ with respect to $R^x_{t-1}$ using

$$\frac{\partial R_{ta_t}}{\partial R^x_{t-1,a_{t-1}}} = \begin{cases} 1 & \text{if } a_t = a^M(a_{t-1}, d) \\ 0 & \text{otherwise.} \end{cases}$$

This allows us to write the derivative as

$$\frac{\partial \widetilde{V}_t(S_t)}{\partial R^x_{t-1,a_{t-1}}} = \frac{\partial \widetilde{V}_t(S_t)}{\partial R_{ta_t}}.$$

$\frac{\partial \widetilde{V}_t(S_t)}{\partial R_{ta}}$ can be found quite easily. A natural byproduct from solving the linear program in equation (12.10) is that we obtain a dual variable $\hat{v}^n_{ta}$ for each flow conservation constraint (12.3). This is a significant advantage over strategies where we visit state $R_t$ (or $S_t$) and then update just the value of being in that single state. The dual variable is an estimate of the slope of the decision problem with respect to $R_{ta}$ at the point $R^n_{ta}$, which gives us

$$\left.\frac{\partial \widetilde{V}_t(S_t)}{\partial R^x_{t-1,a_{t-1}}}\right|_{R^{x,n}_{t-1,a_{t-1}}} = \hat{v}^n_{ta_t}.$$

Dual variables are incredibly convenient for estimating value function approximations in the context of resource allocation problems, both because they approximate the derivatives with respect to $R_{ta}$ (which is all we are interested in), but also because we obtain an entire vector of slopes, rather than a single estimate of the value of being in a state. However, it is important to understand exactly which slope the dual is (or is not) giving us.

Figure 12.5 illustrates a nondifferentiable function (any problem that can be modeled as a sequence of linear programs has this piecewise linear shape). If we are estimating the slope of a piecewise linear function $V_{ta}(R_{ta})$ at a point $R_{ta} = R^n_{ta}$ that corresponds to one of these kinks (which is quite common), then we have a left slope, $\hat{v}^-_{ta} = V_{ta}(R_{ta}) - V_{ta}(R_{ta} - 1)$, and a right slope, $\hat{v}^+_{ta} = V_{ta}(R_{ta} + 1) - V_{ta}(R_{ta})$. If $\hat{v}^n_{ta}$ is our dual variable, we have no way of specifying whether we want the left or right slope. In fact, the dual variable can be anywhere in the range $\hat{v}^+_{ta} \leq \hat{v}^n_{ta} \leq \hat{v}^-_{ta}$.

If we are comfortable with the approximation implicit in the dual variables, then we get the vector $(\hat{v}^n_{ta})_{a\in\mathcal{A}}$ for free from our simplex algorithm. If we specifically want the left or right derivative, then we can perform a numerical derivative by perturbing each element $R^n_{ta}$ by plus or minus 1 and reoptimizing. This sounds clumsy when we get dual variables for free, but it is surprisingly fast.

Once we have computed $\hat{v}^n_{ta}$ for each $a \in \mathcal{A}$, we now have to update our value function approximation. Remember that we have to update the value function at time $t-1$ at the previous post-decision state. For this problem, attributes evolve deterministically (for example, AB+ blood that is 3 weeks old becomes AB+ blood that is 4 weeks old). Let

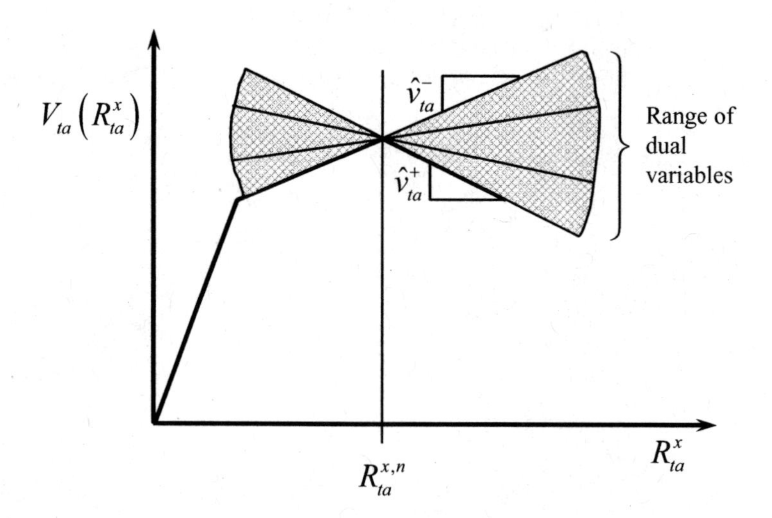


**Figure 12.5** Illustration of range of dual variables for a nondifferentiable function.

$a_t = a^M(a_{t-1}, d)$ describe this evolution (here, $d$ is the decision to hold the blood). Let $\hat{v}^n_{ta_t}$ be the attribute corresponding to blood with attribute $a_t$ at time $t$, and assume that $R^{x,n}_{t-1,a_{t-1}}$ was the amount of blood of type $a_{t-1}$ at time $t-1$. Let $\bar{v}^{n-1}_{t-1,a_{t-1}}(r)$ be the slope corresponding to the interval $r \leq R^{x,n}_{t-1,a_{t-1}} \leq r+1$ (as we did in (12.11)). A simple update is to use

$$\bar{v}^n_{t-1,a_{t-1}}(r) = \begin{cases} (1-\alpha_{n-1})\bar{v}^{n-1}_{t-1,a_{t-1}}(r) + \alpha_{n-1}\hat{v}^n_{ta_t} & \text{if } r = R^{x,n}_{t-1,a_{t-1}}, \\ \bar{v}^{n-1}_{t-1,a_{t-1}}(r) & \text{otherwise.} \end{cases} \tag{12.12}$$

Here, we are only updating the element $\bar{v}^{n-1}_{t-1,a_{t-1}}(r)$ corresponding to $r = R^{x,n}_{t-1,a_{t-1}}$. We have found that it is much better to update a range of slopes, say in the interval $(r-\delta, r+\delta)$ where $\delta$ has been chosen to reflect the range of possible values of $r$. For example, it is best if $\delta$ is chosen initially so that we are updating the entire range, after which we periodically cut it in half until it shrinks to 1 (or an interval small enough that we are getting the precision we want in the decisions).

There is one problem with equation (12.12) that we have already seen before. We know that the value functions are concave, which means that we should have $\bar{v}^n_{ta}(r) \geq \bar{v}^n_{ta}(r+1)$ for all $r$. This might be true for $\bar{v}^{n-1}_{ta}(r)$, but it is entirely possible that after our update, it is no longer true for $\bar{v}^n_{ta}(r)$. Nothing new here. We just have to apply the fix-up techniques that were presented in section 11.3.

The use of separable, piecewise linear approximations has proven effective in a number of applications, but there are open theoretical questions (how good is the approximation?) as well as unresolved computational issues (what is the best way to discretize functions?). What about the use of low-dimensional basis functions? If we use a continuously differentiable approximation (which requires the use of a nonlinear programming algorithm), we can use our regression techniques to fit the parameters of a statistical model that is continuous in the resource variable.

The best value function approximation does not just depend on the structure of the problem. It depends on the nature of the data itself.

### 12.2.4 Extensions

We can anticipate several ways in which we can make the problem richer.

- Instead of modeling the problem in single week increments, model daily decisions.
- Include the presence of blood that has been frozen, and the decision to freeze blood.
- A hospital might require weekly deliveries of blood from a community blood bank to make up for systematic shortages. Imagine that a fixed quantity (e.g., 100 units) of blood arrive each week, but where the amount of blood of each type and age (the blood may have already been held in inventory for several weeks) might be random.
- We presented a model that focused only on blood inventories at a single hospital. We can handle multiple hospitals and distribution centers by simply adding a location attribute, and providing for a decision to move blood (at a cost) from one location to another.

This model can also be applied to any multiproduct inventory problem where there are different types of product and different types of demands, as long as we have the ability to choose which type of product is assigned to each type of demand. We also assume that products are not reusable; once the product is assigned to a demand, it is lost from the system.

## 12.3 A PORTFOLIO OPTIMIZATION PROBLEM

A somewhat different type of resource allocation problem arises in the design of financial portfolios. We can start with some of the same notation we used to manage blood. Let $a$ be an asset class (more generally, the attributes of an asset, such as the type of investment and how long it has been in the investment), where $R_{ta}$ is how much we have invested in asset class $a$ at time $t$. Let $\mathcal{D}$ be our set of decisions, where each element of $\mathcal{D}$ represents a decision to invest in a particular asset class (there is a one-to-one relationship between $\mathcal{D}$ and the asset classes $\mathcal{A}$).

The resource transition functions are somewhat different. We let the indicator function $\delta_{a'}(a,d)$ be defined as before. However, now we have to account for transaction costs. Let

$c_{ad} =$ The cost of acting on asset class $a$ with a decision of type $d$, expressed as a fraction of the assets being transferred.

The decision $d$ might mean "hold the asset," in which case $c_{ad} = 0$. However, if we are moving from one asset class to another, we have to pay $c_{ad}x_{tad}$ to cover the cost of the transaction, which is then deducted from the proceeds. This means that the post-decision resource vector is given by

$$R^x_{ta} = \sum_{a\in\mathcal{A}}\sum_{d\in\mathcal{D}} \delta_{a'}(a,d)(x_{tad} - c_{ad}x_{tad}). \tag{12.13}$$

We now have to account for the market returns from changes in the value of the asset. This is handled by first defining

$\hat{\rho}_{ta} =$ The relative return on assets of type $a$ between $t-1$ and $t$.

These returns are the only source of noise that we are considering at this point. Unlike the demands for the blood problem (where we tend to assume that demands are independent), returns tend to be correlated both across assets, and possibly across time. Since we work only with sample paths, we have no trouble handling complex interactions. With these data, our resource transition from $R_t^x$ to $R_{t+1}$ is given by

$$R_{t+1,a} = \hat{\rho}_{t+1,a} R_{ta}^x. \tag{12.14}$$

Normally, there is one "riskless" asset class to which we can turn to find nominal, but safe, returns.

Finally, we have to measure how well we are doing. We could simply measure our total wealth, given by

$$\bar{R}_t = \sum_{a \in \mathcal{A}} R_{ta}.$$

We might then set up the problem to maximize our total wealth (actually, the total expected wealth) at the end of our planning horizon. This objective ignores the fact that we tend to be risk averse, and we want to avoid significant losses in each time period. So, we propose instead to use a utility function that measures the relative gain in our portfolio, such as

$$U(\bar{R}_t) = -\left(\frac{\bar{R}_t}{\bar{R}_{t-1}}\right)^{-\beta},$$

where $\beta$ is a parameter that controls the degree of risk aversion.

Regardless of how the utility function is constructed, we approach its solution just as we did with the blood management problem. We again write our objective function as

$$\widetilde{V}_t(R_t) = \max_{x_t \in \mathcal{X}_t^n} \left(U(\bar{R}_t) + \gamma \bar{V}_t^{n-1}(R_t^x)\right). \tag{12.15}$$

where $\mathcal{X}_t$ is the set of $x_t$ that satisfy

$$\sum_{d \in \mathcal{D}} x_{tad} = R_{ta}, \tag{12.16}$$

$$x_{tad} \geq 0. \tag{12.17}$$

$R_t^x$ is given by equation (12.13) (which is a function of $x_t$), while $\bar{R}_t$ is a constant at time $t$ (it is determined by $R_{t-1}^x$ and $\hat{\rho}_t$). This allows us to write (12.15) as

$$\widetilde{V}_t(R_t) = U(\bar{R}_t) + F_t(R_t), \tag{12.18}$$

where

$$F_t(R_t) = \max_{x_t \in \mathcal{X}_t^n} \left(\gamma \bar{V}_t^{n-1}(R_t^x)\right).$$

Let $\hat{v}_{ta}^n$ be the dual variable of the flow conservation constraint (12.16). We can use $\hat{v}_{ta}^n$ as an estimate of the derivative of $F_t(R_t)$, giving us

$$\left.\frac{\partial F_t(R_t)}{\partial R_{ta}}\right|_{R_t^n = R_t} = \hat{v}_{ta}^n. \tag{12.19}$$

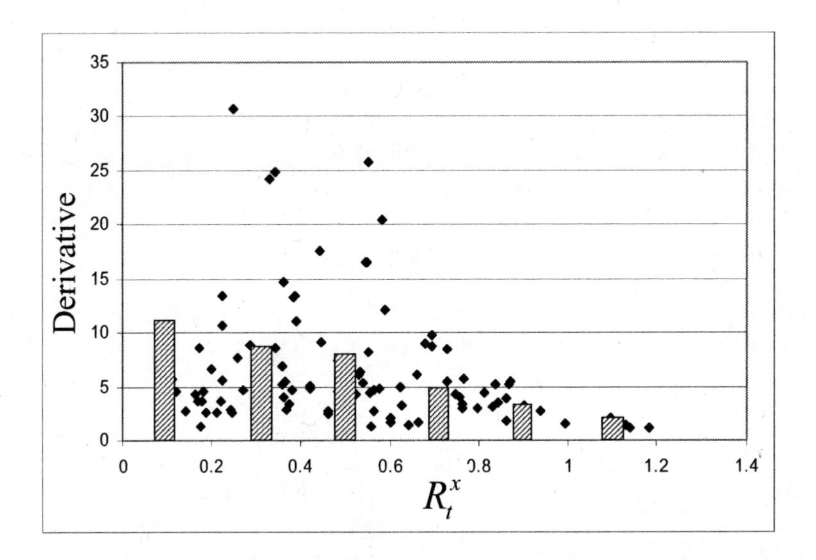


**Figure 12.6** Sample realizations for derivatives of the objective function as a function of $R^x_{ta}$ for a particular asset class $a$. Vertical bars show the average within a range.

Differentiating $U(\bar{R}_t)$ gives

$$\frac{\partial U(\bar{R}_t)}{\partial R_{ta}} = \frac{\beta}{\bar{R}_{t-1}} \left( \frac{\bar{R}_t}{\bar{R}_{t-1}} \right)^{-\beta-1}, \tag{12.20}$$

where we used the fact that $\partial \bar{R}_t / \partial R_{ta} = 1$. We next apply the chain rule to obtain

$$\frac{\partial \widetilde{V}_t(R_t)}{\partial R^x_{t-1,a}} = \sum_{a' \in \mathcal{A}} \frac{\partial \widetilde{V}_t(R_t)}{\partial R_{ta'}} \frac{\partial R_{ta'}}{\partial R^x_{t-1,a}}. \tag{12.21}$$

Finally, we observe that

$$\frac{\partial R_{ta'}}{\partial R^x_{t-1,a}} = \begin{cases} \hat{\rho}_{ta} & \text{If } a' = a, \\ 0 & \text{Otherwise.} \end{cases} \tag{12.22}$$

Combining (12.18) - (12.22) gives us

$$\frac{\partial \widetilde{V}_t(R_t)}{\partial R^x_{t-1,a}} = \left( \left. \frac{\partial U(\bar{R}_t)}{\partial R_{ta}} \right|_{R_t = R^n_t} + \hat{v}^n_{ta} \right) \hat{\rho}_{ta}. \tag{12.23}$$

Figure 12.6 shows sample realizations of the derivatives computed in (12.23) for a problem using the utility function $U(\bar{R}_t) = -0.1(\bar{R}_t/\bar{R}_{t-1})^{-4}$. We would use these sample realizations to estimate a nonlinear function of $R^x_{t-1,a}$ for each $a$. We expect the derivatives to decline, on average, with $R^x_{t-1,a}$, which is a pattern we see in the figure. However, it is apparent that there is quite a bit of noise in this relationship.

Now that we have our derivatives in hand, we can use it to fit a value function approximation. Again, we face a number of strategies. One is to use the same piecewise linear approximations that we implemented for our blood distribution problem, which has the advantage of being independent of the underlying distributions. However, a low-order polynomial would also be expected to work quite well, and we might even be able to estimate a nonseparable one (for example, we might have terms such as $R^x_{t-1,a_1} R^x_{t-1,a_2}$).

Identifying and testing different approximation strategies is one of the most significant challenges of approximate dynamic programming. In the case of the portfolio problem, there is a considerable amount of research into different objective functions and the properties of these functions. For example, a popular class of models minimizes a quadratic function of the variance, which suggests a particular quadratic form for the value function. Even if these functions are not perfect, they would represent a good starting point.

As with the blood management problem, we can anticipate several extensions we could make to our basic model:

- Short-term bonds - Our assets could have a fixed maturity which means we would have to capture how long we have held the bond.

- Sales loads that depend on how long we have held an asset. An asset might be a mutual fund, and some mutual funds impose different loads (sales charges) depending on how long you have held the asset before selling.

- Transaction times - Our model already handles transaction costs. Some equities (e.g., real estate) also incur significant transaction times. We can handle this by including an attribute indicating that the asset is in the process of being sold, as well as how much time has elapsed since the transaction was initiated (or the time until the transaction is expected to be completed).

- Our assets might represent stocks in different industries. We could design value functions that capture not only the amount we have invested in each stock, but also in each industry (we just aggregate the stocks into industry groupings).

The first three of these extensions can be handled by introducing an attribute that measures age in some way. In the case of fixed maturities, or the time until the asset can be sold without incurring a sales load, we can introduce an attribute that specifies the time at which the status of the asset changes.

## 12.4 A GENERAL RESOURCE ALLOCATION PROBLEM

The preceding sections have described different instances of resource allocation problems. While it is important to keep the context of an application in mind (every problem has unique features), it is useful to begin working with a general model for dynamic resource allocation so that we do not have to keep repeating the same notation. This problem class allows us to describe a basic algorithmic strategy that can be used as a starting point for more specialized problems.

### 12.4.1 A basic model

The simplest model of a resource allocation problem involves a "resource" that we are managing (people, equipment, blood, money) to serve "demands" (tasks, customers, jobs).

We describe the resources and demands using

$$R_{ta} \;=\; \text{The number of resources with attribute } a \in \mathcal{A} \text{ in the system at time } t,$$
$$R_t \;=\; (R_{ta})_{a\in\mathcal{A}},$$
$$D_{tb} \;=\; \text{The number of demands of type } b \in \mathcal{B} \text{ in the system at time } t,$$
$$D_t \;=\; (D_{tb})_{b\in\mathcal{B}}.$$

Both $a$ and $b$ are vectors of attributes of resources and demands. In addition to the resources and demands, we sometimes have to model a set of parameters (the price of a stock, the probability of an equipment failure, the price of oil) that govern how the system evolves over time. We model these parameters using

$\rho_t$ = A generic vector of parameters that affects the behavior of costs and the transition function.

The state of our system is given by

$$S_t \;=\; (R_t, D_t, \rho_t).$$

New information is represented as exogenous changes to the resource and demand vectors, as well as to other parameters that govern the problem. These are modeled using

$\hat{R}_{ta}$ = Exogenous changes to $R_{ta}$ from information that arrives during time interval $t$ (between $t-1$ and $t$),

$\hat{D}_{tb}$ = Exogenous changes to $D_{ta}$ from information that arrives during time interval $t$ (between $t-1$ and $t$),

$\hat{\rho}_t$ = Exogenous changes to a vector of parameters (costs, parameters governing the transition).

Our information process, then, is given by

$$W_t = (\hat{R}_t, \hat{D}_t, \hat{\rho}_t).$$

In the blood management problem, $\hat{R}_t$ included blood donations. In a model of complex equipment such as aircraft or locomotives, $\hat{R}_t$ would also capture equipment failures or delays. In a product inventory setting, $\hat{R}_t$ could represent theft of product. $\hat{D}_t$ usually represents new customer demands, but can also represent changes to an existing demand or cancellations. $\hat{\rho}_t$ could represent random costs, or the random returns in our stock portfolio.

Decisions are modeled using

$\mathcal{D}^D$ = Decision to satisfy a demand with attribute $b$ (each decision $d \in \mathcal{D}^D$ corresponds to a demand attribute $b_d \in \mathcal{B}$).

$\mathcal{D}^M$ = Decision to modify a resource (each decision $d \in \mathcal{D}^M$ has the effect of modifying the attributes of the resource). $\mathcal{D}^M$ includes the decision to "do nothing,"

$\mathcal{D}$ = $\mathcal{D}^D \cup \mathcal{D}^M$,

$x_{tad}$ = The number of resources that initially have attribute $a$ that we act on with decision $d$,

$x_t$ = $(x_{tad})_{a\in\mathcal{A},d\in\mathcal{D}}$.

The decisions have to satisfy constraints such as

$$\sum_{d \in \mathcal{D}} x_{tad} = R_{ta}, \tag{12.24}$$

$$\sum_{a \in \mathcal{A}} x_{tad} \leq D_{tb_d}, \quad d \in \mathcal{D}^D, \tag{12.25}$$

$$x_{tad} \geq 0. \tag{12.26}$$

We let $\mathcal{X}_t$ be the set of $x_t$ that satisfy (12.24)-(12.26). As before, we assume that decisions are determined by a class of decision functions

$X_t^\pi(S_t)$ = A function that returns a decision vector $x_t \in \mathcal{X}_t$, where $\pi \in \Pi$ is an element of the set of functions (policies) $\Pi$.

The transition function is given generically by

$$S_{t+1} = S^M(S_t, x_t, W_{t+1}).$$

We now have to deal with each dimension of our state variable. The most difficult, not surprisingly, is the resource vector $R_t$. This is handled primarily through the attribute transition function

$$a_t^x = a^{M,x}(a_t, d_t),$$

where $a_t^x$ is the post-decision attribute (the attribute produced by decision $d$ before any new information has become available). For algebraic purposes, we define the indicator function

$$\delta_{a'}(a, d) = \begin{cases} 1 & \text{if } a' = a_t^x = a^{M,x}(a_t, d_t), \\ 0 & \text{otherwise.} \end{cases}$$

Using matrix notation, we can write the post-decision resource vector $R_t^x$ using

$$R_t^x = \Delta R_t,$$

where $\Delta$ is a matrix in which $\delta_{a'}(a, d)$ is the element in row $a'$ and column $(a, d)$. We emphasize that the function $\delta_{a'}(a, d)$ and matrix $\Delta$ are used purely for notational convenience; in a real implementation, we work purely with the transition function $a^{M,x}(a_t, d_t)$. The pre-decision resource state vector is given by

$$R_{t+1} = R_t^x + \hat{R}_{t+1}.$$

We model demands in a simple way. If a resource is assigned to a demand, then it is "served" and vanishes from the system. Otherwise, it is held to the next time period. Let

$$\begin{aligned} \delta D_{tb_d} &= \text{The number of demands of type } b_d \text{ that are served at time } t. \\ &= \sum_{a \in \mathcal{A}} x_{tad} \quad d \in \mathcal{D}^D, \\ \delta D_t &= (\delta D_{tb})_{b \in \mathcal{B}}. \end{aligned}$$

The demand transition function can be written

$$\begin{aligned} D_t^x &= D_t - \delta D_t, \\ D_{t+1} &= D_t^x + \hat{D}_t. \end{aligned}$$

Finally, we are going to assume that our parameters evolve in a purely exogenous manner such as

$$\rho_{t+1} = \rho_t + \hat{\rho}_{t+1}.$$

We can assume any structure (additive, rule-based). The point is that $\rho_t$ evolves purely exogenously.

The last dimension of our model is the objective function. For our resource allocation problem, we define a contribution for each decision given by

$c_{tad}$ = Contribution earned (negative if it is a cost) from using decision $d$ acting on resources with attribute $a$.

The contribution function for time period $t$ is assumed to be linear, given by

$$C_t(S_t, x_t) = \sum_{a \in \mathcal{A}} \sum_{d \in \mathcal{D}} c_{tad} x_{tad}.$$

The objective function is now given by

$$\max_{\pi \in \Pi} \mathbb{E} \left\{ \sum_{t=0}^{T} C_t(S_t, X_t^\pi(S_t)) \right\}.$$

### 12.4.2 Approximation strategies

The sections on blood management and portfolio optimization both follow our general strategy of solving problems of the form

$$X_t^\pi(S_t^n) = \arg\max_{x_t \in \mathcal{X}_t^n} \left( C_t(S_t^n, x_t) + \gamma \bar{V}_t^{n-1}(S_t^x) \right), \qquad (12.27)$$

where $S_t^n$ is our state at time $t$, iteration $n$, and $\bar{V}_t^{n-1}(S_t^x)$ is our value function approximation computed in iteration $n-1$, evaluated at the post-decision state $S_t^x = S^{M,x}(S_t^n, x_t)$. In the previous sections, our state variable took the form $S_t = (R_t, D_t, \rho_t)$ (for the portfolio problem, $\rho_t$ captured market returns), and we used a value function approximation of the form

$$V_t(S_t) \approx \sum_{a \in \mathcal{A}} \bar{V}_{ta}(R_{ta}^x),$$

where $R_{ta}^x$ is the (post-decision) number of assets with attribute $a$. A particularly powerful approximation strategy takes advantage of the fact that when we solve (12.27), we can easily obtain gradients of the objective function directly from the dual variables of the resource conservation constraints (contained in $\mathcal{X}_t^n$) which state that $\sum_{d \in \mathcal{D}} x_{tad} = R_{ta}$. If we do not feel that the duals are sufficiently accurate, we may compute numerical derivatives fairly easily. Section 12.2.3 shows how we can use these derivatives to estimate piecewise linear approximations.

#### *Linear value function approximation*

The simplest approximation strategy outside of a myopic policy (equivalent to $\bar{V}_t(R_t) = 0$), is one that is linear in the resource state, as in

$$\bar{V}_t(R_t) = \sum_{a' \in \mathcal{A}} \bar{v}_{ta'} R_{ta'}.$$

With a linear value function approximation, the decision function becomes unusually simple. We start by writing $R^x_{ta}$ using

$$R^x_{ta'} = \sum_{a \in \mathcal{A}} \sum_{d \in \mathcal{D}} \delta_{a'}(a,d) x_{tad}. \tag{12.28}$$

Here, a decision $d \in \mathcal{D}$ represents acting on a resource, which might involve serving a demand or simply modifying the resource. Substituting this in our linear value function approximation gives us

$$\bar{V}_t(R^x_t) = \sum_{a' \in \mathcal{A}} \bar{v}_{ta'} \sum_{a \in \mathcal{A}} \sum_{d \in \mathcal{D}} \delta_{a'}(a,d) x_{tad}. \tag{12.29}$$

The decision function is now given by

$$\begin{aligned} X^\pi_t(S_t) &= \arg\max_{x_t \in \mathcal{X}_t} \left( \sum_{a \in \mathcal{A}} \sum_{d \in \mathcal{D}} c_{tad} x_{tad} + \gamma \sum_{a' \in \mathcal{A}} \bar{v}_{ta'} \sum_{a \in \mathcal{A}} \sum_{d \in \mathcal{D}} \delta_{a'}(a,d) x_{tad} \right) \\ &= \arg\max_{x_t \in \mathcal{X}_t(\omega)} \sum_{a \in \mathcal{A}} \sum_{d \in \mathcal{D}} \left( c_{tad} + \gamma \sum_{a' \in \mathcal{A}} \bar{v}_{ta'} \delta_{a'}(a,d) \right) x_{tad}. \end{aligned} \tag{12.30}$$

Recognizing that $\sum_{a' \in \mathcal{A}} \delta_{a'}(a,d) = \delta_{a^M(a_t,d_t)}(a,d) = 1$, we can write (12.30) as

$$X^\pi_t(S_t) = \arg\max_{x_t \in \mathcal{X}_t(\omega)} \sum_{a \in \mathcal{A}} \sum_{d \in \mathcal{D}} \left( c_{tad} + \gamma \bar{v}^{n-1}_{t,a^M(a,d)} \right) x_{tad}. \tag{12.31}$$

So, this is nothing more than a network problem where we are assigning resources with attribute $a$ to demands (for decisions $d \in \mathcal{D}^D$) or we are simply modifying resources ($d \in \mathcal{D}^M$). If we were using a myopic policy, we would just use the contribution $c_{tad}$ to determine which decisions we should make now. When we use a linear value function approximation, we add in the value of the resource after the decision is completed, given by $\bar{v}^{n-1}_{t,a^M(a,d)}$.

Solving $X^\pi(S_t)$ is a linear program, and therefore returns a dual variable $\hat{v}_{ta}$ for each flow conservation constraint. We can use these dual variables to update our linear approximation using

$$\bar{v}^n_{ta} = (1 - \alpha_{n-1}) \bar{v}^{n-1}_{ta} + \alpha_{n-1} \hat{v}^n_{ta}.$$

Linear (in the resource state) approximations are especially easy to develop and use, but they do not always provide good results. They are particularly useful when managing complex assets such as people, locomotives or aircraft. For such problems, the attribute space is quite large and as a result $R_{ta}$ tends to be small (say, 0 or 1). In general, a linear approximation will work well if the size of the attribute space is much larger than the number of discrete resources being managed, although even these problems can have pockets where there are a lot of resources of the same type.

Linear value function approximations are particularly easy to solve. With a linear value function approximation, we are solving network problems with the structure shown in figure 12.7a, which can be easily transformed to the equivalent network shown in figure 12.7b where all we have done is to take the slope of the linear value function for each downstream resource type (for example, the slope $v_1$ for resources that have attribute $a'_1$ in the future) and add this value to the arc assigning the resource to the demand. So, if the cost of assigning resources with attribute $a_1$ to the demand has a cost of $c_1$, then we would use a modified cost of $c_1 + v_1$.

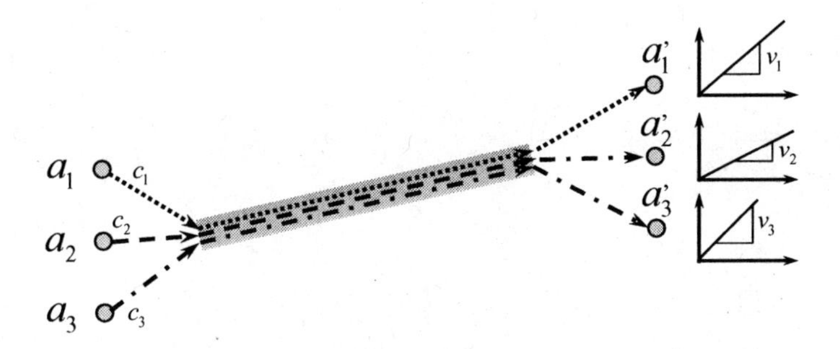


12.7a: A multicommodity flow subproblem with linear value functions.

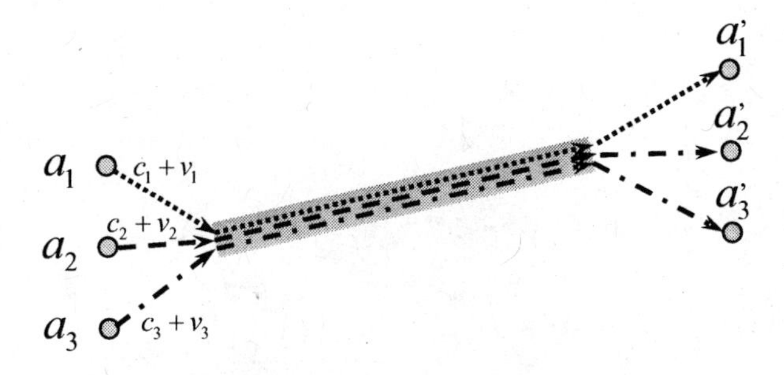


12.7b: Conversion to an equivalent problem which is a pure network.

**Figure 12.7** If linear value functions are used (12.7a) the problem can be converted into an equivalent pure network (12.7b).

### *Piecewise linear, separable approximation*

We solved both the blood management and the portfolio allocation problems using separable, piecewise linear value function approximations. This technique produces results that are more stable when the value function is truly nonlinear in the amount of resources that are provided.

The algorithm is depicted in figure 12.8. At each time period, we solve a linear program similar to what we used in the blood management problem (figure 12.3) using separable, piecewise linear approximations of the value of resources in the future. The process steps forward in time, using value function approximations for assets in the future. The dual variables from each linear program can be used to update the value function for the previous time period, using the techniques described in section 11.3.

A critical assumption in this algorithm is that we obtain $\hat{v}_{ta}$ for all $a \in \mathcal{A}$. If we were managing a single asset, this would be equivalent to sampling the value of being in all possible states at every iteration. In chapter 4, we referred to this strategy as *synchronous* approximate dynamic programming (see section 4.7.2). Below (in section 12.6) we describe an application where this is not possible, and we also describe strategies for overcoming this problem.

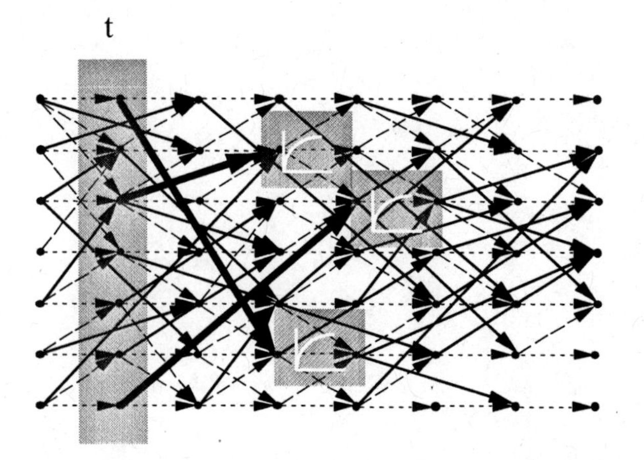


12.8a: The approximate linear program for time $t$.

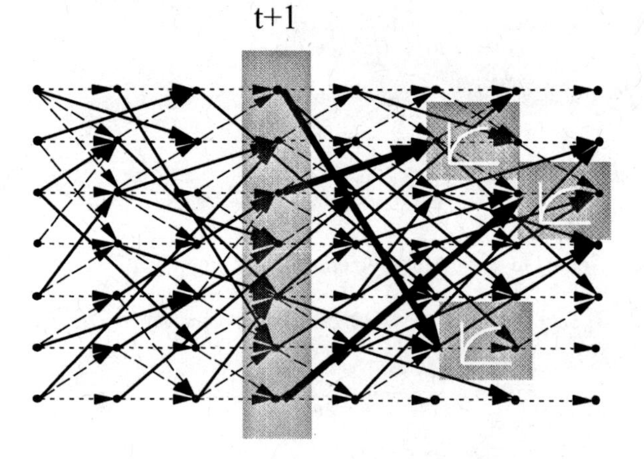


12.8b: The approximate linear program for time $t + 1$.

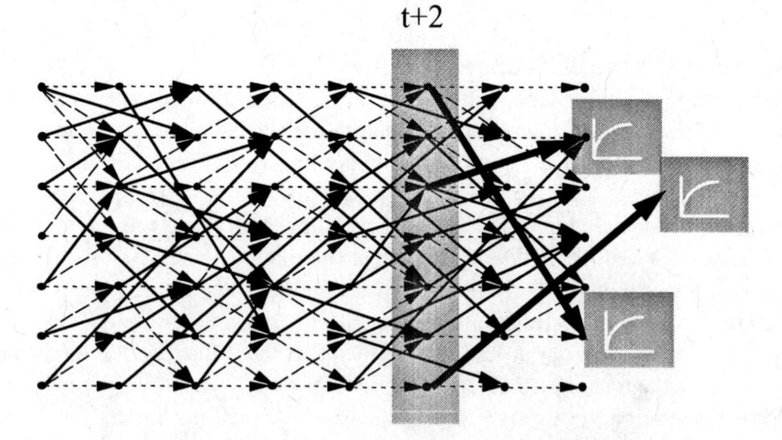


12.8c: The approximate linear program for time $t + 2$.

**Figure 12.8** ADP for general, multistage resource allocation problems using separable value function approximations.

### 12.4.3 Extensions and variations

Piecewise linear, separable approximations have worked quite well for a number of applications, but it is important to realize that this is just one of many potential approximations that could be used. The power of piecewise linear functions is that they are extremely flexible and can be used to handle virtually any distribution for demands (or other parameters). But separable approximations will not work for every problem.

A popular approximation strategy is to use simple polynomials such as

$$\bar{V}_t(S_t) \approx \sum_{a \in \mathcal{A}} \sum_{a' \in \mathcal{A}} \theta_{aa'} R_{ta} R_{ta'}.$$

This is a linear-in-the-parameters regression problem which can be estimated using the techniques of section 7.3 for recursive least squares. This approximation can be a challenge if the set $\mathcal{A}$ is fairly large. Neural networks (section 7.4) offer another avenue of investigation which, as of this writing, has not received any attention for this problem class.

We can retain the use of separable, piecewise linear approximations but partially overcome the separability approximation by writing the value function on an aggregated state space. For example, there may be subsets of resources which substitute fairly easily (e.g., people or equipment located nearby). We can aggregate the attribute space $\mathcal{A}$ into an aggregated attribute space $\mathcal{A}^{(g)}$, where $a^g \in \mathcal{A}^{(g)}$ represents an aggregated attribute. We can then write

$$V_t(S_t) \approx \sum_{a^g \in \mathcal{A}^{(g)}} \bar{V}_{ta^g}(R^x_{ta^g}),$$

where $R^x_{ta^g}$ is the number of resources with aggregated attribute $a^g$. When we use aggregated value functions, we may have several gradients $\hat{v}^n_a$, giving the derivative with respect to $R_{ta}$, which need to be combined in some way to update a single aggregated value function $\bar{V}_{ta^g}(R^x_{ta^g})$.

We may also consider using the techniques of section 7.1.3 and use multiple levels of aggregation, which would produce an approximation that looks like

$$V_t(S_t) \approx \sum_{g \in \mathcal{G}} \sum_{a^g \in \mathcal{A}^{(g)}} \theta^{(g)}_{a^g} \bar{V}_{ta^g}(R^x_{ta^g})$$

where $\mathcal{G}$ is a family of aggregations, and $\theta^{(g)}_{a^g}$ is the weight we give to the function for attribute $a^g$ for aggregation level $g$.

Finally, we should not forget that we can use a linear or piecewise linear function of one variable such as $R_{ta}$ which is indexed by one or more other variables. For example, we might feel that the value of resources of type $a_1$ are influenced by the number of resources of type $a_2$. We can estimate a piecewise linear function of $R_{ta_1}$ that is indexed by $R_{ta_2}$, producing a family of functions as illustrated in figure 12.9. This is a powerful strategy in that it does not assume any specific functional relationship between $R_{ta_1}$ and $R_{ta_2}$, but the price is fairly steep if we want to have the value function indexed by more than just one additional element such as $R_{ta_2}$. We might decide that there are five or ten elements that might affect the behavior of the function with respect to $R_{ta_1}$. Instead of estimating one piecewise linear value function approximation for each $R_{ta_1}$, we might find ourselves estimating thousands, so there is a limit to this strategy.

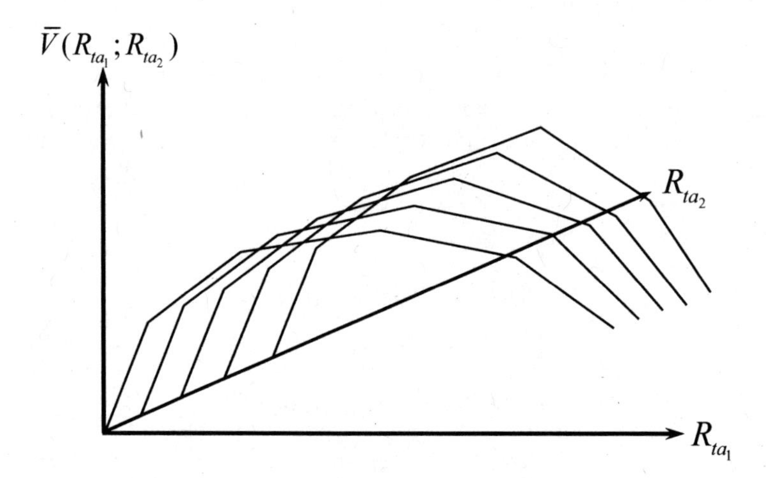


**Figure 12.9** A family of piecewise linear value function approximations.

A powerful idea which has been suggested (but we have not seen it tested in this setting) is the use of ridge regression. Create a scalar *feature resource variable* given by

$$\bar{R}_{tf} = \sum_{a \in \mathcal{A}} \theta_{fa} R_{ta}, \quad f \in \mathcal{F}.$$

Now estimate a value function approximation

$$V_t(S_t) \approx \sum_{f \in \mathcal{F}} \bar{V}_{tf}(\bar{R}_{tf}).$$

This approximation still consists of a series of piecewise linear, scalar functions, but now we are using specially constructed variables $\bar{R}_{tf}$ made up of linear combinations of $R_{ta}$.

### 12.4.4 Value function approximations and problem structure

Linear value functions have a number of nice features. They are easy to estimate (one parameter per attribute). They offer nice opportunities for decomposition. If city $i$ and city $j$ both send equipment to city $k$, a linear value function means that the decisions of city $i$ and city $j$ do not interact (at least not through the value function). Finally, it is often the case that the optimization problem is a special type of linear program known as a *pure network*, which means that it can be completely described in terms of nodes, links, and upper bounds on the flows. Pure networks have a nice property: if all the data (supplies, demands, upper bounds) are integer, then the optimal solution (in terms of flows) will also be integer. This is especially useful if our resources are discrete (people or equipment). It means if we solve the problem as a linear program (ignoring the need for integer solutions), then the optimal solution is still integer.

Unfortunately, linear value function approximations often do not work very well.

Nonlinear value function approximations can improve the quality and stability of the solution considerably. We have already seen with our blood management and portfolio problems that separable, piecewise linear value function approximations can be particularly

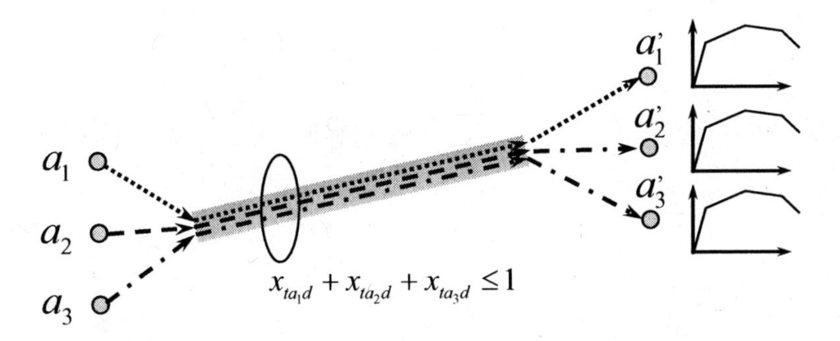


**Figure 12.10** If we can use different types of resources to serve a demand and we use nonlinear value functions to capture the value of each type of resource in the future, we have to use bundle constraints to capture the type of flow serving the demand.

easy to estimate and use. However, if we are managing discrete resources, then integrality becomes an issue. Recall that our decision problem looks like

$$X^\pi(R_t) = \arg\max_{x_t \in \mathcal{X}_t} \left\{ \sum_{a \in \mathcal{A}} \sum_{d \in \mathcal{D}_a} c_{tad} x_{tad} + \sum_{a \in \mathcal{A}} \bar{V}_{ta}^{n-1}(R_{ta}^x(x_t)) \right\}. \tag{12.32}$$

If our only constraints are the flow conservation constraint ($\sum_d x_{tad} = R_{ta}$), nonnegativity and possibly upper bounds on individual flows ($x_{tad} \leq u_{tad}$), then equation (12.32) is a pure network, which means we can ignore integrality (by this we mean that we can solve it as a continuous linear program), but still obtain integer solutions. What may destroy this structure are constraints of the form

$$\sum_{a \in \mathcal{A}} x_{tad} \leq R_{tb_d}^D, \quad d \in \mathcal{D}^D.$$

Constraints such as this are known as *bundle constraints* since they bundle different decisions together ("we can use resources of type 1 or 2 or 3 to cover this demand"). The problem is illustrated in figure 12.10. When we introduce bundle constraints, we are no longer able to solve the problem as a continuous linear program and count on getting integer solutions. Instead, we have to use an integer programming algorithm which can be much slower (it depends on the structure of the problem).

We avoid this situation if our problem exhibits what we might refer to as the *Markov property* which is defined as follows:

**Definition 12.4.1** *We say that a resource allocation problem has the Markov property if*

$$a^M(a, d, W_{t+1}) = a^M(d, W_{t+1}),$$

*which is to say that the attributes of a transition resulting from a decision d acting on a resource with attribute a is independent of a (the transition is "memoryless").*

An example of a problem which exhibits the Markov property is the portfolio problem. If we decide to invest money in asset class $k$, we do not care if the money came from asset

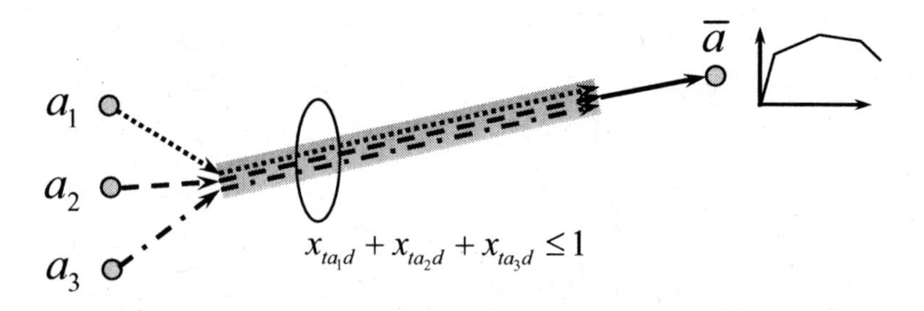


**Figure 12.11** If flow, after serving a demand, has an attribute that depends only on the attributes of the demand, then all the value ends in the same value function.

class $i$ or $j$. Another problem with the Markov property arises when we are managing a fleet of identical vehicles, where the attribute vector $a$ captures only the location of the vehicle. A decision to "move the vehicle to location $j$" determines the attribute of the vehicle after the decision is completed (the vehicle is now at location $j$).

If a problem exhibits the Markov property, then we obtain a network shown in figure 12.11. If a linear program with this structure is solved using a linear programming package without imposing integrality constraints, the solution will still be integer.

If we have a problem that does not exhibit the Markov property, we can restore this property by aggregating resources in the future. To illustrate, imagine that we are managing different types of medical technicians to service different types of medical equipment in a region. The technician might be characterized by his training, his current location, and how many hours he has been working that day. After servicing a machine at a hospital, he has changed his location and the number of hours he has been working, but not his training. Now assume that when we approximate the value of the technician in the future, we ignore his training and how many hours he has been working (leaving only the location). Now his attributes are purely a function of the decision (or equivalently, the job he has just completed).

The good news about this problem is that commercial solvers seem to easily produce optimal (or near optimal) integer solutions even when we combine piecewise linear, separable value function approximations in the presence of bundle constraints. Integer programs come in different flavors, and we have found that this problem class is one of the easy ones. However, it is important to allow the optimization solver to return "near optimal" solutions (all codes allow you to specify an epsilon-tolerance in the solution). Obtaining provably optimal solutions for integer programs can be extremely expensive (the code spends most of the time verifying optimality), while allowing an epsilon tolerance can produce run times that are almost as fast as if we ignore integrality. The reason is that if we ignore integrality, the solution will typically be 99.99 percent integer.

### 12.4.5 Applications

There is a vast range of resource allocation problems that can be modeled with this framework. We began the chapter with three examples (asset acquisition, blood inventories, and portfolio optimization). Some examples of other applications include:

- Inspecting passengers at airports - The Transportation Safety Administration (TSA) assigns each arriving passenger to a risk category that determines the level of inspection for passenger (simple X-ray and magnetic detection, an inspector using the hand wand, more detailed interrogation). Each level requires more time from selected resources (inspectors, machines). For each arriving passenger, we have to determine the right level of inspection taking into account future arrivals, the capacity of each inspection resource, and the probability of detecting a real problem.

- Multi-skill call centers - People phoning in for technical support for their computer have to progress through a menu that provides some information about the nature of their problem. The challenge then is to match a phone call characterized by a vector of attributes $a$ which captures the phone options, with a technician (each of which has different skills). Assigning a fairly simple request to a technician with special skills might tie up that technician for what could be a more important call in the future.

- Energy planning - A national energy model might have to decide how much capacity for different types of energy we should have (coal, ethanol, wind mills, hydro, oil, natural gas) at each time period to meet future demands. Many forms of capacity have lifetimes of several decades, but these decisions have to be made under uncertainty about commodity prices and the evolution of different energy technologies.

- Planning flu vaccines - The Center for Disease Control (CDC) has to decide each year which flu vaccine (or vaccines) should be manufactured to respond to potential outbreaks in the upcoming flu season. It can take six months to manufacture a vaccine for a different strain, while flu viruses in the population evolve continuously.

- Fleet planning for charter jet operators - A company has to purchase different types of jets to serve a charter business. Customers might request a particular type of jet. If the company does not have jets of that type available, the customer might be persuaded to upgrade to a larger jet (possibly with the inducement of "no extra charge."). The company has to decide the proper mix of jets, taking into account not only the uncertainty in the customer demands but also their willingness to accept upgrades.

- Work force planning - Large companies, and the military, have to develop the skills in their employees to meet anticipated needs. The skills may be training in a specific activity (filling out types of forms, inspecting machinery, or working in a particular industry as a consultant) or could represent broader experiences (spending time abroad, working in accounting or marketing, working in the field). It may take time to develop a skill in an employee which previously did not have the skill. The company has to decide which employees to develop to meet uncertain demands in the future.

- Hospital operations - There are numerous problems requiring the management of resources for hospitals. How should different types of patients be assigned to beds? How should nurses be assigned to different shifts? How should doctors and interns be managed? Which patients should be scheduled for surgery and when?

- Queueing networks - Queueing networks arise when we are allocating a resource to serve customers who move through the system according to a set of exogenous rules (that is, we are not optimizing how the customer moves through the network). These applications often arise in manufacturing where the customer is a job (for example, a computer circuit board) which has to go through a series of steps at different stations. At each station is a machine that might perform several tasks, but can only perform one task at a time. The resource allocation problem involves determining what type of task a machine should be working on at a point in time. Given a set of machines (and how they are configured) customers move from one machine to another according to fixed rules (that may depend on the state of the system).

Needless to say, the applications are nearly endless, and cut across different dimensions of society (corporate applications, energy, medical, homeland security). All of these problems can be reasonably approximated using the types of techniques we describe in this chapter.

## 12.5 A FLEET MANAGEMENT PROBLEM

Consider a freight transportation company that moves loads of freight in trailers or containers. Assume each load fills the container (we are not consolidating small shipments into larger containers). After the company moves a load of freight from one location to another, the container becomes empty and may have to be repositioned empty to a new location where it is needed. Applications such as these arise in railroads, truckload motor carriers, and intermodal container shipping.

These companies face two types of decisions (among others): 1) If there are too many customer demands (a fairly common problem), the company has to decide which customers to serve (that is, which loads of freight to move) by balancing the contribution earned by moving the load, plus the value of the container when it becomes empty at the destination. 2) The company may have more containers in one location than are needed, making it necessary to move containers empty to a location where they may be needed (possibly to serve demands that are not yet known).

A model based on the algorithmic strategies described in this chapter was implemented at a major railroad to manage their freight cars. This section briefly describes the modeling and algorithmic strategy used for this problem.

### 12.5.1 Modeling and algorithmic strategy

This problem can be perfectly modeled using the notation in section 12.4. The attribute vector $a$ will include elements such as the location of the container, the type of container, and, because it usually takes multiple time periods for a container to move from one location to the next, the expected time of arrival (or, the number of time periods until it will arrive). The time to travel is usually modeled deterministically, but exercise 12.2 asks you to generalize this basic model to handle random travel times. An important characteristic of this problem is that while the attribute space is not small, it is also not too large, and in fact we will depend on our ability to estimate values for each attribute in the entire attribute space.

The decision set $\mathcal{D}^D$ represents decisions to assign a container to a particular type of load. The load attributes $b$ typically include the origin and destination of the load, but may also include attributes such as the customer (this may determine which types of containers are allowed to be used), the type of freight, and the service requirements.

The problem is very similar to our blood management problem, with two major exceptions. First, in the blood problem, we could assign blood to a demand (after which it would vanish from the system) or hold it. With the fleet management problem, we also have the option of "modifying" a car, captured by the decision set $\mathcal{D}^M$. The most common "modification" is to move a car empty from one location to another (we are modifying the location of the car). Other forms of modification include cleaning a car (freight cars may be too dirty for certain types of freight), repair a car, or change its "ownership" (cars may be assigned to shipper pools, where they are dedicated to the needs of a particular shipper).

The second exception is that while used blood vanishes from the system, a car usually remains in the system after it is assigned to an order. We say usually because a railroad in the eastern United States may pick up a car that has to be taken to the west coast. To complete this move, the car has to be transferred to a railroad that serves the west coast. Although the car will eventually return to the first railroad, for planning purposes the car is considered lost. At a later time, it will appear as if it is a random arrival (similar to our random blood donations).

The most widely used model in practice is an optimization model that assigns individual cars to individual orders at a point in time (figure 12.12a). These models do not consider the downstream impact of a decision now on the future, although they are able to consider cars that are projected to become available in the future, as well as forecasted orders. However, they cannot handle a situation where an order has to be served 10 days from now which might be served with a car after it finishes an order that has to be served three days from now. Figure 12.12b shows the decision function when using value function approximations, where we not only assign known cars to known orders (similar to the myopic model in figure 12.12a), but we also capture the value of cars in the future (through piecewise linear, value function approximations).

Railroads which use the basic assignment model (several major North American railroads do this) only solve them using known cars and orders. If we use approximate dynamic programming, we would first solve the problem at time $t = 0$ using known cars and orders, but we would then step forward in time and use simulated cars and orders. By simulating over a three-week horizon, ADP not only produces a better decision at time $t = 0$, but also produces a forecast of car activities in the future. This proved very valuable in practice.

In section 12.4.2, we mentioned that we can use the dual variable $\hat{v}_{ta}^n$ (or a numerical derivative) for the resource constraint

$$\sum_{d \in \mathcal{D}} x_{tad} = R_{ta}$$

to update the value function approximation $\bar{V}_{t-1,a}^{n-1}(R_{t-1,a}^x)$. When the attribute space is not too large (something that we assume for this problem class), we can assume that we are calculating $\hat{v}_{ta}^n$ for each $a \in \mathcal{A}$. It turns out that this is an extremely powerful capability, and not one that we can do for more complex problems where the attribute space is too large to enumerate.

### 12.5.2 Laboratory experiments

There are two types of laboratory experiments that help provide insights into the quality of the solution provided by ADP: (a) those performed on a deterministic dataset and (b) those on a stochastic dataset. Deterministic experiments are useful because these are problems that can be solved to optimality, providing a tight bound to compare against. Stochastic datasets allow us to evaluate how well the method performs under uncertainty, but we have to compare against heuristic solutions since tight bounds are not available.

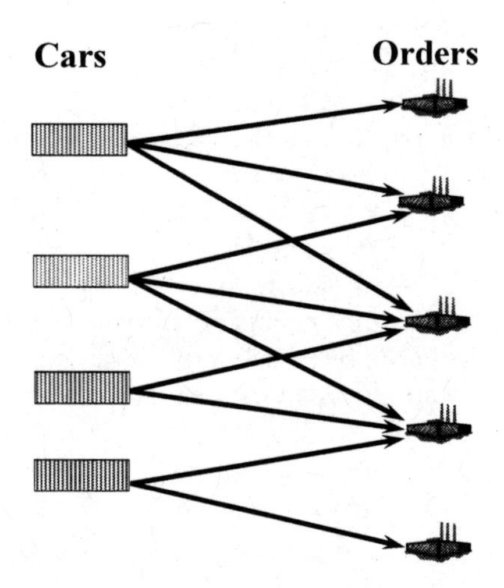


12.12a Basic car-to-order assignment problem.

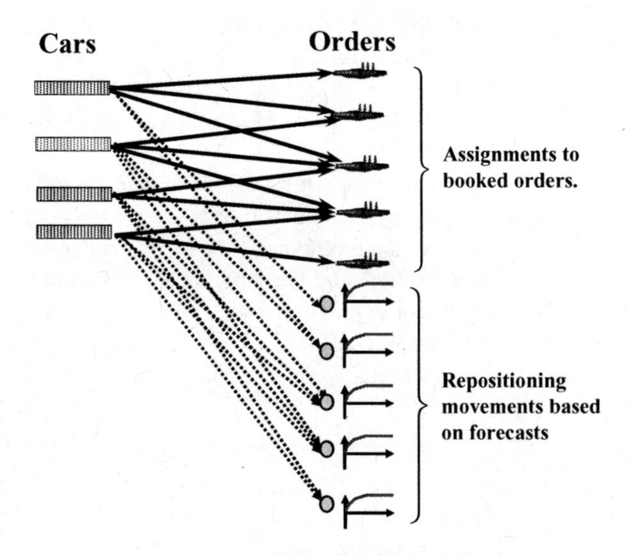


12.12b Car assignment problem using ADP.

**Figure 12.12** (a) A basic assignment model for assigning cars to orders used in engineering practice. (b) A model which uses piecewise linear, separable value function approximations estimated using approximate dynamic programming.

### *Deterministic experiments*

There are many variations of the fleet management problem. In the simplest, we have a single type of container, and we might even assume that the travel time from one location to another is a single time period (in this case, the attribute vector $a$ consists purely of location). We assume that there is no randomness in the resource transition function (which means $\hat{R}_t = 0$), and that if a demand is not satisfied at time $t$, it is lost from the system (which means that $D_t = \hat{D}_t$). We may additionally assume that a demand originating from location

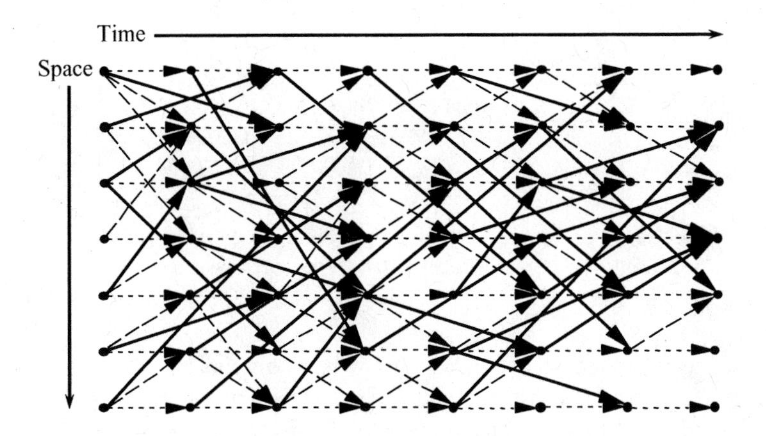


**Figure 12.13** Illustration of a pure network for time-staged, single commodity flow problems.

$i$ can only be served by containers that are already located at location $i$. In section 12.6 below, we relax a number of these assumptions.

If the problem is deterministic (which is to say that all the demands $\hat{D}_t$ are known in advance), then the problem can be modeled as a network such as the one shown in figure 12.13. This problem is easily solved as a linear program. Since the resulting linear program has the structure of what is known as a *pure network*, it is also the case that if the flows (and upper bounds) are integer, then the resulting solution is integer.

We can solve the problem using approximate dynamic programming for a deterministic application by pretending that $\hat{D}_t$ is stochastic (that is, we are not allowed to use $\hat{D}_{t+1}$ when we solve the problem at time $t$), but where every time we "observe" $\hat{D}_t$ we obtain the same realization. Using approximate dynamic programming, our decision function would be given by

$$X_t^\pi(S_t) = \arg\max_{x_t \mathcal{X}_t} \left( C(S_t, x_t) + \bar{V}_t^{n-1}(R_t^x) \right),$$

where we use a separable, piecewise linear value function approximation (as we did in the previous sections).

Experiments (reported in Godfrey & Powell (2002)) were run on problems with 20, 40, and 80 locations and with 15, 30, and 60 time periods. Since the problem is deterministic, we can optimize the entire problem (that is, over all time periods) as a single linear program to obtain the true optimal solution, allowing us to see how well our approximate solution compares to optimal (on a deterministic problem). The results are reported in table 12.3 as percentages of the optimal solution produced by the linear programming algorithm. It is not hard to see that the results are very near optimal. We know that separable, piecewise linear approximations do not produce provably optimal solutions (even in the limit) for this problem class, but it appears that the error is extremely small.

### *Stochastic experiments*

We now consider what happens when the demands are truly stochastic, which is to say that we obtain different values for $\hat{D}_t$ each time we sample information. For this problem, we do not have an optimal solution. Although this problem is relatively small (compared to true industrial applications), if we formulate it as a Markov decision process we would

| | Simulation Horizon | | |
|---|---|---|---|
| **Locations** | **15** | **30** | **60** |
| 20 | 100.00% | 100.00% | 100.00% |
| 40 | 100.00% | 99.99% | 100.00% |
| 80 | 99.99% | 100.00% | 99.99% |

**Table 12.3** Percentage of the optimal deterministic solution produced by separable, piecewise linear value function approximations (from Godfrey & Powell (2002))

produce a state space that is far larger than anything we could hope to solve using the techniques of chapter 3. The standard approach used in engineering (which we have also found works quite well) is to use a rolling horizon procedure where at each time $t$ we combine the demands that are known at time $t$ with expectations of any random demands for future time periods. A deterministic problem is then solved over a *planning horizon* of length $T^{ph}$ which typically has to be chosen experimentally.

For a specific sample realization of the demands, we can still find an optimal solution using a linear programming solver, but this solution "cheats" by being able to use information about what is happening in the future. This solution is known as the *posterior bound* since it uses information that only becomes known after the fact. Although it cheats, it provides a nice benchmark against which we can compare our solutions from the ADP algorithm.

Figure 12.14 compares the results produced by a rolling horizon procedure to those produced using approximate dynamic programming (with separable, piecewise linear value function approximations). All results are shown as a percentage of the posterior bound. The experiments were run on problems with 20, 40, and 80 locations and with 100, 200, and 400 vehicles in our fleet. Problems with 100 vehicles were able to cover roughly half of the demands, while a fleet of 200 vehicles could cover over 90 percent. The fleet of 400 vehicles was much larger than would have been necessary. The ADP approximation produced better results across all the problems, although the difference was most noticeable for problems where the fleet size was not large enough to cover all the demands. Not surprisingly, this was also the problem class where the posterior solution was relatively the best.

These experiments demonstrate that piecewise linear, separable value function approximations work quite well for this problem class. However, as with all experimental evidence, care has to be used when generalizing these results to other datasets. This technique has been used in several industrial applications with success, but without provable bounds, it will be necessary to reevaluate the approximation for different problem classes. We have found, however, that the technique scales quite well to very large scale problems.

### 12.5.3 A case implementation

This strategy was implemented in an industrial application for managing freight cars for a major railroad. Two aspects of the problem proved to be much more complex than anticipated. First, in addition to the usual attributes of location and type of freight car, we also had to consider estimated time of arrival (for cars that are currently moving from one location to another), cleanliness (some shippers reject dirty cars), repair status, and ownership. The second aspect was the complexity of the information process. Random demands turned out to be only one of the sources of uncertainty. When a customer calls in an order, he will specify the origin of the order (this is where we send the car to), but

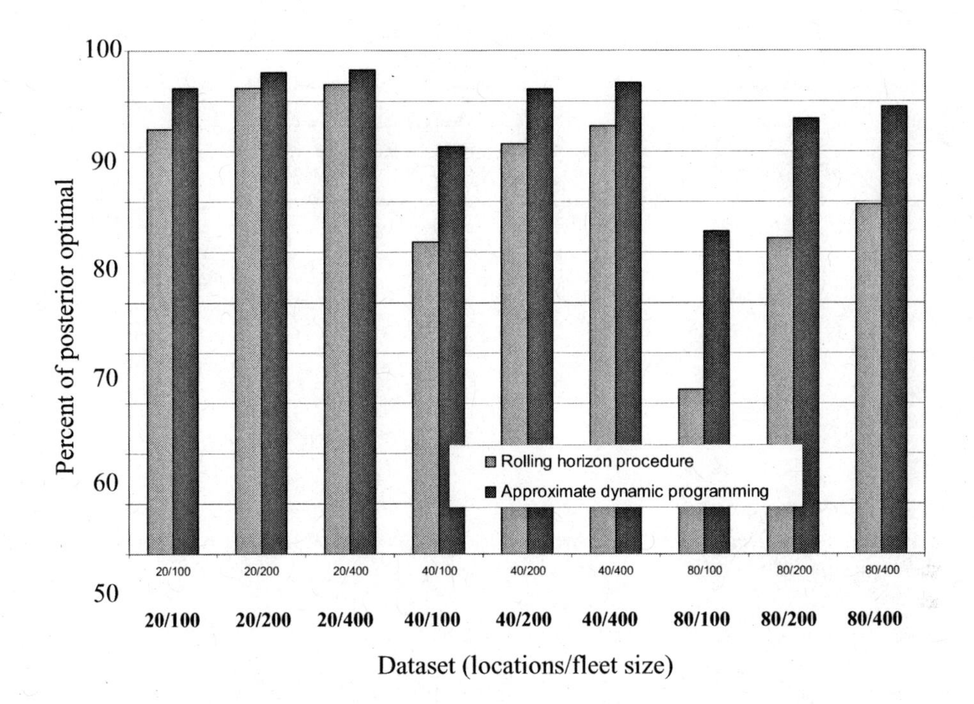


**Figure 12.14** Percentage of posterior bound produced by a rolling horizon procedure using a point forecast of the future versus an approximate dynamic programming approximation.

not the destination (this is only revealed after the car is loaded). Also, a customer might reject a car as being too dirty only after it is delivered to the customer. Finally, the time to load and unload the car, as well as transit times, are all uncertain. None of these issues presented a significant modeling or algorithmic challenge.

Figure 12.12 described two models: (a) a basic car-to-order assignment problem widely used in industry and (b) a model which uses approximate dynamic programming to measure the value of cars in the future. So this raises the question: How do these two methods compare? Figure 12.15 compares the performance of the myopic model (figure 12.12a) to the model using approximate dynamic programming (figure 12.12b) against historical performance, using empty miles as a percent of total as the performance measure. For this dataset, 54 percent of the total mileage generated by the cars were spent moving empty (a surprising statistic). Using the simple assignment model (which we simulated forward in time), this number dropped to about 48 percent empty, which is a significant improvement. Using approximate dynamic programming (where we show the performance after each iteration of the algorithm), we were able to drop empty miles to approximately 35 percent of the total.

## 12.6 A DRIVER MANAGEMENT PROBLEM

In the previous section, we managed a set of "containers" that were described by a fairly simple set of attributes. For this problem, we assumed that we could compute gradients $\hat{v}_{ta}^{n}$

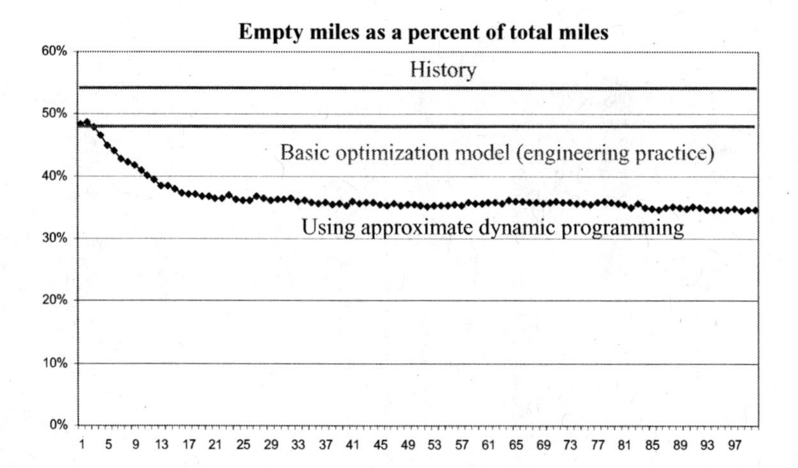


**Figure 12.15** Empty miles as a percent of total from an industrial application, showing historical performance (54 percent of total miles were empty), results using a myopic assignment model and when using approximate dynamic programming.

for each attribute $a \in \mathcal{A}$. We lose this ability when we make the transition from managing simple resources such as a container to complex resources such as drivers.

We use as our motivating application the problem of managing a fleet of drivers for a truckload motor carrier. The mathematical model is identical to what we used for managing containers (which is the model we presented in section 12.4). The only difference now is that the attribute vector $a$ is large enough that we can no longer enumerate the attribute space $\mathcal{A}$. A second difference we are going to introduce is that if we do not handle a demand (to move a load of freight) at time $t$, then the demand is held over to the next time period. This means that if we cannot handle all the demands, we need to think about which demands we want to satisfy first.

This summary is based on an actual industrial project that required modeling drivers (and loads) at a very high level of detail. The company ran a fleet of over 10,000 drivers, and our challenge was to design a model that closely matched actual historical performance. It turns out that experienced dispatchers actually do a fairly good job of thinking about the downstream impact of decisions. For example, a dispatcher might want to put a team of two drivers (which are normally assigned to a long load, which takes advantage of the ability of a team to move constantly) on a load going to Boston, but it might be the case that most of the loads out of Boston are quite short. It might be preferable to put the team on a load going to Atlanta if there are more long loads going out of Atlanta.

### 12.6.1 Working with complex attributes

When we were managing simple resources ("containers") we could expect that the resource state variable $R_{ta}$ would take on integer values greater than 1 (in some applications in transportation, they could number in the hundreds). When the attribute vector becomes complex, then it is generally going to be the case (with rare exception) that $R_{ta}$ is going to be zero (most of the time) or 1.

In an actual project with a truckload motor carrier, the attribute vector was given by

$$a = \begin{pmatrix} a_1 \\ a_2 \\ a_3 \\ a_4 \\ a_5 \\ a_6 \\ a_7 \\ a_8 \\ a_9 \\ a_{10} \end{pmatrix} = \begin{pmatrix} \text{Location} \\ \text{Domicile} \\ \text{Capacity type} \\ \text{Scheduled time-at-home} \\ \text{Days away from home} \\ \text{Available time} \\ \text{Geographical constraints} \\ \text{DOT road hours} \\ \text{DOT duty hours} \\ \text{Eight day duty hours} \end{pmatrix}.$$

The presence of complex attributes introduces one nice simplification: We can use linear (in the resource state) value function approximations (rather than the piecewise linear functions). This means that our value function approximation can be written

$$\bar{V}_t(R_t^x) = \sum_{a \in \mathcal{A}} \bar{v}_{ta} R_{ta}^x.$$

Instead of estimating a piecewise linear function, we have only to estimate $\bar{v}_{ta}$ for each $a$.

Now we have to update the value of a driver with attribute $a$. In theory, we would again compute the dual variable $\hat{v}_{ta}^n$ for the resource constraint $\sum_{d \in \mathcal{D}} x_{tad} = R_{ta}$ for each $a$. This is where we run into problems. The attribute space can be huge, so we can only find $\hat{v}_{ta}^n$ for some of the attributes. A common strategy is to find a dual only for the attributes where $R_{ta} > 0$. Even for a very large problem (the largest trucking companies have thousands of drivers), we have no problem solving linear programs with thousands of rows. But this means that we are not obtaining $\hat{v}_{ta}^n$ when $R_{ta} = 0$.

This issue is very similar to the classical problem in ADP when we do not estimate the value of states we do not visit. Imagine if we were managing a single driver. In this case, $a_t$ would be the "state" of our driver at time $t$. Assuming that we get $\hat{v}_{ta}$ for each $a$ is like assuming that we visit every state at every iteration. This is fine for small state spaces (that is, resources with simple attributes) but causes a problem when the attributes become complicated.

Fortunately, we already have the tools to solve this problem. In section 7.1.3, we saw that we could estimate $\bar{v}_{ta}$ at different levels of aggregation. Let $\mathcal{A}^{(g)}$ be the attribute space at the $g^{th}$ level of aggregation, and let $\bar{v}_{ta}^g$ be an estimate of a driver with attribute $a \in \mathcal{A}^{(g)}$. We can then estimate the value of a driver with (disaggregate) attribute $a$ using

$$\bar{v}_{ta} = \sum_{g \in \mathcal{G}} w_a^{(g)} \bar{v}_{ta}^{(g)}.$$

Section 7.1.3 provides a simple way of estimating the weights $w_a^{(g)}$.

### 12.6.2 Backlogging demands

We next address the problem of deciding which demand to serve when it is possible to hold unserved demands to future time periods. Without backlogging, the post decision state variable was given by $S_t^x = (R_t^x)$. Equation (12.28) determines $R_t^x$ as a function of $R_t$ and $x_t$ (the resource transition function). With backlogging, the post-decision state is given by

$S_t^x = (R_t^x, D_t^x)$ where $D_t^x = (D_{tb}^x)_{b \in \mathcal{B}}$ is the vector of demands that were not served at time $t$. $D_t^x$ is given simply using

$$
\begin{aligned}
D_{tb}^x &= \text{The number of loads with attribute vector } b \text{ which have not been served after decisions were made at time } t \\
&= D_{tb} - \sum_{a \in \mathcal{A}} x_{tad} \quad \text{where } b = b_d,\ d \in \mathcal{D}^D. \qquad (12.33)
\end{aligned}
$$

In equation (12.33), for a decision $d \in \mathcal{D}^D$ to move a load, each element in $\mathcal{D}^D$ references an element in the set of load attributes $\mathcal{B}$. So, for $d \in \mathcal{D}^D$, $\sum_{a \in \mathcal{A}} x_{tad}$ is the number of loads of type $b_d$ that were moved at time $t$.

If we use a linear value function approximation, we would write

$$\bar{V}_t(R_t) = \sum_{a \in \mathcal{A}} \bar{v}_{ta}^R R_{ta}^x + \sum_{b \in \mathcal{B}} \bar{v}_{tb}^D D_{tb}^x.$$

As before, $\bar{v}_{ta}^R$ is the value of a driver with attribute $a$, while $\bar{v}_{tb}^D$ is the value of holding an order with attribute $b$. Just as $\bar{v}_{ta}^R$ is estimated using the dual variable of the resource constraint

$$\sum_{d \in \mathcal{D}} x_{tad} = R_{ta},$$

we would estimate the value of a load in the future, $\bar{v}_{tb}^D$ using the dual variable for the demand constraint

$$\sum_{a \in \mathcal{A}} x_{tad} \leq D_{tb_d} \quad d \in \mathcal{D}^D.$$

Recall that each element in the set $\mathcal{D}^D$ corresponds to a type of demand $\mathcal{B}$. We refer to $\bar{v}_t^R$ as *resource gradients* and $\bar{v}_t^D$ as *task gradients*.

Whenever we propose a value function approximation, we have to ask the following questions: (1) Can we solve the resulting decision function? (2) Can we design an effective updating strategy (to estimate $\bar{V}$)? (3) Does the approximation work well (does it provide high quality decisions)?

For our problem, the decision function is still a linear program. The value functions can be updated using dual variables (or numerical derivatives) just as we did for the single-layer problem. So the only remaining question is whether it works well. Figure 12.16 shows the results of a simulation of a driver-management problem where we make decisions using a purely myopic policy ($\bar{v}^R = \bar{v}^D = 0$), a policy which ignored the value of demands (tasks) in the future ($\bar{v}^D = 0$), and a policy which used value functions for both resources and demands. Simulations were run for a number of datasets of increasing size. All of the runs are evaluated as a percent of the posterior solution (which we can solve optimally by assuming we knew everything that happened in the future). For these datasets, the myopic policy produced an average performance that was 84 percent of the posterior bound. Using value functions for just the first resource layer produced a performance of almost 88 percent, while including value functions for both resource layers produced solutions that were 92.6 percent of the posterior optimal.

### 12.6.3 Extensions

This basic model is capable of handling a number of extensions and generalizations with minor modifications.

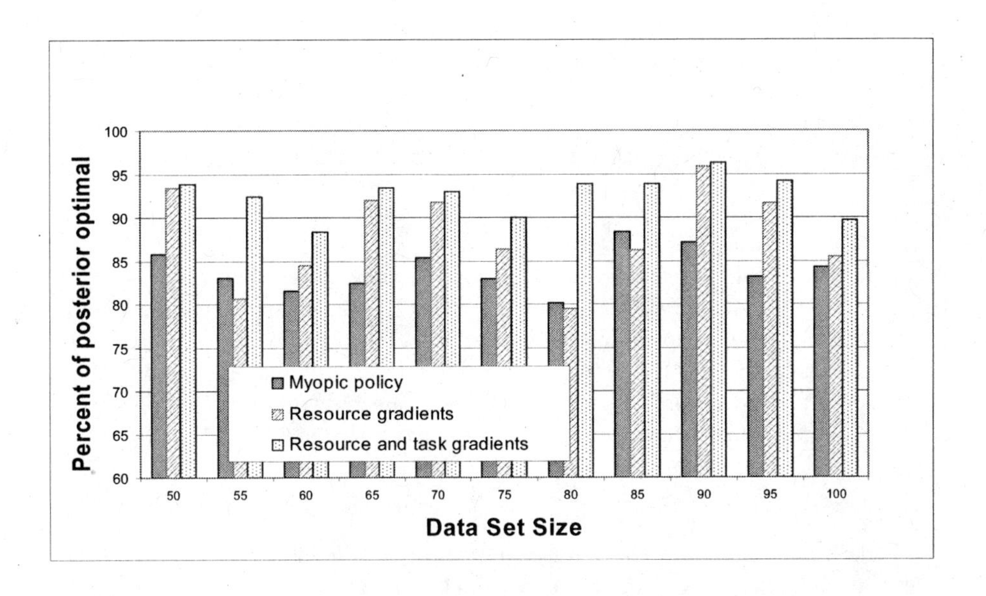


**Figure 12.16** The value of including value functions for resources alone, and resource and tasks, compared to a myopic policy (from Spivey & Powell (2004)).

### *Continuous actionable times vs. discrete decision epochs*

Perhaps one of the most powerful features of approximate dynamic programming is its ability to model activities using a very detailed state variable while maintaining a computationally compact value function approximation. One dimension where this is particularly important is the modeling of time. In our transportation applications where we are often managing thousands of resources (drivers, equipment), it is typically necessary to solve decision subproblems in discrete time, possibly with a fairly coarse discretization (e.g., every six or twelve hours). However, we can still model the fact that real activities happen in continuous time.

A common mistake in modeling is to assume that if we make decisions in discrete time, then every activity has to be modeled using the same level of discretization. This could not be farther from the truth. We can introduce in our attribute vector $a$ an attribute called the *actionable time* which is the time at which we can actually change the status of a resource (we can do the same for the demands, by adding an "actionable time" field to the demand attribute vector $b$). Assume that we are making decisions at time $t = 0, 100, 200, \ldots$ (times are in minutes). Further assume that at time $t = 100$ we wish to consider assigning a resource (driver) with actionable time of $a_{actionable} = 124$ to a demand (load) with actionable time $b_{actionable} = 162$, and assume that it takes $\tau = 50$ minutes for the driver to move to the load. If we make this decision, then the earliest time that the load can actually be picked up is at time $t = 124 + 50 = 174$. We can model these activities as taking place down to the nearest minute, even though we are making decisions at 100-minute intervals.

### *Relaying loads*

The simplest resource allocation problem involves a single *resource layer* (blood, money, containers, people) satisfying demands. In our driver management problem we allowed, for the first time, unsatisfied demands to be held for the future, but we still assumed that if we assign a driver to a load, then the load vanishes from the system.

It is also possible to assume that if we act on a load with attribute $b$ with decision $d$ that we produce a load with attribute $b'$. Previously, we described the effect of a decision $d$ on a resource with attribute $a$ using the attribute transition function

$$a_{t+1} = a^M(a_t, d, W_{t+1}).$$

We could use a similar transition function to describe how the attributes of a load evolve. However, we need some notational sleight of hand. When modeling the evolution of resources, the attribute transition function has access to both the attribute of the driver, $a_t$, and the attribute of the load which is captured implicitly in the decision $d$ which tells us what type of load we are moving (the attributes are given by $b_d$). To know how the attributes of the load change, we can define a load transition function

$$b_{t+1} = b^M(a_t, d, W_{t+1}).$$

This looks a little unusual because we include $a_t$ in the arguments but not $b_t$. In fact, $b_t$ is determined by the decision $d$, and we need to know both the driver attributes $a_t$ as well as the load attributes $b_t = b_d$. Algebraically, we used the function $\delta_{a'}(a, d)$ to indicate the attribute produced by the decision $d$ acting on the attribute $a$. When we have drivers and loads, we might use $\delta^D_{a'}(a, d)$ as the indicator function for drivers and use $\delta^L_{b'}(b, d)$ as the indicator function for loads. All the algebra would carry through symmetrically.

#### *Random travel times*

We can handle random travel times as follows. Assume that at time $t = 100$, we have a driver who is currently headed to Cleveland. We expect him to arrive at time $t = 115$ at which point he will be empty (status E) and available to be reassigned. Although he has not arrived yet, we make the decision to assign him to pick up a load that will take him to Seattle, where we expect him to arrive at time $t = 218$. This decision needs to reflect the value of the driver in Seattle at at time $t = 218$.

By time 200, the driver has not yet arrived but we have learned that he was delayed (our random travel time) and we now expect him to arrive at time $t = 233$. The sequence of pre- and post-decision states is given below.

| Label | t=100 pre-decision | t=100 post-decision | t=200 pre-decision |
|---|---|---|---|
| $\begin{pmatrix} Location \\ Status \\ ETA \end{pmatrix}$ | $\begin{pmatrix} Cleveland \\ \text{E} \\ 115 \end{pmatrix}$ | $\begin{pmatrix} Seattle \\ \text{E} \\ 218 \end{pmatrix}$ | $\begin{pmatrix} Seattle \\ \text{E} \\ 233 \end{pmatrix}.$ |

At time $t = 200$, we use his state (at the time) that he will be in Seattle at time 233 to make a new assignment. We use the value of the driver at this time to update the value of the driver in Seattle at time 218 (the previous post-decision state).

#### *Random destinations*

There are numerous applications in transportation where we have to send a vehicle (container, taxi) to serve a customer before we know where the customer is going. People routinely get into taxi cabs and then tell the driver where they are going. Customers of railroads often ask for "50 box cars" without specifying where they are going.

We can handle random destinations much as we handled random travel times. Assume we have the same driver headed to Cleveland, with an expected time of arrival of 115. Assume that, at time 100, we decide to move the driver empty up to nearby Evanston to pick up a new load, but we do not know the destination of the load. The post-decision state would be "loaded in Evanston." By time $t = 200$, we have learned that the load was headed to Seattle, with an expected arrival time (at time 200) of 233. The sequence of pre- and post-decision states is given by

$$\begin{array}{cccc} \text{Label} & \begin{array}{c} t = 100 \\ \text{pre-decision} \end{array} & \begin{array}{c} t = 100 \\ \text{post-decision} \end{array} & \begin{array}{c} t = 200 \\ \text{pre-decision} \end{array} \\ \left( \begin{array}{c} Location \\ Status \\ ETA \end{array} \right) & \left( \begin{array}{c} Cleveland \\ E \\ 115 \end{array} \right) & \left( \begin{array}{c} Evanston \\ L \\ 132 \end{array} \right) & \left( \begin{array}{c} Seattle \\ E \\ 233 \end{array} \right). \end{array}$$

We again make a new decision to assign the driver, and use the value of the driver when we make this decision to update the value of the driver when he was loaded in Evanston (the previous post-decision state).

## 12.7 BIBLIOGRAPHIC REFERENCES

Section 12.1 - Piecewise linear function approximations have been proven to be convergent for scalar asset acquisition problems using piecewise linear, value function approximations. See Nascimento & Powell (2007*b*,*a*), and Nascimento et al. (2007). The numerical work in this section is based on Nascimento & Powell (2007*a*).

Section 12.2 - The material in this section is based on the undergraduate senior thesis of Lindsey Cant (Cant (2006)). The use of separable, piecewise linear approximations was developed in a series of papers arising in transportation and logistics (see, in particular, Godfrey & Powell (2001) and Powell et al. (2004)).

Section 12.3 - The material in this section is based on Basler (2006).

Section 12.4.2 - The notation in this section is based on a very general model for resource allocation given in Powell et al. (2001). The use of separable, piecewise linear value function approximations has been invested in Godfrey & Powell (2001), Godfrey & Powell (2002), and most thoroughly in Topaloglu & Powell (2006). The idea of using ridge regression to create value function approximations is due to Klabjan & Adelman (to appear).

Section 12.5 - This section is based on Godfrey & Powell (2002), Powell & Topaloglu (2004) and Topaloglu & Powell (2006). For a discussion of approximate dynamic programming for fleet management, using the vocabulary of stochastic programming, see Powell & Topaloglu (2003).

Section 12.6 - This section is based on Powell et al. (2002) and Spivey & Powell (2004).

## PROBLEMS

**12.1** Revise the notation for the blood management problem if we include the presence of testing that may show that inventories that have not yet reached the six week limit have gone bad. State any assumptions and introduce notation as needed. Be sure to model both pre- and post-decision states. Carefully describe the transition function.

**12.2** Consider the problem of managing a fleet of vehicles as described in section 12.5. There, we assumed that the travel time required to move a container from one location to another required one time period. Now we are going to ask you to generalize this problem.

(a) Assume that the time required to move from location $i$ to location $j$ is $\tau_{ij}$, where $\tau_{ij}$ is deterministic and integer. Give two different ways for modeling this by modifying the attribute $a$ in a suitable way (recall that if the travel time is one period, $a = \{location\}$).

(b) How does your answer to (a) change if $\tau_{ij}$ is allowed to take on noninteger values?

(c) Now assume that $\tau_{ij}$ is random, following a discrete uniform distribution (actually, it can be any general discrete distribution). How do you have to define the attribute $a$?

**12.3** **Orange futures project** - We have the opportunity to purchase contracts which allow us to buy frozen concentrated orange juice (FCOJ) at some point in the future. When we sign the contract in year $t$, we commit to purchasing FCOJ during year $t'$ at a specified price. In this project $t' = 10$, while $1 \leq t \leq 10$. To model the problem, let

$x_{tt'}$ = The quantity of FCOJ ordered at time $t$ (more specifically, with the information up through time $t$), to be used during year $t'$,

$x_t$ = $(x_{tt'})_{t' \geq t}$,

$R_{tt'}$ = The FCOJ that we know about at time $t$ that can be used during year $t'$,

$R_t$ = $(R_{tt'})_{t' \geq t}$,

$p_{tt'}$ = The price paid for FCOJ purchased at time $t$ that can be used during year $t'$,

$p_t$ = $(p_{tt'})_{t' \geq t}$,

$D_t$ = The demand for FCOJ during year $t$.

For this problem, $t' = 10$. The demand for FCOJ ($D_t$) does not become known until year 10, and is uniformly distributed between 1000 and 2000 tons. If you have not ordered enough to cover the demand, you must make up the difference by purchasing on the spot market, with price given by $p_{10,10}$.

We are going to specify a stochastic model for prices. For this purpose, let

$$\rho_0^u = \text{Initial upper range for prices} = 2.0,$$

$$\rho_0^\ell = \text{Initial lower range for prices} = 1.0,$$

$$\begin{aligned}
\rho^u &= \text{Upper range for prices} \\
&= 1.2, \\
\rho^\ell &= \text{Lower range for prices} \\
&= 0.9, \\
\beta &= \text{Mean reversion parameter} \\
&= 0.5, \\
\rho^{u,s} &= \text{Upper range for spot prices} \\
&= 1.15, \\
\rho^{\ell,s} &= \text{Lower range for spot prices.} \\
&= 1.03.
\end{aligned}$$

Let $U$ represent a random variable that is uniformly distributed between 0 and 1 (in Excel, this is computed using *RAND()*). Let $\omega = 0$ be the initial sample, and let $\omega = 1, 2, \ldots$ be the remaining samples. Let $p_{t,t'}(\omega)$ be a particular sample realization. Prices are randomly generated using

$$\begin{aligned}
p_{0,10}(\omega) &= 1.7(\rho_0^\ell + (\rho_0^u - \rho_0^\ell)U), \quad \omega \geq 0, \\
p_{t,10}(0) &= p_{t-1,10}(\rho^\ell + (\rho^u - \rho^\ell)U), \quad t = 1, 2, \ldots, 9, \\
p_{10,10}(\omega) &= p_{9,10}(\rho^{\ell,s} + (\rho^{u,s} - \rho^{\ell,s})U), \quad \omega \geq 0, \\
\bar{p}_{t,10}(0) &= p_{t,10}(0), \quad t \geq 0, \\
\bar{p}_{0,10}(\omega) &= .9\bar{p}_{0,10}(\omega - 1) + .05p_{0,10}(\omega), \quad \omega \geq 1, \\
\bar{p}_{t,10}(\omega) &= .9\bar{p}(t, 10)(\omega - 1) + .05p_{t-1,10}(\omega), \quad \omega \geq 1, \\
p_{t,10}(\omega) &= p_{t-1,10}(\rho^u - \rho^\ell)U + \beta(p_{t-1,10}(\omega) - \bar{p}_{t,10}(\omega)), \; t = 1, 2, \ldots, 9, \; \omega \geq 1.
\end{aligned}$$

This process should produce a random set of prices that tend to trend upwards, but which will move downward if they get too high (known as a mean reversion process).

(a) Prepare a detailed model of the problem.

(b) Design an approximate dynamic programming algorithm, including a value function approximation and an updating strategy.

(c) Implement your algorithm and describe the algorithmic tuning steps you had to go through to obtain what appears to be a useful solution. Be careful that your stepsize is not too small.

(d) Compare your results with different mean reversion parameters, such as $\beta = 0$ and 1.

(e) How does your answer change if the upper range for spot prices is increased to 2.0?

**12.4 Blood management project** - Implement the blood management model described in section 12.2. Assume that total demand averages 200 units of blood each week, and total supply averages 150 units of blood per week. Assume that the actual supplies and demands are normally distributed with a standard deviation equal to 30 percent of the mean (set any negative realizations to zero). Use the table below for the distribution of supply and demand (these are actuals for the United States). Use table 12.2 for your costs and contributions.

| Blood type | $AB+$ | $AB-$ | $A+$ | $A-$ | $B+$ | $B-$ | $O+$ | $O-$ |
|---|---|---|---|---|---|---|---|---|
| Percentage of supply | 3.40 | 0.65 | 27.94 | 5.17 | 11.63 | 2.13 | 39.82 | 9.26 |
| Percentage of demand | 3.00 | 1.00 | 34.00 | 6.00 | 9.00 | 2.00 | 38.00 | 7.00 |

(a) Develop a program that simulates blood inventories in steady state, using a discount factor of 0.80, where decisions are made myopically (the value function is equal to zero). Simulate 1000 weeks, and produce a plot showing your shortages each week. This is easiest to implement using a software environment such as Matlab. You will need access to a linear programming solver.

(b) Next implement an ADP strategy using separable, piecewise linear value function approximations. Clearly state any algorithmic choices you have to make (for example, the stepsize rule). Determine how many iterations appear to be needed in order for the solution to stabilize. Then run 1000 testing iterations and plot the shortages each week. Compare your results against the myopic policy. What types of behaviors do you notice compared to the myopic policy?

(c) Repeat part (b) assuming first that the average supply is 200 units of blood each week, and then again with the average set to 250. Compare your results.

**12.5** For the blood management problem, compare the results obtained using value functions trained for an infinite horizon problem against those trained for a 10-week horizon using a fixed set of starting inventories (make reasonable assumptions about these inventories). Compare your results for the finite horizon case using both sets of value functions (when using the infinite horizon value functions, you will be using the same value function for each time period). Experiment with different discount factors.

**12.6** **Transformer replacement project** - There are many industries which have to acquire expensive equipment to meet needs over time. This problem is drawn from the electric power industry (a real project), but the issues are similar to airlines that have to purchase new aircraft or manufacturers that have to purchase special machinery.

Electric power is transported from power sources over high-voltage lines, which drops to successively lower voltages before entering your house by using devices (familiar to all of us) called transformers. In the 1960's, the industry introduced transformers that could step power from 750,000 volts down to 500,000 volts. These high capacity transformers have a lifetime which they expected to be somewhere in the 50- to 80-year range. As of this writing, the oldest units are just reaching 40 years, which means there is no data on the actual lifetime. As a result, there is uncertainty about the lifetime of the units.

The problem is that the transformers cost several million dollars each, can take a year or more to build, and weigh over 200 tons (so they are hard to transport). If a transformer fails, it creates bottlenecks in the grid. If enough fail a blackout can occur, but the more common problem is that failures force the grid operators (these are distinct from the utilities themselves) to purchase power from more expensive plants.

| Age | $R_{0a}$ | Age | $R_{0a}$ | Age | $R_{0a}$ | Age | $R_{0a}$ |
|---|---|---|---|---|---|---|---|
| 1 | 2 | 11 | 2 | 21 | 1 | 31 | 12 |
| 2 | 1 | 12 | 2 | 22 | 1 | 32 | 0 |
| 3 | 0 | 13 | 1 | 23 | 2 | 33 | 8 |
| 4 | 9 | 14 | 0 | 24 | 7 | 34 | 10 |
| 5 | 2 | 15 | 1 | 25 | 2 | 35 | 16 |
| 6 | 4 | 16 | 1 | 26 | 3 | 36 | 6 |
| 7 | 2 | 17 | 0 | 27 | 1 | 37 | 2 |
| 8 | 7 | 18 | 4 | 28 | 1 | 38 | 6 |
| 9 | 1 | 19 | 2 | 29 | 2 | 39 | 10 |
| 10 | 6 | 20 | 1 | 30 | 7 | 40 | 0 |

**Table 12.4** Initial age distribution of the equipment

To model the problem, define

$x_t$ = The number of transformers ordered at the end of year $t$ that will arrive at the beginning of year $t+1$,

$R_{ta}$ = The number of transformers at time $t$ with age $a$,

$R_t^A$ = Total number of active transformers at the end of year $t$ $= \sum_a R_{ta}$,

$F_{ta}$ = A random variable giving the number of failures that occur during year $t$ of transformers with age $a$.

For \$2.5 million, the company can order a transformer in year $t$ to arrive in year $t+1$.

Table 12.4 gives the current age distribution (2 transformers are only 1 year old, while 10 are 39 years old). The probability $f(a)$ that a unit that is $a$ years old will fail is given by

$$f(a) = \begin{cases} \rho^f & \text{if age } a \leq \tau, \\ \rho^f(a-\tau) & \text{if age } a > \tau. \end{cases}$$

$\rho^f$ is the base failure rate for transformers with age $a \leq \tau$, where we assume $\rho^f = 0.01$ and $\tau = 40$. For transformers with age $a > \tau$ ($a$ is an integer number of years), the failure rate rises according to the polynomial given above.

At an aggregate level, the total network congestion costs (in millions of dollars per year) are given approximately by

$$C(R_t) = 5\left(\max(145 - R_t^A), 0\right)^2. \tag{12.34}$$

So if we have $R_t^A \geq 145$ active transformers, then congestion costs are zero.

Your challenge is to use approximate dynamic programming to determine how many transformers the company should purchase in each year. Your performance will be measured based on total purchase and congestion costs over a 50-year period averaged over 1000 samples. For this project, do the following

(a) Write up a complete model of the dynamic program.

(b) Describe at least one value function approximation strategy.

(c) Describe an updating method for your value function approximation. Specify your stepsize rule.

(d) Implement your algorithm. Describe the results of steps to tune your algorithm. Compare different strategies (stepsize rules, approximation strategies) and present your conclusion of the strategy that works the best.

(e) Now assume that you are not allowed to purchase more than 4 transformers in any given year. Compare the acquisition strategy to what you obtained without this constraint.

**12.7** We now extend the previous project by introducing uncertainty in the lifetime. Repeat exercise 12.6, but now assume that the lifetime $\tau$ is a random variable which might be 40, 45, 50, 55, or 60 with equal probability. Describe how this changes the performance of the model.

**12.8** We are going to revisit the transformer problem (holding the lifetime $\tau$ fixed at 40 years), but this time we are going to change the dynamics of the purchasing process. In exercise 12.6 we could purchase a transformer for \$2.5 million which would arrive the following year. Now, we are going to be able to purchase transformers for \$1.5 million, but they arrive in three years. That is, a transformer purchased at the end of year $t$ is available to be used in year $t+3$.

This is a significantly harder problem. For this version, it will be necessary to keep track of transformers which have been ordered but which have not yet arrived. This can be handled by allowing the age $a$ to be negative. So, $R_{t,-2}$ would be the number of transformers we know about at time $t$ which will first be available in 2 time periods.

In addition, it will be necessary to choose between ordering a transformer quickly for a higher price, or more slowly at a lower price. You will need to design an approximation strategy that handles the different types of transformers (consider the use of separable, piecewise linear value function approximations as we did for the blood management problem).

Repeat exercise 12.6, placing special care on the modeling of the state variable, the decision variable and the transition function. Instead of transformers arriving to a single inventory, they now arrive to the system with different ages. Try starting with a separable, piecewise-linear value function approximation which captures the value of transformers of different ages. However, try experimenting with other functional approximations. For example, you might have a single value function that uses the total number of transformers that are not only available, but which also have yet to arrive. Report your experience, and compare your results to the original problem when a transformer arrived in the next time period.

# CHAPTER 13

# IMPLEMENTATION CHALLENGES

So you are finally ready to solve your own problem with approximate dynamic programming. Of course, it is nice if you have a problem that closely matches one that we have already discussed in this volume. But you don't. As is often the case, you have to fall back on general principles to solve a problem that does not look like any of the examples that we have presented.

In this chapter, we describe a number of issues that tend to arise in real applications of approximate dynamic programming. These are intended to help as a rough guide to what can be expected in the path to a successful ADP application.

## 13.1 WILL ADP WORK FOR YOUR PROBLEM?

The first question you have to ask is: do you need approximate dynamic programming to solve your problem? There are many problems where myopic policies are going to work quite well (in special cases, they may be provably optimal). For example, the optimal solution to steady-state batch replenishment problems are known to have the structure where if the amount of product on hand is less than $s$, then we should order an amount that brings our inventory up to $S$ (these are known as $(s, S)$ policies). If we are managing a group of identical repairmen serving jobs that pop up randomly (and uniformly) over a region, then it is also unlikely that a value function that helps us look into the future will noticeably improve the solution.

In short, we have to ask the question, Does a value function add value? If we estimate a value function, how do we think our solution will improve?

Problems where myopic policies work well (and in particular where they are provably optimal) tend to be quite simple (how much product to order, when to sell an asset). There are, of course, more complex problems where myopic policies are known to be optimal. For example, the problem of allocating funds among a set of asset classes to balance risk can be solved with a myopic policy if we assume that there are no transaction costs (or transaction times). In fact, it should not be surprising to find out that if it is possible to move from any state to any other state (instantly and with no cost), then a myopic policy will be optimal. Not surprisingly, ADP is not going to contribute very much to these problems.

By contrast, the hardest ADP problems tend to be applications where it takes numerous steps to obtain a reward. The easiest examples of such problems are games (chess, checkers, Connect-4) where many steps have to be made before we know if we won or lost. ADP works particularly well when we have access to a policy that works "pretty well" which we can use to train value functions. A well designed approximate value function can make a good policy even better. But if we have an intractably hard problem, we may have to start with naive rules which work poorly. We may avoid a decision to visit a state if we do not know how to properly behave once we reach the state. It is these problems where it is often necessary to turn to an outside supervisor who can guide the search algorithm during initial learning stages.

## 13.2 DESIGNING AN ADP ALGORITHM FOR COMPLEX PROBLEMS

There are numerous problems in discrete optimization which have to be solved over time, under uncertainty. For example, it may be necessary to decide which jobs should be assigned to different machines, and in which order, in the presence of random arrivals of jobs in the future. Service vehicles have to be assigned to customer demands (picking up packages, or providing household services such as plumbing) which arise randomly over time. Transportation companies have to dispatch trucks with shipments which arrive randomly to a terminal.

Deterministic versions of these problems (where all customer demands are known in advance) can be extremely difficult, and are generally solved using heuristics. Stochastic, dynamic versions of these problems are generally solved using simulation, where at time $t$ we would solve an optimization problem using only what is known at time $t$. As new information becomes known, the problem would be re-optimized. Solving difficult integer programs in real-time can be quite challenging, primarily because of the time constraints on solution times. But the solution quality can be relatively poor since the decisions do not reflect what might happen in the future.

The first step in developing an ADP strategy for these problems is to begin by implementing a myopic policy in the form of a simulation. Consider a problem faced by the military in the scheduling of unmanned aerial vehicles (UAV's) which have to be routed to collect information about various targets. A target is a request to visit an area to determine its status (Was a bridge destroyed? Is a missile battery operational? Are there people near a building?). Targets arise randomly, and sometimes it is necessary to plan the path of a UAV through a sequence of targets in order to determine which one should be visited first. The problem is illustrated in figure 13.1, which shows UAV's both on the ground and in the air, with paths planned through a series of known targets. In addition, there is a series of potential targets (shown as question marks) that might arise.

One way to model this problem is to assume that we are going to optimally route a fleet of UAV's through a sequence of known targets. This problem can be formulated as an integer

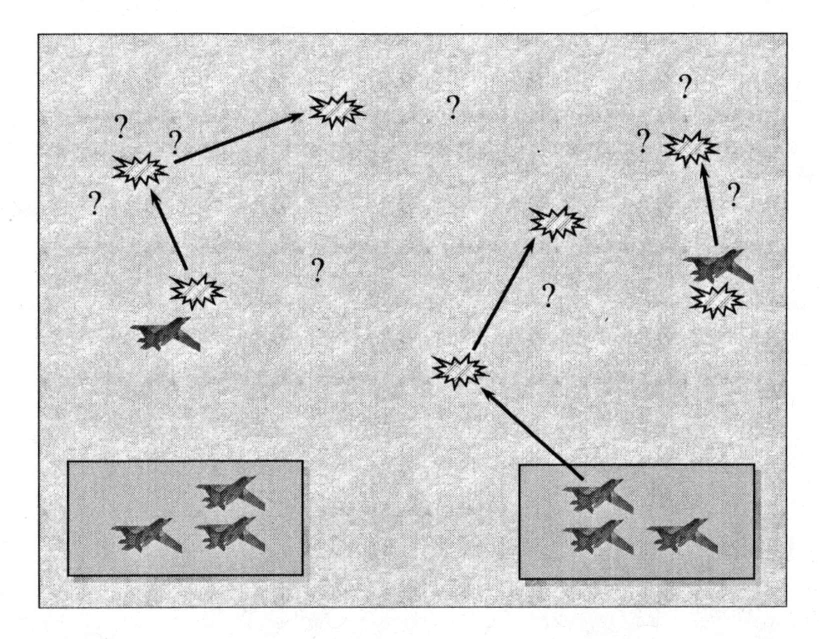

**Figure 13.1** Dynamic routing of UAV's, showing known targets and potential targets.

program and solved using commercial optimization packages as long as the problem is not too large. The biggest complication arises if the tours (the number of targets that a single UAV might cover in a single solution) become long. For example, it is possible to handle problems with thousands of UAV's if they are restricted to covering at most one target at a time. However, the problem becomes much more complex with tours of as few as three or four targets. If we have tours of five or ten targets, we may have to turn to heuristic algorithms that give good, but not necessarily optimal, solutions. These heuristics typically involve starting with one solution, and then testing alternative solutions to try to discover improvements. The simplest are local search algorithms (try making one or more changes, if the solution improves, keep the new solution and proceed from there). These algorithms fall under names such as local search, neighborhood search and tabu search.

Assume that we have devised a strategy for solving the problem myopically. It is quite likely that we will not be entirely satisfied with the behavior of the solution. For example, the system might assign two UAV's to serve two targets in an area that is not likely to produce additional targets in the future. Alternatively, the system might try to assign one UAV to cover two targets in an area that is likely to produce a number of additional targets in the future (too many to be covered by a single UAV). We might also want to consider attributes such as fuel level. We might prefer a UAV with a high fuel level if we expect other demands to arise. If there is little chance of new demands, we might be satisfied assigning a UAV that is low on fuel with the expectation that it will finish the task and then return home.

It is very important when designing an ADP strategy to make sure that we have a clear idea of the behaviors we are trying to achieve which are not being produced with a myopic policy. With this in mind, the next step is to design a value function approximation. The choice of function depends on the type of algorithmic strategy we are using for the myopic problem. If we wish to solve the myopic problem using integer programming, then a nonlinear value function approximation is going to cause complications (now we have a nonlinear, integer programming problem). If we use a local search procedure, then the value function approximation can take on almost any form.

Once we have designed a value function approximation, we have to make sure that we are going to be able to estimate it using information available from the solution procedure. For example, in chapter 11 we showed how some value function approximations can be approximated using gradient information, which means we need to be able to estimate the marginal value of, say, a UAV in a particular region. If we constrain ourselves to assigning a UAV to at most one target at a time, our myopic problem is the same as the dynamic assignment problem we considered in section 12.6, where it is quite easy to obtain marginal values from dual variables. Local search heuristics, on the other hand, do not yield gradient information very easily. As an alternative, we might approximate the value of a UAV as the contribution earned over the tour to which it has been assigned. This can be thought of as the average value of a UAV instead of the marginal value. Not as useful, but perhaps better than nothing.

## 13.3 DEBUGGING AN ADP ALGORITHM

Imagine that you now have your algorithm up and running. There are several strategies for deciding whether it works well:

1. Plotting the objective function over the iterations.

2. Evaluating one or more performance statistics over the iterations.

3. Subjectively evaluating the behavior of your system after the algorithm has completed.

Ideally, the objective function will show generally steady improvement (keep in mind the behavior in figure 13.2). However, you may find that the objective function gets steadily worse, or wanders around with no apparent improvement. Alternatively, you may find that important performance measures that you are interested in are not improving as you hoped or expected. Explanations for this behavior can include:

1. If you are using a forward pass, you may be seeing evidence of a "bounce." You may need to simply run more iterations.

2. You may have a problem where the value function is not actually improving the decisions. Your myopic policy may be doing the best that is possible. You may not have a problem that benefits from looking into the future.

3. Be careful that you are not using a stepsize that is going to zero too quickly.

4. Your value function may be a poor approximation of the true value function. You may not be properly capturing the structure of the problem.

5. You may have a reasonable value function, but the early iterations may have produced a poor estimate, and you are having difficulty recovering from this initial estimate.

6. You may not be updating the value function properly. Carefully verify that $\hat{v}_t^n$ (whether it be the value of being in a state, or a vector of derivatives) is being correctly calculated. This is a common error.

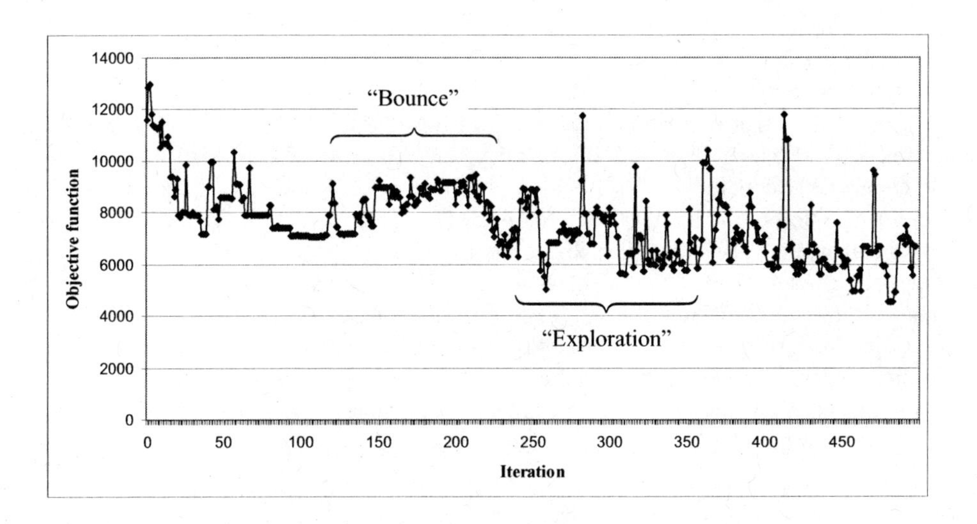


**Figure 13.2** Objective function of an approximate dynamic programming algorithm.

It is often easiest when you can see specific decisions that you simply feel are wrong. If you are solving your decision problems properly (this is not always the case) then presumably the solution is due to an incorrect value function approximation. One problem is that $\bar{V}_t(S_t^x)$ may reflect a number of updates over many iterations. For this reason, it is best to look at $\hat{v}_t^n$. If it takes the algorithm to a poor state, is $\hat{v}_t^n$ small? If not, why do you think it is a good state? If it is clearly a bad state, why does it appear that $\hat{v}_t^n$ is large? If $\hat{v}_t^n$ is reasonable, perhaps the problem is in how the value function approximation is being updated (or the structure of the value function itself).

## 13.4 CONVERGENCE ISSUES

One of the most significant challenges with approximate dynamic programming is determining when to stop. In section 6.7, we already encountered the problem of "apparent convergence" when the objective function appeared to have leveled off after 100 iterations, suggesting that we could stop the algorithm. In fact, the algorithm made considerably more progress when we allowed it to go 1000 iterations. This is one reason that we have to be very careful not to allow the stepsizes to decline too quickly. Also, before we decide that "100 iterations is enough" we should first do a few runs with substantially more iterations.

Apparent convergence, however, is not our only headache. Figure 13.2 shows an example of the objective function (total costs, which we are minimizing) for a problem using a single-pass approximate dynamic programming algorithm. This graph shows a behavior that is fairly common when using a single-pass algorithm which we refer to as the "bounce." When using a single-pass algorithm, the value function for time $t$, iteration $n$ reflects activities from time $t + s$ at iteration $n - s$, since values are passed backward one time period for each iteration. The result can be value function approximations that are simply incorrect, producing solutions that steadily get worse for a period of time.

Figure 13.2 also illustrates some other behaviors that are not uncommon. After the bounce, the algorithm goes through a period of exploration where little progress is made. While this does not always happen, it is not uncommon for algorithms to go through periods of trying different options and slowly learning the effectiveness of these options. This behavior can be particularly pronounced for problems where there is little structure to guide the solution.

Finally, the algorithm starts to settle down in the final 100 iterations, but even here there are short bursts where the solution becomes much worse. At this stage, we might be comfortable concluding that the algorithm has probably progressed as far as it can, but we have to be careful not to stop on one of the poor solutions. This can be avoided in several ways. The most expensive option is to stop every $K$ iterations and perform repeated samples. This strategy allows us to make strong statistical statements about the quality of the solution, but this can be expensive. Another approach is to simply smooth the objective function using exponential smoothing, and simultaneously compute an estimate of the standard deviation of the solution. Then, simple rules can be designed to stop the algorithm when the solution is, for example, one standard deviation below the estimated mean (if we want to hope for a good solution).

## 13.5 MODELING YOUR PROBLEM

There is a long history in the optimization community to first present a complete optimization model of a problem, after which different algorithms may be discussed. A mathematical programming problem is not considered properly stated unless we have properly defined all the variables, constraints and the objective function. The same cannot be said of the approximate dynamic programming community, where a more casual presentation style has evolved. It is not unusual for an author to describe the size of the state space without actually defining a state variable. Transition functions are often ignored, and there appear to be many even in the research community who are unable to write down an objective function properly.

The culture of approximate dynamic programming is closer to that of simulation where it is quite common to see papers without any notation at all. Such papers are often describing fairly complex problems of physical processes that are well understood by those working in the same area of application. Mathematical models of such processes can be quite cumbersome, and have not proved necessary for people who are familiar with the problem class. There are many papers in approximate dynamic programming which have evolved out of the simulation community, and have apparently adopted this minimalist culture for modeling.

ADP should be viewed as a potentially powerful took for the simulation community. However, in a critical way it should be viewed quite differently. Most simulation papers are trying to mimic a physical process rather than optimize it. The most important part of simulation models are captured by the transition function, the details of which may not be particularly interesting. But when we use approximate dynamic programming, we are often trying to optimize (over time) complex problems that might otherwise be solved using simulation.

To properly describe a dynamic program (in preparation for using approximate dynamic programming), we need to specify the value function approximation and the updating strategies used to estimate it. We need to evaluate the quality of a solution by comparing objective functions. These steps require a proper description of the state variable (by far the most important quantity in a dynamic program), the decision variables, exogenous information and the contribution function. But as with many simulation problems, it is not always clear that we need the transition function in painstaking detail. If we formulate a math programming problem, the transition function would be buried in the constraints. When modeling a dynamic system, it is well known that the transition function captures the potentially complex physics of a problem. For simple problems, there is no reason not to completely specify these equations, especially if we need to take advantage of the structure of the transition function. For more complex problems, modeling the transition function can be cumbersome, and the physics tends to be well known by people working with an application class.

If you wish to develop an approximate dynamic programming algorithm, it is important to learn to express your problem mathematically. There is a strong culture in the deterministic optimization community to write out the problem in its entirety using linear algebra. By contrast, there is a vast array of simulation problems which are expressed without any mathematics at all. Approximate dynamic programming sits between these two fields. It is simply inappropriate to solve a problem using ADP without expressing the problem mathematically. At the same time, ADP allows us to address problems with far more complexity than are traditionally handled using tools such as linear programming. For this reason, we suggest the following guidelines when modeling a problem:

- The state variable. Clearly identify the dimensions of the state variable using explicit notation. This is the mechanism by which we can understand the complexity and structure of your problem.

- The decision variable. We need to know what dimensions of the problem we are controlling. It is often convenient to describe constraints on decisions that apply at a point in time (rather than over time).

- The exogenous information process. What information is arriving to our system from exogenous sources? What do we know (if anything) about the probability law describing what information might arrive?

- The contribution function. Also known as the cost function (minimizing), the reward function (maximizing), or the utility function (usually maximizing). We need to understand what goal we are trying to achieve.

- The transition function. If the problem is relatively simple, this should be stated mathematically in its entirety. But as the problem becomes more complicated, a more descriptive presentation may be appropriate. For many problems, the details of the transition function are not central to the development of an approximate dynamic programming algorithm. Also, as problems become more complex, a complete mathematical model of the transition function can become tedious. Obviously, the right level of detail depends on the context and what we are trying to accomplish.

- The objective function. This is where you specify how you are going to evaluate contributions over time (finite or infinite horizon? Discounted or average reward?).

Thus, we feel that every component should be stated explicitly using mathematical notation with the single major exception of the transition function (which captures all the physics of the problem). Some problems are simply too complex, and for these problems, the structure of the transition function may not be relevant to the design of the solution algorithm. A transition function might be a single, simple equation, but it can also be thousands of lines of code. The details of the transition function are generally not that interesting if the focus is on the development of an ADP algorithm.

When writing up your model, pay particular attention to the modeling of time. Keep in mind that dynamic programs are solved at discrete points in time that we call decision epochs. Variables indexed at time $t$ are being measured at time $t$, which determines the information that is available when we make a decision. It does *not* imply that physical activities are also taking place at these points in time. We can decide at time 100 to assign a technician to perform a task at time 117. It may be that we will not make the next decision until time 200 (or time 143, when the next phone call comes in). If you are modeling a deterministic problem, these issues are less critical. But for stochastic problems, the modeling of information (what we know when we make a decision) needs to be separated from the modeling of physical processes (when something happens).

## 13.6 ON-LINE VS. OFF-LINE MODELS

When testing approximate dynamic programming algorithms, it is most common to perform the testing in an "off-line" setting. In this mode (which describes virtually every algorithm in this book) we start at time $t = 0$ and step forward in time, using information realizations (which we have denoted $W(\omega)$) that are either generated randomly, drawn from a file of sample paths (for example, prices of stocks), or sampled from a known probability distribution.

In an on-line setting, the model is running in real time alongside an actual physical process. The process might be a stock price, an inventory at a store, or a problem in engineering such as flying an aircraft or controlling the temperature of a chemical reaction. In such settings, we may observe the exogenous information process (such as a stock price) or we may even depend on physical observations to observe state transitions. When we observe exogenous information, we do not need to estimate probability distributions for the random variables (such as the change in the stock price). These are sometimes referred to as distribution-free models. If we make a decision and then observe the state transition, then we do not need a mathematical transition function. This is referred to as model-free (or sometimes "direct") approximate dynamic programming.

There are two important cases of on-line applications: steady state and transient. In a steady-state situation, we do not have to forecast future information, but instead need only to use live information to update the process. We can use live information to drive our ADP algorithm. For example, consider a steady-state version of the asset acquisition problem we solved in section 12.1. Assume that our asset is a type of DVD player (a popular theft item). If we have $R_t$ DVD's in inventory, we use some rule to decide to order an amount $x_t$. Assume now that we observe that we sold $\hat{D}_t$ units, allowing us to compute a profit

$$C_t(x_t) = p\hat{D}_t - cx_t,$$

where $p$ is the price at which we sell the units, and $c$ is the wholesale cost of new units purchased at time $t$. Note that $\hat{D}_t$ is sales (rather than actual demand). $\hat{D}_t$ is limited by the measured inventory $R_t$. We let $R_t^x = R_t - \hat{D}_t + x_t$ be our post-decision state, which is how much we think we have left over (but excluding loss due to theft). After this, we measure the inventory $R_{t+1}$ which differs from $R_t^x$ because of potential theft.

We can use the methods in chapter 11 to help estimate an approximate value function $\bar{V}(R_t^x)$. We would then make decisions using

$$x_t = \arg\max \big(C_t(x_t) + \gamma\bar{V}(R_t^x)\big).$$

Now assume that we have a nonstationary problem, where demands might be steadily increasing or decreasing, or might follow some sort of cyclic pattern (for example, hourly, weekly or monthly cycles, depending on the application). We might use a time series model with the form

$$F_{tt'} = \theta_{t0} + \sum_{c \in \mathcal{C}} \theta^{cal}_{tc} X_{t'c} + \sum_{\tau=1}^{\tau^M} \theta^{time}_{t\tau} p_{t-\tau}$$

where $\mathcal{C}$ is a set of calendar effects (for example, the seven days of the week), $X_{t'c}$ is an indicator variable (for example, which is equal to 1 if $t'$ falls on a Monday), $\theta^{cal}$ is the calendar parameters, $p_{t-\tau}$ is the price $\tau$ time periods ago, and $\theta^{time}_{t\tau}$ is the impact of the price $\tau$ time periods ago on the model. The precise specification of our forecast model is not important right now, but we observe that it is a function of a series of past prices, so we can define a price state variable $S^p = (p_{t-1}, \ldots, p_{t-\tau^M})$ and a series of parameters $\theta_t = (\theta_{t0}, (\theta^{cal}_{tc})_{c \in \mathcal{C}}, (\theta^{time}_{t\tau})_\tau)$ which are updated after each time period. Assume that we have devised a system for updating the parameter vector $\theta_t$ after we observe the demands $\hat{D}_t$. Further assume that in addition to our point estimate $F_{tt'}$, we update our estimate of the variance $\bar{\sigma}^2_{t,\tau}$ which is an estimate of the variance of the difference between $\hat{D}_t$ and $F_{t'',t}$ for each $\tau = t - t''$ (presumably, for a certain number of time periods into the past).

In the presence of transient demands, to determine the decision $x_t$ we have to set up and solve a dynamic program over the time periods $t, t+1, \ldots, t+\tau^{ph}$, where $\tau^{ph}$ is a specified planning horizon. This means we have to use our forecast model $F_{tt'}$ over $t' = t, t+1, \ldots, t+\tau^{ph}$, along with our estimate of the variance of $F_{tt'}$, given by $\bar{\sigma}^2_{t,\tau}$ for $\tau = 1, \ldots, \tau^{ph}$. To run an ADP algorithm, we have to assume a distribution (for example, normal with mean $F_{tt'}$ and variance $\bar{\sigma}^2_{t,t'-t}$) to generate random observations of demands.

## 13.7 IF IT WORKS, PATENT IT!

A successful ADP algorithm for an important problem class is, we believe, a patentable invention. While we continue to search for the holy grail of a general class of algorithms that work reliably on all problems, we suspect the future will involve fairly specific problem classes with a well-defined structure. For such problems, we can, with some confidence, perform a set of experiments that can determine whether an ADP algorithm solves the problems that are likely to come up in that class. Of course, we encourage developers to keep these results in the public domain, but our point is that these represent specific technological breakthroughs which could form the basis of a patent application.

# Bibliography


Adelman, D. (2003), 'Price-directed replenishment of subsets: Methodology and its application to inventory routing', *Manufacturing & Service Operations Management* **5**(4), 348–371.

Adelman, D. (2004), 'A price-directed approach to stochastic inventory/routing', *Operations Research* **52**(4), 499–514.

Adelman, D. (to appear), 'On approximate dynamic programming and bid-price controls in revenue management', *Operations Research.*

Andreatta, G. & Romeo, L. (1988), 'Stochastic shortest paths with recourse', *Networks* **18**, 193–204.

Arrow, K., Harris, T. & Marschak, T. (1951), 'Optimal inventory policy', *Econometrica.*

Banks, J., J. S. Carson, I. & Nelson, B. L. (1996), *Discrete-Event System Simulation*, second edn, Prentice-Hall, Inc., Englewood Cliffs, N.J.

Barto, A. G., Bradtke, S. J. & Singh, S. P. (1995), 'Learning to act using real-time dynamic programming', *Artificial Intelligence, Special Volume on Computational Research on Interaction and Agency* **72**, 81–138.

Basler, J. T. (2006), 'Optimal portfolio rebalancing: an approximate dynamic programming approach', Senior thesis, Department of Operations Research and Financial Engineering, Princeton University.

Bean, J., Birge, J. & Smith, R. (1987), 'Aggregation in dynamic programming', *Operations Research* **35**, 215–220.

Bellman, R. (1957), *Dynamic Programming*, Princeton University Press, Princeton.

Bellman, R. (1971), *Introduction to the Mathematical Theory of Control Processes, Vol. II*, Academic Press, New York.

Bellman, R. & Dreyfus, S. (1959), 'Functional approximations and dynamic programming', *Mathematical Tables and Other Aids to Computation* **13**, 247–251.

Bellman, R. & Kalaba, R. (1959), 'On adaptive control processes', *OIRE Trans.* **4**, 1–9.

Bellman, R., Glicksberg, I. & Gross, O. (1955), 'On the optimal inventory equation', *Management Science* **1**, 83–104.

Benveniste, A., Metivier, M. & Priouret, P. (1990), *Adaptive Algorithms and Stochastic Approximations*, Springer-Verlag, New York.

Berry, D. A. & Fristedt, B. (1985), *Bandit Problems*, Chapman and Hall, London.

Bertsekas, D. (2000), *Dynamic Programming and Optimal Control*, Athena Scientific, Belmont, MA.

Bertsekas, D. & Castanon, D. (1989), 'Adaptive aggregation methods for infinite horizon dynamic programming', *IEEE Transactions on Automatic Control* **34**(6), 589–598.

Bertsekas, D. & Castanon, D. (1999), 'Rollout algorithms for stochastic scheduling problems', *J. Heuristics* **5**, 89–108.

Bertsekas, D. & Tsitsiklis, J. (1991), 'An analysis of stochastic shortest path problems', *Mathematics of Operations Research* **16**, 580–595.

Bertsekas, D. & Tsitsiklis, J. (1996), *Neuro-Dynamic Programming*, Athena Scientific, Belmont, MA.

Bertsekas, D. P. (1982), 'Distributed dynamic programming', *IEEE Transactions on Automatic Control* **27**(3), 610–616.

Bertsekas, D. P. & Tsitsiklis, J. N. (1989), *Parallel and Distributed Computation: Numerical Methods*, Prentice-Hall, Inc., Englewood Cliffs.

Bertsekas, D. P., Borkar, V. S. & Nedic, A. (2004), Improved temporal difference methods with linear function approximation, *in* J. Si, A. G. Barto, W. B. Powell & D. Wunsch, eds, 'Handbook of Learning and Approximate Dynamic Programming', IEEE Press, New York, pp. 233–257.

Birge, J. (1985), 'Decomposition and partitioning techniques for multistage stochastic linear programs', *Operations Research* **33**(5), 989–1007.

Birge, J. & Louveaux, F. (1997), *Introduction to Stochastic Programming*, Springer-Verlag, New York.

Blum, J. (1954*a*), 'Approximation methods which converge with probability one', *Annals of Mathematical Statistics* **25**, 382–386.

Blum, J. (1954*b*), 'Multidimensional stochastic approximation methods', *Annals of Mathematical Statistics* **25**, 737–744.

Boutilier, C., Dean, T. & Hanks, S. (1999), 'Decision-theoretic planning: Structural assumptions and computational leverage', *Journal of Artificial Intelligence Research* **11**, 1–94.

Boyan, J. (2002), 'Technical update: Least-squares temporal difference learning', *Machine Learning* **49**, 1–15.

Bradtke, S. J. & Barto, A. G. (1996), 'Linear least-squares algorithms for temporal difference learning', *Machine Learning* **22**, 33–57.

Brown, R. (1959), *Statistical Forecasting for Inventory Control*, McGraw-Hill, New York.

Brown, R. (1963), *Smoothing, Forecasting and Prediction of Discrete Time Series*, Prentice-Hall, Englewood Cliffs, N.J.

Cant, L. (2006), 'Life saving decisions: A model for optimal blood inventory management', Senior thesis, Department of Operations Research and Financial Engineering, Princeton University.

Chang, H., Fu, M., Hu, J. & Marcus, S. (2007), *Simulation-Based Algorithms for Markov Decision Processes*, Springer, New York.

Chen, Z.-L. & Powell, W. (1999), 'A convergent cutting-plane and partial-sampling algorithm for multistage linear programs with recourse', *Journal of Optimization Theory and Applications* **103**(3), 497–524.

Cheung, R. K.-M. & Powell, W. B. (2000), 'SHAPE: A stochastic hybrid approximation procedure for two-stage stochastic programs', *Operations Research* **48**(1), 73–79.

Choi, D. P. & Van Roy, B. (2006), 'A generalized Kalman filter for fixed point approximation and efficient temporal-difference learning', *Discrete Event Dynamic Systems* **16**, 207–239.

Chong, E. (1991), On-line stochastic optimization of queueing systems, Ph.D. dissertation, Department of Electrical Engineering, Princeton University.

Chow, G. (1997), *Dynamic Economics*, Oxford University Press, New York.

Chung, K. (1974), *A Course in Probability Theory*, Academic Press, New York.

Clark, C. (1961), 'The greatest of a finite set of random variables', *Operations Research* **9**, 145–163.

Clement, E., Lamberton, D. & Protter, P. (2002), 'An analysis of a least squares regression method for American option pricing', *Finance and Stochastics* **17**, 448–471.

Coles, S. G. (2001), *An introduction to statistical modeling of extreme values*, Springer, New York.

Crites, R. & Barto, A. (1994), 'Elevator group control using multiple reinforcement learning agents', *Machine Learning* **33**, 235–262.

Dantzig, G. & Ferguson, A. (1956), 'The allocation of aircrafts to routes: An example of linear programming under uncertain demand', *Management Science* **3**, 45–73.

Darken, C. & Moody, J. (1991), Note on learning rate schedules for stochastic optimization, *in* R. P. Lippmann, J. Moody & D. S. Touretzky, eds, 'Advances in Neural Information Processing Systems 3', pp. 1009–1016.

Darken, C. & Moody, J. (1992), Towards faster stochastic gradient search, *in* J. Moody, D. L. Hanson & R. P. Lippmann, eds, 'Advances in Neural Information Processing Systems 4', pp. 1009–1016.

Darken, C., Chang, J. & Moody, J. (1992), 'Learning rate schedules for faster stochastic gradient search', *Neural Networks for Signal Processing 2 - Proceedings of the 1992 IEEE Workshop*.

Dayan, P. (1992), 'The convergence of TD($\lambda$) for general $\lambda$', *Machine Learning* **8**, 341–362.

de Farias, D. & Van Roy, B. (2003), 'The linear programming approach to approximate dynamic programming', *Operations Research* **51**(6), 850–865.

de Farias, D. & Van Roy, B. (2004), 'On constraint sampling in the linear programming approach to approximate dynamic programming', *Mathematics of Operations Research* **29**(3), 462–478.

de Farias, D. P. & Van Roy, B. (2000), 'On the existence of fixed points for approximate value iteration and temporal-difference learning', *Journal of Optimization Theory and Applications* **105**, 1201–1225.

DeGroot, M. H. (1970), *Optimal Statistical Decisions*, John Wiley and Sons.

Denardo, E. V. (1982), *Dynamic Programming*, Prentice-Hall, Englewood Cliffs, NJ.

Derman, C. (1962), 'On Sequential Decisions and Markov Chains', *Management Science* **9**(1), 16–24.

Derman, C. (1970), *Finite State Markovian Decision Processes*, Academic Press, New York.

Doob, J. L. (1953), *Stochastic Processes*, John Wiley & Sons, New York.

Douglas, S. & Mathews, V. (1995), 'Stochastic gradient adaptive step size algorithms for adaptive filtering', *Proc. International Conference on Digital Signal Processing, Limassol, Cyprus* **1**, 142–147.

Drake Jr., D. E. (2005), Overcoming overestimation error in approximate adaptive dynamic programming algorithms for the dynamic vehicle allocation problem with driver domiciles, Master's thesis in industrial engineering, Lehigh University.

Dreyfus, S. & Law, A. M. (1977), *The Art and Theory of Dynamic Programming*, Academic Press, New York.

Dvoretzky, A. (1956), On stochastic approximation, *in* J. Neyman, ed., 'Proceedings $3^{rd}$ Berkeley Symposium on Mathematical Statistics and Probability', University of California Press, pp. 39–55.

Dvoretzky, A., Kiefer, J. & Wolfowitz, J. (1952), 'The inventory problem I, II', *Econometrica* **20**, 187–222.

Dynkin, E. B. (1979), *Controlled Markov Processes*, Springer-Verlag, New York.

Ermoliev, Y. (1983), 'Stochastic quasigradient methods and their application to system optimization', *Stochastics* **9**, 1–36.

Ermoliev, Y. (1988), Stochastic quasigradient methods, *in* Y. Ermoliev & R. Wets, eds, '*Numerical Techniques for Stochastic Optimization*', Springer-Verlag, Berlin.

Even-Dar, E. & Mansour, Y. (2004), 'Learning rates for q-learning', *Journal of Machine Learning Research* **5**, 1–25.

Even-Dar, E. & Mansour, Y. (2007), 'Learning rates for Q-learning', *Journal of Machine Learning Research* **5**, 1–25.

Fan, J. (1996), *Local Polynomial Modelling and Its Applications*, Chapman & Hall/CRC.

Ford, L. & Fulkerson, D. (1962), *Flows in Networks*, Princeton University Press, Princeton, N.J.

Frank, H. (1969), 'Shortest paths in probabilistic graphs', *Operations Research* **17**, 583–599.

Frazier, P. & Powell, W. B. (2007), Approximate value iteration converges slowly when smoothed with a $1/n$ stepsize, Technical report, Princeton University.

Frazier, P., Powell, W. B. & Dayanik, S. (2007), A knowledge gradient policy for sequential information collection, Working Paper.

Frieze, A. & Grimmet, G. (1985), 'The shortest path problem for graphs with random arc lengths', *Discrete Applied Mathematics* **10**, 57–77.

Gaivoronski, A. (1988), Stochastic quasigradient methods and their implementation, *in* Y. Ermoliev & R. Wets, eds, 'Numerical Techniques for Stochastic Optimization', Springer-Verlag, Berlin.

Gardner, E. S. (1983), 'Automatic monitoring of forecast errors', *Journal of Forecasting* **2**, 1–21.

George, A. (2005), Optimal Learning Strtegies for Multi-Attribute Resource Allocation Problems, PhD thesis, Princeton University.

George, A. & Powell, W. B. (2006), 'Adaptive stepsizes for recursive estimation with applications in approximate dynamic programming', *Machine Learning* **65**(1), 167–198.

George, A., Powell, W. B. & Kulkarni, S. (2005), Value function approximation using hierarchical aggregation for multi-attribute resource management, Technical report, Princeton University, Department of Operations Research and Financial Engineering.

Giffin, W. (1971), *Introduction to Operations Engineering*, R. D. Irwin, Inc., Homewood, IL.

Gittins, J. (1979), 'Bandit processes and dynamic allocation indices', *Journal of the Royal Statistical Society, Series B* **14**, 148–177.

Gittins, J. (1981), 'Multiserver scheduling of jobs with increasing completion times', *Journal of Applied Probability* **16**, 321–324.

Gittins, J. (1989), *Multi-Armed Bandit Allocation Indices*, John Wiley and Sons, New York.

Gittins, J. C. & Jones, D. M. (1974), A dynamic allocation index for the sequential design of experiments, *in* J. Gani, ed., 'Progress in Statistics', pp. 241–266.

Gladyshev, E. G. (1965), 'On Stochastic Approximation', *Theory of Probability and its Applications* **10**, 275–278.

Godfrey, G. & Powell, W. B. (2002), 'An adaptive, dynamic programming algorithm for stochastic resource allocation problems I: Single period travel times', *Transportation Science* **36**(1), 21–39.

Godfrey, G. A. & Powell, W. B. (2001), 'An adaptive, distribution-free approximation for the newsvendor problem with censored demands, with applications to inventory and distribution problems', *Management Science* **47**(8), 1101–1112.

Golub, G. H. & Loan, C. F. V. (1996), *Matrix Computations*, John Hopkins University Press, Baltimore, MD.

Goodwin, G. C. & Sin, K. S. (1984), *Adaptive Filtering and Control*, Prentice-Hall, Englewood Cliffs, NJ.

Guestrin, C., Koller, D. & Parr, R. (2003), 'Efficient solution algorithms for factored MDPs', *Journal of Artificial Intelligence Research* **19**, 399–468.

Hastie, T., Tibshirani, R. & Friedman, J. (2001), *The Elements of Statistical Learning*, Springer series in Statistics, New York, NY.

Haykin, S. (1999), *Neural Networks: A Comprehensive Foundation*, Prentice Hall.

Heuberger, P. S. C., den Hov, P. M. J. V. & Wahlberg, B., eds (2005), *Modeling and Identification with Rational Orthogonal Basis Functions*, Springer, New York.

Heyman, D. & Sobel, M. (1984), *Stochastic Models in Operations Research, Volume II: Stochastic Optimization*, McGraw Hill, New York.

Higle, J. & Sen, S. (1991), 'Stochastic decomposition: An algorithm for two stage linear programs with recourse', *Mathematics of Operations Research* **16**(3), 650–669.

Holt, C., Modigliani, F., Muth, J. & Simon, H. (1960), *Planning, Production, Inventories and Work Force*, Prentice-Hall, Englewood Cliffs, NJ.

Howard, R. (1971), *Dynamic Probabilistic Systems, Volume II: Semimarkov and Decision Processes*, John Wiley and Sons, New York.

Infanger, G. (1994), *Planning under Uncertainty: Solving Large-Scale Stochastic Linear Programs*, The Scientific Press Series, Boyd & Fraser, New York.

Jaakkola, T., Jordan, M. I. & Singh, S. P. (1994), Convergence of stochastic iterative dynamic programming algorithms, *in* J. D. Cowan, G. Tesauro & J. Alspector, eds, 'Advances in Neural Information Processing Systems', Vol. 6, Morgan Kaufmann Publishers, San Francisco, pp. 703–710.

Judd, K. (1998), *Numerical Methods in Economics*, MIT Press.

Kaelbling, L. P. (1993), *Learning in Embedded Systems*, MIT Press, Cambridge, MA.

Kaelbling, L. P., Littman, M. L. & Moore, A. W. (1996), 'Reinforcement learning: A survey', *Journal of Artifcial Intelligence Research* **4**, 237–285.

Kall, P. & Mayer, J. (2005), *Stochastic Linear Programming: Models, Theory, and Computation*, Springer, New York.

Kall, P. & Wallace, S. (1994), *Stochastic Programming*, John Wiley & Sons, New York.

Kesten, H. (1958), 'Accelerated stochastic approximation', *The Annals of Mathematical Statistics* **29**(4), 41–59.

Kiefer, J. & Wolfowitz, J. (1952), 'Stochastic estimation of the maximum of a regression function', *Annals Mathematical Statistics* **23**, 462–466.

Kirk, D. E. (1998), *Optimal Control Theory: An Introduction*, Dover, New York.

Klabjan, D. & Adelman, D. (to appear), 'A convergent infinite dimensional linear programming algorithm for deterministic semi-Markov decision processes on Borel spaces', *Mathematics of Operations Research*.

Kmenta, J. (1997), *Elements of Econometrics*, second edn, University of Michigan Press, Ann Arbor, MI.

Konda, V. R. & Tsitsiklis, J. (2003), 'On actor-critic algorithms', *SIAM J. Control and Optimization* **42**(4), 1143–1166.

Kushner, H. J. & Clark, S. (1978), *Stochastic Approximation Methods for Constrained and Unconstrained Systems*, Springer-Verlag, New York.

Kushner, H. J. & Yin, G. G. (1997), *Stochastic Approximation Algorithms and Applications*, Springer-Verlag, New York.

Lagoudakis, M. G., Parr, R. & Littman, M. L. (2002), Least-squares methods in reinforcement learning for control, *in* I. Vlahavas & C. Spyropoulos, eds, 'Methods and Applications of Artificial Intelligence : Second Hellenic Conference on AI', Springer-Verlag, New York.

Lai, T. L. & Robbins, H. (1985), 'Asymptotically efficient adaptive allocation rules', *Advances in Applied Mathematics* **6**, 4–22.

Lambert, T., Smith, R. & Epelman, M. (2002), 'Aggregation in stochastic dynamic programming', *Working Paper*.

Law, A. M. & Kelton, W. D. (2000), *Simulation Modeling and Analysis*, third edn, McGraw-Hill, New York.

LeBlanc, M. & Tibshirani, R. (1996), 'Combining estimates in regression and classification', *Journal of the American Statistical Association* **91**, 1641–1650.

Lendaris, G. G. & Neidhoefer, J. C. (2004), Guidance in the use of adaptive critics for control, *in* J. Si, A. G. Barto, W. B. Powell & D. W. II, eds, 'Handbook of Learning and Approximate Dynamic Programming', IEEE Press, New York.

Ljung, l. & Soderstrom, T. (1983), *Theory and Practice of Recursive Identification*, MIT Press, Cambridge, MA.

Longstaff, F. & Schwartz, E. S. (2001), 'Valuing American options by simulation: A simple least squares approach', *The Review of Financial Studies* **14**, 113–147.

Manne, A. (1960), 'Linear Programming and Sequential Decisions', *Management Science* **6**(3), 259–267.

Mathews, V. J. & Xie, Z. (1993), 'A stochastic gradient adaptive filter with gradient adaptive step size', *IEEE Transactions on Signal Processing* **41**, 2075–2087.

Mayer, J. (1998), *Stochastic linear programming algorithms: A comparison based on a model management system*, Springer.

McClain, J. (1974), 'Dynamics of exponential smoothing with trend and seasonal terms', *Management Science* **20**, 1300–1304.

Menache, I., Mannor, S. & Shimkin, N. (2005), Basis function adaptation in temporal difference reinforcement learning, *in* D. P. Kroese & R. Y. Rubinstein, eds, 'Annals of Operations Research', J.C. Baltzer AG, pp. 215–238.

Mendelssohn, R. (1982), 'An iterative aggregation procedure for Markov decision processes', *Operations Research* **30**(1), 62–73.

Miller, W. T. I., Sutton, R. S. & Werbos, P. J., eds (1990), *Neural Networks for Control*, MIT Press, Cambridge, MA.

Mirozahmedov, F. & Uryasev, S. P. (1983), 'Adaptive stepsize regulation for stochastic optimization algorithm', *Zurnal vicisl. mat. i. mat. fiz. 23* **6**, 1314–1325.

Mulvey, J. M. & Ruszczyński, A. J. (1995), 'A new scenario decomposition method for large-scale stochastic optimization', *Operations Research* **43**(3), 477–490.

Nascimento, J. & Powell, W. B. (2007*a*), Dynamic programming models and algorithms for the mutual fund cash balance problem, Technical report, Princeton University.

Nascimento, J. & Powell, W. B. (2007*b*), An optimal approximate dynamic programming algorithm for the lagged asset acquisition problem, Technical report, Princeton University.

Nascimento, J., Powell, W. B. & Ruszczyński, A. (2007), Optimal approximate dynamic programming algorithms for a general class of storage problems, Technical report, Princeton University.

Nediç, A. & Bertsekas, D. P. (2003), 'Least-squares policy evaluation algorithms with linear function approximation', *Journal of Discrete Event Systems* **13**, 79–110.

Nemhauser, G. L. (1966), *Introduction to dynamic programming*, John Wiley & Sons, New York.

Neveu, J. (1975), *Discrete Parameter Martingales*, North Holland, Amsterdam.

Ormoneit, D. & Glynn, P. (2002), 'Kernel-based reinforcement learning in average-cost problems', *IEEE Transactions on Automatic Control* **47**, 1624–1636.

Ormoneit, D. & Sen, S. (2002), 'Kernel-based reinforcement learning', *Machine Learning* **49**, 161–178.

Papadaki, K. & Powell, W. B. (2002), 'A monotone adaptive dynamic programming algorithm for a stochastic batch service problem', *European Journal of Operational Research* **142**(1), 108–127.

Papadaki, K. & Powell, W. B. (2003), 'An adaptive dynamic programming algorithm for a stochastic multiproduct batch dispatch problem', *Naval Research Logistics* **50**(7), 742–769.

Pearl, J. (1984), *Heuristics: Intelligent Search Strategies for Computer Problem Solving*, Addison-Wesley.

Pereira, M. & Pinto, L. (1991), 'Multistage stochastic optimization applied to energy planning', *Mathematical Programming* **52**, 359–375.

Pflug, G. C. (1988), Stepsize rules, stopping times and their implementation in stochastic quasi-gradient algorithms, *in* 'Numerical Techniques for Stochastic Optimization', Springer-Verlag, New York, pp. 353–372.

Pflug, G. C. (1996), *Optimization of Stochastic Models: The Interface Between Simulation and Optimization*, Kluwer International Series in Engineering and Computer Science: Discrete Event Dynamic Systems, Kluwer Academic Publishers, Boston.

Pollard, D. (2002), *A User's Guide to Measure Theoretic Probability*, Cambridge University Press, Cambridge.

Porteus, E. L. (1990), *Handbooks in Operations Research and Management Science: Stochastic Models*, Vol. 2, North Holland, Amsterdam, chapter Stochastic Inventory Theory, pp. 605–652.

Powell, W. B. & Topaloglu, H. (2003), Stochastic programming in transportation and logistics, *in* A. Ruszczyński & A. Shapiro, eds, '*Handbooks in Operations Research and Management Science: Stochastic Programming*', Vol. 10, Elsevier, Amsterdam, pp. 555–635.

Powell, W. B. & Topaloglu, H. (2004), Fleet management, *in* S. Wallace & W. Ziemba, eds, 'Applications of Stochastic Programming', Math Programming Society - SIAM Series in Optimization, Philadelphia.

Powell, W. B. & Van Roy, B. (2004), Approximate dynamic programming for high dimensional resource allocation problems, *in* J. Si, A. G. Barto, W. B. Powell & D. W. II, eds, 'Handbook of Learning and Approximate Dynamic Programming', IEEE Press, New York.

Powell, W. B., Ruszczyński, A. & Topaloglu, H. (2004), 'Learning algorithms for separable approximations of stochastic optimization problems', *Mathematics of Operations Research* **29**(4), 814–836.

Powell, W. B., Shapiro, J. A. & Simão, H. P. (2001), A representational paradigm for dynamic resource transformation problems, *in* R. F. C. Coullard & J. H. Owens, eds, 'Annals of Operations Research', J.C. Baltzer AG, pp. 231–279.

Powell, W. B., Shapiro, J. A. & Simão, H. P. (2002), 'An adaptive dynamic programming algorithm for the heterogeneous resource allocation problem', *Transportation Science* **36**(2), 231–249.

Prokhorov, D. V. & Wunsch, D. C. (1997), 'Adaptive critic designs', *IEEE Transactions on Neural Networks* **8**(5), 997–1007.

Psaraftis, H. & Tsitsiklis, J. (1993), 'Dynamic Shortest Paths in Acyclic Networks with Markovian Arc Costs', *Operations Research* **41**(1), 91–101.

Puterman, M. L. (1994), *Markov Decision Processes*, John Wiley & Sons, New York.

Robbins, H. & Monro, S. (1951), 'A stochastic approximation method', *Annals of Math. Stat.* **22**, 400–407.

Rockafellar, R. & Wets, R. (1991), 'Scenarios and policy aggregation in optimization under uncertainty', *Mathematics of Operations Research* **16**(1), 119–147.

Rogers, D., Plante, R., Wong, R. & Evans, J. (1991), 'Aggregation and disaggregation techniques and methodology in optimization', *Operations Research* **39**(4), 553–582.

Ross, A. (2005), Useful bounds on the expected maximum of correlated normal random variables, Technical report, Lehigh University.

Ross, S. (1983), *Introduction to Stochastic Dynamic Programming*, Academic Press, New York.

Ross, S. R. (2002), *Simulation*, third edn, Academic Press, New York.

Ruszczyński, A. (1980), 'Feasible direction methods for stochastic programming problems', *Math. Programming* **19**, 220–229.

Ruszczyński, A. (1987), 'A linearization method for nonsmooth stochastic programming problems', *Mathematics of Operations Research* **12**(1), 32–49.

Ruszczyński, A. (1993), 'Parallel decomposition of multistage stochastic programming problems', *Mathematical Programming* **58**(2), 201–228.

Ruszczyński, A. (2003), Decomposition methods, *in* A. Ruszczyński & A. Shapiro, eds, '*Handbook in Operations Research and Management Science*, Volume on *Stochastic Programming*', Elsevier, Amsterdam.

Ruszczyński, A. & Shapiro, A. (2003), *Handbooks in Operations Research and Management Science: Stochastic Programming*, Vol. 10, Elsevier, Amsterdam.

Ruszczyński, A. & Syski, W. (1986), 'A method of aggregate stochastic subgradients with on-line stepsize rules for convex stochastic programming problems', *Mathematical Programming Study* **28**, 113–131.

Samuel, A. L. (1959), 'Some studies in machine learning using the game of checkers', *IBM Journal of Research and Development* **3**, 211–229.

Samuel, A. L. (1967), 'Some studies in machine learning using the game of checkers II: Recent progress', *IBM Journal of Research and Development* **11**, 601–617.

Schweitzer, P. & Seidmann, A. (1985), 'Generalized polynomial approximations in Markovian decision processes', *Journal of Mathematical Analysis and Applications* **110**, 568–582.

Sen, S. & Higle, J. (1999), 'An introductory tutorial on stochastic linear programming models', *Interfaces* **29**(2), 33–61.

Si, J., Barto, A. G., Powell, W. B. & Wunsch, D., eds (2004*a*), *Handbook of Learning and Approximate Dynamic Programming*, IEEE Press, New York.

Si, J., Yang, L. & Liu, D. (2004*b*), Direct neural dynamic programming, *in* J. Si, A. Barto, W. B. Powell & D. Wunsch, eds, 'Learning and Approximate Dynamic Programming', IEEE Press, New York, pp. 95–122.

Sigal, C., Pritsker, A. & Solberg, J. (1980), 'The stochastic shortest route problem', *Operations Research* **28**(5), 1122–1129.

Singh, S., Jaakkola, T. & Jordan, M. I. (1995), Reinforcement learning with soft state aggregation, *in* G. Tesauro, D. Touretzky & T. K. Leen, eds, 'Advances in Neural Information Processing Systems 7', MIT Press, Cambridge, MA.

Singh, S., Jaakkola, T., Littman, M. & Szepesvari, C. (2000), 'Convergence results for single-step on-policy reinforcement-learning algorithms', *Machine Learning* **38**(3), 287–308.

Soderstrom, T., Ljung, l. & Gustavsson, I. (1978), 'A theoretical analysis of recursive identification methods', *Automatica* **78**(9), 231–244.

Spall, J. C. (2003), *Introduction to Stochastic Search and Optimization: Estimation, Simulation and Control*, John Wiley & Sons, Hoboken, NJ.

Spivey, M. & Powell, W. B. (2004), 'The dynamic assignment problem', *Transportation Science* **38**(4), 399–419.

Stengel, R. (1994), *Optimal Control and Estimation*, Dover Publications, New York.

Stokey, N. L. & R. E. Lucas, J. (1989), *Recursive Methods in Dynamic Economics*, Harvard University Press, Cambridge, MA.

Sutton, R. (1984), Temporal credit assignment in reinforcement learning, PhD thesis, University of Massachusetts, Amherst, MA.

Sutton, R. (1988), 'Learning to predict by the methods of temporal differences', *Machine Learning* **3**, 9–44.

Sutton, R. & Barto, A. (1998), *Reinforcement Learning*, The MIT Press, Cambridge, Massachusetts.

Sutton, R. S. & Singh, S. P. (1994), On step-size and bias in temporal-difference learning, *in* C. for System Science, ed., 'Eight Yale Workshop on Adaptive and Learning Systems', Yale University, pp. 91–96.

Szepesvari, C. (1998), The asymptotic convergence-rate of Q-learning, *in* M. J. K. M. I. Jordan & S. A. Solla, eds, 'Advances in neural information processing systems 10'.

Taylor, H. (1967), 'Evaluating a call option and optimal timing strategy in the stock market', *Management Science* **12**, 111–120.

Taylor, H. M. (1990), *Martingales and Random Walks*, Vol. 2, Elsevier Science Publishers B.V.,, Amsterdam, chapter 3.

Thrun, S. (1992), The role of exploration in learning control, *in* D. White & D. Sofge, eds, 'Handbook for Intelligent Control: Neural, Fuzzy and Adaptive Approaches', Van Nostrand Reinhold, Florence, Kentucky 41022.

Topaloglu, H. (2001), Dynamic Programming Approximations for Dynamic Resource Allocation Problems, PhD thesis, Princeton University.

Topaloglu, H. & Powell, W. B. (2003), 'An algorithm for approximating piecewise linear concave functions from sample gradients', *Operations Research Letters* **31**(1), 66–76.

Topaloglu, H. & Powell, W. B. (2006), 'Dynamic programming approximations for stochastic, time-staged integer multicommodity flow problems', *Informs Journal on Computing* **18**(1), 31–42.

Topkins, D. M. (1978), 'Minimizing a submodular function on a lattice', *Operations Research* **26**, 305–321.

Trigg, D. (1964), 'Monitoring a forecasting system', *Operations Research Quarterly* **15**(3), 271–274.

Trigg, D. & Leach, A. (1967), 'Exponential smoothing with an adaptive response rate', *Operations Research Quarterly* **18**(1), 53–59.

Tsitsiklis, J. (2002), 'On the convergence of optimistic policy iteration', *Journal of Machine Learning Research* **3**, 59–72.

Tsitsiklis, J. & Van Roy, B. (1997), 'An analysis of temporal-difference learning with function approximation', *IEEE Transactions on Automatic Control* **42**, 674–690.

Tsitsiklis, J. N. (1994), 'Asynchronous stochastic approximation and Q-learning', *Machine Learning* **16**, 185–202.

Tsitsiklis, J. N. & Van Roy, B. (1996), 'Feature-based methods for large scale dynamic programming', *Machine Learning* **22**, 59–94.

Tsitsiklis, J. N. & Van Roy, B. (2001), 'Regression methods for pricing complex American-style options', *IEEE Transactions on Neural Networks* **12**(4), 694–703.

Van Roy, B. (2001), Neuro-dynamic programming: Overview and recent trends, *in* E. Feinberg & A. Shwartz, eds, 'Handbook of Markov Decision Processes: Methods and Applications', Kluwer, Boston.

Van Roy, B., Bertsekas, D. P., Lee, Y. & Tsitsiklis, J. N. (1997), A neuro-dynamic programming approach to retailer inventory management, *in* 'Proceedings of the IEEE Conference on Decision and Control', Vol. 4, pp. 4052–4057.

Van Slyke, R. & Wets, R. (1969), 'L-shaped linear programs with applications to optimal control and stochastic programming', *SIAM Journal of Applied Mathematics* **17**(4), 638–663.

Venayagamoorthy, G. K., Harley, R. G. & Wunsch, D. C. (2002), 'Comparison of heuristic dynamic programming and dual heuristic programming adaptive critics for neurocontrol of a turbogenerator', *IEEE Transactions on Neural Networks* **13**(3), 764–773.

Wallace, S. (1986*a*), 'Decomposing the requirement space of a transportation problem', *Math. Prog. Study* **28**, 29–47.

Wallace, S. (1986*b*), 'Solving stochastic programs with network recourse', *Networks* **16**, 295–317.

Wallace, S. W. (1987), 'A piecewise linear upper bound on the network recourse function', *Mathematical Programming* **38**, 133–146.

Wasan, M. (1969), *Stochastic approximation*, Cambridge University Press, Cambridge.

Watkins, C. (1989), Learning from delayed rewards, Ph.d. thesis, Cambridge University, Cambridge, UK.

Watkins, C. & Dayan, P. (1992), 'Q-learning', *Machine Learning* **8**, 279–292.

Weber, R. (1992), 'On the Gittins index for multiarmed bandits', *The Annals of Applied Probability* **2**(4), 1024–1033.

Werbos, P. (1974), Beyond regression: new tools for prediction and analysis in the behavioral sciences, PhD thesis, Harvard University.

Werbos, P. (1990), A menu of designs for reinforcement learning over time, *in* R. S. W.T. Miller & P. Werbos, eds, 'Neural Networks for Control', MIT PRess, Cambridge, MA, pp. 67–96.

Werbos, P. (1994), *The Roots of Backpropagation: From Ordered Derivatives to Neural Networks and Political Forecasting*, John Wiley & Sons, New York.

Werbos, P. J. (1987), 'Building and understanding adaptive systems: A statistical/numerical approach to factory automation and brain research', *IEEE Transactions on Systems, Man and Cybernetics* **17**(1), 7–20.

Werbos, P. J. (1992*a*), Approximate dynamic programming for real-time control and neural modelling, *in* D. J. White & D. A. Sofge, eds, 'Handbook of Intelligent Control: Neural, Fuzzy, and Adaptive Approaches'.

Werbos, P. J. (1992*b*), Neurocontrol and supervised learning: an overview and evaluation, *in* D. A. White & D. A. Sofge, eds, 'Handbook of Intelligent Control', Von Nostrand Reinhold, New York, pp. 65–86.

White, C. C. (1991), 'A survey of solution techniques for the partially observable Markov decision process', *Annals of operations research* **32**, 215–230.

White, D. A. & Sofge, D. A. (1992), *Handbook of Intelligent Control*, Von Nostrand Reinhold, New York, NY.

White, D. J. (1969), *Dynamic Programming*, Holden-Day, San Francisco.

Whitin, T. M. (1953), *The Theory of Inventory Management*, Princeton University Press, Princeton, NJ.

Whitt, W. (1978), 'Approximations of dynamic programs I', *Mathematics of Operations Research* **3**, 231–243.

Whittle, P. (1982*a*), *Optimization Over Time: Dynamic Programming and Stochastic Control*, Vol. II, John Wiley and Sons, New York.

Whittle, P. (1982*b*), *Optimization Over Time: Dynamic Programming and Stochastic Control*, Vol. I, John Wiley & Sons, New York.

Widrow, B., Gupta, N. & Maitra, S. (1973), 'Punish/reward: Learning with a critic in adaptive threshold systems', *IEEE Transactions on Systems, Man and Cybernetics* **5**, 455–465.

Yang, Y. (2001), 'Adaptive regression by mixing', *Journal of the American Statistical Association*.

Young, P. (1984), *Recursive Estimation and Time-Series Analysis*, Springer-Verlag, Berlin, Heidelberg.

Zipkin, P. (1980*a*), 'Bounds for row-aggregation in linear programming', *Operations Research* **28**, 903–916.

Zipkin, P. (1980*b*), 'Bounds on the effect of aggregating variables in linear programming', *Operations Research* **28**, 403–418.

# INDEX

## WILEY SERIES IN PROBABILITY AND STATISTICS

ESTABLISHED BY WALTER A. SHEWHART AND SAMUEL S. WILKS



The ***Wiley Series in Probability and Statistics*** is well established and authoritative. It covers many topics of current research interest in both pure and applied statistics and probability theory. Written by leading statisticians and institutions, the titles span both state-of-the-art developments in the field and classical methods.

Reflecting the wide range of current research in statistics, the series encompasses applied, methodological and theoretical statistics, ranging from applications and new techniques made possible by advances in computerized practice to rigorous treatment of theoretical approaches.

This series provides essential and invaluable reading for all statisticians, whether in academia, industry, government, or research.

† ABRAHAM and LEDOLTER · Statistical Methods for Forecasting
AGRESTI · Analysis of Ordinal Categorical Data
AGRESTI · An Introduction to Categorical Data Analysis, *Second Edition*
AGRESTI · Categorical Data Analysis, *Second Edition*
ALTMAN, GILL, and McDONALD · Numerical Issues in Statistical Computing for the Social Scientist
AMARATUNGA and CABRERA · Exploration and Analysis of DNA Microarray and Protein Array Data
ANDĚL · Mathematics of Chance
ANDERSON · An Introduction to Multivariate Statistical Analysis, *Third Edition*
* ANDERSON · The Statistical Analysis of Time Series
ANDERSON, AUQUIER, HAUCK, OAKES, VANDAELE, and WEISBERG · Statistical Methods for Comparative Studies
ANDERSON and LOYNES · The Teaching of Practical Statistics
ARMITAGE and DAVID (editors) · Advances in Biometry
ARNOLD, BALAKRISHNAN, and NAGARAJA · Records
* ARTHANARI and DODGE · Mathematical Programming in Statistics
* BAILEY · The Elements of Stochastic Processes with Applications to the Natural Sciences
BALAKRISHNAN and KOUTRAS · Runs and Scans with Applications
BALAKRISHNAN and NG · Precedence-Type Tests and Applications
BARNETT · Comparative Statistical Inference, *Third Edition*
BARNETT · Environmental Statistics
BARNETT and LEWIS · Outliers in Statistical Data, *Third Edition*
BARTOSZYNSKI and NIEWIADOMSKA-BUGAJ · Probability and Statistical Inference
BASILEVSKY · Statistical Factor Analysis and Related Methods: Theory and Applications
BASU and RIGDON · Statistical Methods for the Reliability of Repairable Systems
BATES and WATTS · Nonlinear Regression Analysis and Its Applications

*Now available in a lower priced paperback edition in the Wiley Classics Library.
†Now available in a lower priced paperback edition in the Wiley–Interscience Paperback Series.

BECHHOFER, SANTNER, and GOLDSMAN · Design and Analysis of Experiments for Statistical Selection, Screening, and Multiple Comparisons
BELSLEY · Conditioning Diagnostics: Collinearity and Weak Data in Regression
† BELSLEY, KUH, and WELSCH · Regression Diagnostics: Identifying Influential Data and Sources of Collinearity
BENDAT and PIERSOL · Random Data: Analysis and Measurement Procedures, *Third Edition*
BERRY, CHALONER, and GEWEKE · Bayesian Analysis in Statistics and Econometrics: Essays in Honor of Arnold Zellner
BERNARDO and SMITH · Bayesian Theory
BHAT and MILLER · Elements of Applied Stochastic Processes, *Third Edition*
BHATTACHARYA and WAYMIRE · Stochastic Processes with Applications
BILLINGSLEY · Convergence of Probability Measures, *Second Edition*
BILLINGSLEY · Probability and Measure, *Third Edition*
BIRKES and DODGE · Alternative Methods of Regression
BLISCHKE AND MURTHY (editors) · Case Studies in Reliability and Maintenance
BLISCHKE AND MURTHY · Reliability: Modeling, Prediction, and Optimization
BLOOMFIELD · Fourier Analysis of Time Series: An Introduction, *Second Edition*
BOLLEN · Structural Equations with Latent Variables
BOLLEN and CURRAN · Latent Curve Models: A Structural Equation Perspective
BOROVKOV · Ergodicity and Stability of Stochastic Processes
BOULEAU · Numerical Methods for Stochastic Processes
BOX · Bayesian Inference in Statistical Analysis
BOX · R. A. Fisher, the Life of a Scientist
BOX and DRAPER · Response Surfaces, Mixtures, and Ridge Analyses, *Second Edition*
* BOX and DRAPER · Evolutionary Operation: A Statistical Method for Process Improvement
BOX and FRIENDS · Improving Almost Anything, *Revised Edition*
BOX, HUNTER, and HUNTER · Statistics for Experimenters: Design, Innovation, and Discovery, *Second Editon*
BOX and LUCEÑO · Statistical Control by Monitoring and Feedback Adjustment
BRANDIMARTE · Numerical Methods in Finance: A MATLAB-Based Introduction
BROWN and HOLLANDER · Statistics: A Biomedical Introduction
BRUNNER, DOMHOF, and LANGER · Nonparametric Analysis of Longitudinal Data in Factorial Experiments
BUCKLEW · Large Deviation Techniques in Decision, Simulation, and Estimation
CAIROLI and DALANG · Sequential Stochastic Optimization
CASTILLO, HADI, BALAKRISHNAN, and SARABIA · Extreme Value and Related Models with Applications in Engineering and Science
CHAN · Time Series: Applications to Finance
CHARALAMBIDES · Combinatorial Methods in Discrete Distributions
CHATTERJEE and HADI · Regression Analysis by Example, *Fourth Edition*
CHATTERJEE and HADI · Sensitivity Analysis in Linear Regression
CHERNICK · Bootstrap Methods: A Practitioner's Guide
CHERNICK and FRIIS · Introductory Biostatistics for the Health Sciences
CHILÈS and DELFINER · Geostatistics: Modeling Spatial Uncertainty
CHOW and LIU · Design and Analysis of Clinical Trials: Concepts and Methodologies, *Second Edition*
CLARKE and DISNEY · Probability and Random Processes: A First Course with Applications, *Second Edition*
* COCHRAN and COX · Experimental Designs, *Second Edition*
CONGDON · Applied Bayesian Modelling
CONGDON · Bayesian Models for Categorical Data
CONGDON · Bayesian Statistical Modelling

*Now available in a lower priced paperback edition in the Wiley Classics Library.
†Now available in a lower priced paperback edition in the Wiley–Interscience Paperback Series.

CONOVER · Practical Nonparametric Statistics, *Third Edition*
COOK · Regression Graphics
COOK and WEISBERG · Applied Regression Including Computing and Graphics
COOK and WEISBERG · An Introduction to Regression Graphics
CORNELL · Experiments with Mixtures, Designs, Models, and the Analysis of Mixture Data, *Third Edition*
COVER and THOMAS · Elements of Information Theory
COX · A Handbook of Introductory Statistical Methods
* COX · Planning of Experiments
CRESSIE · Statistics for Spatial Data, *Revised Edition*
CSÖRGŐ and HORVÁTH · Limit Theorems in Change Point Analysis
DANIEL · Applications of Statistics to Industrial Experimentation
DANIEL · Biostatistics: A Foundation for Analysis in the Health Sciences, *Eighth Edition*
* DANIEL · Fitting Equations to Data: Computer Analysis of Multifactor Data, *Second Edition*
DASU and JOHNSON · Exploratory Data Mining and Data Cleaning
DAVID and NAGARAJA · Order Statistics, *Third Edition*
* DEGROOT, FIENBERG, and KADANE · Statistics and the Law
DEL CASTILLO · Statistical Process Adjustment for Quality Control
DEMARIS · Regression with Social Data: Modeling Continuous and Limited Response Variables
DEMIDENKO · Mixed Models: Theory and Applications
DENISON, HOLMES, MALLICK and SMITH · Bayesian Methods for Nonlinear Classification and Regression
DETTE and STUDDEN · The Theory of Canonical Moments with Applications in Statistics, Probability, and Analysis
DEY and MUKERJEE · Fractional Factorial Plans
DILLON and GOLDSTEIN · Multivariate Analysis: Methods and Applications
DODGE · Alternative Methods of Regression
* DODGE and ROMIG · Sampling Inspection Tables, *Second Edition*
* DOOB · Stochastic Processes
DOWDY, WEARDEN, and CHILKO · Statistics for Research, *Third Edition*
DRAPER and SMITH · Applied Regression Analysis, *Third Edition*
DRYDEN and MARDIA · Statistical Shape Analysis
DUDEWICZ and MISHRA · Modern Mathematical Statistics
DUNN and CLARK · Basic Statistics: A Primer for the Biomedical Sciences, *Third Edition*
DUPUIS and ELLIS · A Weak Convergence Approach to the Theory of Large Deviations
EDLER and KITSOS · Recent Advances in Quantitative Methods in Cancer and Human Health Risk Assessment
* ELANDT-JOHNSON and JOHNSON · Survival Models and Data Analysis
ENDERS · Applied Econometric Time Series
† ETHIER and KURTZ · Markov Processes: Characterization and Convergence
EVANS, HASTINGS, and PEACOCK · Statistical Distributions, *Third Edition*
FELLER · An Introduction to Probability Theory and Its Applications, Volume I, *Third Edition,* Revised; Volume II, *Second Edition*
FISHER and VAN BELLE · Biostatistics: A Methodology for the Health Sciences
FITZMAURICE, LAIRD, and WARE · Applied Longitudinal Analysis
* FLEISS · The Design and Analysis of Clinical Experiments
FLEISS · Statistical Methods for Rates and Proportions, *Third Edition*
† FLEMING and HARRINGTON · Counting Processes and Survival Analysis
FULLER · Introduction to Statistical Time Series, *Second Edition*
† FULLER · Measurement Error Models

*Now available in a lower priced paperback edition in the Wiley Classics Library.
†Now available in a lower priced paperback edition in the Wiley–Interscience Paperback Series.

GALLANT · Nonlinear Statistical Models
GEISSER · Modes of Parametric Statistical Inference
GELMAN and MENG · Applied Bayesian Modeling and Causal Inference from Incomplete-Data Perspectives
GEWEKE · Contemporary Bayesian Econometrics and Statistics
GHOSH, MUKHOPADHYAY, and SEN · Sequential Estimation
GIESBRECHT and GUMPERTZ · Planning, Construction, and Statistical Analysis of Comparative Experiments
GIFI · Nonlinear Multivariate Analysis
GIVENS and HOETING · Computational Statistics
GLASSERMAN and YAO · Monotone Structure in Discrete-Event Systems
GNANADESIKAN · Methods for Statistical Data Analysis of Multivariate Observations, *Second Edition*
GOLDSTEIN and LEWIS · Assessment: Problems, Development, and Statistical Issues
GREENWOOD and NIKULIN · A Guide to Chi-Squared Testing
GROSS and HARRIS · Fundamentals of Queueing Theory, *Third Edition*
* HAHN and SHAPIRO · Statistical Models in Engineering
HAHN and MEEKER · Statistical Intervals: A Guide for Practitioners
HALD · A History of Probability and Statistics and their Applications Before 1750
HALD · A History of Mathematical Statistics from 1750 to 1930
† HAMPEL · Robust Statistics: The Approach Based on Influence Functions
HANNAN and DEISTLER · The Statistical Theory of Linear Systems
HEIBERGER · Computation for the Analysis of Designed Experiments
HEDAYAT and SINHA · Design and Inference in Finite Population Sampling
HEDEKER and GIBBONS · Longitudinal Data Analysis
HELLER · MACSYMA for Statisticians
HINKELMANN and KEMPTHORNE · Design and Analysis of Experiments, Volume 1: Introduction to Experimental Design
HINKELMANN and KEMPTHORNE · Design and Analysis of Experiments, Volume 2: Advanced Experimental Design
HOAGLIN, MOSTELLER, and TUKEY · Exploratory Approach to Analysis of Variance
* HOAGLIN, MOSTELLER, and TUKEY · Exploring Data Tables, Trends and Shapes
* HOAGLIN, MOSTELLER, and TUKEY · Understanding Robust and Exploratory Data Analysis
HOCHBERG and TAMHANE · Multiple Comparison Procedures
HOCKING · Methods and Applications of Linear Models: Regression and the Analysis of Variance, *Second Edition*
HOEL · Introduction to Mathematical Statistics, *Fifth Edition*
HOGG and KLUGMAN · Loss Distributions
HOLLANDER and WOLFE · Nonparametric Statistical Methods, *Second Edition*
HOSMER and LEMESHOW · Applied Logistic Regression, *Second Edition*
HOSMER and LEMESHOW · Applied Survival Analysis: Regression Modeling of Time to Event Data
† HUBER · Robust Statistics
HUBERTY · Applied Discriminant Analysis
HUBERTY and OLEJNIK · Applied MANOVA and Discriminant Analysis, *Second Edition*
HUNT and KENNEDY · Financial Derivatives in Theory and Practice, *Revised Edition*
HUSKOVA, BERAN, and DUPAC · Collected Works of Jaroslav Hajek—with Commentary
HUZURBAZAR · Flowgraph Models for Multistate Time-to-Event Data
IMAN and CONOVER · A Modern Approach to Statistics

*Now available in a lower priced paperback edition in the Wiley Classics Library.
†Now available in a lower priced paperback edition in the Wiley–Interscience Paperback Series.

† JACKSON · A User's Guide to Principle Components
JOHN · Statistical Methods in Engineering and Quality Assurance
JOHNSON · Multivariate Statistical Simulation
JOHNSON and BALAKRISHNAN · Advances in the Theory and Practice of Statistics: A Volume in Honor of Samuel Kotz
JOHNSON and BHATTACHARYYA · Statistics: Principles and Methods, *Fifth Edition*
JOHNSON and KOTZ · Distributions in Statistics
JOHNSON and KOTZ (editors) · Leading Personalities in Statistical Sciences: From the Seventeenth Century to the Present
JOHNSON, KOTZ, and BALAKRISHNAN · Continuous Univariate Distributions, Volume 1, *Second Edition*
JOHNSON, KOTZ, and BALAKRISHNAN · Continuous Univariate Distributions, Volume 2, *Second Edition*
JOHNSON, KOTZ, and BALAKRISHNAN · Discrete Multivariate Distributions
JOHNSON, KEMP, and KOTZ · Univariate Discrete Distributions, *Third Edition*
JUDGE, GRIFFITHS, HILL, LÜTKEPOHL, and LEE · The Theory and Practice of Econometrics, *Second Edition*
JUREČKOVÁ and SEN · Robust Statistical Procedures: Aymptotics and Interrelations
JUREK and MASON · Operator-Limit Distributions in Probability Theory
KADANE · Bayesian Methods and Ethics in a Clinical Trial Design
KADANE AND SCHUM · A Probabilistic Analysis of the Sacco and Vanzetti Evidence
KALBFLEISCH and PRENTICE · The Statistical Analysis of Failure Time Data, *Second Edition*
KARIYA and KURATA · Generalized Least Squares
KASS and VOS · Geometrical Foundations of Asymptotic Inference
† KAUFMAN and ROUSSEEUW · Finding Groups in Data: An Introduction to Cluster Analysis
KEDEM and FOKIANOS · Regression Models for Time Series Analysis
KENDALL, BARDEN, CARNE, and LE · Shape and Shape Theory
KHURI · Advanced Calculus with Applications in Statistics, *Second Edition*
KHURI, MATHEW, and SINHA · Statistical Tests for Mixed Linear Models
KLEIBER and KOTZ · Statistical Size Distributions in Economics and Actuarial Sciences
KLUGMAN, PANJER, and WILLMOT · Loss Models: From Data to Decisions, *Second Edition*
KLUGMAN, PANJER, and WILLMOT · Solutions Manual to Accompany Loss Models: From Data to Decisions, *Second Edition*
KOTZ, BALAKRISHNAN, and JOHNSON · Continuous Multivariate Distributions, Volume 1, *Second Edition*
KOVALENKO, KUZNETZOV, and PEGG · Mathematical Theory of Reliability of Time-Dependent Systems with Practical Applications
KVAM and VIDAKOVIC · Nonparametric Statistics with Applications to Science and Engineering
LACHIN · Biostatistical Methods: The Assessment of Relative Risks
LAD · Operational Subjective Statistical Methods: A Mathematical, Philosophical, and Historical Introduction
LAMPERTI · Probability: A Survey of the Mathematical Theory, *Second Edition*
LANGE, RYAN, BILLARD, BRILLINGER, CONQUEST, and GREENHOUSE · Case Studies in Biometry
LARSON · Introduction to Probability Theory and Statistical Inference, *Third Edition*
LAWLESS · Statistical Models and Methods for Lifetime Data, *Second Edition*
LAWSON · Statistical Methods in Spatial Epidemiology
LE · Applied Categorical Data Analysis
LE · Applied Survival Analysis

*Now available in a lower priced paperback edition in the Wiley Classics Library.
†Now available in a lower priced paperback edition in the Wiley–Interscience Paperback Series.

LEE and WANG · Statistical Methods for Survival Data Analysis, *Third Edition*
LePAGE and BILLARD · Exploring the Limits of Bootstrap
LEYLAND and GOLDSTEIN (editors) · Multilevel Modelling of Health Statistics
LIAO · Statistical Group Comparison
LINDVALL · Lectures on the Coupling Method
LIN · Introductory Stochastic Analysis for Finance and Insurance
LINHART and ZUCCHINI · Model Selection
LITTLE and RUBIN · Statistical Analysis with Missing Data, *Second Edition*
LLOYD · The Statistical Analysis of Categorical Data
LOWEN and TEICH · Fractal-Based Point Processes
MAGNUS and NEUDECKER · Matrix Differential Calculus with Applications in Statistics and Econometrics, *Revised Edition*
MALLER and ZHOU · Survival Analysis with Long Term Survivors
MALLOWS · Design, Data, and Analysis by Some Friends of Cuthbert Daniel
MANN, SCHAFER, and SINGPURWALLA · Methods for Statistical Analysis of Reliability and Life Data
MANTON, WOODBURY, and TOLLEY · Statistical Applications Using Fuzzy Sets
MARCHETTE · Random Graphs for Statistical Pattern Recognition
MARDIA and JUPP · Directional Statistics
MASON, GUNST, and HESS · Statistical Design and Analysis of Experiments with Applications to Engineering and Science, *Second Edition*
McCULLOCH and SEARLE · Generalized, Linear, and Mixed Models
McFADDEN · Management of Data in Clinical Trials, *Second Edition*
* McLACHLAN · Discriminant Analysis and Statistical Pattern Recognition
McLACHLAN, DO, and AMBROISE · Analyzing Microarray Gene Expression Data
McLACHLAN and KRISHNAN · The EM Algorithm and Extensions
McLACHLAN and PEEL · Finite Mixture Models
McNEIL · Epidemiological Research Methods
MEEKER and ESCOBAR · Statistical Methods for Reliability Data
MEERSCHAERT and SCHEFFLER · Limit Distributions for Sums of Independent Random Vectors: Heavy Tails in Theory and Practice
MICKEY, DUNN, and CLARK · Applied Statistics: Analysis of Variance and Regression, *Third Edition*
* MILLER · Survival Analysis, *Second Edition*
MONTGOMERY, PECK, and VINING · Introduction to Linear Regression Analysis, *Fourth Edition*
MORGENTHALER and TUKEY · Configural Polysampling: A Route to Practical Robustness
MUIRHEAD · Aspects of Multivariate Statistical Theory
MULLER and STOYAN · Comparison Methods for Stochastic Models and Risks
MURRAY · X-STAT 2.0 Statistical Experimentation, Design Data Analysis, and Nonlinear Optimization
MURTHY, XIE, and JIANG · Weibull Models
MYERS and MONTGOMERY · Response Surface Methodology: Process and Product Optimization Using Designed Experiments, *Second Edition*
MYERS, MONTGOMERY, and VINING · Generalized Linear Models. With Applications in Engineering and the Sciences
† NELSON · Accelerated Testing, Statistical Models, Test Plans, and Data Analyses
† NELSON · Applied Life Data Analysis
NEWMAN · Biostatistical Methods in Epidemiology
OCHI · Applied Probability and Stochastic Processes in Engineering and Physical Sciences
OKABE, BOOTS, SUGIHARA, and CHIU · Spatial Tesselations: Concepts and Applications of Voronoi Diagrams, *Second Edition*

*Now available in a lower priced paperback edition in the Wiley Classics Library.
†Now available in a lower priced paperback edition in the Wiley–Interscience Paperback Series.

OLIVER and SMITH · Influence Diagrams, Belief Nets and Decision Analysis
PALTA · Quantitative Methods in Population Health: Extensions of Ordinary Regressions
PANJER · Operational Risk: Modeling and Analytics
PANKRATZ · Forecasting with Dynamic Regression Models
PANKRATZ · Forecasting with Univariate Box-Jenkins Models: Concepts and Cases
* PARZEN · Modern Probability Theory and Its Applications
PEÑA, TIAO, and TSAY · A Course in Time Series Analysis
PIANTADOSI · Clinical Trials: A Methodologic Perspective
PORT · Theoretical Probability for Applications
POURAHMADI · Foundations of Time Series Analysis and Prediction Theory
POWELL · Approximate Dynamic Programming: Solving the Curses of Dimensionality
PRESS · Bayesian Statistics: Principles, Models, and Applications
PRESS · Subjective and Objective Bayesian Statistics, *Second Edition*
PRESS and TANUR · The Subjectivity of Scientists and the Bayesian Approach
PUKELSHEIM · Optimal Experimental Design
PURI, VILAPLANA, and WERTZ · New Perspectives in Theoretical and Applied Statistics
† PUTERMAN · Markov Decision Processes: Discrete Stochastic Dynamic Programming
QIU · Image Processing and Jump Regression Analysis
* RAO · Linear Statistical Inference and Its Applications, *Second Edition*
RAUSAND and HØYLAND · System Reliability Theory: Models, Statistical Methods, and Applications, *Second Edition*
RENCHER · Linear Models in Statistics
RENCHER · Methods of Multivariate Analysis, *Second Edition*
RENCHER · Multivariate Statistical Inference with Applications
* RIPLEY · Spatial Statistics
* RIPLEY · Stochastic Simulation
ROBINSON · Practical Strategies for Experimenting
ROHATGI and SALEH · An Introduction to Probability and Statistics, *Second Edition*
ROLSKI, SCHMIDLI, SCHMIDT, and TEUGELS · Stochastic Processes for Insurance and Finance
ROSENBERGER and LACHIN · Randomization in Clinical Trials: Theory and Practice
ROSS · Introduction to Probability and Statistics for Engineers and Scientists
ROSSI, ALLENBY, and McCULLOCH · Bayesian Statistics and Marketing
† ROUSSEEUW and LEROY · Robust Regression and Outlier Detection
* RUBIN · Multiple Imputation for Nonresponse in Surveys
RUBINSTEIN · Simulation and the Monte Carlo Method
RUBINSTEIN and MELAMED · Modern Simulation and Modeling
RYAN · Modern Experimental Design
RYAN · Modern Regression Methods
RYAN · Statistical Methods for Quality Improvement, *Second Edition*
SALEH · Theory of Preliminary Test and Stein-Type Estimation with Applications
* SCHEFFE · The Analysis of Variance
SCHIMEK · Smoothing and Regression: Approaches, Computation, and Application
SCHOTT · Matrix Analysis for Statistics, *Second Edition*
SCHOUTENS · Levy Processes in Finance: Pricing Financial Derivatives
SCHUSS · Theory and Applications of Stochastic Differential Equations
SCOTT · Multivariate Density Estimation: Theory, Practice, and Visualization
† SEARLE · Linear Models for Unbalanced Data
† SEARLE · Matrix Algebra Useful for Statistics
† SEARLE, CASELLA, and McCULLOCH · Variance Components
SEARLE and WILLETT · Matrix Algebra for Applied Economics
SEBER and LEE · Linear Regression Analysis, *Second Edition*
† SEBER · Multivariate Observations

*Now available in a lower priced paperback edition in the Wiley Classics Library.
†Now available in a lower priced paperback edition in the Wiley–Interscience Paperback Series.

† SEBER and WILD · Nonlinear Regression
SENNOTT · Stochastic Dynamic Programming and the Control of Queueing Systems
* SERFLING · Approximation Theorems of Mathematical Statistics
SHAFER and VOVK · Probability and Finance: It's Only a Game!
SILVAPULLE and SEN · Constrained Statistical Inference: Inequality, Order, and Shape Restrictions
SMALL and McLEISH · Hilbert Space Methods in Probability and Statistical Inference
SRIVASTAVA · Methods of Multivariate Statistics
STAPLETON · Linear Statistical Models
STAUDTE and SHEATHER · Robust Estimation and Testing
STOYAN, KENDALL, and MECKE · Stochastic Geometry and Its Applications, *Second Edition*
STOYAN and STOYAN · Fractals, Random Shapes and Point Fields: Methods of Geometrical Statistics
STREET and BURGESS · The Construction of Optimal Stated Choice Experiments: Theory and Methods
STYAN · The Collected Papers of T. W. Anderson: 1943–1985
SUTTON, ABRAMS, JONES, SHELDON, and SONG · Methods for Meta-Analysis in Medical Research
TAKEZAWA · Introduction to Nonparametric Regression
TANAKA · Time Series Analysis: Nonstationary and Noninvertible Distribution Theory
THOMPSON · Empirical Model Building
THOMPSON · Sampling, *Second Edition*
THOMPSON · Simulation: A Modeler's Approach
THOMPSON and SEBER · Adaptive Sampling
THOMPSON, WILLIAMS, and FINDLAY · Models for Investors in Real World Markets
TIAO, BISGAARD, HILL, PEÑA, and STIGLER (editors) · Box on Quality and Discovery: with Design, Control, and Robustness
TIERNEY · LISP-STAT: An Object-Oriented Environment for Statistical Computing and Dynamic Graphics
TSAY · Analysis of Financial Time Series, *Second Edition*
UPTON and FINGLETON · Spatial Data Analysis by Example, Volume II: Categorical and Directional Data
VAN BELLE · Statistical Rules of Thumb
VAN BELLE, FISHER, HEAGERTY, and LUMLEY · Biostatistics: A Methodology for the Health Sciences, *Second Edition*
VESTRUP · The Theory of Measures and Integration
VIDAKOVIC · Statistical Modeling by Wavelets
VINOD and REAGLE · Preparing for the Worst: Incorporating Downside Risk in Stock Market Investments
WALLER and GOTWAY · Applied Spatial Statistics for Public Health Data
WEERAHANDI · Generalized Inference in Repeated Measures: Exact Methods in MANOVA and Mixed Models
WEISBERG · Applied Linear Regression, *Third Edition*
WELSH · Aspects of Statistical Inference
WESTFALL and YOUNG · Resampling-Based Multiple Testing: Examples and Methods for *p*-Value Adjustment
WHITTAKER · Graphical Models in Applied Multivariate Statistics
WINKER · Optimization Heuristics in Economics: Applications of Threshold Accepting
WONNACOTT and WONNACOTT · Econometrics, *Second Edition*
WOODING · Planning Pharmaceutical Clinical Trials: Basic Statistical Principles
WOODWORTH · Biostatistics: A Bayesian Introduction
WOOLSON and CLARKE · Statistical Methods for the Analysis of Biomedical Data, *Second Edition*

*Now available in a lower priced paperback edition in the Wiley Classics Library.
†Now available in a lower priced paperback edition in the Wiley–Interscience Paperback Series.

WU and HAMADA · Experiments: Planning, Analysis, and Parameter Design Optimization
WU and ZHANG · Nonparametric Regression Methods for Longitudinal Data Analysis
YANG · The Construction Theory of Denumerable Markov Processes
YOUNG, VALERO-MORA, and FRIENDLY · Visual Statistics: Seeing Data with Dynamic Interactive Graphics
ZELTERMAN · Discrete Distributions—Applications in the Health Sciences
* ZELLNER · An Introduction to Bayesian Inference in Econometrics
ZHOU, OBUCHOWSKI, and McCLISH · Statistical Methods in Diagnostic Medicine

*Now available in a lower priced paperback edition in the Wiley Classics Library.
†Now available in a lower priced paperback edition in the Wiley–Interscience Paperback Series.